Classics in Mathematics

Ludvig D. Faddeev · Leon A. Takhtajan

Hamiltonian Methods in the Theory of Solitons

Ludwig D. Faddeev was born in Leningrad, USSR in 1934. He graduated from the Leningrad State University in 1956 and received his Ph.D. from there in 1959. Since 1959 he has been affiliated with the Leningrad branch of Steklov Mathematical Institute and was its Director from 1976 to 2000. Currently Faddeev is Director of the Euler International Mathematical Institute in St. Petersburg, Russia, and Academician-Secretary of the Mathematics Division of the Russian Academy of Sciences. He was President of the International Mathematical Union during 1986–1990.

Faddeev's principal interests and contributions cover the large area of mathematical physics. They include, in chronological order, quantum scattering theory, spectral theory of automorphic functions, quantization of Yang-Mills theories, Hamiltonian methods in classical and quantum integrable systems, quantum groups and quantum integrable systems on a lattice. Faddeev's work laid a mathematical foundation for functional methods in quantum gauge theories. A great deal of his work was directed towards development of Hamiltonian methods in classical and quantum field theories.

Leon A. Takhtajan was born in Erevan, Republic of Armenia of the USSR, in 1950. He was schooled in Leningrad, graduated from the Leningrad State University in 1973, and received his Ph.D. from the Leningrad branch of Steklov Mathematical Institute in 1975, with which he was affiliated during 1973-1998. Since 1992 he has been Professor of Mathematics at Stony Brook University, USA.

Takhtajan's principal interests and contributions are in the area of mathematical physics. They include classical and quantum integrable systems, quantum groups, Weil-Petersson geometry of moduli spaces of Riemann surfaces and moduli spaces of vector bundles, and application of quantum methods to algebraic and complex analysis. His work, together with L.D. Faddeev and E.K. Sklyanin, led to the development of the quantum inverse scattering method from which the theory of quantum groups was born.

Ludvig D. Faddeev · Leon A. Takhtajan

Hamiltonian Methods in the Theory of Solitons

Translated from the Russian
by Alexey G. Reyman

Reprint of the 1987 Edition

Springer

Authors

Ludvig D. Faddeev
Russian Academy of Sciences
Steklov Institute of Mathematics
Fontanka 27
191011 St. Petersburg
Russia
e-mail: faddeev@pdmi.ras.ru

Leon A. Takhtajan
State University New York
Department Mathematics
11794-3651 Stony Brook, NY
USA
e-mail: leontak@math.sunysb.edu

Translator

Alexey G. Reyman

Title of the Russian edition: *Gamiltonov podkhod v teorii solitonov*
Publisher Nauka, Moscow 1986

Originally published in the *Springer Series in Soviet Mathematics*

Mathematics Subject Classification (2000): 35, 45, 58

Library of Congress Control Number: 2007924346

ISSN 1431-0821

ISBN 978-3-540-69843-2 Springer Berlin Heidelberg New York

Springer is a part of Springer Science+Business Media

springer.com

Production: LE-TEX Jelonek, Schmidt & Vöckler GbR, Leipzig
Cover-design: WMX Design GmbH, Heidelberg

SPIN 11944317 41/3100YL - 5 4 3 2 1 0 Printed on acid-free paper

L. D. Faddeev L. A. Takhtajan

Hamiltonian Methods in the Theory of Solitons

Translated from the Russian
by A. G. Reyman

Reprint of the 1st edition

Springer-Verlag Berlin Heidelberg GmbH

Ludwig D. Faddeev   Leon A. Takhtajan
Alexey G. Reyman
Steklov Mathematical Institute
Fontanka 27, 191011 Leningrad, USSR

Title of the Russian edition:
Gamiltonov podkhod v teorii solitonov
Publisher Nauka, Moscow 1986

This volume is part of the *Springer Series in Soviet Mathematics*
Advisers: L. D. Faddeev (Leningrad), R. V. Gamkrelidze (Moscow)

Mathematics Subject Classification (1980): 35, 45, 58

ISBN 978-3-540-69843-2

Library of Congress Cataloging in Publication Data
Faddeev, L. D. Hamiltonian methods in the theory of solitons.
(Springer series in Soviet mathematics)
Translation of: Gamiltonov podkhod v teorii solitonov.
Includes bibliographies.
1. Solitons. 2. Inverse scattering transform.
3. Hamiltonian systems. 4. Mathematical physics.
I. Takhtajan, L. A. (Leon A.), 1950–.
II. Title. III. Series.
QC174.26.W28F3313   1987   530.1'4   86-31410
ISBN 978-3-540-69843-2   ISBN 978-3-540-69969-9 (eBook)
DOI 10.1007/978-3-540-69969-9

Originally published by Springer-Verlag Berlin Heidelberg New York in 1987
Softcover reprint of the hardcover 1st edition 1987

2141/3140-543210

Preface

This book presents the foundations of the inverse scattering method and its applications to the theory of solitons in such a form as we understand it in Leningrad.

The concept of soliton was introduced by Kruskal and Zabusky in 1965. A soliton (a solitary wave) is a localized particle-like solution of a nonlinear equation which describes excitations of finite energy and exhibits several characteristic features: propagation does not destroy the profile of a solitary wave; the interaction of several solitary waves amounts to their elastic scattering, so that their total number and shape are preserved. Occasionally, the concept of the soliton is treated in a more general sense as a localized solution of finite energy. At present this concept is widely spread due to its universality and the abundance of applications in the analysis of various processes in nonlinear media. The inverse scattering method which is the mathematical basis of soliton theory has developed into a powerful tool of mathematical physics for studying nonlinear partial differential equations, almost as vigorous as the Fourier transform.

The book is based on the Hamiltonian interpretation of the method, hence the title. Methods of differential geometry and Hamiltonian formalism in particular are very popular in modern mathematical physics. It is precisely the general Hamiltonian formalism that presents the inverse scattering method in its most elegant form. Moreover, the Hamiltonian formalism provides a link between classical and quantum mechanics. So the book is not only an introduction to the classical soliton theory but also the groundwork for the quantum theory of solitons, to be discussed in another volume.

The book is addressed to specialists in mathematical physics. This has determined the choice of material and the level of mathematical rigour. We hope that it will also be of interest to mathematicians of other specialities and to theoretical physicists as well. Still, being a mathematical treatise it does not contain applications of soliton theory to specific physical phenomena.

While the book was written in Leningrad, the contents passed through several revisions caused by new developments of the method. We hope that in its present version the text has reached sufficient steadiness. At the same time, we do not claim to give an exhaustive account of the current state of the subject. In this sense the book is an introduction to the subject rather than an outline of all modern constructions connected with multi-dimensional generalizations and representations of infinite-dimensional algebraic structures.

We would like to thank our colleagues at the laboratory of mathematical problems of physics at the Leningrad branch of V. A. Steklov Mathematical Institute: V. E. Korepin, P. P. Kulish, A. G. Reyman, N. Yu. Reshetikhin, M. A. Semenov-Tian-Shansky, E. K. Sklyanin, F. A. Smirnov. The book undoubtedly gained from our contacts. We are also grateful to V. O. Tarasov for his careful reading of the manuscript.

Contents

Introduction

Over the past fifteen years the theory of solitons and the related theory of integrable nonlinear evolution equations in two space-time dimensions has attracted a large number of research workers of different orientations ranging from algebraic geometry to applied hydrodynamics. Modern mathematical physics has witnessed the development of a vast new area of research devoted to this theory and called the inverse scattering method of solving nonlinear equations (other names are: the inverse spectral transform, the method of isospectral deformations and, more colloquially, the L-A pair method).

The method was initiated by the pioneering work of the Princeton group. In 1967 in the paper "Method for solving the Korteweg-de Vries equation" [GGKM 1967] Gardner, Greene, Kruskal and Miura introduced a remarkable nonlinear change of variables which made the equation linear and explicitly solvable. The change of variables involves the direct and inverse scattering problems for the one-dimensional Schrödinger equation, which accounts for the name of the method.

The formation of the theory was greatly influenced by the following two contributions. In "Integrals of nonlinear equations of evolution and solitary waves" [L 1968] Lax formalized the results of the Princeton group and introduced the concept of an L-A pair. Next, in "Exact theory of two-dimensional self-focusing and one-dimensional self-modulation of waves in nonlinear media" [ZS 1971] Zakharov and Shabat showed that the concept of an L-A pair is not necessarily tied to the Korteweg-de Vries equation but can also be used for the nonlinear Schrödinger equation, thus opening perspectives for treating other equations.

Since then the increasingly fast development of the inverse scattering method and its applications has created a large new domain of mathematical physics. Characteristically, most of the work in this field is collective. Several long-standing groups can be listed besides the one in Princeton (of course, some of the people have subsequently moved to other locations). They are:

1. The group in Moscow represented by Zakharov, Manakov, Novikov, Krichever, Dubrovin and Mikhailov. Later they were joined by Gelfand, Manin and Perelomov with their collaborators.

2. The group in Potsdam represented by Ablowitz, Kaup, Newell, Segur and their collaborators.

3. The group in Arizona which includes Flaschka, Lamb and McLaughlin.

More recently a group appeared in Kyoto (Sato, Miwa, Jimbo, Kashiwara et al.). There are also other centers: in New York (Lax, Moser, Kac, McKean, Case, Deift and Trubowitz), in Rome (Calogero and Degasperis), in Manchester (Bullough with collaborators), in Freiburg (Pohlmeyer and Honerkamp). There is a group in Leningrad too, which includes the authors of the present book and also Korepin, Kulish, Reyman, Sklyanin, Semenov-Tian-Shansky, Izergin, Its and Matveev. Besides the groups some single contributors should be mentioned, Shabat, Kostant, Adler and van Moerbeke among them.

So far we have only listed mathematical physicists without mentioning the large army of specialists engaged in applications of soliton theory in quantum field theory, solid state physics, nonlinear optics, plasma physics, hydrodynamics, biology and other natural sciences. This impressive list of people and topics is indicative of the range of interests and geographical spread of those involved.

At present soliton theory is believed to have reached maturity. The increasingly prominent role of this theory was an impetus for the appearance of many monographs in which the schools mentioned above made known their particular views on the subject. They are the following:

1. Zakharov, Manakov, Novikov, Pitaievski, Theory of Solitons. The Inverse Problem Method [ZMNP 1980].

2. Lamb, Elements of Soliton Theory [L 1980].

3. Ablowitz, Segur, Solitons and the Inverse Scattering Transform [AS 1981].

4. Calogero, Degasperis, Spectral Transform and Solitons [CD 1982].

5. Dodd, Eilbeck, Gibbon, Morris, Solitons and Nonlinear waves [DEGM 1982].

There are also the following collections of papers:

1. Solitons in Action, Lonngren and Scott, eds. [LS 1978].

2. Solitons, Bullough and Caudrey, eds. [BC 1980].

3. Bäcklund Transformations, Miura, ed. [M 1976].

4. Proceedings of the Joint US–USSR symposium on Soliton Theory, Manakov and Zakharov, eds. [MZ 1981].

5. Nonlinear Evolution equations Solvable by the Spectral Transform, Calogero, ed. [C 1978].

There is also a textbook of Eilenberger "Solitons: Mathematical Methods for Physicists" [E 1981].

After publishing a number of reviews devoted to the quantum theory of solitons and its applications in quantum field theory [KF 1977], [FK 1978], [F 1980a], [F 1980b] the Leningrad specialists think it timely to voice their attitude towards the inverse scattering method as a whole. Naturally, the attitude presented is influenced by the orientation towards the quantum formulation of soliton theory. The quantum version of the inverse scattering method which has been developed since 1978 and reviewed in a series of papers [TF 1979], [KS 1980], [F 1980b], [F 1981], [F 1982a], [F 1982b], [IK 1982], [F 1983], [T 1983] forced us to look afresh at the basic tools and devices of the classical version of the method. Particularly, this concerns the language of Hamiltonian dynamics closely associated with quantum applications.

As a matter of fact, most integrable models (including all of applied importance) possess a Hamiltonian structure, that is, the equations defining them are infinite-dimensional analogues of Hamilton's equations in classical mechanics. The inverse scattering transform can be interpreted as a canonical transformation with respect to this structure so that the variables which linearize the equation have the meaning of action-angle variables.

For the example of the Korteweg-de Vries equation this programme was formulated and carried out in the paper of Zakharov and Faddeev "Korteweg-de Vries equation, a completely integrable Hamiltonian system" [ZF 1971] published in 1971, which was the formative period of the theory. Later the same was done for other interesting models.

The treatises cited above often mention the Hamiltonian approach but never assign to it a principal methodological role. The main point in which our book differs from others is the emphasis on the Hamiltonian structure and the ensuing choice and arrangement of the material (see the Preface). At the same time the text is self-contained enough to serve as an independent introduction to the subject.

At first we planned to devote the book mainly to the quantum version, with a suitable introduction to the classical method. However, as often happens, the project expanded in the course of writing and the book will appear in two volumes. The present volume is devoted entirely to the classical theory.

The pedagogical novelties of the book are clearly noticeable. In contrast to other authors, we have chosen the nonlinear Schrödinger (NS) equation

$$i\frac{\partial\psi}{\partial t}=-\frac{\partial^2\psi}{\partial x^2}+2\varkappa|\psi|^2\psi,$$

where $\psi(x,t)$ is a complex-valued function, to be our principal representative example, instead of the Korteweg-de Vires (KdV) equation

$$\frac{\partial u}{\partial t}=6u\frac{\partial u}{\partial x}-\frac{\partial^3 u}{\partial x^3}.$$

For this there are several reasons:

2. Alongside the usual analysis of the direct and inverse scattering problems for the auxiliary linear system on an infinite interval we also consider the finite interval $-L \leqslant x \leqslant L$ with quasi-periodic boundary conditions. However, the associated inverse problem involves analysis on Riemann surfaces and goes beyond the scope of our book.

3. Our treatment of the inverse problem is based on the matrix Riemann problem of analytic factorization of matrix-valued functions, rather then on the traditional Gelfand-Levitan-Marchenko equation. As has now become clear, this method is more universal and technically more transparent. For the prototype NS equation we explain how the Gelfand-Levitan-Marchenko method can be naturally incorporated into the Riemann problem.

4. The Hamiltonian structure is defined in terms of the so-called r-matrix. This construction originated in the quantum spectral transform method and later was adapted to the classical case. We believe it to be most adequate and universal and hope to demonstrate this.

5. A comprehensive classification of integrable models based on the concept of an r-matrix is presented. The Lie-bracket formalism for (infinite-dimensional) current algebras turns out to provide an adequate language for continuous models. We also discuss an extension of the classification to lattice models.

We emphasize again that the above characteristic features have their natural counterparts in the quantum version of the method.

Now, a word about the level of mathematical rigour. Our presentation, mostly elementary, is based on techniques of classical analysis. Proofs are given of all results on direct and inverse problems for the auxiliary linear system for the NS model in the rapidly decreasing case. This is not done when other models are discussed in order to avoid overloading the text with tiresome details. We believe that the NS model is treated in a sufficiently invariant manner, so that the reader will be able to fill in the gaps.

However, a rigorous proof of the assertions concerning the Hamiltonian formalism should make use of analysis on infinite-dimensional manifolds. We consider this level of rigour superfluous for our subject and therefore do not hesitate to use differential-geometric terminology in the infinite-dimensional case without complete justification. This is done deliberately because in our view rigorous proofs in this situation do not reveal the heart of the matter; so we leave the job to specialists in global analysis. We believe that this agrees with the state of affairs in modern mathematical physics to which the present text belongs.

The inverse scattering method is now developed to such an extent that it can be presented from the very beginning in its most general form. However, this does not seem to be the best way of introducing the subject. As an alternative we have chosen to introduce its basic concepts by means of a particular example and to illustrate its generality by other models, so that the reader is led gradually to the fairly natural and general construction underlying the method. In our opinion, this agrees with the spirit of modern mathematical physics.

With this the main part of the introduction is ended. We have omitted a formal summary of the text confining ourselves to basic historical comments and methodological principles. It is hoped that the table of contents gives sufficient information about the contents of the book.

Finally, a few words about the structure of the text. It consists of two parts, each divided into chapters and sections. The main text contains no references to original papers. These are given in special sections at the end of each chapter, which also contain various remarks and comments. Thus, other aspects of the theory are mentioned there which are not included in the main text, with appropriate references.

Each formula has a number composed of the number of the section and its own number within the section. References to formulae in other chapters have triple numerations, the first entry being the number of the chapter. References to a different part of the volume are specified explicitly.

References

[AS 1981] Ablowitz, M. J., Segur, H.: Solitons and the Inverse Scattering Transform. SIAM, Philadelphia 1981

[BC 1980] Bullough, R. K., Caudrey, P. J. (ed.): Solitons, Topics in Current Physics *17*, Berlin–Heidelberg–New York, Springer 1980

[C 1978] Calogero, F. (ed.): Nonlinear Evolution Equations Solvable by the Spectral Transform. Research Notes in Math. *26*, London, Pitman 1978

[CD 1982] Calogero, F., Degasperis, A.: Spectral Transform and Solitons. Vol. I. Amsterdam, North-Holland 1982

[DEGM 1982] Dodd, R. K., Eilbeck, J. C., Gibbon, J. D., Morris, H. C.: Solitons and Nonlinear Waves. New York, Academic Press 1982

[E 1981] Eilenberger, G.: Solitons. Mathematical Method for Physicists. Berlin, Springer 1981

[F 1980a] Faddeev, L. D.: A Hamiltonian interpretation of the inverse scattering method. In: Solitons, edited by Bullough R. K., Caudrey P. J., Topics in Current Physics *17*, 339–354, Berlin–Heidelberg–New York, Springer 1980

[F 1980b] Faddeev, L. D.: Quantum completely integrable models in field theory. In: Mathematical Physics Review. Sect. C.: Math. Phys. Rev. *1*, 107–155, Harwood Academic 1980

[F 1981] Faddeev, L. D.: Two-dimensional integrable models in quantum field theory. Physica Scripta *24*, 832–835 (1981)

[F 1982a] Faddeev, L. D.: Recent development of QST. In: Recent development in gauge theory and integrable systems. Kyoto, Kyoto Univ. Research Inst. for Math. Sci., 53–71, 1982

[F 1982b] Faddeev, L. D.: Integrable models in 1+1-dimensional quantum field theory. In: Les Houches, Session XXXIX, 1982, Recent Advances in Field Theory and Statistical Mechanics, Zuber, J.-B., Stora, R. (editors), 563–608. Elsevier Science Publishers 1984

[F 1983] Faddeev, L. D.: Quantum scattering transformation. In: Structural Elements in Particle Physics and Statistical Mechanics. (Freiburg Summer Inst. on Theor. Physics 1981) vol. 82, 93-114, New York-London, Plenum Press 1983

[FK 1978] Faddeev, L. D., Korepin, V. E.: Quantum Theory of Solitons. Physics Reports *42C* (1), 1-87 (1978)

[GGKM 1967] Gardner, C. S., Greene, J. M., Kruskal, M. D., Miura, R. M.: Method for solving the Korteweg-de Vries equation. Phys. Rev. Lett. *19*, 1095-1097 (1967)

[IK 1982] Izergin, A. G., Korepin, V. E.: The quantum inverse scattering method. Physics of elementary particles and atomic nuclei, v. 13, N 3, 501-541 (1982) [Russian]; English transl. in Soviet J. Particles and Nuclei *13* (3), 207-223 (1982)

[KF 1977] Korepin, V. E., Faddeev, L. D.: Quantization of solitons. In: Physics of elementary particles (Proceedings of the XII winter school of the Leningrad institute of nuclear physics), 130-146, Leningrad 1977 [Russian]

[KS 1980] Kulish, P. P., Sklyanin, E. K.: Solutions of the Yang-Baxter equation. In: Differential geometry, Lie groups and mechanics. III. Zapiski Nauchn. Semin. LOMI *95*, 129-160 (1980) [Russian]; English transl. in J. Sov. Math. *19* (5), 1596-1620 (1982)

[KS 1982] Kulish, P. P., Sklyanin, E. K.: Quantum spectral transform method. Recent Developments. Lecture Notes in Physics, vol. 151, 61-119, Berlin-Heidelberg-New York, Springer 1982

[L 1968] Lax, P. D.: Integrals of nonlinear equations of evolution and solitary waves. Comm. Pure Appl. Math. *21*, 467-490 (1968)

[L 1980] Lamb, G. L., Jr.: Elements of Soliton Theory. New York, Wiley 1980

[LS 1978] Lonngren, K., Scott, A. (eds.): Solitons in Action. New York, Academic Press 1978

[M 1976] Miura, R. (ed.): Bäcklund transformations. Lecture Notes in Mathematics, vol. 515, Berlin-Heidelberg-New York, Springer 1976

[MZ 1981] Manakov, S. L., Zakharov, V. E. (eds.): Soliton Theory. Proceedings of the Soviet-American Symposium on Soliton Theory. Physica D, *3D*, no. 1+2 (1981)

[T 1983] Takhtajan, L. A.: Integrable models in classical and quantum field theory. In: Proceedings of the International Congress of Mathematicians 1983, 1331-1340, Warszawa, North-Holland 1984

[TF 1979] Takhtajan, L. A., Faddeev, L. D.: The quantum inverse problem method and the XYZ Heisenberg model. Uspekhi Mat. Nauk *34* (5), 13-63 (1979) [Russian]; English transl. in Russian Math. Surveys *34* (5), 11-68 (1979)

[ZF 1971] Zakharov, V. E., Faddeev, L. D.: Korteweg-de Vries equation, a completely integrable Hamiltonian system. Funk. Anal. Prilož. *5* (4) 18-27 (1971) [Russian]: English transl. in Funct. Anal. Appl. *5*, 280-287 (1971)

[ZMNP 1980] Zakharov, V. E., Manakov, S. V., Novikov, S. P., Pitaievski, L. P.: Theory of Solitons. The Inverse Problem Method. Moscow, Nauka 1980 [Russian]; English transl.: New York, Plenum 1984

[ZS 1971] Zakharov, V. E., Shabat, A. B.: Exact theory of two-dimensional self-focusing and one-dimensional self-modulation of waves in non-linear media. Zh. Exp. Teor. Fiz. *61*, 118-134 (1971) [Russian]; English transl. in Soviet Phys. JETP *34*, 62-69 (1972)

Part One
The Nonlinear Schrödinger Equation (NS Model)

Chapter I
Zero Curvature Representation

§ 1. Formulation of the NS Model

The dynamical system to be considered is generated by the nonlinear equation

$$i\frac{\partial \psi}{\partial t} = -\frac{\partial^2 \psi}{\partial x^2} + 2\varkappa |\psi|^2 \psi \tag{1.1}$$

with the initial condition

$$\psi(x,t)|_{t=0} = \psi(x). \tag{1.2}$$

Here $\psi(x,t)$ is a complex-valued function (the classical charged field) and $|\psi|^2 = \psi\bar{\psi}$, the bar denoting complex conjugation. A real parameter $\varkappa$ in (1.1) is the coupling constant. The domain of the variable x is the whole real line $-\infty < x < \infty$ and the initial data $\psi(x)$ are supposed to be sufficiently smooth.

In the linear limit, $\varkappa = 0$, (1.1) goes into the Schrödinger equation for the wave function of a free one-dimensional particle of mass $m = \frac{1}{2}$. For this reason (1.1) is colloquially called the *nonlinear Schrödinger (NS) equation*, though its physical meaning lies far from one-particle quantum mechanics. Its most significant physical applications are in nonlinear optics. At the same time, (1.1) provides a fairly universal model of a nonlinear equation.

The initial value problem (1.1)–(1.2) must be supplemented with some boundary conditions. We shall consider the following three types.

1. *Rapidly decreasing type:*

$$\psi(x,t) \to 0 \quad \text{as} \quad |x| \to \infty \tag{1.3}$$

sufficiently rapidly; for instance, ψ is in the Schwartz space $\mathscr{S}(\mathbb{R}^1)$ i.e. ψ is infinitely differentiable and together with all its derivatives decays faster than any power of $|x|^{-1}$ as $|x| \to \infty$. Weaker conditions will also be used.

2. *Finite density type:*

$$\psi(x,t)\to\varrho\, e^{i\varphi_\pm(t)} \quad \text{as} \quad x\to\pm\infty \tag{1.4}$$

where $\varrho>0$ and $0\leqslant\varphi_\pm<2\pi$. The constant ϱ^2 plays the role of density and $\varphi_\pm$ are called *asymptotic phases.*

The boundary conditions (1.4) are said to be *satisfied in the sense of Schwartz* if $\psi-\varrho\, e^{i\varphi_\pm}$ is of Schwartz type in the vicinity of $\pm\infty$. This terminology will be frequently used in what follows.

The boundary conditions (1.4) are compatible with (1.1) in the sense that ϱ and $\theta=\varphi_+-\varphi_-$ are time-independent. It is more convenient, however, to force both φ_+ and φ_- to be time-independent. To this end (1.1) should be modified by adding a linear term $-2\varkappa\varrho^2\psi$, so that it becomes

$$i\frac{\partial\psi}{\partial t}=-\frac{\partial^2\psi}{\partial\varkappa^2}+2\varkappa(|\psi|^2-\varrho^2)\psi. \tag{1.5}$$

3. *Quasi-periodic boundary conditions.* Here ψ is a smooth function satisfying

$$\psi(x+2L,t)=e^{i\theta}\psi(x,t), \tag{1.6}$$

where $0\leqslant\theta<2\pi$ and θ does not depend on t. Just as before, (1.6) is consistent with (1.1). Clearly, in this case it is sufficient to consider (1.1) in the fundamental domain for the group of translations generated by the shift $x\mapsto x+2L$. For definiteness, let the fundamental domain be the interval $-L\leqslant x<L$.

Condition 3 is the most general of all listed above. Other types are obtained from (1.6) as subsequent limits $L\to\infty$, $\varrho\to 0$.

Equation (1.1) supplied with the aforementioned boundary conditions determines a dynamical system called the NS model.

Let us show that this is a Hamiltonian system for all the three types of boundary conditions. We shall assume that the reader is familiar with the basic concepts of Hamiltonian mechanics, at least in the finite-dimensional case. Therefore we discuss below only what is specific for infinite-dimensional systems.

We begin with the rapidly decreasing case. Here the phase space $\mathscr{M}_0$ is an infinite-dimensional real linear space with complex coordinates defined by pairs of functions $\psi(x)$, $\bar{\psi}(x)$ in $\mathscr{S}(\mathbb{R}^1)$. By analogy with finite-dimensional coordinates labelled by a discrete parameter, the variable x may be thought of as a coordinate label; for x fixed, $\psi(x)$ and $\bar{\psi}(x)$ range over the two-dimensional real space $\mathbb{R}^2$ with the real coordinates $\operatorname{Re}\psi(x)=\dfrac{\psi(x)+\bar{\psi}(x)}{2}$, $\operatorname{Im}\psi(x)=\dfrac{\psi(x)-\bar{\psi}(x)}{2i}$.

We shall define the *algebra of observables* on the phase space $\mathscr{M}_0$. To this end consider real-valued functionals of the form

$$F(\psi,\bar\psi)=c+\sum_{\substack{n,m=0\\(n,m)\neq(0,0)}}^{\infty}\int_{-\infty}^{\infty}\dots\int_{-\infty}^{\infty}c_{nm}(y_1,\dots,y_n|z_1,\dots,z_m)$$
$$\times\psi(y_1)\dots\psi(y_n)\bar\psi(z_1)\dots\bar\psi(z_m)\,dy_1\dots dy_n\,dz_1\dots dz_m, \tag{1.7}$$

where $c_{nm}(y_1,\dots,y_n|z_1,\dots,z_m)$ are tempered generalized functions on $\mathbb{R}^{n+m}$ symmetric in $y_1,\dots,y_n$ and, separately, in $z_1,\dots,z_m$ and satisfying the reality condition

$$c_{nm}(y_1,\dots,y_n|z_1,\dots,z_m)=\overline{c_{mn}(z_1,\dots,z_m|y_1,\dots,y_n)}\,. \tag{1.8}$$

Suppose in addition that (1.7) is absolutely convergent for all $\psi(x)$, $\bar\psi(x)$ in $\mathcal{S}(\mathbb{R}^1)$. Such functionals will be naturally referred to as *real-analytic* ones.

According to the general definition of the *variational derivative*,

$$\delta F(\psi,\bar\psi)=F(\psi+\delta\psi,\bar\psi+\delta\bar\psi)-F(\psi,\bar\psi)$$
$$=\int_{-\infty}^{\infty}\left(\frac{\delta F}{\delta\psi(x)}\delta\psi(x)+\frac{\delta F}{\delta\bar\psi(x)}\delta\bar\psi(x)\right)dx \tag{1.9}$$

up to terms of higher order in $\delta\psi$ and $\delta\bar\psi$. Then for functionals such as (1.7) we have

$$\frac{\delta F}{\delta\psi(x)}=\sum_{\substack{n,m=0\\(n,m)\neq(0,0)}}^{\infty}n\int_{-\infty}^{\infty}\dots\int_{-\infty}^{\infty}c_{nm}(x,y_1,\dots,y_{n-1}|z_1,\dots,z_m)$$
$$\times\psi(y_1)\dots\psi(y_{n-1})\bar\psi(z_1)\dots\bar\psi(z_m)\,dy_1\dots dy_{n-1}\,dz_1\dots dz_m, \tag{1.10}$$

$$\frac{\delta F}{\delta\bar\psi(x)}=\sum_{\substack{n,m=0\\(n,m)\neq(0,0)}}^{\infty}m\int_{-\infty}^{\infty}\dots\int_{-\infty}^{\infty}c_{nm}(y_1,\dots,y_n|x,z_1,\dots,z_{m-1})$$
$$\times\psi(y_1)\dots\psi(y_n)\bar\psi(z_1)\dots\bar\psi(z_{m-1})\,dy_1\dots dy_n\,dz_1\dots dz_{m-1}. \tag{1.11}$$

Thus in general the variational derivatives $\dfrac{\delta F}{\delta\psi(x)}$, $\dfrac{\delta F}{\delta\bar\psi(x)}$ are generalized functions. A functional is said to be *smooth* if these derivatives are usual functions in Schwartz space.

Smooth real-analytic functionals make up the *algebra of observables on the phase space* $\mathcal{M}_0$. We define a *Poisson structure* on this algebra by the following Poisson bracket

$$\{F,G\}=i\int_{-\infty}^{\infty}\left(\frac{\delta F}{\delta\psi(x)}\frac{\delta G}{\delta\bar\psi(x)}-\frac{\delta F}{\delta\bar\psi(x)}\frac{\delta G}{\delta\psi(x)}\right)dx\,. \tag{1.12}$$

Obviously, the bracket (1.12) possesses the basic properties of a Poisson bracket: it is *skew-symmetric*

$$\{F, G\} = -\{G, F\} \tag{1.13}$$

and satisfies the *Jacobi identity*

$$\{F, \{G, H\}\} + \{H, \{F, G\}\} + \{G, \{H, F\}\} = 0. \tag{1.14}$$

The bracket (1.12) is the infinite-dimensional generalization of the usual Poisson bracket in the phase space $\mathbb{R}^{2n}$ with real coordinates $p_k, q_k, k=1, \ldots, n$,

$$\{f, g\} = \sum_{k=1}^{n} \left(\frac{\partial f}{\partial p_k} \frac{\partial g}{\partial q_k} - \frac{\partial f}{\partial q_k} \frac{\partial g}{\partial p_k} \right), \tag{1.15}$$

written in terms of complex coordinates $z_k = \frac{q_k + i p_k}{\sqrt{2}}$, $\bar{z}_k = \frac{q_k - i p_k}{\sqrt{2}}$:

$$\{f, g\} = i \sum_{k=1}^{n} \left(\frac{\partial f}{\partial z_k} \frac{\partial g}{\partial \bar{z}_k} - \frac{\partial f}{\partial \bar{z}_k} \frac{\partial g}{\partial z_k} \right). \tag{1.16}$$

The coordinates $\psi(x)$ and $\bar{\psi}(x)$ themselves may be considered as functionals on $\mathscr{M}_0$. However, their variational derivatives are generalized functions,

$$\frac{\delta \psi(x)}{\delta \psi(y)} = \delta(x-y), \qquad \frac{\delta \bar{\psi}(x)}{\delta \bar{\psi}(y)} = \delta(x-y), \tag{1.17}$$

where $\delta(x-y)$ is the Dirac δ-function, and $\frac{\delta \psi(x)}{\delta \bar{\psi}(y)}$ and $\frac{\delta \bar{\psi}(x)}{\delta \psi(y)}$ vanish. Substituting (1.17) formally into (1.12) gives

$$\begin{gathered} \{\psi(x), \psi(y)\} = \{\bar{\psi}(x), \bar{\psi}(y)\} = 0, \\ \{\psi(x), \bar{\psi}(y)\} = i\delta(x-y), \end{gathered} \tag{1.18}$$

which can serve as a definition of the Poisson structure in the sense that

$$\{F, G\} = \int_{-\infty}^{\infty} \int_{-\infty}^{\infty} \left(\frac{\delta F}{\delta \psi(x)} \frac{\delta G}{\delta \psi(y)} \{\psi(x), \psi(y)\} + \frac{\delta F}{\delta \psi(x)} \frac{\delta G}{\delta \bar{\psi}(y)} \{\psi(x), \bar{\psi}(y)\} + \frac{\delta F}{\delta \bar{\psi}(x)} \frac{\delta G}{\delta \psi(y)} \{\bar{\psi}(x), \psi(y)\} + \frac{\delta F}{\delta \bar{\psi}(x)} \frac{\delta G}{\delta \bar{\psi}(y)} \{\bar{\psi}(x), \bar{\psi}(y)\} \right) dx\, dy. \tag{1.19}$$

In fact, substituting (1.18) into (1.19) gives (1.12). These formulae also yield

$$\frac{\delta F}{\delta \psi(x)} = -i\{F, \bar{\psi}(x)\}, \qquad \frac{\delta F}{\delta \bar{\psi}(x)} = i\{F, \psi(x)\}. \tag{1.20}$$

It is clear from the definition that the Poisson bracket is *non-degenerate* on the algebra of observables, i.e. if

$$\{F, G\} = 0 \tag{1.21}$$

for each observable G, then $F(\psi, \bar{\psi}) = \text{const}$. Indeed, (1.21) implies that the variational derivatives $\dfrac{\delta F}{\delta \psi(x)}$ and $\dfrac{\delta F}{\delta \bar{\psi}(x)}$ vanish, so that the coefficient functions $c_{nm}(y_1, \ldots, y_n | z_1, \ldots, z_m)$ in (1.7) also vanish. Hence, the phase space $\mathcal{M}_0$ acquires a *symplectic structure*. The corresponding closed 2-form Ω *(symplectic form)* is

$$\Omega = i \int_{-\infty}^{\infty} d\bar{\psi}(x) \wedge d\psi(x)\, dx. \tag{1.22}$$

Each observable H gives rise to a one-parameter group of transformations on the phase space $\mathcal{M}_0$ defined by *Hamilton's equations of motion*

$$\begin{aligned} \frac{\partial \psi}{\partial t} &= \{H, \psi\} = -i \frac{\delta H}{\delta \bar{\psi}}, \\ \frac{\partial \bar{\psi}}{\partial t} &= \{H, \bar{\psi}\} = i \frac{\delta H}{\delta \psi}. \end{aligned} \tag{1.23}$$

In this case H is called the *Hamiltonian*.

In particular, the NS equation of motion (1.1) is represented in the form (1.23) by choosing the Hamiltonian H to be

$$H = \int_{-\infty}^{\infty} \left(\left| \frac{\partial \psi}{\partial x} \right|^2 + \varkappa |\psi|^4 \right) dx . \tag{1.24}$$

The Hamiltonian H (also called the *energy integral*) is the generator of the group of time displacements.

Along with H we consider the functionals

$$N = \int_{-\infty}^{\infty} |\psi|^2 \, dx \tag{1.25}$$

and

$$P = \frac{1}{2i} \int_{-\infty}^{\infty} \left(\frac{\partial \psi}{\partial x} \bar{\psi} - \frac{\partial \bar{\psi}}{\partial x} \psi \right) dx . \tag{1.26}$$

The Hamiltonian transformations generated by N and P are the phase shift

$$\psi(x) \rightarrow e^{i\varphi} \psi(x) \tag{1.27}$$

and the x-displacement

$$\psi(x) \rightarrow \psi(x+a), \tag{1.28}$$

respectively. The physical meaning of N and P is that of *charge (number of particles) and momentum*, respectively.

It is easily verified that

$$\{H, P\} = \{H, N\} = 0 \tag{1.29}$$

and

$$\{N, P\} = 0 . \tag{1.30}$$

Indeed, the NS equation is invariant under the transformations (1.27) and (1.28). By (1.29) N and P are *integrals of the motion*, i.e. they are constant along the trajectories of (1.23), since for every observable F one has

$$\frac{dF}{dt} = \int_{-\infty}^{\infty} \left(\frac{\delta F}{\delta \psi(x)} \frac{\partial \psi(x)}{\partial t} + \frac{\delta F}{\delta \bar{\psi}(x)} \frac{\partial \bar{\psi}(x)}{\partial t} \right) dx = \{H, F\} . \tag{1.31}$$

Two observables are said to be in involution if their Poisson bracket vanishes. Thus, (1.30) shows that the integrals of the motion N and P *are in involution*. Later we shall see that there is an infinite set of integrals of the motion in involution for the NS model implying that the model is completely integrable.

Let us now consider the case of quasi-periodic boundary conditions. Coordinates in the phase space $\mathscr{M}_{L,\theta}$ are given by pairs of smooth functions $\psi(x)$, $\bar{\psi}(x)$ subject to (1.6). Naturally, functionals on $\mathscr{M}_{L,\theta}$ depend only on the values of $\psi(x)$ and $\bar{\psi}(x)$ in the fundamental domain of the group of translations $x \mapsto x+2nL$, where n is an integer.

The definition of admissible functionals associated with observables differs from the one given above in only two points: first, integration over the y_i, z_j in (1.7) is restricted to the fundamental domain; second, the coefficient functions $c_{nm}(y_1,\dots,y_n|z_1,\dots,z_m)$ must satisfy the *quasi-periodicity conditions*

$$\begin{aligned} &c_{nm}(y_1,\dots,y_i+2L,\dots,y_n|z_1,\dots,z_m)\\ &\quad=e^{-i\theta}c_{nm}(y_1,\dots,y_i,\dots,y_n|z_1,\dots,z_m); \qquad i=1,\dots,n, \end{aligned} \tag{1.32}$$

$$\begin{aligned} &c_{nm}(y_1,\dots,y_n|z_1,\dots,z_j+2L,\dots,z_m)\\ &\quad=e^{i\theta}c_{nm}(y_1,\dots,y_n|z_1,\dots,z_j,\dots,z_m); \qquad j=1,\dots m, \end{aligned} \tag{1.33}$$

understood in the sense of generalized functions. The integrands in (1.7) are then periodic in each variable separately so that the integral does not depend on the choice of the fundamental domain. As before, the variational derivatives are given by (1.10)–(1.11) and are supposed to be smooth functions of x.

In terms of variational derivatives, conditions (1.32)–(1.33) become

$$\begin{aligned} \frac{\delta F}{\delta\psi(x)}&=e^{i\theta}\frac{\delta F}{\delta\psi(y)}\bigg|_{y=x+2L},\\ \frac{\delta F}{\delta\bar{\psi}(x)}&=e^{-i\theta}\frac{\delta F}{\delta\bar{\psi}(y)}\bigg|_{y=x+2L}. \end{aligned} \tag{1.34}$$

The Poisson bracket of two observables is defined as in (1.12) and has the form

$$\{F,G\}=i\int_{-L}^{L}\left(\frac{\delta F}{\delta\psi(x)}\frac{\delta G}{\delta\bar{\psi}(x)}-\frac{\delta F}{\delta\bar{\psi}(x)}\frac{\delta G}{\delta\psi(x)}\right)dx, \tag{1.35}$$

where the value of the integral does not depend on the choice of the fundamental domain. This Poisson bracket is well defined and non-degenerate on the algebra of observables.

Formally, the Poisson brackets of the coordinates $\psi(x)$ and $\bar{\psi}(x)$ are

$$\begin{aligned} \{\psi(x), \psi(y)\} &= \{\bar{\psi}(x), \bar{\psi}(y)\} = 0, \\ \{\psi(x), \bar{\psi}(y)\} &= i\delta_{L,\theta}(x-y), \end{aligned} \tag{1.36}$$

where $\delta_{L,\theta}(x)$ is the averaged δ-function,

$$\delta_{L,\theta}(x) = \sum_{n=-\infty}^{\infty} e^{i\theta n}\,\delta(x-2nL), \tag{1.37}$$

which satisfies the quasi-periodicity condition with respect to x.

Here the NS equation can also be represented in the Hamiltonian form (1.23). The Hamiltonian H is given by (1.24) as before, with integration over the fundamental domain. The observables N and P are defined in a similar way and have the same physical interpretation as above.

Besides the functionals associated with observables, in Chapter III we shall also need *compactly supported functionals*. Fix a fundamental domain, say $-L \leqslant x < L$. The definition requires that with respect to each variable the coefficients $c_{nm}(y_1, \ldots, y_n | z_1, \ldots, z_m)$ have support in this interval. The variational derivatives are supposed to be smooth functions within the support. For such functionals the Poisson bracket (1.35) is well defined and non-degenerate. The algebra of admissible functionals is the completion of compactly supported functionals after imposing the quasi-periodicity condition. The Poisson bracket for observables is the corresponding limit of the Poisson bracket for compactly supported functionals.

Finally, let us discuss the finite density case. The phase space $\mathscr{M}_{\varrho,\theta}$ is obtained from $\mathscr{M}_{L,\theta}$ in the limit $L \to \infty$ if the values of $\psi(x)$ and $\bar{\psi}(x)$ at $x = -L$ are kept fixed and equal to ϱ. Thus, $\mathscr{M}_{\varrho,\theta}$ is parametrized by two real parameters ϱ and θ, $0 < \varrho < \infty$, $0 \leqslant \theta < 2\pi$ and consists of pairs of functions $\psi(x)$, $\bar{\psi}(x)$ satisfying the boundary conditions (1.4) in the sense of Schwartz, with $\varphi_- = 0$, $\varphi_+ = \theta$. Notice that, in contrast with the previous examples, $\mathscr{M}_{\varrho,\theta}$ is not a linear space.

Functionals on $\mathscr{M}_{\varrho,\theta}$ are obtained as limiting values, as $L \to \infty$, of the admissible functionals for the quasi-periodic case. However, *admissible functionals* associated with observables are subject to the additional condition that their variational derivatives be of Schwartz type. In fact, $\dfrac{\delta F}{\delta \psi(x)}$ and $\dfrac{\delta F}{\delta \bar{\psi}(x)}$ enter into Hamilton's equations (1.23) and their decay guarantees that the Hamiltonian transformations leave the phase space $\mathscr{M}_{\varrho,\theta}$ invariant.

The Poisson bracket for observables is again given by (1.12), and the formal Poisson bracket for the coordinates $\psi(x)$ and $\bar{\psi}(x)$ has the same form as (1.18), since $\delta_{L,\theta}(x)$ goes into the usual δ-function as $L\to\infty$. The Poisson structure is non-degenerate because it is the limit, as $L\to\infty$, of the non-degenerate quasi-periodic structure with the following two non-commuting constraints

$$\psi(-L)=\bar{\psi}(-L)=\varrho. \tag{1.38}$$

The simplest example of an inadmissible functional is given by

$$N_\varrho=\int_{-\infty}^{\infty}(|\psi|^2-\varrho^2)\,dx \tag{1.39}$$

which could have been a natural analogue of charge in the rapidly decreasing case. Indeed, the variational derivatives

$$\frac{\delta N_\varrho}{\delta\psi(x)}=\bar{\psi}(x), \qquad \frac{\delta N_\varrho}{\delta\bar{\psi}(x)}=\psi(x) \tag{1.40}$$

do not vanish as $|x|\to\infty$ by virtue of the boundary conditions (1.4). This is related to the fact that the phase shift (1.27) cannot be defined on $\mathcal{M}_{\varrho,\theta}$ because the phase of $\psi(x)$ has a fixed limit as $x\to-\infty$. Another example of an inadmissible functional is provided by a naive regularization of the quasi-periodic Hamiltonian

$$\tilde{H}_\varrho=\int_{-\infty}^{\infty}\left(\left|\frac{\partial\psi}{\partial x}\right|^2+\varkappa(|\psi|^4-\varrho^4)\right)dx. \tag{1.41}$$

On the other hand,

$$H_\varrho=\int_{-\infty}^{\infty}\left(\left|\frac{\partial\psi}{\partial x}\right|^2+\varkappa(|\psi|^2-\varrho^2)^2\right)dx \tag{1.42}$$

and

$$P=\frac{1}{2i}\int_{-\infty}^{\infty}\left(\frac{\partial\psi}{\partial x}\bar{\psi}-\frac{\partial\bar{\psi}}{\partial x}\psi\right)dx \tag{1.43}$$

are admissible functionals on $\mathcal{M}_{\varrho,\theta}$ and play the role of energy and momentum, respectively. The modified equations of motion (1.5) are precisely those generated by H_ϱ.

The basic characteristic of the problem is its *monodromy matrix* which will be defined here in the quasi-periodic case. In terms of U, V the quasi-periodicity conditions can be expressed as

$$U(x+2L,t,\lambda)=Q^{-1}(\theta)\,U(x,t,\lambda)\,Q(\theta), \tag{2.23}$$

$$V(x+2L,t,\lambda)=Q^{-1}(\theta)\,V(x,t,\lambda)\,Q(\theta), \tag{2.24}$$

with a diagonal matrix $Q(\theta)$,

$$Q(\theta)=\begin{pmatrix} e^{\frac{i\theta}{2}} & 0 \\ 0 & e^{-\frac{i\theta}{2}} \end{pmatrix} = \exp\left\{\frac{i\theta\sigma_3}{2}\right\}. \tag{2.25}$$

The monodromy matrix T_L is the matrix of parallel transport along the contour $t=t_0$, $-L\leqslant x\leqslant L$ oriented in the positive x-direction,

$$T_L(\lambda,t_0)=\overset{\curvearrowleft}{\exp}\int_{-L}^{L} U(x,t_0,\lambda)\,dx. \tag{2.26}$$

The zero curvature condition leads to a remarkable relationship between monodromy matrices for different values of t. To derive it, consider a closed rectilinear curve γ presented in Fig. 1.

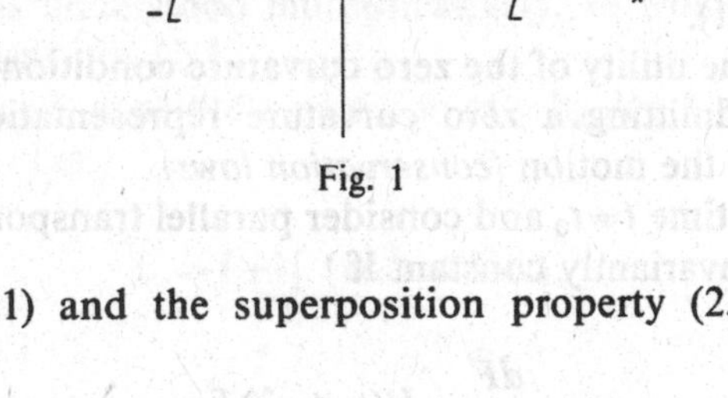

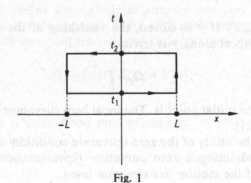

Fig. 1

By virtue of (2.21) and the superposition property (2.18) we have for such γ

$$S_-^{-1}T_L^{-1}(t_2)S_+T_L(t_1)=I, \tag{2.27}$$

where

$$S_\pm(\lambda,t_1,t_2)=\overset{\curvearrowleft}{\exp}\int_{t_1}^{t_2} V(\pm L,t,\lambda)\,dt. \tag{2.28}$$

By the quasi-periodicity (2.24) and the definition of the ordered exponential we see that S_+ is conjugate to S_-,

$$S_+ = Q^{-1}(\theta) S_- Q(\theta). \tag{2.29}$$

Thus (2.27) becomes

$$T_L(\lambda, t_2) Q(\theta) = S_+(t_1, t_2) T_L(\lambda, t_1) Q(\theta) S_+^{-1}(t_1, t_2), \tag{2.30}$$

that is, $T_L(\lambda, t) Q(\theta)$ *are conjugate to each other for different values of* t. In particular, (2.30) implies an important relation

$$\operatorname{tr} T_L(\lambda, t_2) Q(\theta) = \operatorname{tr} T_L(\lambda, t_1) Q(\theta), \tag{2.31}$$

where tr denotes matrix trace in $\mathbb{C}^2$. From this we conclude that the trace of $T_L(\lambda, t) Q(\theta)$ does not depend on t.

So, with the zero curvature representation as a starting point we have shown that the functional $F_L(\lambda)$ given by

$$F_L(\lambda) = \operatorname{tr} T_L(\lambda) Q(\theta) \tag{2.32}$$

is a *generating function for the conservation laws for (1.1).*

The particular choice of the fundamental domain $-L \leqslant x < L$ in the definition of the monodromy matrix is not essential. For a different domain $x_0 - L \leqslant x < x_0 + L$ we set

$$T_{L,x_0}(\lambda, t) = \overset{\curvearrowleft}{\exp} \int_{x_0 - L}^{x_0 + L} U(x, t, \lambda)\, dx \tag{2.33}$$

and show that $\operatorname{tr} T_{L,x_0}(\lambda, t) Q(\theta)$ does not depend on x_0. To this end we will prove that $T_L(\lambda, t) Q(\theta)$ and $T_{L,x_0}(\lambda, t) Q(\theta)$ are conjugate to each other. In fact, from (2.33) we have

$$T_{L,x_0}(\lambda, t) = P_+ T_L(\lambda, t) P_-^{-1}, \tag{2.34}$$

where

$$P_\pm(x_0) = \overset{\curvearrowleft}{\exp} \int_{\pm L}^{x_0 \pm L} U(x, t, \lambda)\, dx. \tag{2.35}$$

Using the quasi-periodicity condition (2.23) we find

$$P_+ = Q^{-1}(\theta) P_- Q(\theta), \tag{2.36}$$

whence using (2.34) we conclude that

$$T_{L,x_0}(\lambda, t) Q(\theta) = P_+(x_0) T_L(\lambda, t) Q(\theta) P_+^{-1}(x_0). \tag{2.37}$$

We thus see that the monodromy matrix $T_L(\lambda)$ is a useful tool for describing the dynamics in our model. In the following sections we shall investigate it further and obtain some new dynamical applications.

§ 3. Properties of the Monodromy Matrix in the Quasi-Periodic Case

In this section we shall analyze the monodromy matrix, i.e. the matrix of parallel transport along the fundamental domain $-L \leqslant x \leqslant L$,

$$T_L(\lambda) = \overset{\curvearrowleft}{\exp} \int_{-L}^{L} U(x, \lambda)\, dx, \tag{3.1}$$

where $U(x, \lambda)$ is given by (2.3)–(2.5) and satisfies the quasi-periodicity condition

$$U(x + 2L, \lambda) = Q^{-1}(\theta)\, U(x, \lambda)\, Q(\theta). \tag{3.2}$$

Here we have suppressed the t-dependence assuming that t is fixed.

Along with the monodromy matrix we shall consider a more general object, the parallel transport matrix from y to x along the x-axis,

$$T(x, y, \lambda) = \overset{\curvearrowleft}{\exp} \int_{y}^{x} U(z, \lambda)\, dz, \tag{3.3}$$

which will be called the *transition matrix*. The monodromy matrix $T_L(\lambda)$ is a special case of the transition matrix:

$$T_L(\lambda) = T(L, -L, \lambda). \tag{3.4}$$

The basic properties of $T(x, y, \lambda)$ are the following.

It satisfies the differential equation (2.22) of the auxiliary linear problem

$$\frac{\partial}{\partial x} T(x, y, \lambda) = U(x, \lambda)\, T(x, y, \lambda) \tag{3.5}$$

with the initial condition

$$T(x, y, \lambda)|_{x=y} = I. \tag{3.6}$$

This property can serve as an alternative definition of the transition matrix. The well-known theorems for ordinary differential equations assure that the solution $T(x, y, \lambda)$ exists for all x, y, is unique, and is an entire function of λ. The latter follows because $U(x, \lambda)$ and the initial data (3.6) are analytic in λ.

The *superposition property*

$$T(x, z, \lambda)\, T(z, y, \lambda) = T(x, y, \lambda) \tag{3.7}$$

is a consequence of either the more general relation (2.18) or the differential equation (3.5), (3.6). In particular, one has the relation

$$T(x, y, \lambda) = T^{-1}(y, x, \lambda), \tag{3.8}$$

consistent with the differential equation for $T(x, y, \lambda)$ with respect to y,

$$\frac{\partial}{\partial y} T(x, y, \lambda) = -T(x, y, \lambda)\, U(y, \lambda), \tag{3.9}$$

which follows directly from (3.3).

The matrix $T(x, y, \lambda)$ is unimodular

$$\det T(x, y, \lambda) = 1 \tag{3.10}$$

because $U(x, \lambda)$ is traceless,

$$\operatorname{tr} U(x, \lambda) = 0. \tag{3.11}$$

Indeed, from (3.5) we have

$$\frac{\partial}{\partial x} \det T(x, y, \lambda) = \operatorname{tr} U(x, \lambda) \det T(x, y, \lambda) = 0. \tag{3.12}$$

This computation works for any matrix solution of (3.5) thus proving that its determinant does not depend on x.

The matrix $U(x, \lambda)$ is of a rather special form and satisfies the *involution relation*

$$\bar{U}(x, \lambda) = \sigma\, U(x, \bar{\lambda})\, \sigma, \tag{3.13}$$

where $\sigma = \sigma_1$ for $\varkappa > 0$, $\sigma = \sigma_2$ for $\varkappa < 0$, and $\bar{U}$ denotes a matrix whose elements are the complex conjugates of those of U.

The involution relation naturally extends to the transition matrix, so that

$$\bar{T}(x, y, \lambda) = \sigma T(x, y, \bar{\lambda}) \sigma. \tag{3.14}$$

In particular, the monodromy matrix can be written as

$$T_L(\lambda) = \begin{pmatrix} a_L(\lambda) & \varepsilon \bar{b}_L(\bar{\lambda}) \\ b_L(\lambda) & \bar{a}_L(\bar{\lambda}) \end{pmatrix}, \tag{3.15}$$

with $\varepsilon = \operatorname{sign} \varkappa$. We shall call $a_L(\lambda)$ and $b_L(\lambda)$ *transition coefficients*. They are entire functions of λ satisfying for real λ the *normalization condition*

$$|a_L(\lambda)|^2 - \varepsilon |b_L(\lambda)|^2 = 1, \tag{3.16}$$

which follows from the unimodularity of $T_L(\lambda)$.

This completes the list of elementary properties of the transition and monodromy matrices.

Let us now discuss the time dependence of the monodromy matrix. In § 2 we used its geometric interpretation to obtain the result in the integral form (2.30). As an alternative, it is possible to derive a differential equation for $T_L(\lambda, t)$ with respect to t. In the derivation which follows we restore the t-dependance in our notation.

First we shall obtain the corresponding equation for the transition matrix. By differentiating (3.5) with respect to t

$$\frac{\partial^2 T}{\partial x \partial t} = U \frac{\partial T}{\partial t} + \frac{\partial U}{\partial t} T \tag{3.17}$$

and using the zero curvature condition we can write it as

$$\begin{aligned} \frac{\partial^2 T}{\partial x \partial t} &= \left(\frac{\partial V}{\partial x} + VU - UV \right) T + U \frac{\partial T}{\partial t} \\ &= \frac{\partial V}{\partial x} T + V \frac{\partial T}{\partial x} - UVT + U \frac{\partial T}{\partial t} \\ &= \frac{\partial}{\partial x} (VT) + U \left(\frac{\partial T}{\partial t} - VT \right) \end{aligned} \tag{3.18}$$

or

$$\frac{\partial}{\partial x} \left(\frac{\partial T}{\partial t} - VT \right) = U \left(\frac{\partial T}{\partial t} - VT \right), \tag{3.19}$$

whence we conclude that

$$\frac{\partial}{\partial t} T(x, y) = V(x) T(x, y) + T(x, y) C, \tag{3.20}$$

where the matrix C does not depend on x. Using the initial condition (3.6) we get $C = -V(y)$. As a result, we find the *evolution equation for the transition matrix*,

$$\frac{\partial}{\partial t} T(x, y) = V(x) T(x, y) - T(x, y) V(y). \tag{3.21}$$

For the monodromy matrix this equation simplifies due to the quasi-periodicity conditions. In fact, from (3.21) we get for $T_L(\lambda, t) Q(\theta)$ an *evolution equation of Heisenberg type*

$$\frac{\partial}{\partial t} T_L(\lambda, t) Q(\theta) = [V(L, t, \lambda), T_L(\lambda, t) Q(\theta)]. \tag{3.22}$$

Its solution in terms of ordered exponentials is given by (2.30).

From (3.22) we deduce that

$$\frac{\partial}{\partial t} \operatorname{tr} T_L(\lambda, t) Q(\theta) = 0, \tag{3.23}$$

and we see once again that the functional $F_L(\lambda)$ (see (2.32)) is a generating function for the motion integrals of (1.1).

We conclude this section with a discussion of some deeper analytic properties of the monodromy matrix. Assuming that $\psi(x)$, $\bar{\psi}(x)$ are infinitely differentiable we will prove that $a_L(\lambda)$ and $b_L(\lambda)$ *are entire functions of exponential type L with the following asymptotic expansion for large real* λ:

$$a_L(\lambda) = e^{-i\lambda L} + e^{-i\lambda L} \sum_{n=1}^{\infty} \frac{a_n}{\lambda^n} + e^{i\lambda L} \sum_{n=1}^{\infty} \frac{\tilde{a}_n}{\lambda^n} + O(|\lambda|^{-\infty}) \tag{3.24}$$

and

$$b_L(\lambda) = e^{-i\lambda L} \sum_{n=1}^{\infty} \frac{b_n}{\lambda^n} + e^{i\lambda L} \sum_{n=1}^{\infty} \frac{\tilde{b}_n}{\lambda^n} + O(|\lambda|^{-\infty}). \tag{3.25}$$

Here $O(|\lambda|^{-\infty})$ indicates a function whose asymptotic expansion in powers of λ^{-1} vanishes identically.

The proof is based on an integral representation for the transition matrix which will also be important later. Therefore we shall give the details of the derivation.

We shall start with the integral equations for $T(x,y,\lambda)$ equivalent to the differential problem (3.5)–(3.6)

$$T(x,y,\lambda)=E(x-y,\lambda)+\int_y^x T(x,z,\lambda)\,U_0(z)\,E(z-y,\lambda)\,dz \tag{3.26}$$

and

$$T(x,y,\lambda)=E(x-y,\lambda)+\int_y^x E(x-z,\lambda)\,U_0(z)\,T(z,y,\lambda)\,dz\,, \tag{3.27}$$

where for the sake of definiteness it is assumed that $y\leqslant x$. The matrix $U_0(x)$ is given by (cf. (2.3)–(2.4))

$$U_0(x)=U(x,\lambda)+\frac{i\lambda}{2}\sigma_3=\sqrt{\varkappa}\,(\bar{\psi}\sigma_+ +\psi\sigma_-)\,, \tag{3.28}$$

and $E(x-y,\lambda)$ is the solution of (3.5)–(3.6) for $U_0=0$,

$$E(x-y,\lambda)=\exp\left\{\frac{\lambda}{2i}(x-y)\sigma_3\right\}. \tag{3.29}$$

From

$$\sigma_3\sigma_\pm=-\sigma_\pm\sigma_3 \tag{3.30}$$

we have a useful relation

$$E(x,\lambda)\,U_0(y)=U_0(y)\,E(-x,\lambda)\,. \tag{3.31}$$

Equations (3.26), (3.27) are Volterra integral equations, so that their iterations are absolutely convergent. The analysis of the iterations shows that for $y\leqslant x$, $T(x,y,\lambda)$ can be represented as

$$T(x,y,\lambda)=E(x-y,\lambda)+\int_{2y-x}^{x}\Gamma(x,y,z)\,E(z-y,\lambda)\,dz \tag{3.32}$$

or

$$T(x,y,\lambda)=E(x-y,\lambda)+\int_{y}^{2x-y}E(x-z,\lambda)\,\tilde{\Gamma}(x,y,z)\,dz\,. \tag{3.33}$$

Indeed, using (3.31) we can assemble in each iteration all the factors $E(\cdot,\lambda)$ either on the left or on the right of the product of the $U_0(\cdot)$. After that we use the superposition property of $E(x,\lambda)$.

The matrix-valued kernels Γ, $\tilde{\Gamma}$ satisfy

$$\Gamma(x,y,z)=\frac{1}{2}\,U_0\left(\frac{x+z}{2}\right)+\int_y^{\frac{x+z}{2}}\Gamma(x,s,2s-z)\,U_0(s)\,ds\,, \tag{3.34}$$

where $y\leqslant\frac{x+z}{2}\leqslant x$ and

$$\tilde{\Gamma}(x,y,z)=\frac{1}{2}\,U_0\left(\frac{y+z}{2}\right)+\int_{\frac{y+z}{2}}^{x}U_0(s)\,\tilde{\Gamma}(s,y,2s-z)\,ds\,, \tag{3.35}$$

where $y\leqslant\frac{y+z}{2}\leqslant x$.

For example, in order to derive (3.34) insert (3.32) into (3.26), interchange the integrals and use (3.31). Equating the terms with the common factor $E(z-y,\lambda)$, $2y-x\leqslant z\leqslant x$, gives the desired equation. A similar argument leads to (3.35).

Obviously, the above arguments are reversible, so that the integral equations (3.26), (3.27) are equivalent to (3.34), (3.35), respectively. The integral representations (3.32), (3.33) for the transition matrix as well as the integral equations (3.34), (3.35) for the kernels Γ, $\tilde{\Gamma}$ will be needed later in §§ 5, 6.

The iterations of (3.34) and (3.35) are absolutely convergent. This assertion relies only on the fact that $\psi(x)$, $\bar{\psi}(x)$ are bounded. Let $\|\cdot\|$ denote some matrix norm and let

$$C=\max_{-L\leqslant x\leqslant L}\|U_0(x)\|\,. \tag{3.36}$$

From (3.34), (3.35) we easily get the estimates

$$\|\Gamma(x,y,z)\|\leqslant\frac{c}{2}\left(1+c\left(\frac{x+z}{2}-y\right)\right)I_0(c\sqrt{(x-z)(x+z-2y)}) \tag{3.37}$$

and

$$\|\tilde{\Gamma}(x,y,z)\|\leqslant\frac{c}{2}\left(1+c\left(x-\frac{y+z}{2}\right)\right)I_0(c\sqrt{(z-y)(2x-y-z)}), \tag{3.38}$$

where $I_0(x)$ is the modified Bessel function.

We point out that the estimates (3.37), (3.38) become too rough for large values of the arguments in Γ, $\tilde{\Gamma}$. More accurate estimates will be obtained in § 5.

The involution property extends to Γ and $\tilde{\Gamma}$,

$$\bar{\Gamma}(x,y,z)=\sigma\Gamma(x,y,z)\sigma, \quad \bar{\tilde{\Gamma}}(x,y,z)=\sigma\tilde{\Gamma}(x,y,z)\sigma \tag{3.39}$$

and allows us to write them as

$$\Gamma=\begin{pmatrix}\alpha & \varepsilon\bar{\beta}\\ \beta & \bar{\alpha}\end{pmatrix}, \quad \tilde{\Gamma}=\begin{pmatrix}\tilde{\alpha} & \varepsilon\bar{\tilde{\beta}}\\ \tilde{\beta} & \bar{\tilde{\alpha}}\end{pmatrix}, \tag{3.40}$$

with $\varepsilon=\operatorname{sign}\varkappa$.

The integral representations (3.32) and (3.33) determine the relationship between the scalar kernels α,β and $\tilde{\alpha},\tilde{\beta}$, respectively. In fact, $E(x,\lambda)$ commutes with the diagonal parts of Γ, $\tilde{\Gamma}$ and is replaced by its inverse after permutation with their off-diagonal parts. Therefore by carrying $E(z-y,\lambda)$ to the left in (3.32) we obtain a representation such as (3.33); then, by comparing the coefficients we have

$$\alpha(x,y,z)=\tilde{\alpha}(x,y,x+y-z), \tag{3.41}$$

$$\beta(x,y,z)=\tilde{\beta}(x,y,x-y+z). \tag{3.42}$$

Now observe that, as is clear from (3.34), (3.35), the smoothness properties of $\Gamma(x,y,z)$ and $\tilde{\Gamma}(x,y,z)$ are the same as those of $\psi(x)$, $\bar{\psi}(x)$. In particular, if $\psi(x)$, $\bar{\psi}(x)$ are infinitely differentiable, so are the kernels Γ and $\tilde{\Gamma}$ with respect to each of their arguments. Hence we may successively integrate by parts in the integral representations (3.32) and (3.33). Using the differential equation for $E(x,\lambda)$ we then obtain an asymptotic expansion for $T(x,y,\lambda)$ for large real λ,

$$T(x,y,\lambda)= E(x-y,\lambda)+\sum_{n=1}^{\infty}\frac{T_n(x,y)}{\lambda^n}E(x-y,\lambda)$$
$$+\sum_{n=1}^{\infty}\frac{\tilde{T}_n(x,y)}{\lambda^n}E(y-x,\lambda)+O(|\lambda|^{-\infty}). \tag{3.43}$$

Let us now return to the monodromy matrix $T_L(\lambda)$. We have established the representations

$$T_L(\lambda)=E(2L,\lambda)+\int_{-2L}^{2L}\Gamma(L,-L,x-L)E(x,\lambda)dx \tag{3.44}$$

and

$$T_L(\lambda)=E(2L,\lambda)+\int_{-2L}^{2L}E(x,\lambda)\tilde{\Gamma}(L,-L,L-x)dx. \tag{3.45}$$

For the transition coefficients $a_L(\lambda)$ and $b_L(\lambda)$ we then find

$$a_L(\lambda)=e^{-i\lambda L}+\int_{-L}^{L}\alpha_L(x)e^{-i\lambda x}dx \tag{3.46}$$

and

$$b_L(\lambda)=\int_{-L}^{L}\beta_L(x)e^{i\lambda x}dx, \tag{3.47}$$

where

$$\alpha_L(x)=2\alpha(L,-L,2x-L)=2\tilde{\alpha}(L,-L,L-2x), \tag{3.48}$$

$$\beta_L(x)=2\beta(L,-L,2x-L)=2\tilde{\beta}(L,-L,L+2x). \tag{3.49}$$

Thus the entire functions $a_L(\lambda)$ and $b_L(\lambda)$ are of exponential type L and if $\psi(x)$, $\bar{\psi}(x)$ are infinitely differentiable, allow the asymptotic expansions (3.24) and (3.25).

This completes our discussion of the analytic properties.

The generating function $F_L(\lambda)$ has the following expression in terms of $a_L(\lambda)$:

$$F_L(\lambda)=\operatorname{tr}T_L(\lambda)Q(\theta)=a_L(\lambda)e^{\frac{i\theta}{2}}+\bar{a}_L(\bar{\lambda})e^{-\frac{i\theta}{2}}. \tag{3.50}$$

Thus the coefficients a_n, $\tilde{a}_n$ are involved in the construction of the integrals of the motion. An explicit procedure for expressing them in terms of $\psi(x)$, $\bar{\psi}(x)$ is developed in the following section.

§ 4. Local Integrals of the Motion

So far, the set of conservation laws produced by the generating function

$$F_L(\lambda)=\operatorname{tr}T_L(\lambda)Q(\theta) \tag{4.1}$$

has not been described explicitly enough. *Here we will show that*

$$p_L(\lambda)=\arccos\tfrac{1}{2}F_L(\lambda) \tag{4.2}$$

is the generating function for the local integrals of the motion and develope an explicit recursion procedure for computing them in terms of $\psi(x)$ and $\bar{\psi}(x)$. By a local functional on the phase space $\mathcal{M}_{L,\theta}$ we mean a functional of the form

$$F(\psi,\bar{\psi})=\int_{-L}^{L} P(x)\,dx, \tag{4.3}$$

where $P(x)$ is a polynomial in $\psi(x)$, $\bar{\psi}(x)$ and their x-derivatives. Of course, $\psi(x)$, $\bar{\psi}(x)$ are supposed to be infinitely differentiable.

The procedure is based on the large real λ expansion of $p_L(\lambda)$,

$$p_L(\lambda)=-\lambda L+\frac{\theta}{2}+\varkappa\sum_{n=1}^{\infty}\frac{I_n}{\lambda^n}+O(|\lambda|^{-\infty}), \tag{4.4}$$

which is a consequence of (3.24), (3.50) and (4.2). We are going to show that the coefficients I_n are local functionals of $\psi(x)$, $\bar{\psi}(x)$.

Let us first consider the transition matrix $T(x,y,\lambda)$.

In the previous section we proved that it has the asymptotic expansion (3.43). Let us show that *by a suitable transformation this expansion can be reduced to*

$$T(x,y,\lambda)=(I+W(x,\lambda))\exp Z(x,y,\lambda)\cdot(I+W(y,\lambda))^{-1}, \tag{4.5}$$

where W and Z are an off-diagonal and a diagonal matrix, respectively, with the following asymptotic representations as $|\lambda|\to\infty$:

$$W(x,\lambda)=\sum_{n=1}^{\infty}\frac{W_n(x)}{\lambda^n}+O(|\lambda|^{-\infty}), \tag{4.6}$$

$$Z(x,y,\lambda)=\frac{(x-y)\lambda\sigma_3}{2i}+\sum_{n=1}^{\infty}\frac{Z_n(x,y)}{\lambda^n}+O(|\lambda|^{-\infty}). \tag{4.7}$$

Clearly, the expansion associated with the right hand side of (4.5) has the same structure as (3.43). Therefore, in order to prove (4.5) it is sufficient to show that the coefficients $W_n(x)$ and $Z_n(x,y)$ are uniquely determined by $T(x,y,\lambda)$. To do so we shall use the differential equation (3.5) with the initial condition (3.6) which characterize $T(x,y,\lambda)$ uniquely.

Geometrically (4.5) may be interpreted as a gauge transformation defined by the matrix $G(x,\lambda)=(I+W(x,y))^{-1}$ (see (2.19)). This transformation asymptotically reduces the transition matrix to diagonal form $\exp Z(x,y,\lambda)$. Alternatively one can say that this gauge transformation asymptotically reduces the potential $U(x,\lambda)$ of the differential equation (3.5) to diagonal form.

Let us now come back to the problem (3.5)–(3.6) for $T(x,y,\lambda)$ and determine $W(x,\lambda)$ and $Z(x,y,\lambda)$.

For that insert (4.5) into (3.5), cancel out the x-independent matrix $(I+W(y,\lambda))^{-1}$ and split the result into diagonal and off-diagonal parts. We then obtain the equations

$$\frac{dW}{dx}+W\frac{\partial Z}{\partial x}=U_0+\lambda U_1 W, \tag{4.8}$$

$$\frac{\partial Z}{\partial x}=U_0 W+\lambda U_1, \tag{4.9}$$

where we have used the decomposition $U(x,\lambda)=U_0(x)+\lambda U_1$, $U_1=\frac{1}{2i}\sigma_3$. By eliminating $\partial Z/\partial x$ from (4.8) we find for W a nonlinear equation of Riccati type

$$\frac{dW}{dx}+i\lambda\sigma_3 W+WU_0 W-U_0=0, \tag{4.10}$$

where we have used that U_1 anticommutes with W.

The differential equation (4.9) with the initial condition $Z(x,y,\lambda)|_{x=y}=0$ implied by (3.6) can easily be solved,

$$Z(x,y,\lambda)=\frac{\lambda(x-y)}{2i}\sigma_3+\int_y^x U_0(z)\,W(z,\lambda)\,dz, \tag{4.11}$$

which gives the asymptotic expansion (4.7) in terms of the expansion (4.6) for $W(x,\lambda)$.

Substituting (4.6) into the differential equation (4.10) gives the following recursion relations for $W_n(x)$:

$$W_{n+1}(x)=i\sigma_3\left(\frac{dW_n(x)}{dx}+\sum_{k=1}^{n-1}W_k(x)\,U_0(x)\,W_{n-k}(x)\right) \tag{4.12}$$

with the initial condition

$$W_1(x)=-i\sigma_3 U_0(x)=i\sqrt{\varkappa}\begin{pmatrix}0 & -\bar{\psi}(x)\\ \psi(x) & 0\end{pmatrix}$$
$$=i\sqrt{\varkappa}\,(\psi(x)\sigma_- -\bar{\psi}(x)\sigma_+). \tag{4.13}$$

The coefficients $W_n(x)$ are uniquely determined by (4.12), (4.13) and can be expressed locally in terms of $U_0(x)$ and its derivatives. By (4.12) and (4.13) the asymptotic series $W(x,\lambda)$ satisfies the involution relation

$$\bar{W}(x,\lambda)=\sigma W(x,\bar{\lambda})\sigma \tag{4.14}$$

and is quasi-periodic

$$W(x+2L,\lambda)=Q^{-1}(\theta)\,W(x,\lambda)\,Q(\theta). \tag{4.15}$$

So, $W(x,\lambda)$ can be represented in the form

$$W(x,\lambda)=i\sqrt{\varkappa}\,(w(x,\lambda)\sigma_{-}-\bar{w}(x,\bar{\lambda})\sigma_{+}), \tag{4.16}$$

where

$$w(x,\lambda)=\sum_{n=1}^{\infty}\frac{w_n(x)}{\lambda^n} \tag{4.17}$$

with the quasi-periodic $w_n(x)$,

$$w_n(x+2L)=e^{i\theta}w_n(x). \tag{4.18}$$

In terms of $w_n(x)$, the recursion relations and the initial condition become

$$w_{n+1}(x)=-i\frac{dw_n}{dx}(x)+\varkappa\bar{\psi}(x)\sum_{k=1}^{n-1}w_k(x)\,w_{n-k}(x) \tag{4.19}$$

and

$$w_1(x)=\psi(x). \tag{4.20}$$

Recalling the representation for Z we see that the diagonal matrix $U_0 W$ involved in (4.11) has the form

$$U_0(x)\,W(x,\lambda)=i\varkappa\begin{pmatrix}\bar{\psi}(x)\,w(x,\lambda) & 0\\ 0 & -\psi(x)\,\bar{w}(x,\bar{\lambda})\end{pmatrix}, \tag{4.21}$$

where the asymptotic series $\bar{\psi}(x)\,w(x,\lambda)$ and $\psi(x)\,\bar{w}(x,\bar{\lambda})$ are periodic in x. This completes our discussion of the modified asymptotic expansion for the transition matrix $T(x,y,\lambda)$.

We now turn to the monodromy matrix. From (4.5) we have the representation

$$T_L(\lambda)=(I+W(L,\lambda))\exp Z_L(\lambda)\,(I+W(-L,\lambda))^{-1}, \tag{4.22}$$

where

$$Z_L(\lambda) = -i\lambda L\sigma_3 + \int_{-L}^{L} U_0(x)\, W(x,\lambda)\, dx. \tag{4.23}$$

Since $U_0(x)\, W(x,\lambda)$ is periodic, the integral in (4.23) is independent of the choice of the fundamental domain.

Taking the product of the asymptotic expansions for $I + W(\pm L, \lambda)$ and $Z_L(\lambda)$ we get from (4.22) an asymptotic representation for $T_L(\lambda)$ which has the same form as (3.24)–(3.25). Moreover, we also have a method for computing the coefficients a_n, $\tilde{a}_n$ and b_n, $\tilde{b}_n$ occurring in (3.24)–(3.25).

In particular, for the coefficients I_n in the expansion of $p_L(\lambda)$ the procedure simplifies considerably. Indeed, using quasi-periodicity and (4.22) we find

$$T_L(\lambda)\, Q(\theta) = (I + W(L,\lambda)) \exp(Z_L(\lambda))\, Q(\theta)\, (I + W(L,\lambda))^{-1}, \tag{4.24}$$

so that

$$F_L(\lambda) = \operatorname{tr} T_L(\lambda)\, Q(\theta) = \operatorname{tr} \exp\left\{ Z_L(\lambda) + \frac{i\theta}{2}\sigma_3 \right\}. \tag{4.25}$$

Since $T_L(\lambda)\, Q(\theta)$ is unimodular, we have

$$\operatorname{tr} Z_L(\lambda) = O(|\lambda|^{-\infty}). \tag{4.26}$$

From (4.11) and (4.21) we then conclude that

$$\varphi_L(\lambda) = \varkappa \int_{-L}^{L} \bar{\psi}(x)\, w(x,\lambda)\, dx \tag{4.27}$$

is an asymptotic series with real coefficients,

$$\varphi_L(\lambda) = \bar{\varphi}_L(\bar{\lambda}). \tag{4.28}$$

As a result, $Z_L(\lambda)$ can be expressed as

$$Z_L(\lambda) = i\sigma_3(\varphi_L(\lambda) - \lambda L), \tag{4.29}$$

so that

$$F_L(\lambda) = 2\cos\left(\varphi_L(\lambda) + \frac{\theta}{2} - \lambda L\right). \tag{4.30}$$

Thus it is natural to work with the function $p_L(\lambda)=\arccos\frac{1}{2}F_L(\lambda)$ used in (4.2).

In this notation we have

$$p_L(\lambda)=-\lambda L+\frac{\theta}{2}+\varphi_L(\lambda), \tag{4.31}$$

where the function $\varphi_L(\lambda)$ defined by (4.27) has an asymptotic expansion

$$\varphi_L(\lambda)=\varkappa\sum_{n=1}^{\infty}\frac{I_n}{\lambda^n}+O(|\lambda|^{-\infty}). \tag{4.32}$$

Here

$$I_n(\psi,\bar\psi)=\int_{-L}^{L}P_n(x)\,dx, \tag{4.33}$$

where

$$P_n(x)=\bar\psi(x)\,w_n(x) \tag{4.34}$$

is a polynomial in $\psi(x)$, $\bar\psi(x)$ and their derivatives at x. Since $P_n(x)$ is periodic, this implies that I_n is an admissible functional on $\mathscr{M}_{L,\theta}$, i.e. satisfies (1.34).

The first four densities $P_n(x)$ have the form

$$P_1(x)=|\psi(x)|^2,\qquad P_2(x)=-i\bar\psi(x)\frac{d\psi}{dx}(x),$$

$$P_3(x)=-\bar\psi(x)\frac{d^2\psi(x)}{dx^2}+\varkappa|\psi(x)|^4,$$

$$P_4(x)=i\left(\bar\psi(x)\frac{d^3\psi(x)}{dx^3}-\varkappa|\psi(x)|^2\left(\psi(x)\frac{d\bar\psi}{dx}(x)+4\bar\psi(x)\frac{d\psi(x)}{dx}\right)\right). \tag{4.35}$$

The functionals I_n, $n=1,2,\ldots$, are the promised local motion integrals for the NS model in the quasi-periodic case. As it follows from (4.35), the first three of them, I_1, I_2 and I_3, coincide with the functionals N, P and H introduced in § 1. Later we shall see that all the I_n are in involution with respect to the Poisson bracket defined in § 1.

The results of this and earlier sections cover all the essential elementary properties of the NS model and its monodromy matrix for the quasi-periodic boundary conditions. A complete description of the corresponding dynamics requires a more sophisticated machinary which in beyond the scope of this book. Considerable simplifications occur in the limit $L\to\infty$ for the boundary conditions of rapid decrease or finite density which will be examined in the following sections.

§ 5. The Monodromy Matrix in the Rapidly Decreasing Case

This section is of auxiliary character. Here we shall analyze the properties of the transition matrix $T(x, y, \lambda)$ on the whole axis $-\infty < x, y < \infty$ under the assumption that $\psi(x)$, $\bar{\psi}(x)$ vanish as $|x| \to \infty$. More precisely, these functions will be supposed absolutely integrable on $\mathbb{R}^1$, i.e. $\psi(x)$ lies in $L_1(-\infty, \infty)$. In terms of $U_0(x)$ this amounts to

$$\int_{-\infty}^{\infty} \|U_0(x)\| \, dx < \infty . \tag{5.1}$$

In the following the space of 2×2 matrix functions satisfying (5.1) will be denoted by $L_1^{(2\times 2)}(-\infty, \infty)$. The off-diagonal matrix $U_0(x)$ is a specific element of $L_1^{(2\times 2)}(-\infty, \infty)$.

Under this assumption we will show that *the limits*

$$T_\pm(x, \lambda) = \lim_{y \to \pm\infty} T(x, y, \lambda) E(y, \lambda) \tag{5.2}$$

exist for real λ, where $E(x, \lambda)$ was defined in § 3. Next we will examine the properties of $T_\pm(x, \lambda)$ and, in particular, their asymptotic behaviour for large x and λ.

The proof will make use of the integral representations (3.32), (3.33). To be more specific, let us consider the limit $y \to -\infty$ and write (3.32) in the form

$$T(x, y, \lambda) E(y, \lambda) = E(x, \lambda) + \int_{2y-x}^{x} \Gamma(x, y, z) E(z, \lambda) \, dz . \tag{5.3}$$

Let us show that Γ is absolutely integrable over the interval $2y - x \leqslant z \leqslant x$ uniformly in y. To this end consider the function

$$\Phi(x, y) = \int_{2y-x}^{x} \|\Gamma(x, y, z)\| \, dz . \tag{5.4}$$

By integrating (3.34) over z in the interval indicated above and interchanging the integrals we obtain the estimate

$$\Phi(x, y) \leqslant \int_{y}^{x} \|U_0(z)\| \, dz + \int_{y}^{x} \|U_0(s)\| \, \Phi(x, s) \, ds . \tag{5.5}$$

By iterating this estimate we find

$$\Phi(x,y) \leqslant \exp \int_y^x \|U_0(z)\|\, dz - 1. \tag{5.6}$$

We will now show that the limit

$$\Gamma_-(x,z) = \lim_{y \to -\infty} \Gamma(x,y,z) \tag{5.7}$$

exists, where the expression $\Gamma_-(x,z)$ belongs to $L_1^{(2\times 2)}(-\infty, x)$ as a function of z for each fixed x and convergence is taken in the L_1 norm. For this it is sufficient to show that the representation

$$\Gamma_-(x,z) = \frac{1}{2} U_0\left(\frac{x+z}{2}\right) + \int_{-\infty}^{\frac{x+z}{2}} \Gamma(x,s,2s-z)\, U_0(s)\, ds, \tag{5.8}$$

which results from (3.34) by formally taking the limit $y \to -\infty$, defines a function in $L_1^{(2\times 2)}(-\infty, x)$. The latter follows immediately from (5.6):

$$\begin{aligned}\int_{-\infty}^x \|\Gamma_-(x,z)\|\, dz &\leqslant \int_{-\infty}^x \|U_0(z)\|\, dz + \int_{-\infty}^x \|U_0(s)\| \cdot \Phi(x,s)\, ds \\ &\leqslant \int_{-\infty}^x \|U_0(z)\|\, dz \exp \int_{-\infty}^x \|U_0(z)\|\, dz.\end{aligned} \tag{5.9}$$

Now, using (5.7) we see that the limit (5.2), as $y \to -\infty$, does exist and there is an integral representation for $T_-(x,\lambda)$,

$$T_-(x,\lambda) = E(x,\lambda) + \int_{-\infty}^x \Gamma_-(x,z) E(z,\lambda)\, dz. \tag{5.10}$$

The existence of the limit (5.2) as $y \to +\infty$ is established by a similar argument using (3.33). Recalling that

$$T(x,y,\lambda) = T^{-1}(y,x,\lambda), \tag{5.11}$$

we see that the limit exists and there is an integral representation

$$T_+^{-1}(x,\lambda) = E(-x,\lambda) + \int_x^{\infty} E(-z,\lambda) \tilde{\Gamma}_+(x,z)\, dz. \tag{5.12}$$

The kernel $\tilde{\Gamma}_+$ is given by

$$\tilde{\Gamma}_+(x,z) = \lim_{y\to\infty} \tilde{\Gamma}(y,x,z) = \frac{1}{2} U_0\left(\frac{x+z}{2}\right)$$

$$+ \int_{\frac{x+z}{2}}^{\infty} U_0(s)\tilde{\Gamma}(s,x,2s-z)\,ds\,, \tag{5.13}$$

and satisfies

$$\int_x^\infty \|\tilde{\Gamma}_+(x,z)\|\,dz \leqslant \int_x^\infty \|U_0(z)\|\,dz \exp \int_x^\infty \|U_0(z)\|\,dz\,. \tag{5.14}$$

It is not difficult to obtain an integral representation for $T_+(x,\lambda)$ itself. Observe that, along with $T(x,y,\lambda)$ and $E(x,\lambda)$, the matrix $T_+(x,\lambda)$ is also unimodular. Using the general formula

$$A^{-1} = \sigma_2 A^\tau \sigma_2\,, \tag{5.15}$$

which is valid for any 2×2 unimodular matrix and relates the inverse, A^{-1}, with the transposed matrix A^τ, we find the integral representation for $T_+(x,\lambda)$,

$$T_+(x,\lambda) = E(x,\lambda) + \int_x^\infty \Gamma_+(x,z)E(z,\lambda)\,dz\,, \tag{5.16}$$

where

$$\Gamma_+(x,z) = \sigma_2 \tilde{\Gamma}_+^\tau(x,z)\sigma_2\,. \tag{5.17}$$

The involution property (3.39) extends naturally to $\Gamma_\pm(x,z)$,

$$\bar{\Gamma}_\pm(x,z) = \sigma\Gamma_\pm(x,z)\sigma\,, \tag{5.18}$$

and also to $T_\pm(x,\lambda)$,

$$\bar{T}_\pm(x,\lambda) = \sigma T_\pm(x,\lambda)\sigma\,. \tag{5.19}$$

In particular, for $\Gamma_\pm$ we have the representation

$$\Gamma_\pm = \begin{pmatrix} \alpha_\pm & \varepsilon\bar{\beta}_\pm \\ \beta_\pm & \bar{\alpha}_\pm \end{pmatrix}, \qquad \varepsilon = \operatorname{sign}\varkappa\,. \tag{5.20}$$

Just as $T(x,y,z)$, the matrix functions $T_\pm(x,\lambda)$ satisfy the differential equation

$$\frac{dF}{dx} = U(x,\lambda)F. \tag{5.21}$$

The initial conditions are now replaced by the asymptotic conditions

$$T_\pm(x,\lambda) = E(x,\lambda) + o(1) \quad \text{if} \quad x \to \pm\infty, \tag{5.22}$$

which follow immediately from (5.10), (5.16) and the estimates (5.9), (5.14).

Here it would be appropriate to comment on the relationship with scattering theory. Since (5.21) contains the spectral parameter λ linearly, multiplying by $i\sigma_3$ on the left reduces it to the usual form of the eigenvalue problem

$$\mathscr{L}F = \frac{\lambda}{2}F \tag{5.23}$$

for the first order matrix differential operator

$$\mathscr{L} = i\sigma_3 \frac{d}{dx} + i\sqrt{\varkappa}\,(\psi(x)\sigma_- - \bar{\psi}(x)\sigma_+). \tag{5.24}$$

If $\varkappa > 0$, $\mathscr{L}$ is a formally self-adjoint operator. The corresponding spectral problem with the coefficients $\psi(x)$, $\bar{\psi}(x)$ stabilizing as $|x| \to \infty$ is the main object of scattering theory. In particular, a significant role in the theory is played by the solutions $T_\pm(x,\lambda)$ which are called the *Jost solutions*.

Let us now examine the analytic properties of the matrix elements of $T_\pm(x,\lambda)$ considered as functions of λ for a fixed x. Recall that $T(x,y,\lambda)$ is an entire function of λ. However, $T_\pm(x,\lambda)$ are not such in general, because their definition involves a passage to the limit. Nevertheless, the integral representations (5.10) and (5.16) combined with the absolute integrability of $\Gamma_\pm(x,z)$ in z imply that the first column of $T_-(x,\lambda)$ and the second column of $T_+(x,\lambda)$ may be analytically extended into the upper half-plane, while the first column of $T_+(x,\lambda)$ and the second column of $T_-(x,\lambda)$ may be analytically extended into the lower half-plane. Indeed, for the corresponding λ the exponentials, $\exp\left\{\pm\frac{i\lambda x}{2}\right\}$, involved in the integral representations (5.10) and (5.16) decay as the integration variable z goes to $+\infty$ or $-\infty$, respectively.

We shall denote the above columns by $T_\pm^{(1,2)}(x,\lambda)$, so that

$$T_\pm(x,\lambda) = (T_\pm^{(1)}(x,\lambda),\, T_\pm^{(2)}(x,\lambda)). \tag{5.25}$$

The integral representations combined with the Riemann-Lebesgue lemma give, at a fixed x, the following asymptotic behaviour:

$$e^{\frac{i\lambda x}{2}} T_{-}^{(1)}(x,\lambda) = \begin{pmatrix} 1 \\ 0 \end{pmatrix} + o(1), \tag{5.26}$$

$$e^{-\frac{i\lambda x}{2}} T_{+}^{(2)}(x,\lambda) = \begin{pmatrix} 0 \\ 1 \end{pmatrix} + o(1) \tag{5.27}$$

for $\operatorname{Im}\lambda \geqslant 0$, $|\lambda| \to \infty$, and

$$e^{\frac{i\lambda x}{2}} T_{+}^{(1)}(x,\lambda) = \begin{pmatrix} 1 \\ 0 \end{pmatrix} + o(1), \tag{5.28}$$

$$e^{-\frac{i\lambda x}{2}} T_{-}^{(2)}(x,\lambda) = \begin{pmatrix} 0 \\ 1 \end{pmatrix} + o(1) \tag{5.29}$$

for $\operatorname{Im}\lambda \leqslant 0$, $|\lambda| \to \infty$.

The involution property extends to complex values of λ and takes the form

$$\bar{T}_{+}^{(1)}(x,\lambda) = \tilde{\sigma} T_{+}^{(2)}(x,\bar{\lambda}), \tag{5.30}$$

where $\operatorname{Im}\lambda \geqslant 0$ and

$$\bar{T}_{-}^{(1)}(x,\lambda) = \tilde{\sigma} T_{-}^{(2)}(x,\bar{\lambda}), \tag{5.31}$$

where $\operatorname{Im}\lambda \leqslant 0$ and $\tilde{\sigma} = \sigma_1$ when $\varkappa > 0$, $\tilde{\sigma} = i\sigma_2$ when $\varkappa < 0$.

In the case when $\psi(x)$, $\bar{\psi}(x)$ vanish outside the interval $-q \leqslant x \leqslant q$, $T_{\pm}(x,\lambda)E(-x,\lambda)$ are entire functions of exponential type q. In fact, from (5.8) and (5.13) it is easily seen that the associated kernels $\Gamma_{+}(x,z)$ and $\Gamma_{-}(x,z)$ vanish for $z > 2q - x$ or $z < -2q - x$, respectively.

Later we shall need the following relationship between $\Gamma_{\pm}(x,z)$ restricted to the diagonal $z = x$ and $U_0(x)$,

$$[\sigma_3, \Gamma_{-}(x,x)] = \sigma_3 U_0(x) \tag{5.32}$$

and

$$[\sigma_3, \Gamma_{+}(x,x)] = -\sigma_3 U_0(x). \tag{5.33}$$

In order to prove (5.32) it is sufficient to observe that the integral equation (3.34) implies

$$\Gamma(x, y, 2y-x) = \tfrac{1}{2} U_0(y), \tag{5.34}$$

so that by (5.8) we have

$$\Gamma_-(x, x) = \frac{1}{2}\left(U_0(x) + \int_{-\infty}^{x} U_0^2(s)\, ds\right). \tag{5.35}$$

Since the diagonal matrix $U_0^2(s)$ commutes with σ_3, this implies (5.32). The proof of (5.33) is quite similar.

To conclude our discussion of the Jost solutions we note that a more traditional method for studying them in scattering theory is based on the integral equations

$$T_-(x, \lambda) = E(x, \lambda) + \int_{-\infty}^{x} E(x-z, \lambda)\, U_0(z)\, T_-(z, \lambda)\, dz \tag{5.36}$$

and

$$T_+(x, \lambda) = E(x, \lambda) - \int_{x}^{\infty} E(x-z, \lambda)\, U_0(z)\, T_+(z, \lambda)\, dz, \tag{5.37}$$

which result from (3.26) and (3.27) in the limit $y \to \pm\infty$ for real λ. We prefer the method used here because the dependence on λ in the integral representations (5.10) and (5.16) is located entirely in the elementary functions $\exp\left\{\pm \frac{i\lambda x}{2}\right\}$.

Now we shall introduce an analogue of $T_L(\lambda)$, the *reduced monodromy matrix* $T(\lambda)$. For real λ it is defined by

$$T(\lambda) = \lim_{\substack{x \to \infty \\ y \to -\infty}} E(-x, \lambda)\, T(x, y, \lambda)\, E(y, \lambda). \tag{5.38}$$

To prove that the limit in (5.38) exists we observe that the transition matrix can be expressed as

$$T(x, y, \lambda) = T_+(x, \lambda)\, T_+^{-1}(y, \lambda) = T_-(x, \lambda)\, T_-^{-1}(y, \lambda), \tag{5.39}$$

because the right hand sides in (5.39) satisfy the differential equation (3.5) with the initial condition (3.6). From (5.39) it is clear that $T_+^{-1}(x, \lambda)\, T_-(x, \lambda)$ does not depend on x. We will show that it coincides with the limit (5.38).

Indeed, let us set

$$T(\lambda) = T_+^{-1}(x, \lambda)\, T_-(x, \lambda), \tag{5.40}$$

then

$$T_-(x,\lambda)=T_+(x,\lambda)\,T(\lambda). \tag{5.41}$$

By inserting (5.41) into (5.39) we find

$$T(x,y,\lambda)=T_+(x,\lambda)\,T(\lambda)\,T_-^{-1}(y,\lambda). \tag{5.42}$$

By virtue of the boundary conditions (5.22) we then conclude that the limit in (5.38) exists and coincides with (5.40).

Putting $x=L$ and $y=-L$ in (5.38) we obtain, as a special case of this formula,

$$T(\lambda)=\lim_{L\to\infty} E(-L,\lambda)\,T(L,-L,\lambda)\,E(-L,\lambda). \tag{5.43}$$

The factor $T(L,-L,\lambda)$ on the right can be interpreted as the monodromy matrix, $T_L(\lambda)$, of the periodic problem with $\psi(x)$, $\bar{\psi}(x)$ extended periodically outside the interval $(-L,L)$ (possibly with dicontinuities). In this sense $T(\lambda)$ can be regarded as the infinite period limit, as $L\to\infty$, of the periodic monodromy matrix $T_L(\lambda)$ with the trivial oscillating factors reduced out.

The reduced monodromy matrix $T(\lambda)$ possesses the same involution property as $T_L(\lambda)$,

$$\overline{T(\lambda)}=\sigma\,T(\lambda)\,\sigma, \tag{5.44}$$

so that it can be written in the form

$$T(\lambda)=\begin{pmatrix} a(\lambda) & \varepsilon\bar{b}(\lambda) \\ b(\lambda) & \bar{a}(\lambda)\end{pmatrix}, \quad \varepsilon=\operatorname{sign}\varkappa. \tag{5.45}$$

We retain for $a(\lambda)$ and $b(\lambda)$ the name of *transition coefficients*. They satisfy the *normalization relation*

$$|a(\lambda)|^2-\varepsilon\,|b(\lambda)|^2=1. \tag{5.46}$$

In terms of transition coefficients the limiting relation (5.43) can be written as

$$a(\lambda)=\lim_{L\to\infty} e^{i\lambda L}a_L(\lambda), \qquad b(\lambda)=\lim_{L\to\infty} b_L(\lambda). \tag{5.47}$$

Some deeper properties of these coefficients will be examined in the next section.

$$\beta(x)=\lim_{L\to\infty}\beta_L(x)=\sqrt{\varkappa}\,\psi(x)+2\sqrt{\varkappa}\int_{-\infty}^{x}\alpha_+(s,2x-s)\,\psi(s)\,ds \tag{6.14}$$

exists and

$$\int_{-\infty}^{\infty}|\beta(x)|\,dx<\infty\,. \tag{6.15}$$

Finally, we obtain the desired representation for $b(\lambda)$,

$$b(\lambda)=\int_{-\infty}^{\infty}\beta(x)\,e^{i\lambda x}\,dx\,, \tag{6.16}$$

where $\beta(x)$ is in $L_1(-\infty,\infty)$.

The set of all functions of the form

$$F(\lambda)=\int_{-\infty}^{\infty}f(x)\,e^{i\lambda x}\,dx\,, \tag{6.17}$$

where $f(x)$ is in $L_1(-\infty,\infty)$, constitutes a complete normed ring $\mathfrak{R}_0$ well known in the mathematical literature. So, under our assumptions on $\psi(x)$, $\bar{\psi}(x)$ the coefficient $b(\lambda)$ belongs to $\mathfrak{R}_0$. The function $a(\lambda)$ in turn belongs to the ring $\mathfrak{R}_+$ composed of functions of the form

$$F_+(\lambda)=c+\int_{0}^{\infty}f_+(x)\,e^{i\lambda x}\,dx\,, \tag{6.18}$$

where $f_+(x)$ lies in $L_1(0,\infty)$. Functions belonging to $\mathfrak{R}_+$ extend analytically into the upper half-plane and tend to a constant as $|\lambda|\to\infty$.

The aforementioned analytic properties of $a(\lambda)$ and $b(\lambda)$ can be deduced directly from their integral representations which also give a complete characterization of the transition coefficients in terms of their Fourier transforms.

The smoothness properties of $\alpha(x)$ and $\beta(x)$ are the same as those of $\psi(x)$ and $\bar{\psi}(x)$. If the latter are in Schwartz space, so is $\beta(x)$ and hence $b(\lambda)$, whereas $\alpha(x)$ is infinitely differentiable and of Schwartz type at $+\infty$.

Let us now investigate the zeros of $a(\lambda)$ in the upper half-plane $\operatorname{Im}\lambda\geqslant 0$.

First we will show that $a(\lambda)$ has no zeros if $\varkappa>0$. In fact, there are no zeros on the real line by virtue of the normalization relation. Suppose that $a(\lambda_0)=0$ with $\operatorname{Im}\lambda_0>0$. It follows from (6.1) that the column vectors $T_-^{(1)}(x,\lambda_0)$ and $T_+^{(2)}(x,\lambda_0)$ are linearly dependent. Since $\operatorname{Im}\lambda_0>0$, the integral representations (5.10), (5.16) imply that $T_-^{(1)}(x,\lambda_0)$ and $T_+^{(2)}(x,\lambda_0)$ decay exponentially as $x\to-\infty$ or $x\to+\infty$, respectively. Thus, for $\lambda=\lambda_0$ equation

(5.21) has a column vector solution decaying exponentially as $|x| \to \infty$. However, (5.21) is equivalent to the spectral problem (5.23) for a formally self-adjoint operator $\mathscr{L}$ (5.24) which would then have a non-real eigenvalue λ_0. This contradiction shows that $a(\lambda)$ has no complex zeros.

If $\varkappa<0$, $\mathscr{L}$ is not self-adjoint, and $a(\lambda)$ may have zeros. The properties of $a(\lambda)$ discussed impose only mild restrictions on them. Analyticity and the asymptotic behaviour (6.3) imply that the zeros are located in a bounded region of the half-plane $\operatorname{Im}\lambda \geqslant 0$ and may only accumulate towards the real line.

To simplify our analysis we shall assume the following *condition (A):*

1) *no zeros occur on the real axis;*

2) *all the zeros are simple.*

In particular, it follows that the total number of zeros is finite and there is a strict inequality for $b(\lambda)$,

$$|b(\lambda)|<1. \tag{6.19}$$

It is hard to formulate the corresponding sufficient conditions in terms of $\psi(x)$, $\bar{\psi}(x)$. Here one is faced with difficult problems of spectral analysis of non-self-adjoint differential operators. This is, however, of little importance for our investigation of the NS dynamical system. Actually, the set of functions $\psi(x)$, $\bar{\psi}(x)$ satisfying condition (*A*) is, in a natural sense, open and dense in the phase space $\mathscr{M}_0$. Later in Chapter III we shall give an alternative description of $\mathscr{M}_0$ which will clarify this statement.

Let $\lambda_1, \ldots, \lambda_n$ be the complete list of zeros of $a(\lambda)$, $\operatorname{Im}\lambda_j>0$, $j=1, \ldots, n$. As noted above, for $\lambda=\lambda_j$ the column $T_-^{(1)}(x, \lambda)$ is proportional to $T_+^{(2)}(x, \lambda)$. Let γ_j be the proportionality coefficient,

$$T_-^{(1)}(x, \lambda_j)=\gamma_j T_+^{(2)}(x, \lambda_j), \quad j=1, \ldots, n, \tag{6.20}$$

$\gamma_j \neq 0$. The set of complex numbers γ_j is one of the characteristics of the auxiliary linear problem and will play an important role in what follows.

It is clear from (6.4) that $\bar{\lambda}_1, \ldots, \bar{\lambda}_n$ are the zeros of $a^*(\lambda)$ in the lower half-plane. Using the involution property we find that

$$T_-^{(2)}(x, \bar{\lambda}_j)=-\bar{\gamma}_j T_+^{(1)}(x, \bar{\lambda}_j), \quad j=1, \ldots, n. \tag{6.21}$$

If $\psi(x)$, $\bar{\psi}(x)$ have compact support, relation (5.41) characterizing the reduced monodromy matrix makes sense for all complex λ, so that substituting $\lambda=\lambda_j$ or $\lambda=\bar{\lambda}_j$ into (5.41) gives $\gamma_j=b(\lambda_j)$, $\bar{\gamma}_j=\bar{b}(\lambda_j)$, $j=1, \ldots, n$. We emphasize that this holds only for compactly supported $\psi(x)$, $\bar{\psi}(x)$.

The set λ_j, $\bar{\lambda}_j$, $j=1, \ldots, n$, is the *discrete part of the spectrum* of (5.23) for $\varkappa<0$. Furthermore, for any $\varkappa$, $\mathscr{L}$ has *continuous spectrum* of multiplicity two on the whole real line, according to the existence for real λ of two linearly independent solutions to (5.23) which are bounded in x. These are, for instance, the columns of $T_-(x,\lambda)$ or of $T_+(x,\lambda)$. With this in mind we shall call $a(\lambda)$ and $b(\lambda)$ *transition coefficients for the continuous spectrum*, and γ_j, $\bar{\gamma}_j$, $j=1,\ldots,n$, will be called *transition coefficients for the discrete spectrum.*

To conclude this section we will show that *the analyticity of $a(\lambda)$ and the normalization relation can be used to express $a(\lambda)$ through its zeros (if there are any) and $b(\lambda)$.* Namely, for $\operatorname{Im}\lambda>0$

$$a(\lambda)=\exp\left\{\frac{1}{2\pi i}\int_{-\infty}^{\infty}\frac{\log(1+|b(\mu)|^2)}{\mu-\lambda}\,d\mu\right\} \tag{6.22}$$

if $\varkappa>0$, and

$$a(\lambda)=\exp\left\{\frac{1}{2\pi i}\int_{-\infty}^{\infty}\frac{\log(1-|b(\mu)|^2)}{\mu-\lambda}\,d\mu\right\}\prod_{j=1}^{n}\frac{\lambda-\lambda_j}{\lambda-\bar{\lambda}_j} \tag{6.23}$$

if $\varkappa<0$. These formulae can be extended up to the real line according to the *Sochocki-Plemelj formula*

$$\frac{1}{\mu-\lambda}\rightarrow\frac{1}{\mu-\lambda-i0}=\text{p.v.}\,\frac{1}{\mu-\lambda}+\pi i\delta(\mu-\lambda), \tag{6.24}$$

where p.v. indicates principal value.

To prove (6.23) consider the function

$$\tilde{a}(\lambda)=a(\lambda)\prod_{j=1}^{n}\frac{\lambda-\bar{\lambda}_j}{\lambda-\lambda_j}, \tag{6.25}$$

which differs from $a(\lambda)$ by a product of elementary Blaschke factors and is analytic in the upper half-plane. The function $\tilde{a}(\lambda)$ has no zeros for $\operatorname{Im}\lambda\geqslant 0$ and, as before, satisfies the asymptotic condition (6.3). On the real line we have

$$|\tilde{a}(\lambda)|^2=|a(\lambda)|^2=1-|b(\lambda)|^2. \tag{6.26}$$

The function $\eta(\lambda)=\log\tilde{a}(\lambda)$ is also analytic for $\operatorname{Im}\lambda>0$, is continuous down to the real line by virtue of the condition (A), and vanishes as $|\lambda|\rightarrow\infty$. Therefore, for real λ its imaginary and real parts are related by

$$\operatorname{Im}\eta(\lambda)=-\frac{1}{\pi}\,\text{p.v.}\int_{-\infty}^{\infty}\frac{\operatorname{Re}\eta(\mu)}{\mu-\lambda}\,d\mu, \tag{6.27}$$

which follows immediately from Cauchy's theorem. In the physical literature (6.27) is referred to as a *dispersion relation*. Using (6.24) it can be written as

$$\eta(\lambda) = \frac{1}{\pi i} \int_{-\infty}^{\infty} \frac{\operatorname{Re}\eta(\mu)}{\mu - \lambda}\, d\mu, \quad \operatorname{Im}\lambda > 0. \tag{6.28}$$

The representation (6.23) now follows from (6.28) and the obvious relation

$$\operatorname{Re}\eta(\lambda) = \log|\tilde{a}(\lambda)| = \tfrac{1}{2}\log(1 - |b(\lambda)|^2). \tag{6.29}$$

The proof of (6.22) is similar. This concludes our discussion of the analytic properties of transition coefficients.

To sum up, in §§ 5–7 we have defined the mapping

$$(\psi(x), \bar{\psi}(x)) \to (b(\lambda), \bar{b}(\lambda); \lambda_j, \bar{\lambda}_j, \gamma_j, \bar{\gamma}_j) \tag{6.30}$$

and described its image for various functional classes of $\psi(x)$, $\bar{\psi}(x)$. Thus, if $\psi(x)$, $\bar{\psi}(x)$ are in $L_1(-\infty, \infty)$ then the corresponding $b(\lambda)$, $\bar{b}(\lambda)$ are in $\mathfrak{R}_0$; if $\psi(x)$, $\bar{\psi}(x)$ are of Schwartz type, then $b(\lambda)$, $\bar{b}(\lambda)$ are also of Schwartz type.

This mapping will be essential in describing the dynamics of our model. So, in the following section we shall see that the equations of motion become trivial in the new variables, and in the following chapter we shall study the inverse mapping to (6.30).

§ 7. The Dynamics of Transition Coefficients

In § 3 the transition matrix, $T(x, y, \lambda)$, was shown to satisfy the evolution equation

$$\frac{\partial T}{\partial t}(x, y, \lambda) = V(x, \lambda)\, T(x, y, \lambda) - T(x, y, \lambda)\, V(y, \lambda), \tag{7.1}$$

if $\psi(x)$, $\bar{\psi}(x)$ satisfy the NS equation. For $\psi(x)$, $\bar{\psi}(x)$ rapidly decreasing one can take the limit of (7.1) as $y \to -\infty$, $x \to +\infty$ to obtain simple evolution equations for the transition coefficients.

For this purpose notice that

$$V(x, \lambda) \to V(\lambda) = \frac{i\lambda^2}{2}\sigma_3, \tag{7.2}$$

as $|x|\to\infty$, so that $V(\lambda)$ commutes with $E(x,\lambda)$. Now multiply (7.1) by $E(y,\lambda)$ from the right and take the limit as $y\to\pm\infty$ for real λ. Recalling the definition of the Jost solutions, $T_\pm(x,\lambda)$, given by (5.2) we find

$$\frac{\partial T_\pm}{\partial t}(x,\lambda)=V(x,\lambda)\,T_\pm(x,\lambda)-\frac{i\lambda^2}{2}\,T_\pm(x,\lambda)\sigma_3. \tag{7.3}$$

Performing the same operation with respect to x we obtain the equation for the reduced monodromy matrix,

$$\frac{\partial}{\partial t}\,T(\lambda,t)=\frac{i\lambda^2}{2}\,[\sigma_3,T(\lambda,t)]. \tag{7.4}$$

This equation is remarkable in that the dependence on $\psi(x)$, $\bar\psi(x)$ is completely eliminated. In terms of transition coefficients for the continuous spectrum (7.4) can be written as

$$\frac{\partial}{\partial t}\,a(\lambda,t)=0,\qquad \frac{\partial}{\partial t}\,b(\lambda,t)=-i\lambda^2 b(\lambda,t). \tag{7.5}$$

In particular, we deduce that $a(\lambda)$ is time-independent for real λ,

$$a(\lambda,t)=a(\lambda,0). \tag{7.6}$$

By virtue of analyticity, the same holds for $\operatorname{Im}\lambda>0$, so that the zeros, λ_j, of $a(\lambda)$ are time-independent as well. Thus *in the rapidly decreasing case the generating function for the conservation laws is just* $a(\lambda)$.

Let us now determine the evolution of transition coefficients for the discrete spectrum. By (7.3) we have for the column-vectors $T_-^{(1)}(x,\lambda)$ and $T_+^{(2)}(x,\lambda)$

$$\frac{\partial T_-^{(1)}}{\partial t}(x,\lambda)=V(x,\lambda)\,T_-^{(1)}(x,\lambda)-\frac{i\lambda^2}{2}\,T_-^{(1)}(x,\lambda) \tag{7.7}$$

and

$$\frac{\partial T_+^{(2)}}{\partial t}(x,\lambda)=V(x,\lambda)\,T_+^{(2)}(x,\lambda)+\frac{i\lambda^2}{2}\,T_+^{(2)}(x,\lambda). \tag{7.8}$$

These equations also hold for $\operatorname{Im}\lambda>0$. They are compatible with (6.20) for $\lambda=\lambda_j$,

$$T_-^{(1)}(x,\lambda_j)=\gamma_j\,T_+^{(2)}(x,\lambda_j), \tag{7.9}$$

only if

$$\frac{d}{dt}\gamma_j(t) = -i\lambda_j^2\gamma_j(t), \quad j=1,\ldots,n. \tag{7.10}$$

Equations (7.5) and (7.10) can easily be solved so that *the time dependence of transition coefficients is given by the remarkably simple formulae*

$$\begin{aligned} b(\lambda,t) &= e^{-i\lambda^2 t} b(\lambda,0), \\ \gamma_j(t) &= e^{-i\lambda_j^2 t}\gamma_j(0), \quad j=1,\ldots,n. \end{aligned} \tag{7.11}$$

This is the simplification of the dynamics due to the mapping (6.30) promised at the end of the preceding section. In the new variables the equations of motion can be solved explicitly. Up to the assumption that (6.30) has an inverse we can affirm that (7.11) provides a complete solution of the initial-value problem (1.1)–(1.2) in the rapidly decreasing case.

Let us now discuss the local integrals of the motion. We assume that $\psi(x)$, $\bar\psi(x)$ are of Schwartz type. In order to exploit the results obtained earlier, suppose that $\psi(x)$, $\bar\psi(x)$ are the limits, as $L\to\infty$, of the $2L$-periodic functions $\psi_L(x)$, $\bar\psi_L(x)$. In this case the densities, $P_n(x)$, of the local integrals of the motion defined by (4.19)–(4.20) and (4.34) have limits, as $L\to\infty$, which are also of Schwartz type. Therefore we can take the limit, as $L\to\infty$, in (4.33) to obtain

$$I_n = \int_{-\infty}^{\infty} P_n(x)\,dx. \tag{7.12}$$

Here $P_n(x)$ is constructed from $\psi(x)$, $\bar\psi(x)$ according to (4.19)–(4.20) and (4.34).

Let us now consider the limit, as $L\to\infty$, in the generating function $p_L(\lambda)$,

$$p_L(\lambda) = \arccos \tfrac{1}{2}\operatorname{tr} T_L(\lambda). \tag{7.13}$$

Notice that, in contrast to (4.1)–(4.2), we have set $\theta=0$. The definition of the reduced monodromy matrix $T(\lambda)$ and (5.47) imply that for real λ

$$\begin{aligned} \operatorname{tr} T_L(\lambda) &= e^{-i\lambda L}a(\lambda) + e^{i\lambda L}\bar a(\lambda) + o(1) \\ &= 2|a(\lambda)|\cos(\arg a(\lambda) - \lambda L) + o(1), \end{aligned} \tag{7.14}$$

as $L\to\infty$.

Since $b(\lambda)$ is of Schwartz type, the normalization relation yields

$$|a(\lambda)| = 1 + O(|\lambda|^{-\infty}). \tag{7.15}$$

Hence, up to terms of order $O(|\lambda|^{-\infty})$,

$$p_L(\lambda) = -\lambda L + \arg a(\lambda) + o(1)$$
$$= -\lambda L + \frac{1}{i}\log a(\lambda) + o(1), \tag{7.16}$$

as $L \to \infty$, where we have used

$$\log a(\lambda) = i \arg a(\lambda) + O(|\lambda|^{-\infty}), \tag{7.17}$$

which is a consequence of (7.15).

Thus, the generating function for the conservation laws in the limit $L \to \infty$ coincides with $\log a(\lambda)$,

$$\lim_{L\to\infty} (p_L(\lambda) + \lambda L) = \frac{1}{i}\log a(\lambda), \tag{7.18}$$

up to terms of order $O(|\lambda|^{-\infty})$. By comparing this with (4.4) we deduce that $\log a(\lambda)$ is the generating function for local integrals of the motion,

$$\log a(\lambda) = i\varkappa \sum_{n=1}^{\infty} \frac{I_n}{\lambda^n} + O(|\lambda|^{-\infty}). \tag{7.19}$$

The uniform behaviour with respect to L, as $L \to \infty$, of the asymptotic expansion (4.4) for $p_L(\lambda) + \lambda L$ is an obvious consequence of the existence of the limit in the integral representation (3.44) for $E(-L, \lambda)\, T_L(\lambda)\, E(-L, \lambda)$, as $L \to \infty$, proved in § 6.

The coefficients in (7.19) can be determined from (6.22)–(6.23). The densities $\log(1 + \varepsilon |b(\mu)|^2)$ in the integrals (6.22) and (6.23) are of Schwartz type; expanding $\frac{1}{\mu - \lambda}$ in a geometric progression leads to the asymptotic expansion

$$\log a(\lambda) = i\varkappa \sum_{n=1}^{\infty} \frac{c_n}{\lambda^n} + O(|\lambda|^{-\infty}), \tag{7.20}$$

where

$$c_k = \frac{1}{2\pi\varkappa} \int_{-\infty}^{\infty} \log(1 + \varepsilon |b(\lambda)|^2) \lambda^{k-1} d\lambda$$
$$+ \frac{1}{i\varkappa k} \sum_{j=1}^{n} (\bar{\lambda}_j^k - \lambda_j^k), \quad k = 1, 2, \ldots \tag{7.21}$$

Here $\varepsilon = \operatorname{sign} \varkappa$; if $\varepsilon = 1$, the sum over the zeros on the right hand side of (7.21) disappears.

Comparing the asymptotic expansion (7.19) with (7.20) gives the identities

$$c_n = I_n = \int_{-\infty}^{\infty} P_n(x)\, dx \tag{7.22}$$

relating certain functionals of $\psi(x)$, $\bar{\psi}(x)$ with functionals of $b(\lambda)$, $\bar{b}(\lambda)$ and λ_j, $\bar{\lambda}_j$. In spectral theory such formulae are called *trace identities*.

What is essential for our purposes is that we have managed to represent the integrals of the motion, I_n, as functionals of the new variables $(b(\lambda), \bar{b}(\lambda), \lambda_j, \bar{\lambda}_j, \gamma_j, \bar{\gamma}_j;\ j = 1, \ldots, n)$ introduced in (6.30). Noteworthy, only half of these variables enter into I_n, namely, $|b(\lambda)|^2$ and λ_j, $\bar{\lambda}_j$. An interpretation of this phenomenon in terms of Hamiltonian mechanics will be given in Chapter III.

§ 8. The Case of Finite Density. Jost Solutions

The finite density boundary conditions have meaningful applications only when $\varkappa > 0$, hence we shall confine ourselves to this case. We shall assume that $\psi(x)$, $\bar{\psi}(x)$ take their boundary values

$$\lim_{x \to \pm\infty} \psi(x) = \varrho e^{i\varphi_\pm}, \qquad \varphi_+ - \varphi_- = \theta \tag{8.1}$$

in the sense of Schwartz. With no loss of generality we can fix $\varphi_- = 0$, so that $\varphi_+ = \theta$.

In terms of $U(x, \lambda)$ these boundary conditions take the form

$$\lim_{x \to +\infty} U(x, \lambda) = \frac{1}{2} \begin{pmatrix} -i\lambda & \omega e^{-i\theta} \\ \omega e^{i\theta} & i\lambda \end{pmatrix} = U_+(\lambda) \tag{8.2}$$

and

$$\lim_{x \to -\infty} U(x, \lambda) = \frac{1}{2} \begin{pmatrix} -i\lambda & \omega \\ \omega & i\lambda \end{pmatrix} = U_-(\lambda), \tag{8.3}$$

where

$$\omega = 2\sqrt{\varkappa}\, \varrho. \tag{8.4}$$

The matrices $U_\pm(\lambda)$ are related by

$$U_+(\lambda) = Q^{-1}(\theta)\, U_-(\lambda)\, Q(\theta). \tag{8.5}$$

In this section we shall introduce appropriate solutions of the linear problem

$$\frac{dF}{dx} = U(x, \lambda) F \tag{8.6}$$

with the boundary conditions (8.2)-(8.3) and analyze their properties following the pattern of the rapidly decreasing case and leaving out unessential details.

The role of $E(x, \lambda)$ will now be played by the solution matrix $E_\varrho(x, \lambda)$ of the equation

$$\frac{dE_\varrho}{dx}(x, \lambda) = U_-(\lambda) E_\varrho(x, \lambda), \tag{8.7}$$

which results from (8.6) in the limit $x \to -\infty$. The continuous spectrum associated with (8.7) consists of real λ satisfying

$$\lambda^2 \geqslant \omega^2. \tag{8.8}$$

Let $\mathbb{R}_\omega$ denote the set of these λ. For λ in $\mathbb{R}_\omega$ we choose $E_\varrho(x, \lambda)$ in the form

$$E_\varrho(x, \lambda) = \begin{pmatrix} 1 & \frac{i(k-\lambda)}{\omega} \\ \frac{i(\lambda - k)}{\omega} & 1 \end{pmatrix} e^{-\frac{ikx}{2}\sigma_3}, \tag{8.9}$$

where

$$k(\lambda) = \sqrt{\lambda^2 - \omega^2}, \tag{8.10}$$

the branch of the square root being fixed by

$$\operatorname{sign} k(\lambda) = \operatorname{sign} \lambda. \tag{8.11}$$

The choice of $E_\varrho(x, \lambda)$ is uniquely determined by the requirements of analytic continuation, namely, that for $\operatorname{Im} k > 0$ the first column of $E_\varrho(x, \lambda)$ vanishes as $x \to -\infty$ and the second one vanishes as $x \to +\infty$.

The corresponding solution to the equation resulting from (8.6) in the limit $x \to +\infty$ is $Q^{-1}(\theta) E_\varrho(x, \lambda)$.

Let us now discuss the analytic properties of $E_\varrho(x, \lambda)$ in more detail. Notice that $E_\varrho(x, \lambda)$, unlike $E(x, \lambda)$, is not unimodular,

$$\det E_\varrho(x, \lambda) = \frac{2k(\lambda - k)}{\omega^2} \tag{8.12}$$

and thus degenerates at $\lambda=\pm\omega$. As $\varrho\to 0$, $E_\varrho(x,\lambda)$ goes into $E(x,\lambda)$.

The main difference between (8.6) and the same problem in the rapidly decreasing case is the presence of the *gap* $-\omega<\lambda<\omega$ in the continuous spectrum. The degeneracy points of $E_\varrho(x,\lambda)$ are the boundary points of the continuous spectrum. A natural domain for studying the analytic properties of $E_\varrho(x,\lambda)$ is provided by the Riemann surface Γ of the function $k(\lambda)$. The surface Γ is pieced together of two copies, Γ_+ and Γ_-, of the complex plane $\mathbb{C}^1$ slit along the intervals $(-\infty,-\omega]$ and $[\omega,\infty)$ of the real line (see Fig. 2), the edges of the cuts being suitably identified.

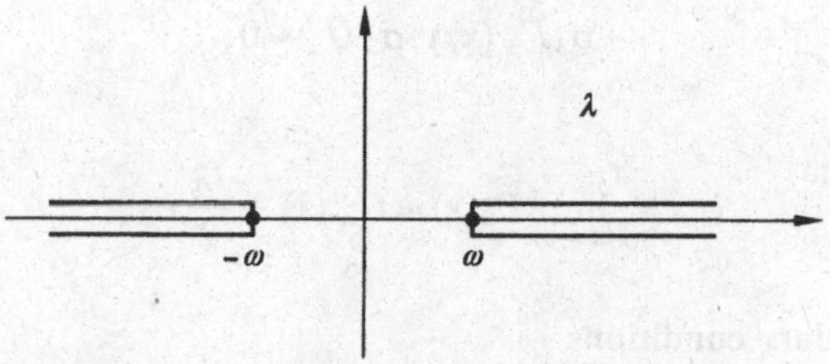

Fig. 2

A point on Γ different from the branch points, $\pm\omega$, will be represented by a pair (λ,ε) where λ is a complex number and $\varepsilon=\pm 1$, so that $\varepsilon=1$ on Γ_+ and $\varepsilon=-1$ on Γ_-. The function $k(\lambda)$ is defined on Γ by (8.10) where $\pm\operatorname{Im}k(\lambda)\geqslant 0$ on $\Gamma_\pm$. Alternatively, the sheet Γ_+ is specified by the condition that $k(\lambda+i0)>0$ for $\operatorname{Im}\lambda=0$ and $\lambda>\omega$; then $k(\lambda+i0)<0$ for $\operatorname{Im}\lambda=0$ and $\lambda<-\omega$. Thus, convention (8.11) is fulfilled for the limiting values of $k(\lambda)$ on the upper banks of the cuts on Γ_+ and on the lower banks of the cuts on Γ_-.

In what follows we shall often suppress the dependence of $k(\lambda)$ on λ. So, if both k and λ occur in a formula, then k is always considered as a function of λ.

For λ outside the cuts, $E_\varrho(x,\lambda)$ is not bounded in x. However, its first column on Γ_+ and its second column on Γ_- decay exponentially as $x\to-\infty$; its second column on Γ_+ and its first column on Γ_- decay exponentially as $x\to+\infty$. In the opposite directions the columns grow exponentially.

For λ in $\mathbb{R}_\omega$ we introduce the matrix Jost solutions of (8.6) through their integral representations

$$T_+(x,\lambda)=Q^{-1}(\theta)E_\varrho(x,\lambda)+\int_x^\infty \Gamma_+(x,y)Q^{-1}(\theta)\cdot E_\varrho(y,\lambda)\,dy \tag{8.13}$$

and

$$T_-(x,\lambda)=E_\varrho(x,\lambda)+\int_{-\infty}^{x}\Gamma_-(x,y)\,E_\varrho(y,\lambda)\,dy\,. \tag{8.14}$$

The derivation will use a method alternative to that of § 3. Substitute (8.13) and (8.14) into (8.6) and put the terms containing the common factor $E_\varrho(x,\lambda)$ together. It results that the kernels $\Gamma_\pm(x,y)$ satisfy the differential equations

$$\frac{\partial}{\partial x}\Gamma_\pm(x,y)+\sigma_3\frac{\partial}{\partial y}\Gamma_\pm(x,y)\sigma_3-U_0(x)\Gamma_\pm(x,y)$$
$$+\sigma_3\Gamma_\pm(x,y)\sigma_3\,U_\pm=0\,, \tag{8.15}$$

where

$$U_\pm=\lim_{x\to\pm\infty}U_0(x)=U_\pm(\lambda)+\frac{i\lambda}{2}\sigma_3\,, \tag{8.16}$$

with the boundary conditions

$$\Gamma_\pm(x,x)-\sigma_3\Gamma_\pm(x,x)\sigma_3=\mp(U_0(x)-U_\pm)\,,\qquad \lim_{y\to\pm\infty}\Gamma_\pm(x,y)=0\,. \tag{8.17}$$

These differential equations defining two Goursat problems may be reduced to systems of integral equations. For instance, the system for $\Gamma_-(x,y)$ is

$$\Gamma_-^{(d)}(x,y)=\int_{-\infty}^{x}\left(U_0(s)\Gamma_-^{(od)}(s,s+y-x)+\Gamma_-^{(od)}(s,s+y-x)\,U_-\right)ds\,, \tag{8.18}$$

$$\Gamma_-^{(od)}(x,y)=\frac{1}{2}\left(U_0\left(\frac{x+y}{2}\right)-U_-\right)$$
$$+\int_{\frac{x+y}{2}}^{x}\left(U_0(s)\Gamma_-^{(d)}(s,x+y-s)-\Gamma_-^{(d)}(s,x+y-s)\,U_-\right)ds\,, \tag{8.19}$$

where $x\geqslant y$ and $\Gamma_-^{(d)}$ and $\Gamma_-^{(od)}$ indicate the diagonal and off-diagonal parts of Γ_-, respectively.

These are Volterra integral equations, and their iterations are absolutely convergent. Under our assumptions on $\psi(x)$, $\bar\psi(x)$ their solution $\Gamma_-(x,y)$ is infinitely differentiable with respect to x and y and of Schwartz type in y as $y\to-\infty$. The kernel $\Gamma_+(x,y)$ is analyzed in a similar way and proves to be of Schwartz type in y as $y\to+\infty$.

Thus we have established the representations (8.13) and (8.14) defining the Jost solutions. Let us now discuss their properties.

1. From (8.13) and (8.14) it follows that the Jost solutions have the following asymptotic behaviour as $|x|\to\pm\infty$:

$$T_+(x,\lambda)=Q^{-1}(\theta)E_\varrho(x,\lambda)+o(1) \quad \text{as} \quad x\to+\infty \tag{8.20}$$

and

$$T_-(x,\lambda)=E_\varrho(x,\lambda)+o(1) \qquad \text{as} \quad x\to-\infty\,. \tag{8.21}$$

2. For λ in $\mathbb{R}_\omega$ the matrices $T_\pm(x,\lambda)$ result from $T(x,y,\lambda)$ in the limit $y\to\pm\infty$:

$$T_+(x,\lambda)=\lim_{y\to+\infty} T(x,y,\lambda)Q^{-1}(\theta)E_\varrho(y,\lambda) \tag{8.22}$$

and

$$T_-(x,\lambda)=\lim_{y\to-\infty} T(x,y,\lambda)E_\varrho(y,\lambda)\,. \tag{8.23}$$

To prove this it is sufficient to write $T(x,y,\lambda)$ as

$$T(x,y,\lambda)=T_\pm(x,\lambda)T_\pm^{-1}(y,\lambda) \tag{8.24}$$

and then use the asymptotics (8.20) and (8.21).

3. The determinant of the Jost solutions coincides with that of $E_\varrho(x,\lambda)$,

$$\det T_\pm(x,\lambda)=\det E_\varrho(x,\lambda)=\frac{2k(\lambda-k)}{\omega^2}\,, \tag{8.25}$$

so that $T_\pm(x,\lambda)$ are degenerate at $\lambda=\pm\omega$.

4. For λ in $\mathbb{R}_\omega$ the involution property holds

$$\bar{T}_\pm(x,\lambda)=\sigma_1 T_\pm(x,\lambda)\sigma_1 \tag{8.26}$$

which coincides with (5.19) when $\varkappa>0$; a similar relation holds for $E_\varrho(x,\lambda)$. The kernels $\Gamma_\pm(x,y)$ also possess this property and hence may be written in the form

$$\Gamma_\pm=\begin{pmatrix}\alpha_\pm & \bar{\beta}_\pm\\ \beta_\pm & \bar{\alpha}_\pm\end{pmatrix}. \tag{8.27}$$

5. The integral representations (8.13), (8.14) and the analytic properties of $E_\varrho(x,\lambda)$ imply the following analytic properties of the Jost solutions. The first column, $T_-^{(1)}(x,\lambda)$, of $T_-(x,\lambda)$ and the second column, $T_+^{(2)}(x,\lambda)$, of

$T_+(x,\lambda)$ extend analytically to the sheet Γ_+ of the Riemann surface Γ, whereas the columns $T_+^{(1)}(x,\lambda)$ and $T_-^{(2)}(x,\lambda)$ extend analytically to the sheet Γ_-. For x fixed, one has, asymptotically as $L\to\infty$,

$$e^{\frac{ikx}{2}}T_-^{(1)}(x,\lambda)=\begin{pmatrix}1\\ \frac{i(\lambda-k)}{\omega}\end{pmatrix}+O\left(\frac{|1+\lambda-k|}{|\lambda|}\right), \tag{8.28}$$

$$e^{-\frac{ikx}{2}}T_+^{(2)}(x,\lambda)=\begin{pmatrix}\frac{i(k-\lambda)}{\omega}e^{-\frac{i\theta}{2}}\\ e^{\frac{i\theta}{2}}\end{pmatrix}+O\left(\frac{|1+\lambda-k|}{|\lambda|}\right), \tag{8.29}$$

for λ on the sheet Γ_+, and

$$e^{\frac{ikx}{2}}T_+^{(1)}(x,\lambda)=\begin{pmatrix}e^{-\frac{i\theta}{2}}\\ \frac{i(\lambda-k)}{\omega}e^{\frac{i\theta}{2}}\end{pmatrix}+O\left(\frac{|1+\lambda-k|}{|\lambda|}\right), \tag{8.30}$$

$$e^{-\frac{ikx}{2}}T_-^{(2)}(x,\lambda)=\begin{pmatrix}\frac{i(k-\lambda)}{\omega}\\ 1\end{pmatrix}+O\left(\frac{|1+\lambda-k|}{|\lambda|}\right), \tag{8.31}$$

for λ on the sheet Γ_-.

Notice that for λ in Γ_+

$$k=\lambda+O\left(\frac{1}{|\lambda|}\right) \tag{8.32}$$

as $|\lambda|\to\infty$, $\operatorname{Im}\lambda>0$, and

$$k=-\lambda+O\left(\frac{1}{|\lambda|}\right) \tag{8.33}$$

as $|\lambda|\to\infty$, $\operatorname{Im}\lambda<0$. So, to estimate the remainders in (8.28)–(8.31) it should be remembered that $1+\lambda-k=O(1)$ in the first case, but $1+\lambda-k=O(\lambda)$ in the second case. On Γ_- these cases are interchanged.

6. The involution property (8.26) extends to the analytically continued columns of $T_{\pm}(x,\lambda)$. To show this let P be an involution on Γ sending $\lambda\to\bar{\lambda}$, $k\to\bar{k}$. A more formal definition is

$$P(\lambda,\varepsilon)=(\bar{\lambda},-\varepsilon), \tag{8.34}$$

so that P permutes Γ_+ and Γ_-. We have

$$\bar{E}_{\varrho}(x,\lambda)=\sigma_1 E_{\varrho}(x,P(\lambda))\sigma_1, \tag{8.35}$$

which gives

$$\sigma_1 \bar{T}_{\pm}^{(1)}(x,\lambda)=T_{\pm}^{(2)}(x,P(\lambda)), \tag{8.36}$$

where λ is in Γ_- for the plus sign and in Γ_+ for the minus sign.

There is another involution on Γ sending $\lambda\to\bar{\lambda}$, $k\to-\bar{k}$. A more formal definition is

$$J(\lambda,\varepsilon)=(\bar{\lambda},\varepsilon), \tag{8.37}$$

so that J leaves $\Gamma_{\pm}$ invariant. For any λ in Γ we have

$$\bar{E}_{\varrho}(x,\lambda)=-\frac{i(\bar{\lambda}-\bar{k})}{\omega}\sigma_1 E_{\varrho}(x,J(\lambda))\sigma_3, \tag{8.38}$$

which also extends to the corresponding columns of the Jost solutions. We have

$$\bar{T}_{\pm}^{(1)}(x,J(\lambda))=\frac{\lambda+k}{i\omega}\sigma_1 T_{\pm}^{(1)}(x,\lambda) \tag{8.39}$$

and

$$\bar{T}_{\pm}^{(2)}(x,J(\lambda))=-\frac{\lambda+k}{i\omega}\sigma_1 T_{\pm}^{(2)}(x,\lambda), \tag{8.40}$$

where λ belongs to the corresponding $\Gamma_{\pm}$. In particular, we deduce a relationship between the values of the columns of $T_{\pm}(x,\lambda)$ on the upper and lower banks of the cuts on the corresponding sheets of analyticity:

$$\bar{T}_{\pm}^{(1)}(x,\lambda-i0)=\frac{\lambda+k}{i\omega}\sigma_1 T_{\pm}^{(1)}(x,\lambda+i0) \tag{8.41}$$

and

$$\bar{T}_{\pm}^{(2)}(x,\lambda-i0) = -\frac{\lambda+k}{i\omega}\sigma_1 T_{\pm}^{(2)}(x,\lambda+i0), \tag{8.42}$$

where λ is real, $|\lambda| \geqslant \omega$.

This completes our list of the properties of $T_{\pm}(x,\lambda)$.

As in the rapidly decreasing case, there exists a *reduced monodromy matrix*, $T_{\varrho}(\lambda)$, relating $T_{+}(x,\lambda)$ to $T_{-}(x,\lambda)$:

$$T_{-}(x,\lambda) = T_{+}(x,\lambda)\,T_{\varrho}(\lambda). \tag{8.43}$$

The matrix $T_{\varrho}(\lambda)$ is well defined and unimodular for λ in $\mathbb{R}_{\omega}$, $\lambda \neq \pm\omega$. For such λ it can also be obtained as the limit

$$T_{\varrho}(\lambda) = \lim_{L\to\infty} E_{\varrho}^{-1}(L,\lambda)\,Q(\theta)\,T(L,-L,\lambda)\,E_{\varrho}(-L,\lambda). \tag{8.44}$$

Once again, for λ in $\mathbb{R}_{\omega}$ the involution property yields

$$\bar{T}_{\varrho}(\lambda) = \sigma_1 T_{\varrho}(\lambda)\sigma_1 \tag{8.45}$$

which allows us to represent $T_{\varrho}(\lambda)$ in the already familiar form

$$T_{\varrho}(\lambda) = \begin{pmatrix} a_{\varrho}(\lambda) & \bar{b}_{\varrho}(\lambda) \\ b_{\varrho}(\lambda) & \bar{a}_{\varrho}(\lambda) \end{pmatrix}. \tag{8.46}$$

As before, the coefficients $a_{\varrho}(\lambda)$ and $b_{\varrho}(\lambda)$ will be called *transition coefficients*. Since $T_{\varrho}(\lambda)$ is unimodular, we have the *normalization relation*

$$|a_{\varrho}(\lambda)|^2 - |b_{\varrho}(\lambda)|^2 = 1. \tag{8.47}$$

Further properties of $a_{\varrho}(\lambda)$ and $b_{\varrho}(\lambda)$ will be discussed in the following section.

§ 9. The Case of Finite Density. Transition Coefficients

We shall begin by enumerating the properties of transition coefficients following the pattern of the rapidly decreasing case. From (8.43) and (8.25) one immediately obtains the representations

$$a_{\varrho}(\lambda) = \frac{\omega^2}{2k(\lambda-k)}\det\left(T_{-}^{(1)}(x,\lambda),\, T_{+}^{(2)}(x,\lambda)\right) \tag{9.1}$$

and

$$b_\varrho(\lambda) = \frac{\omega^2}{2k(\lambda-k)} \det(T_+^{(1)}(x,\lambda), T_-^{(1)}(x,\lambda)), \tag{9.2}$$

which generalize those given by (6.1) and (6.2).

It follows from (9.1) that $a_\varrho(\lambda)$ may be analytically extended to the sheet Γ_+ except the branch points $\lambda = \pm\omega$. From (8.28) and (8.29) we see that $a_\varrho(\lambda)$ has the asymptotic behaviour

$$a_\varrho(\lambda) = \cos\frac{\theta}{2} + i\frac{\lambda}{k}\sin\frac{\theta}{2} + O\left(\frac{1}{|\lambda|}\right), \tag{9.3}$$

as $|\lambda| \to \infty$. In other words,

$$a_\varrho(\lambda) = e^{\frac{i\theta}{2}} + O\left(\frac{1}{|\lambda|}\right) \tag{9.4}$$

when $\operatorname{Im}\lambda > 0$ and

$$a_\varrho(\lambda) = e^{-\frac{i\theta}{2}} + O\left(\frac{1}{|\lambda|}\right) \tag{9.5}$$

when $\operatorname{Im}\lambda < 0$.

In a similar way, $\bar{a}_\varrho(\lambda)$ extends analytically to the sheet Γ_- except $\lambda = \pm\omega$. Let $a_\varrho^*(\lambda)$ denote this analytic continuation. Then (8.36) yields

$$a_\varrho^*(\lambda) = \bar{a}_\varrho(P(\lambda)). \tag{9.6}$$

The representation (9.2) shows that in general $b_\varrho(\lambda)$ need not extend off $\mathbb{R}_\omega$. Such an extension, of course, exists if $\psi(x)$, $\bar{\psi}(x)$ differ from their asymptotic values only in a finite interval. It follows from (8.28) and (8.30) that for λ in $\mathbb{R}_\omega$,

$$b_\varrho(\lambda) = O\left(\frac{1}{|\lambda|}\right), \tag{9.7}$$

as $|\lambda| \to \infty$.

Let us now discuss the possible behaviour of $a_\varrho(\lambda)$ and $b_\varrho(\lambda)$ in the vicinity of $\lambda = \pm\omega$. From (9.1) it follows that if the columns $T_-^{(1)}(x,\lambda)$ and $T_+^{(2)}(x,\lambda)$ are linearly independent at $\lambda = \omega$ or $\lambda = -\omega$ (i.e. at $k=0$), then $a_\varrho(\lambda)$ has a singularity of the form

$$a_\varrho(\lambda)|_{\lambda \sim \pm\omega} = \frac{a_\pm}{k} + O(1), \tag{9.8}$$

with the non-zero $a_\pm$. *This is what happens generically*. In a special situation when $T_-^{(1)}(x,\lambda)$ and $T_+^{(2)}(x,\lambda)$ become linearly dependent at $\lambda=\omega$ or $\lambda=-\omega$, either a_+ or a_- or both of them vanish, so that $a_\varrho(\lambda)$ is non-singular near the corresponding branch point. In scattering theory one usually says in this case that $\lambda=\omega$ or $\lambda=-\omega$ or both are *virtual levels*.

The coefficient $b_\varrho(\lambda)$ is either singular or regular in the vicinity of $\lambda=\pm\omega$ simultaneously with $a_\varrho(\lambda)$. In fact, the matrices $T_\pm(x,\lambda)$ become degenerate at $\lambda=\pm\omega$, so that the columns $T_\pm^{(1)}(x,\pm\omega)$ become proportional to $T_\pm^{(2)}(x,\pm\omega)$. The asymptotic expressions (8.20) and (8.21) and the definition (8.9) of $E_\varrho(x,\lambda)$ imply that

$$T_+^{(1)}(x,\pm\omega)=\pm i T_+^{(2)}(x,\pm\omega). \tag{9.9}$$

Comparing (9.1), (9.2) and (9.9) shows that if a_+ or a_- does not vanish then

$$b_\varrho(\lambda)|_{\lambda\approx\pm\omega}=\mp\frac{ia_\pm}{k}+O(1). \tag{9.10}$$

In particular, under this hypothesis one has

$$\lim_{\lambda\to\pm\omega}\frac{a_\varrho(\lambda)}{b_\varrho(\lambda)}=\pm i. \tag{9.11}$$

It should be emphasized that it is the gap in the continuous spectrum for the finite density boundary conditions which makes the analytic properties of the transition coefficients more complicated than in the rapidly decreasing case.

The involution J on Γ relates the values of $a_\varrho(\lambda)$ in the half-planes $\pm\operatorname{Im}\lambda>0$ of the sheet Γ_+. Specifically, (8.39) and (8.40) yield

$$a_\varrho(\lambda)=\bar{a}_\varrho(J(\lambda)). \tag{9.12}$$

Besides, the involution allows us to relate the values of $a_\varrho(\lambda)$ and $b_\varrho(\lambda)$ on the upper and lower banks of the cuts on Γ_+,

$$a_\varrho(\lambda+i0)=\bar{a}_\varrho(\lambda-i0), \quad b_\varrho(\lambda+i0)=-\bar{b}_\varrho(\lambda-i0), \tag{9.13}$$

λ being real, $|\lambda|>\omega$. The latter formula follows from (8.41). By taking the limit in (9.13) we find that $a_\pm$ are pure imaginary.

As in the rapidly decreasing case, there are integral representations

$$a_\varrho(\lambda)=\cos\frac{\theta}{2}+i\frac{\lambda}{k}\sin\frac{\theta}{2}+\int_0^\infty \alpha_1(x)e^{ikx}\,dx$$
$$+i\frac{\lambda}{k}\int_0^\infty \alpha_2(x)e^{ikx}\,dx+\frac{i}{k}\int_0^\infty \alpha_3(x)e^{ikx}\,dx \tag{9.14}$$

and

$$b_\varrho(\lambda)=i\int_{-\infty}^{\infty}\beta_1(x)e^{ikx}\,dx+\frac{\lambda}{k}\int_{-\infty}^{\infty}\beta_2(x)e^{ikx}\,dx$$

$$+\frac{1}{k}\int_{-\infty}^{\infty}\beta_3(x)e^{ikx}\,dx. \qquad (9.15)$$

The derivation proceeds by taking the limit $x\to+\infty$ in (8.43) and using (8.14) and (8.18), (8.19) (cf. § 6). The functions $\alpha_j(x), \beta_j(x), j=1,2,3$, are real by virtue of (9.13), and of Schwartz type near $+\infty$ or on the whole axis, respectively. Therefore, $b_\varrho(\lambda)$ is of Schwartz type as $|\lambda|\to\infty$.

It follows from (9.14) that $a_\varrho(\lambda)$ has an asymptotic expansion in inverse powers of λ or k as $|\lambda|\to\infty$. To write it down we shall use the following asymptotic expansions on Γ_+:

$$\frac{1}{k(\lambda)}=\pm\left(1-\frac{\omega^2}{\lambda^2}\right)^{-\frac{1}{2}}\cdot\frac{1}{\lambda}=\pm\sum_{n=0}^{\infty}\frac{(-\omega^2)^n}{\lambda^{2n+1}}\binom{-\frac{1}{2}}{n}+O(|\lambda|^{-\infty}) \qquad (9.16)$$

or

$$\lambda=\pm k\left(1+\frac{\omega^2}{k^2}\right)^{\frac{1}{2}}=\pm\sum_{n=0}^{\infty}\frac{\omega^{2n}}{k^{2n-1}}\binom{\frac{1}{2}}{n}+O(|k|^{-\infty}), \qquad (9.17)$$

where the common $\pm$sign is that of $\operatorname{Im}\lambda$, and $\binom{\pm\frac{1}{2}}{n}$ indicates the binomial coefficient. Integration by parts in (9.14) then gives the desired expansions for $a_\varrho(\lambda)$. For instance, one has

$$a_\varrho(\lambda)=e^{\frac{i\theta}{2}}+\sum_{n=1}^{\infty}\frac{a_n}{k^n}+O(|k|^{-\infty}), \qquad (9.18)$$

when $\operatorname{Im}\lambda>0$, and

$$a_\varrho(\lambda)=e^{-\frac{i\theta}{2}}+\sum_{n=1}^{\infty}\frac{(-1)^n\bar{a}_n}{k^n}+O(|k|^{-\infty}), \qquad (9.19)$$

when $\operatorname{Im}\lambda<0$. The expansions agree with the involution (9.12).

As in the rapidly decreasing case, the zeros of $a_\varrho(\lambda)$ for λ outside of $\mathbb{R}_\omega$ correspond to the discrete spectrum of the auxiliary linear problem (8.6) which is equivalent to the spectral problem

$$\mathscr{L} F = \frac{\lambda}{2} F, \tag{9.20}$$

where

$$\mathscr{L} = i\sigma_3 \frac{d}{dx} + i\sqrt{\varkappa}\,(\psi\sigma_- - \bar{\psi}\sigma_+). \tag{9.21}$$

Indeed, if $a_\varrho(\lambda)$ vanishes at $\lambda = \lambda_j$, then the columns $T_-^{(1)}(x, \lambda)$ and $T_+^{(2)}(x, \lambda)$ become linearly dependent,

$$T_-^{(1)}(x, \lambda_j) = \gamma_j T_+^{(2)}(x, \lambda_j), \tag{9.22}$$

and for λ outside of $\mathbb{R}_\omega$ decay exponentially as $x \to -\infty$ or $x \to +\infty$, respectively. Therefore, equation (9.20) has a column-solution decaying exponentially as $|x| \to \infty$.

If $\varkappa > 0$, the operator $\mathscr{L}$ is formally self-adjoint for the boundary conditions in question, so that its eigenvalues, and hence the zeros of $a_\varrho(\lambda)$, are real. By virtue of the normalization relation (8.47), the zeros must be located inside the gap $-\omega < \lambda < \omega$. In fact, we have either $|a_\varrho(\omega)| = \infty$ (the general case) or $|a_\varrho(\omega)| < \infty$ (a virtual level) in which case $|a_\varrho(\omega)| \geqslant 1$ by the normalization relation. The same holds for $\lambda = -\omega$. In particular, it follows that $a_\varrho(\lambda)$ has only finitely many zeros; they will be labelled $\lambda_1, \ldots, \lambda_n$.

Let us show that the zeros are simple. Let λ_j be a zero of $a_\varrho(\lambda)$ lying in the gap. We will show that $\frac{da_\varrho}{d\lambda}$ does not vanish at $\lambda = \lambda_j$. From (9.1) and $a_\varrho(\lambda_j) = 0$ we find

$$\dot{a}_\varrho(\lambda_j) = \frac{\omega^2}{2k_j(\lambda_j - k_j)} \left(\det(\dot{T}_-^{(1)}(x, \lambda_j), T_+^{(2)}(x, \lambda_j)) + \det(T_-^{(1)}(x, \lambda_j), \dot{T}_+^{(2)}(x, \lambda_j))\right), \tag{9.23}$$

where the dot indicates the derivative with respect to λ. Since $T_-^{(1)}(x, \lambda)$ and $T_+^{(2)}(x, \lambda)$ satisfy (8.6), their derivatives, $\dot{T}_-^{(1)}(x, \lambda)$ and $\dot{T}_+^{(2)}(x, \lambda)$, satisfy

$$\frac{d}{dx} \dot{F} = U(x, \lambda) \dot{F} - \frac{i\sigma_3}{2} F. \tag{9.24}$$

Together with

$$U^\tau(x, \lambda) = -\sigma_2 U(x, \lambda) \sigma_2 \tag{9.25}$$

these equations easily lead to

$$\frac{\partial}{\partial x}\det(T_-^{(1)}(x,\lambda),\dot{T}_+^{(2)}(x,\lambda))=\frac{i}{2}\det(\sigma_3 T_-^{(1)}(x,\lambda),T_+^{(2)}(x,\lambda)) \tag{9.26}$$

and

$$\frac{\partial}{\partial x}\det(\dot{T}_-^{(1)}(x,\lambda),T_+^{(2)}(x,\lambda))=-\frac{i}{2}\det(\sigma_3 T_-^{(1)}(x,\lambda),T_+^{(2)}(x,\lambda)). \tag{9.27}$$

Recalling that, for $\lambda=\lambda_j$, the columns $T_-^{(1)}(x,\lambda)$ and $T_+^{(2)}(x,\lambda)$ are proportional to each other and decay exponentially as $|x|\to\infty$ we derive that

$$\det(T_-^{(1)}(x,\lambda_j),\dot{T}_+^{(2)}(x,\lambda_j))=\frac{1}{2i}\gamma_j\int_x^\infty \Delta(x',\lambda_j)\,dx' \tag{9.28}$$

and

$$\det(\dot{T}_-^{(1)}(x,\lambda_j),T_+^{(2)}(x,\lambda_j))=\frac{1}{2i}\gamma_j\int_{-\infty}^x \Delta(x',\lambda_j)\,dx', \tag{9.29}$$

where

$$\Delta(x,\lambda)=\det(\sigma_3 T_+^{(2)}(x,\lambda),T_+^{(2)}(x,\lambda)). \tag{9.30}$$

Now we notice that

$$\frac{\omega}{\lambda-k}=\frac{\lambda+k}{\omega}, \tag{9.31}$$

so that by using the involution (8.40) the expression $\frac{\omega}{\lambda-k}\Delta(x,\lambda)$ for λ inside the gap can be transformed into

$$\frac{\lambda+k}{\omega}\Delta(x,\lambda)=\det(\sigma_2\bar{T}_+^{(2)}(x,\lambda),T_+^{(2)}(x,\lambda))=\frac{1}{i}\|T_+^{(2)}(x,\lambda_j)\|^2, \tag{9.32}$$

where $\|\cdot\|$ denotes the usual vector norm in $\mathbb{C}^2$. Putting together the formulae obtained we get the final expression for $\dot{a}_\varrho(\lambda_j)$,

$$\dot{a}_\varrho(\lambda_j)=-\frac{\omega\gamma_j}{4k_j}\int_{-\infty}^\infty \|T_+^{(2)}(x,\lambda_j)\|^2\,dx, \tag{9.33}$$

which shows that $\dot{a}_\varrho(\lambda_j)$ does not vanish.

The involution (8.40) also shows that the *transition coefficients for the discrete spectrum*, γ_j, involved in (9.22) are pure imaginary. Therefore (9.33) implies that $\dot{a}_\varrho(\lambda_j)$ is real and of the same sign as $i\gamma_j$,

$$\operatorname{sign} \dot{a}_\varrho(\lambda_j) = \operatorname{sign} i\gamma_j\,; \quad j = 1, \ldots, n. \tag{9.34}$$

Let us now show that, just as in the rapidly decreasing case, $a_\varrho(\lambda)$ *is uniquely determined by* $b_\varrho(\lambda)$, *the zeros* $\lambda_j, j=1, \ldots, n$, *inside the gap, and the parameter* θ. For this purpose we will derive an analogue of the dispersion relations (6.22), (6.23). Consider a conformal mapping of the sheet Γ_+ onto the upper half-plane of the variable z defined by

$$z = z(\lambda) = \lambda + k(\lambda), \qquad \operatorname{Im} z \geqslant 0. \tag{9.35}$$

The mapping takes the cuts on Γ_+ into the real axis $-\infty < z < \infty$ and takes a neighbourhood of ∞ for $\operatorname{Im}\lambda < 0$ into a neighbourhood of $z=0$. The inverse mapping is given by

$$\lambda = \lambda(z) = \frac{1}{2}\left(z + \frac{\omega^2}{z}\right), \tag{9.36}$$

where $\lambda(z)$ is sometimes referred to as the *Žukowsky function*.

For $\operatorname{Im} z \geqslant 0$ consider the function

$$f(z) = e^{-\frac{i\theta}{2}} a_\varrho(\lambda(z)). \tag{9.37}$$

It is analytic in the upper half-plane and behaves asymptotically as

$$f(z) = 1 + O\left(\frac{1}{|z|}\right), \tag{9.38}$$

as $|z| \to \infty$. Its zeros, z_j, are

$$z_j = z(\lambda_j) = \lambda_j + i\sqrt{\omega^2 - \lambda_j^2}\,, \quad |z_j| = \omega\,; \quad j = 1, \ldots, n. \tag{9.39}$$

Therefore the following dispersion relation holds:

$$f(z) = \prod_{j=1}^{n} \frac{z - z_j}{z - \bar{z}_j} \exp\left\{\frac{1}{\pi i} \int_{-\infty}^{\infty} \frac{\log|f(s)|}{s - z}\, ds\right\}, \tag{9.40}$$

which is just another variant of (6.28).

By (9.12), $f(z)$ satisfies the involution

$$f(z) = e^{-i\theta} \bar{f}\left(\frac{\omega^2}{\bar{z}}\right). \tag{9.41}$$

In particular, for real s,

$$|f(s)| = \left| f\left(\frac{\omega^2}{s}\right) \right| . \tag{9.42}$$

Hence, the integral in (9.40) can be reduced to an integral over the half-lines $|s| \geqslant \omega$. Coming back to the initial variables λ and k and using the normalization relation we obtain the required representation for $a_\varrho(\lambda)$,

$$a_\varrho(\lambda) = e^{\frac{i\theta}{2}} \prod_{j=1}^{n} \frac{\lambda + k(\lambda) - \lambda_j - k_j}{\lambda + k(\lambda) - \lambda_j - \bar{k}_j}$$
$$\times \exp\left\{ \frac{1}{2\pi i} \int_{\mathbb{R}_\omega} \frac{\log(1 + |b_\varrho(\mu)|^2)}{k(\mu)} \left(1 + \frac{k(\lambda)}{\mu - \lambda}\right) d\mu \right\}, \tag{9.43}$$

where integration goes over the upper banks of the cuts on Γ_+, and λ lies outside of $\mathbb{R}_\omega$.

In contrast with the rapidly decreasing case, the data $b_\varrho(\lambda)$, λ_j and θ are not all independent. Namely, the asymptotic formula (9.5) implies the relation

$$e^{-i\theta} = \prod_{j=1}^{n} \frac{\lambda_j + k_j}{\lambda_j + \bar{k}_j} \exp\left\{ \frac{1}{\pi i} \int_{\mathbb{R}_\omega} \frac{\log(1 + |b_\varrho(\lambda)|^2)}{k(\lambda)} d\lambda \right\}, \tag{9.44}$$

which results from (9.43) by taking the limit as $|\lambda| \to \infty$ in the half-plane $\operatorname{Im}\lambda < 0$ and using that $\lambda + k = O\left(\frac{1}{|\lambda|}\right)$. In what follows, (9.44) will be called the *condition* (θ).

However, this does not exhaust all the constraints on the data $b_\varrho(\lambda)$, λ_j and θ. Generically, $b_\varrho(\lambda)$ is singular in the neighbourhood of $\lambda = \pm\omega$,

$$b_\varrho(\lambda)|_{\lambda \approx \pm\omega} = \frac{b_\pm}{k} + O(1), \tag{9.45}$$

with $b_\pm$ real. On the other hand, $a_\varrho(\lambda)$ satisfies (9.8) with

$$a_\pm = \pm i b_\pm . \tag{9.46}$$

Since (9.43) involves only $|b_\varrho(\lambda)|$, the latter relation imposes some restrictions on $b_\pm$, which have the form

$$\operatorname{sign} b_\pm = (-1)^{N_\pm}, \tag{9.47}$$

where $N_\pm$ are some integers (see (9.58)).

To find $N_{\pm}$ let us consider the integral in (9.43) for λ in the neighbourhood of $\pm\omega$ outside of $\mathbb{R}_{\omega}$. Clearly, for λ near ω the only singular contribution comes from the integral

$$I(\lambda)=\int_{\omega}^{\omega\operatorname{ch}\delta}\frac{\log(1+|b_{\varrho}(\mu)|^{2})k(\lambda)}{k(\mu)(\mu-\lambda)}\,d\mu, \tag{9.48}$$

with $\delta>0$. It follows from (9.45) that

$$I(\lambda)=2\pi i\log|b_{+}|+I_{0}(\lambda)+O(|k(\lambda)|), \tag{9.49}$$

where

$$I_{0}(\lambda)=-2k(\lambda)\int_{\omega}^{\omega\operatorname{ch}\delta}\frac{\log k(\mu)}{k(\mu)(\mu-\lambda)}\,d\mu. \tag{9.50}$$

By a change of variables,

$$\mu(x)=\frac{1}{2}\left(x+\frac{\omega^{2}}{x}\right),\qquad k(\mu(x))=\frac{1}{2}\left(x-\frac{\omega^{2}}{x}\right), \tag{9.51}$$

the last integral is reduced to

$$I_{0}(\lambda)=-2\int_{\omega e^{-\delta}}^{\omega e^{\delta}}\frac{\log|k(\mu(x))|}{x-z}\,dx+\int_{\omega e^{-\delta}}^{\omega e^{\delta}}\frac{\log|k(\mu(x))|}{x}\,dx, \tag{9.52}$$

where $z=z(\lambda)$ (see (9.35)). Now we shall use the formula

$$\int_{\omega}^{\omega e^{\delta}}\frac{\log k(\mu(x))}{x-z}\,dx=-\frac{1}{2}\log^{2}k(\lambda)+\pi i\log k(\lambda)+\varphi(z), \tag{9.53}$$

where $\varphi(z)$ is regular near $z=\omega$, and $\log k$ on the right hand side is the branch of the logarithm with a cut along the positive axis $0\leqslant k<\infty$. To prove (9.53), consider the integral $\int\limits_{|\zeta-\omega|=\omega(e^{\delta}-1)}\frac{\log^{2}k(\mu(\zeta))}{\zeta-z}\,d\zeta$ and apply Cauchy's theorem. From (9.53) we have

$$\int_{\omega e^{-\delta}}^{\omega} \frac{\log(-k(\mu(x)))}{x-z}\,dx = \frac{1}{2}\log^2(-k(\lambda)) - \pi i \log(-k(\lambda))$$

$$+\frac{1}{2}\int_{\omega e^{-\delta}}^{\omega e^{\delta}} \frac{\log|k(\mu(x))|}{x}\,dx - \varphi\left(\frac{\omega^2}{z}\right), \qquad (9.54)$$

which gives a representation for $I_0(\lambda)$ in the neighbourhood of $\lambda=\omega$,

$$I_0(\lambda) = -2\pi i \log k(\lambda) - \pi^2 + O(|k(\lambda)|). \qquad (9.55)$$

It then follows that

$$a_\varrho(\lambda)|_{\lambda\approx\omega} = \frac{i|b_+|}{k(\lambda)} e^{\frac{i\theta}{2}} \prod_{j=1}^{n} \frac{\omega-\lambda_j-k_j}{\omega-\lambda_j-\bar{k}_j}$$

$$\times \exp\left\{\frac{1}{2\pi i}\int_{\mathbb{R}_\omega} \frac{\log(1+|b_\varrho(\mu)|^2)}{k(\mu)}\,d\mu\right\} + O(1). \qquad (9.56)$$

Notice that for $-\omega<\lambda<\omega$ one has

$$\left(\frac{\omega-\lambda-k}{\omega-\lambda-\bar{k}}\right)^2 = \frac{\lambda+k}{\lambda+\bar{k}}. \qquad (9.57)$$

Comparing (9.56), (9.57) and (9.44) shows that (9.46) holds if the integer N_+ is determined by

$$0 \leqslant \sum_{j=1}^{n} \arg\frac{\omega-\lambda_j-\bar{k}_j}{\omega-\lambda_j-k_j} + \frac{1}{2\pi}\int_{\mathbb{R}_\omega} \frac{\log(1+|b_\varrho(\lambda)|^2)}{k(\lambda)}\,d\lambda + \pi N_+ < \pi. \qquad (9.58)$$

Clearly, $(-1)^{N_+}$ does not depend on the choice of the branch of the argument.

The neighbourhood of $\lambda=-\omega$ is treated in a similar manner. The integer N_- is determined by a relation such as (9.58) with ω replaced by $-\omega$. Henceforth conditions (9.47) will be called the *conditions for the determination of signs*. If $b_\varrho(\lambda)$ is regular at $\lambda=\omega$ or $\lambda=-\omega$ then additional constraints of this type do not arise.

The properties of the data $b_\varrho(\lambda), \lambda_j$, $j=1,\ldots,n$, *and* θ *stated above, namely*:

1) *the involution* (see (9.13))
2) *the condition* (θ) (see (9.44))
3) *the conditions for the determination of signs* (see (9.47) and (9.58))

allow us, using (9.43), *to reconstruct a unique coefficient* $a_\varrho(\lambda)$ *with the properties*

1) *the involution* (see (9.12))
2) *the asymptotic behaviour as* $|\lambda|\to\infty$ (see (9.3)–(9.5))
3) *the sign correlations* (see (9.46)).

The verification of 2) and 3) is obvious because their derivation can be reversed. To verify 1) one should use the condition (θ) and the equation

$$\frac{\lambda-k-\lambda_j-\bar{k}_j}{\lambda-k-\lambda_j-k_j}\cdot\frac{\lambda_j+k_j}{\lambda_j+\bar{k}_j}=\frac{\lambda+k-\lambda_j-k_j}{\lambda+k-\lambda_j-\bar{k}_j}, \tag{9.59}$$

which holds for $-\omega<\lambda_j<\omega$.

This closes our list of the properties of transition coefficients.

§ 10. The Case of Finite Density. Time Dynamics and Integrals of the Motion

We shall begin by deriving the evolution equations for the Jost solutions. For that purpose we multiply the equation for the transition matrix (3.21)

$$\frac{\partial T}{\partial t}(x,y,\lambda)=V_\varrho(x,\lambda)\,T(x,y,\lambda)-T(x,y,\lambda)\,V_\varrho(y,\lambda) \tag{10.1}$$

by $E_\varrho(y,\lambda)$ on the right and let $y\to-\infty$. Consider the limit of $E_\varrho^{-1}(y,\lambda)\,V_\varrho(y,\lambda)\,E_\varrho(y,\lambda)$ as $y\to-\infty$, with $V_\varrho(y,\lambda)$ as defined in § 2,

$$V_\varrho=\lambda^2 V_2+\lambda V_1+V_{0,\varrho} \tag{10.2}$$

(see (2.4)–(2.8) and (2.11)). By virtue of the boundary conditions (8.1), the last term in (10.2) vanishes as $y\to-\infty$, while the first two terms turn into $-\lambda U_-(\lambda)$ (see (8.3)). The differential equation (8.7) and the explicit expression (8.9) for $E_\varrho(y,\lambda)$ yield

$$E_\varrho^{-1}(y,\lambda)\,U_-(\lambda)\,E_\varrho(y,\lambda)=E_\varrho^{-1}(y,\lambda)\,\frac{d}{dy}\,E_\varrho(y,\lambda)=-\frac{ik}{2}\sigma_3. \tag{10.3}$$

So, passing to the limit in (10.1) leads to

$$\frac{\partial T_-}{\partial t}(x,\lambda)=V_\varrho(x,\lambda)\,T_-(x,\lambda)-\frac{ik\lambda}{2}\,T_-(x,\lambda)\sigma_3. \tag{10.4}$$

The limit as $y \to +\infty$ is analyzed in a similar manner; it results that $T_+(x, \lambda)$ satisfies the same differential equation as $T_-(x, \lambda)$.

The subsequent limit with respect to x gives the evolution equation for the reduced monodromy matrix,

$$\frac{\partial}{\partial t} T_\varrho(\lambda) = \frac{ik\lambda}{2} [\sigma_3, T_\varrho(\lambda)]. \tag{10.5}$$

Comparing the evolution equations for the columns $T_-^{(1)}(x, \lambda)$ and $T_+^{(2)}(x, \lambda)$ at $\lambda = \lambda_j$ leads to differential equations for the transition coefficients for the discrete spectrum,

$$\frac{d}{dt} \gamma_j = -ik_j \lambda_j \gamma_j, \quad j = 1, \ldots, n. \tag{10.6}$$

Notice that the only difference between these equations and their analogues (7.4) and (7.10) is that λ^2 is replaced by $k\lambda$.

From (10.5) and (10.6) we derive that *the time-dependence of the transition coefficients is given by*

$$a_\varrho(\lambda, t) = a_\varrho(\lambda, 0), \qquad b_\varrho(\lambda, t) = e^{-ik\lambda t} b_\varrho(\lambda, 0),$$
$$\gamma_j(t) = e^{-ik_j \lambda_j t} \gamma_j(0), \quad j = 1, \ldots, n. \tag{10.7}$$

Thus, also in the case of finite density boundary conditions, the dynamics is considerably simplified by passing from $\psi(x)$, $\bar{\psi}(x)$ to the transition coefficients and discrete spectrum:

$$(\psi(x), \bar{\psi}(x)) \to (b_\varrho(\lambda), \bar{b}_\varrho(\lambda); \lambda_j, \gamma_j, j = 1, \ldots, n). \tag{10.8}$$

In the next chapter we shall investigate the reversibility of (10.8) and in Chapter III we shall discuss this mapping from the Hamiltonian point of view.

Let us now proceed to the integrals of the motion. From (10.7) it follows that *their generating function is $a_\varrho(\lambda)$. We will show that, as in the rapidly decreasing case, $\log a_\varrho(\lambda) e^{-\frac{i\theta}{2}}$ is the generating function for the local integrals of the motion.* To be able to use the results of § 4 let us assume that $\psi(x)$, $\bar{\psi}(x)$ are obtained in the limit $L \to \infty$ from the functions $\psi_L(x)$, $\bar{\psi}_L(x)$ satisfying (1.6)

$$\psi_L(x + 2L) = e^{i\theta} \psi_L(x), \qquad \bar{\psi}_L(x + 2L) = e^{-i\theta} \bar{\psi}_L(x) \tag{10.9}$$

with the additional constraint (cf. § 1)

$$\psi_L(x)|_{x=-L} = \bar{\psi}_L(x)|_{x=-L} = \varrho . \tag{10.10}$$

First consider the limit of the generating function

$$p_L(\lambda) = \arccos \tfrac{1}{2} \operatorname{tr} T_L(\lambda) Q(\theta), \tag{10.11}$$

as $L \to \infty$. Recalling that $b_\varrho(\lambda)$ is of Schwartz type for $|\lambda| \to \infty$, from (8.44) and the explicit formula (8.9) for $E_\varrho(x, \lambda)$ we deduce that, for λ in $\mathbb{R}_\omega$,

$$\begin{aligned} \operatorname{tr} T_L(\lambda) Q(\theta) &= e^{-ikL} a_\varrho(\lambda) + e^{ikL} \bar{a}_\varrho(\lambda) + o(1) \\ &= 2\cos(-kL + \arg a_\varrho(\lambda)) + o(1), \end{aligned} \tag{10.12}$$

as $L \to \infty$, up to terms of order $O(|\lambda|^{-\infty})$. The last identity relies upon

$$|a_\varrho(\lambda)| = 1 + O(|\lambda|^{-\infty}) \tag{10.13}$$

(cf. § 7). Thus, up to terms of order $O(|\lambda|^{-\infty})$ we obtain

$$\lim_{L \to \infty} \left(p_L(\lambda) + kL - \frac{\theta}{2} \right) = \frac{1}{i} \log a_\varrho(\lambda) e^{-\frac{i\theta}{2}} . \tag{10.14}$$

Now recall the expansion (4.4)

$$p_L(\lambda) = -\lambda L + \frac{\theta}{2} + \varkappa \sum_{n=1}^{\infty} \frac{I_n}{\lambda^n} + O(|\lambda|^{-\infty}), \tag{10.15}$$

where (see (4.32)–(4.34))

$$I_n = \int_{-L}^{L} P_n(x)\, dx . \tag{10.16}$$

Using the asymptotic expansion

$$\lambda - k(\lambda) = \sum_{n=1}^{\infty} \frac{p_n}{\lambda^n} + O(|\lambda|^{-\infty}), \tag{10.17}$$

valid for λ in $\mathbb{R}_\omega$, where

$$p_{2n-1} = (-1)^{n+1} \omega^{2n} \binom{\frac{1}{2}}{n}; \quad p_{2n} = 0, \tag{10.18}$$

we conclude from (10.14) that in the case of finite density boundary conditions the expressions

$$I_n - \frac{p_n}{\varkappa} L = \int_{-L}^{L} \left(P_n(x) - \frac{p_n}{2\varkappa} \right) dx \tag{10.19}$$

have a limit $I_{n,\varrho}$ as $L \to \infty$. We have thus shown that adding kL to $p_L(\lambda)$ regularizes the functionals $I_n(\psi_L, \bar{\psi}_L)$ so that, as $L \to \infty$, they have finite limits $I_{n,\varrho}$,

$$I_{n,\varrho} = \int_{-\infty}^{\infty} \left(P_n(x) - \frac{1}{2\varkappa} p_n \right) dx. \tag{10.20}$$

It follows from (10.14) and (9.43) that the $I_{n,\varrho}$ may be represented as functionals of $\log(1 + |b_\varrho(\lambda)|^2)$ and λ_j, $j = 1, \ldots, n$. However, if n is odd, $I_{n,\varrho}$ is not an admissible functional on the phase space $\mathscr{M}_{\varrho,\theta}$. Indeed, we have, for example,

$$\frac{\delta I_{3,\varrho}}{\delta \bar{\psi}(x)} = -\frac{d^2 \psi(x)}{dx^2} + 2\varkappa |\psi(x)|^2 \psi(x), \tag{10.21}$$

so that $\dfrac{\delta I_{3,\varrho}}{\delta \bar{\psi}(x)}$ does not vanish as $|x| \to \infty$.

Nevertheless, suitable linear combinations of the $I_{n,\varrho}$ turn out to be admissible. They are obtained by expanding $p_L(\lambda)$ in inverse powers of $k(\lambda)$,

$$p_L(\lambda) = -kL + \frac{\theta}{2} + \varkappa \sum_{n=1}^{\infty} \frac{J_n}{k^n} + O(|k|^{-\infty}), \tag{10.22}$$

which follows from (10.15) by using (10.17) and the asymptotic expansion

$$\frac{1}{\lambda^n} = \sum_{m=0}^{\infty} \frac{\omega^{2m}}{k^{n+2m}} \begin{pmatrix} -\frac{n}{2} \\ m \end{pmatrix}, \tag{10.23}$$

valid for λ in $\mathbb{R}_\omega$. The functionals J_n have the same form as in (10.16) and their limit as $L \to \infty$ is

$$J_{n,\varrho}=\lim_{L\to\infty} J_n=\int_{-\infty}^{\infty} P_{n,\varrho}(x)\,dx. \tag{10.24}$$

By (10.23), there is a simple expression for $J_{n,\varrho}$ in terms of $I_{n,\varrho}$

$$J_{n,\varrho}=\sum_{n=l+2m,\,l>0} \omega^{2m}\begin{pmatrix}-\frac{l}{2}\\ m\end{pmatrix} I_{l,\varrho}. \tag{10.25}$$

In particular, we find

$$J_{1,\varrho}=N_\varrho,\qquad J_{2,\varrho}=P_\varrho,\qquad J_{3,\varrho}=H_\varrho \tag{10.26}$$

(see § 1), so that, in contrast with $J_{1,\varrho}$, the functionals $J_{2,\varrho}$ and $J_{3,\varrho}$ correspond to observables. In Chapter III we shall give a simple proof of the fact that the $J_{n,\varrho}$ are admissible for $n>1$, i.e. their variational derivatives vanish as $|x|\to\infty$. The proof will be based on an explicit formula for the variational derivatives of $p_L(\lambda)$.

As in the rapidly decreasing case (see § 7), the local constants of the motion, $J_{n,\varrho}$, depend on only half of the new variables ($b_\varrho(\lambda), \bar{b}_\varrho(\lambda)$; $\lambda_j, \gamma_j, j=1,\dots,n$). In order to determine them consider the asymptotic expansion of $\log a_\varrho(\lambda)e^{-\frac{i\theta}{2}}$ for λ in $\mathbb{R}_\omega$ as $|\lambda|\to\infty$,

$$\log a_\varrho(\lambda)e^{-\frac{i\theta}{2}}=i\varkappa\sum_{l=1}^{\infty}\frac{c_{l,\varrho}}{k^l}+O(|k|^{-\infty}), \tag{10.27}$$

which results from (9.18). To find the real coefficients $c_{l,\varrho}$ in closed form we use (9.43) and expand the denominator $\dfrac{1}{\mu-\lambda}$ in a geometric progression (cf. § 7). Using (10.23) we find

$$c_{l,\varrho}=\frac{1}{2\pi\varkappa}\int_{\mathbb{R}_\omega}\varphi_l(\lambda)\log(1+|b_\varrho(\lambda)|^2)\,d\lambda+\frac{1}{\varkappa}\sum_{j=1}^{n}\varphi_{l,j}. \tag{10.28}$$

For $\varphi_l(\lambda)$ we have the expression

$$\varphi_l(\lambda)=\frac{1}{k(\lambda)}\sum_{l=p+2q}\lambda^p\,\omega^{2q}\begin{pmatrix}-\frac{p+1}{2}\\ q\end{pmatrix}. \tag{10.29}$$

For odd $l=2m+1$, using the elementary formula

$$\binom{\alpha}{n}=(-1)^n\binom{n-\alpha-1}{n} \tag{10.30}$$

and the binomial theorem, we obtain from (10.29)

$$\varphi_{2m+1}(\lambda)=\lambda k^{2m-1}(\lambda). \tag{10.31}$$

For even $l=2m$, using (10.30) and the simple formula

$$\sum_{l=0}^{n}\binom{\alpha}{n-l}\binom{\beta}{l}=\binom{\alpha+\beta}{n} \tag{10.32}$$

we can rewrite (10.29) as

$$\begin{aligned}\varphi_{2m}(\lambda)&=\frac{1}{k(\lambda)}\sum_{p=0}^{m}\omega^{2p}(k^2(\lambda)+\omega^2)^{m-p}\binom{p-m-\frac{1}{2}}{p}\\ &=k^{2m-1}(\lambda)\sum_{p=0}^{m}\sum_{q=0}^{m-p}\left(\frac{\omega^2}{k^2(\lambda)}\right)^{m-q}\binom{m-p}{q}\binom{p-m-\frac{1}{2}}{p}\\ &=k^{2m-1}(\lambda)\sum_{p=0}^{m}\binom{\frac{1}{2}}{p}\left(\frac{\omega^2}{k^2(\lambda)}\right)^{p}.\end{aligned} \tag{10.33}$$

To compute the coefficients $\varphi_{l,j}$ we shall make use of the relation

$$\log\frac{\lambda+k-\lambda_j-k_j}{\lambda+k-\lambda_j-\bar{k}_j}=-\int_{\lambda_j}^{\omega}\frac{1}{k(\mu)}\left(1+\frac{k(\lambda)}{\mu-\lambda}\right)d\mu, \tag{10.34}$$

which holds for $-\omega<\lambda_j<\omega$ and can be proved either directly or by a change of variable as in § 9. It follows that

$$\varphi_{l,j}=\frac{1}{i}\int_{\lambda_j}^{\omega}\varphi_l(\lambda)\,d\lambda. \tag{10.35}$$

Integration then gives

$$\varphi_{2m+1,j}=\frac{i}{2m+1}k_j^{2m+1} \tag{10.36}$$

and

$$\varphi_{2m,j} = \frac{i\lambda_j k_j^{2m-1}}{2m} \sum_{p=0}^{m-1} \binom{-\frac{1}{2}}{p} \left(\frac{\omega^2}{k_j^2}\right)^p . \tag{10.37}$$

We thus conclude that $\log a_\varrho(\lambda) e^{-\frac{i\theta}{2}}$ has the asymptotic expansion (10.27) with the coefficients $c_{l,\varrho}$ given by (10.28), (10.31), (10.33) and (10.36), (10.37). The first few coefficients are

$$c_{1,\varrho} = \frac{1}{2\pi\varkappa} \int_{\mathbb{R}_\omega} \frac{\log(1+|b_\varrho(\lambda)|^2)}{k(\lambda)} \lambda \, d\lambda + \frac{i}{\varkappa} \sum_{j=1}^{n} k_j , \tag{10.38}$$

$$c_{2,\varrho} = \frac{1}{2\pi\varkappa} \int_{\mathbb{R}_\omega} \frac{\log(1-|b_\varrho(\lambda)|^2)}{k(\lambda)} \left(k^2(\lambda) + \frac{\omega^2}{2}\right) d\lambda + \frac{i}{2\varkappa} \sum_{j=1}^{n} \lambda_j k_j , \tag{10.39}$$

$$c_{3,\varrho} = \frac{1}{2\pi\varkappa} \int_{\mathbb{R}_\omega} \log(1+|b_\varrho(\lambda)|^2) \lambda k(\lambda) \, d\lambda + \frac{i}{3\varkappa} \sum_{j=1}^{n} k_j^3 . \tag{10.40}$$

Comparing (10.14), (10.22), (10.24) with (10.27) leads to the identities

$$J_{n,\varrho} = c_{n,\varrho} = \int_{-\infty}^{\infty} P_{n,\varrho}(x) \, dx \tag{10.41}$$

which are *the trace formulae for the case of finite density*. Their interpretation in terms of Hamiltonian mechanics will be given in Chapter III.

This concludes our discussion of the properties of transition coefficients and their dynamics. In the next chapter we shall investigate the invertibility of the mapping from $\psi(x)$, $\bar{\psi}(x)$ to the transition coefficients and the discrete spectrum for both rapidly decreasing and finite density boundary conditions. The results obtained will be fundamental for the complete solution of the NS initial value problem (1.1)–(1.2) under these boundary conditions.

§ 11. Notes and References

1. The zero curvature representation (2.10) is an alternative to the Lax representation (see the Introduction) exploited at the early stage of the inverse scattering method. A significant contribution to the related analysis was made by the papers [AKNS 1974], [N 1974], [ZM 1978], [ZS 1979]. In our

text we have given preference to this representation because it has a clear geometric meaning emphasized in [ZT 1979].

2. The NS equation supplied with various boundary conditions provides models for a broad class of nonlinear phenomena in physics. We have already mentioned its applications in nonlinear optics; it is encountered in plasma physics as well. Here we point out its role of the Hartree-Fock equation for a multi-particle quantum system (Bose gas) with a pairwise interaction via the potential $2\varkappa\delta(x-y)$. The sign of the coupling constant $\varkappa$ distinguishes between attraction ($\varkappa<0$) and repulsion ($\varkappa>0$) of particles. In the attractive case a physically meaningful problem is that of a finite number of particles and their bound states. Its classical limit is modelled by rapidly decreasing boundary conditions. In the repulsive case a problem of interest is the one corresponding to the gas of particles of finite density. The boundary conditions called here the finite density conditions model this situation.

The first imbedding of the NS model into the framework of the inverse scattering method was given in [ZS 1971], [ZS 1973].

3. The integral representations (3.32) and (3.33) for the transition matrix are a variant of the formulae of M. G. Krein [K 1956]. The triangular representations (5.10) and (5.16) for the Jost solutions are more traditional. They were introduced by B. Ja. Levin [L 1956] starting from the eigenvalue problem for the *one-dimensional Schrödinger operator*

$$-\frac{d^2y}{dx^2}+u(x)y=\lambda y \tag{11.1}$$

with a potential $u(x)$ satisfying

$$\int_{-\infty}^{\infty}(1+|x|)\,|u(x)|\,dx<\infty\,, \tag{11.2}$$

and were used extensively by V. A. Marchenko (see the summarizing treatise [M 1977]). Our derivation of these representations in § 5 differs from the usual one which we follow in § 8 in the case of finite density.

4. The discussion in §§ 5–6 is just a variant of quantum scattering theory for the operator $\mathscr{L}$ (see (5.24)) whose quantum mechanical meaning is that of the Dirac operator with zero mass. In the context of inverse scattering method this operator is sometimes called the *Zakharov-Shabat operator*.

5. In the case of the finite density boundary conditions the operator $\mathscr{L}$ actually coincides with the Dirac operator with non-zero mass. Its scattering theory is closer to the one-dimensional Schrödinger operator (11.1) because in both cases the continuous spectrum has boundary points. The minimal requirements on $\psi(x)$, $\bar{\psi}(x)$ assuring that the general scattering formalism works are more stringent than in the rapidly decreasing case. So, not only

the decaying parts, $\psi(x)-\varrho e^{i\varphi_\pm}$, of $\psi(x)$, but also their first moments must be absolutely integrable in the neighbourhood of $\pm\infty$ (cf. (11.2)) [AK 1981]. The inverse problem was studied in [Fr 1972] and [GK 1978]; in [Fr 1972] the condition (θ) was omitted.

References

[AK 1981] Asano, N., Kato, Y.: Non-self-adjoint Zakharov-Shabat operator with a potential of the finite asymptotic values. I. Direct spectral and scattering problems. J. Math. Phys. *22*, 2780–2793 (1981)

[AKNS 1974] Ablowitz, M. J., Kaup, D. J., Newell, A. C., Segur, H.: The inverse scattering transform-Fourier analysis for nonlinear problems. Stud. Appl. Math. *53*, 249–315 (1974)

[Fr 1972] Frolov, I. S.: Inverse scattering problem for the Dirac system on the whole line. Dokl. Akad. Nauk SSSR *207*, 44–47 (1972) [Russian]; English transl. in Sov. Math. Dokl. *13*, 1468–1472 (1972)

[GK 1978] Gerdjikov, V. C., Kulish, P. P.: Completely integrable Hamiltonian system associated with the non-self-adjoint Dirac operator. Bulg. J. Phys. *5*, 337–348 (1978) [Russian]

[K 1956] Krein, M. G.: On the theory of accelerants and *S*-matrices of canonical differential systems. Dokl. Akad. Nauk SSSR *111*, 1167–1170 (1956) [Russian]

[L 1956] Levin, B. Ja.: Transformations of Fourier and Laplace types by means of solutions of a second order differential equation. Dokl. Akad. Nauk SSSR *106*, 187–190 (1956) [Russian]

[M 1977] Marchenko, V. A.: Sturm-Liouville Operators and their Applications. Kiev, Naukova Dumka 1977 [Russian]

[N 1974] Novikov, S. P.: The periodic problem for the Korteweg-de Vries equation. Funk. Anal. Prilož. *8* (3) 54–66 (1974) [Russian]; English transl. in Funct. Anal. Appl. *8*, 236–246 (1974)

[ZM 1978] Zakharov, V. E., Mikhailov, A. V.: Relativistically invariant two-dimensional models of field theory which are integrable by means of the inverse scattering problem method. Zh. Exp. Teor. Fiz. *74*, 1953–1973 (1978) [Russian]; English transl. in Sov. Phys. JETP *47*, 1017–1027 (1978)

[ZS 1971] Zakharov, V. E., Shabat, A. B.: Exact theory of two-dimensional self-focusing and one-dimensional self-modulation of waves in non-linear media. Zh. Exp. Teor. Fiz. *61*, 118–134 (1971) [Russian]; English transl. in Soviet Phys. JETP *34*, 62–69 (1972)

[ZS 1973] Zakharov, V. E., Shabat, A. B.: Interaction between solitons in a stable medium. Zh. Exp. Teor. Fiz. *64*, 1627–1639 (1973) [Russian]; English transl. in Sov. Phys. JETP *37*, 823–828 (1973)

[ZS 1979] Zakharov, V. E., Shabat, A. B.: Integration of the nonlinear equations of mathematical physics by the method of the inverse scattering problem. II. Funk. Anal. Prilož. *13* (3), 13–22 (1979) [Russian]; English transl. in Funct. Anal. Appl. *13*, 166–174 (1979)

[ZT 1979] Zakharov, V. E., Takhtajan, L. A.: Equivalence of the nonlinear Schrödinger equation and the Heisenberg ferromagnet equation. Teoret. Mat. Fiz. *38* (1), 26–35 (1979) [Russian]; English transl. in Theor. Math. Phys. *38*, 17–23 (1979)

Chapter II
The Riemann Problem

In Chapter I we analyzed the mapping

$$\mathscr{F}: (\psi(x), \bar{\psi}(x)) \to (b(\lambda), \bar{b}(\lambda); \lambda_j, \gamma_j)$$

from the functions $\psi(x)$, $\bar{\psi}(x)$ to the transition coefficients and discrete spectrum of the auxiliary linear problem. We saw that for both rapidly decreasing and finite density boundary conditions this "change of variables" makes the dynamics quite simple because the time evolution of the transition coefficients for the continuous and discrete spectra becomes linear.

In this chapter we shall investigate the inverse mapping $\mathscr{F}^{-1}$. More precisely, we shall explain in what sense $\mathscr{F}$ has an inverse and present a solution of the inverse problem, i.e. the problem of reconstructing $\psi(x)$, $\bar{\psi}(x)$ from the transition coefficients and discrete spectrum. The basic tool for solving the inverse problem is provided by the formalism of conjugation problem in function theory, also called the Riemann problem or analytic factorization problem. There are different variants of this problem depending on the boundary conditions, discrete spectrum, etc. Here we shall specify the Riemann problem for the boundary conditions considered above and analyze it in detail.

§ 1. The Rapidly Decreasing Case. Formulation of the Riemann Problem

The rapidly decreasing case leads to the following Riemann problem. Let $G(\lambda)$ be a matrix function on the real line $-\infty < \lambda < \infty$. *The problem consists in representing it in the factorized form*

$$G(\lambda) = G_+(\lambda) G_-(\lambda), \tag{1.1}$$

where $G_+(\lambda)$ and $G_-(\lambda)$ may be analytically extended into the upper and lower half-planes, respectively. The existence problem for (1.1) under various restrictions on $G(\lambda)$, $G_{\pm}(\lambda)$ has been extensively studied in the mathematical literature. Let us show how (1.1) arises in the framework of the auxiliary linear problem,

$$\frac{dF}{dx} = U(x, \lambda) F, \tag{1.2}$$

and what are the properties of $G(\lambda)$, $G_{\pm}(\lambda)$.

Our starting point is the relationship between the Jost solutions $T_{+}(x, \lambda)$ and $T_{-}(x, \lambda)$ involving the reduced monodromy matrix $T(\lambda)$,

$$T_{-}(x, \lambda) = T_{+}(x, \lambda) T(\lambda). \tag{1.3}$$

This is not yet a relation of the type (1.1) because the columns of $T_{\pm}(x, \lambda)$ are analytic in different half-planes. In fact, from § I.5 we know that, in the column notation for $T_{\pm}(x, \lambda)$

$$T_{\pm}(x, \lambda) = (T_{\pm}^{(1)}(x, \lambda), T_{\pm}^{(2)}(x, \lambda)), \tag{1.4}$$

the columns $T_{-}^{(1)}(x, \lambda)$ and $T_{+}^{(2)}(x, \lambda)$ are analytic in the upper half-plane, whereas $T_{+}^{(1)}(x, \lambda)$ and $T_{-}^{(2)}(x, \lambda)$ are analytic in the lower half-plane. However, (1.3) can easily be transformed to the form (1.1). To this end we define the matrices

$$S_{+}(x, \lambda) = (T_{-}^{(1)}(x, \lambda), T_{+}^{(2)}(x, \lambda)) \tag{1.5}$$

and

$$S_{-}(x, \lambda) = (T_{+}^{(1)}(x, \lambda), T_{-}^{(2)}(x, \lambda)), \tag{1.6}$$

which solve the linear problem (1.2) and extend analytically into the upper and lower half-planes, respectively. In these half-planes they have the following asymptotic behaviour:

$$S_{\pm}(x, \lambda) E^{-1}(x, \lambda) = I + o(1), \tag{1.7}$$

as $|\lambda| \to \infty$, which follows from (I.5.26)-(I.5.29).

By using (1.3), $S_{\pm}(x, \lambda)$ can be written as

$$S_{+}(x, \lambda) = T_{+}(x, \lambda) M_{++}(\lambda) = T_{-}(x, \lambda) M_{-+}(\lambda) \tag{1.8}$$

and

$$S_{-}(x, \lambda) = T_{+}(x, \lambda) M_{+-}(\lambda) = T_{-}(x, \lambda) M_{--}(\lambda), \tag{1.9}$$

where $M_{\pm,\pm}(\lambda)$ are given by

$$M_{++}(\lambda)=\begin{pmatrix} a(\lambda) & 0 \\ b(\lambda) & 1 \end{pmatrix}, \quad M_{-+}(\lambda)=\begin{pmatrix} 1 & -\varepsilon\bar{b}(\lambda) \\ 0 & a(\lambda) \end{pmatrix},$$
$$M_{+-}(\lambda)=\begin{pmatrix} 1 & \varepsilon\bar{b}(\lambda) \\ 0 & \bar{a}(\lambda) \end{pmatrix}, \quad M_{--}(\lambda)=\begin{pmatrix} \bar{a}(\lambda) & 0 \\ -b(\lambda) & 1 \end{pmatrix}, \tag{1.10}$$

with $\varepsilon=\operatorname{sign}\varkappa$. Here $a(\lambda)$ and $b(\lambda)$ are the transition coefficients entering into $T(\lambda)$ (see (I.5.45)). The fact that (1.8) and (1.9) are compatible with (1.3) yields the following factorization of the reduced monodromy matrix:

$$T(\lambda)=M_{+-}(\lambda)M_{--}^{-1}(\lambda)=M_{++}(\lambda)M_{-+}^{-1}(\lambda). \tag{1.11}$$

The special triangular form of $M_{\pm,\pm}(\lambda)$ permits to reconstruct them in a unique way from the given unimodular matrix $T(\lambda)$ of the form (I.5.45).

In terms of $S_{\pm}(x,\lambda)$, (1.3) becomes

$$S_{-}(x,\lambda)=\dot{S}_{+}(x,\lambda)S(\lambda), \tag{1.12}$$

where

$$S(\lambda)=M_{++}^{-1}(\lambda)M_{+-}(\lambda)=M_{-+}^{-1}(\lambda)M_{--}(\lambda)$$
$$=\begin{pmatrix} \dfrac{1}{a(\lambda)} & \dfrac{\varepsilon\bar{b}(\lambda)}{a(\lambda)} \\ -\dfrac{b(\lambda)}{a(\lambda)} & \dfrac{1}{a(\lambda)} \end{pmatrix}. \tag{1.13}$$

The matrix $S(\lambda)$ plays the role of *scattering matrix* for the auxiliary linear problem. In scattering theory the coefficients $\dfrac{1}{a(\lambda)}$ and $\dfrac{b(\lambda)}{a(\lambda)}$ are commonly called *transmission* and *reflection coefficients*, respectively. Unlike $T(\lambda)$, the scattering matrix is not unimodular,

$$\det S(\lambda)=\frac{\bar{a}(\lambda)}{a(\lambda)}. \tag{1.14}$$

It satisfies the involutions

$$\sigma\bar{S}(\lambda)\sigma=S^{-1}(\lambda) \tag{1.15}$$

and

$$\tilde{\sigma}S^{*}(\lambda)\tilde{\sigma}=S^{-1}(\lambda), \tag{1.16}$$

where $\sigma=\sigma_1$ when $\varepsilon=1$, $\sigma=\sigma_2$ when $\varepsilon=-1$ (see 1.2), $\tilde{\sigma}=I$ when $\varepsilon=1$, $\tilde{\sigma}=\sigma_3$ when $\varepsilon=-1$, and * indicates Hermitian conjugation. Formula (1.16) can be interpreted as *$\tilde{\sigma}$-unitarity* of the scattering matrix. Relations (1.15) and (1.16) uniquely determine the form of $S(\lambda)$.

Equation (1.12) has almost the same form as (1.1), so that it might seem sufficient to put $G_+(x,\lambda)=S_+^{-1}(x,\lambda)$ and $G_-(x,\lambda)=S_-(x,\lambda)$. However, (1.8) implies that

$$\det S_+(x,\lambda)=a(\lambda). \tag{1.17}$$

Hence, the zeros of $a(\lambda)$ correspond to singularities of $S_+^{-1}(x,\lambda)$ in the upper half-plane. Therefore we set

$$G_-(x,\lambda)=S_-(x,\lambda)E^{-1}(x,\lambda) \tag{1.18}$$

and

$$G_+(x,\lambda)=a(\lambda)E(x,\lambda)S_+^{-1}(x,\lambda), \tag{1.19}$$

where in addition we have cancelled out the asymptotic part of $S_\pm(x,\lambda)$ as $|\lambda|\to\infty$.

The matrices $G_\pm(x,\lambda)$ solve the Riemann problem

$$G_+(x,\lambda)G_-(x,\lambda)=G(x,\lambda), \tag{1.20}$$

where

$$G(x,\lambda)=E(x,\lambda)G(\lambda)E^{-1}(x,\lambda)=\begin{pmatrix} 1 & \varepsilon\bar{b}(\lambda)e^{-i\lambda x} \\ -b(\lambda)e^{i\lambda x} & 1 \end{pmatrix} \tag{1.21}$$

and

$$G(\lambda)=\begin{pmatrix} 1 & \varepsilon\bar{b}(\lambda) \\ -b(\lambda) & 1 \end{pmatrix}, \tag{1.22}$$

with the usual normalization

$$G_\pm(x,\lambda)=I+o(1), \qquad G(x,\lambda)=I+o(1), \tag{1.23}$$

as $|\lambda|\to\infty$. The variable x is a parameter of the Riemann problem entering through exponential factors in $G(x,\lambda)$.

We shall now list the properties of $G(x,\lambda)$ and $G_\pm(x,\lambda)$ associated with the auxiliary linear problem with $\psi(x)$, $\bar{\psi}(x)$ in $L_1(-\infty,\infty)$. If $\varkappa<0$, the condition (A) concerning the location of the zeros of $a(\lambda)$ will be assumed (see § I.6). We start with $G(\lambda)$ and the related matrix $G(x,\lambda)$.

1. *The involution property*

$$\tau G^*(x,\lambda)\tau=G(x,\lambda), \tag{1.24}$$

where $\tau=\sigma_3$ when $\varepsilon=1$ and $\tau=I$ when $\varepsilon=-1$.

2. *The non-degeneracy property*

$$\det G(x,\lambda)=\det G(\lambda)=1+\varepsilon|b(\lambda)|^2, \tag{1.25}$$

so that by virtue of the condition (*A*)

$$\det G(x,\lambda)>0 \tag{1.26}$$

for each λ.

3. *The integral representations.*

From (I.6.16) it follows that $G(\lambda)$ can be represented as

$$G(\lambda)=I+\int_{-\infty}^{\infty}\Phi(s)e^{i\lambda s}\,ds, \tag{1.27}$$

where $\Phi(s)$ has the specific form

$$\Phi(s)=\begin{pmatrix}0 & \varepsilon\bar\beta(-s)\\ -\beta(s) & 0\end{pmatrix}, \tag{1.28}$$

and $\beta(s)$ is given by

$$\beta(s)=\frac{1}{2\pi}\int_{-\infty}^{\infty}b(\lambda)e^{-i\lambda s}\,d\lambda \tag{1.29}$$

and belongs to $L_1(-\infty,\infty)$. There is a similar expression for $G(x,\lambda)$ with $\Phi(s)$ replaced by $\Phi(x,s)$,

$$\Phi(x,s)=\begin{pmatrix}0 & \varepsilon\bar\beta(-s-x)\\ -\beta(s-x) & 0\end{pmatrix}. \tag{1.30}$$

In other words, $G(\lambda)$ and $G(x,\lambda)$ are specific elements of the *normed ring* $\mathfrak{R}^{(2\times 2)}$ consisting of matrix functions of the form

$$F(\lambda)=cI+\int_{-\infty}^{\infty}\Omega(s)e^{i\lambda s}\,ds, \tag{1.31}$$

where $\Omega(s)$ is in $L_1^{(2\times 2)}(-\infty,\infty)$, c is in $\mathbb{C}^1$ and the norm is defined in the usual way (cf. § I.6)

$$\|F\| = |c| + \int_{-\infty}^{\infty} \|\Omega(s)\|\, ds. \tag{1.32}$$

Later we shall also need the *subrings* $\mathfrak{R}_{\pm}^{(2\times 2)}$ of the ring $\mathfrak{R}^{(2\times 2)}$ consisting of matrix functions of the form

$$F_{\pm}(\lambda) = c_{\pm} I + \int_{0}^{\infty} \Omega_{\pm}(s) e^{\pm i\lambda s}\, ds, \tag{1.33}$$

respectively, where $\Omega_{\pm}(s)$ are in $L_1^{(2\times 2)}(0, \infty)$. The elements of $\mathfrak{R}_{+}^{(2\times 2)}$ and $\mathfrak{R}_{-}^{(2\times 2)}$ are analytic in the upper and lower half-planes, respectively. By the Riemann-Lebesgue lemma they tend to $c_{\pm} I$ as $|\lambda| \to \infty$.

Now we turn to the properties of $G_{\pm}(x, \lambda)$.

1. *The involution property*

$$\tau G_{+}^{*}(x, \lambda) \tau = G_{-}(x, \bar{\lambda}), \tag{1.34}$$

which follows from (I.5.30), (I.5.31) and (1.5), (1.6), (1.17). In particular, if $\varkappa < 0$ we have

$$G_{+}^{*}(x, \lambda) = G_{-}(x, \bar{\lambda}). \tag{1.35}$$

2. *The integral representations*

$$G_{\pm}(x, \lambda) = I + \int_{0}^{\infty} \Phi_{\pm}(x, s) e^{\pm i\lambda s}\, ds, \tag{1.36}$$

which result from (I.5.10), (I.5.16) and (I.6.10) after trivial transformations. Note that the matrix functions $\Phi_{\pm}(x, s)$ are in $L_1^{(2\times 2)}(0, \infty)$, so that $G_{\pm}(x, \lambda)$ are in $\mathfrak{R}_{\pm}^{(2\times 2)}$.

3. *The relationship* (see § I.2)

$$U_0(x) = \tfrac{1}{2}[\sigma_3, \Phi_{\pm}(x, s)]|_{s=0}, \tag{1.37}$$

where

$$U_0(x) = U(x, \lambda) - \frac{\lambda}{2i}\sigma_3 = \sqrt{\varkappa}\begin{pmatrix} 0 & \bar{\psi}(x) \\ \psi(x) & 0 \end{pmatrix}, \tag{1.38}$$

which is an immediate consequence of (I.5.32) and (I.5.33).

4. *The asymptotic behaviour for real λ, as $|x|\to\infty$:*

$$G_+^{-1}(x,\lambda)=\begin{pmatrix}1 & 0\\ \frac{b(\lambda)}{a(\lambda)}e^{i\lambda x} & \frac{1}{a(\lambda)}\end{pmatrix}+o(1), \tag{1.39}$$

$$G_-(x,\lambda)=\begin{pmatrix}1 & \varepsilon\bar{b}(\lambda)e^{-i\lambda x}\\ 0 & \bar{a}(\lambda)\end{pmatrix}+o(1), \tag{1.40}$$

as $x\to+\infty$ and

$$G_+^{-1}(x,\lambda)=\begin{pmatrix}\frac{1}{a(\lambda)} & -\varepsilon\frac{\bar{b}(\lambda)}{a(\lambda)}e^{-i\lambda x}\\ 0 & 1\end{pmatrix}+o(1), \tag{1.41}$$

$$G_-(x,\lambda)=\begin{pmatrix}\bar{a}(\lambda) & 0\\ -b(\lambda)e^{i\lambda x} & 1\end{pmatrix}+o(1), \tag{1.42}$$

as $x\to-\infty$, which follows from (1.8), (1.9) and (1.18), (1.19).

5. *Degeneracy properties for complex λ.*

From (1.8), (1.9) and (1.18), (1.19) we find that $\det G_+(x,\lambda)=a(\lambda)$, $\det G_-(x,\lambda)=a^*(\lambda)$, where $a^*(\lambda)$ is the analytic extension of $\bar{a}(\lambda)$ into the lower half-plane (see § I.6). If $\varkappa>0$, it follows that $G_\pm(x,\lambda)$ are non-degenerate in their domains of analyticity. If $\varkappa<0$ and the condition (*A*) holds, $G_+(x,\lambda)$ and $G_-(x,\lambda)$ become degenerate at $\lambda=\lambda_j$ and $\lambda=\bar{\lambda}_j$, respectively, where λ_j, $j=1,\dots,n$, are the zeros of $a(\lambda)$. More precisely, they have simple zeros, i.e. $G_+(x,\lambda_j)$ and $G_-(x,\bar{\lambda}_j)$ are rank one matrices. Comparing (1.5), (1.19) with (I.6.20) shows that $G_+(x,\lambda_j)$ can be written in the form

$$G_+(x,\lambda_j)=e^{-\frac{i\lambda_j x}{2}}\begin{pmatrix}1\\ -\gamma_j(x)\end{pmatrix}T_+^{(2)^\tau}(x,\lambda_j)\cdot\frac{1}{i}\sigma_2, \tag{1.43}$$

with

$$\gamma_j(x)=e^{i\lambda_j x}\gamma_j,\quad j=1,\dots,n, \tag{1.44}$$

and the column-matrix $\begin{pmatrix}1\\ -\gamma_j(x)\end{pmatrix}$ is multiplied by the row-matrix $-iT_+^{(2)^\tau}(x,\lambda_j)\sigma_2$. By virtue of (1.35) and (I.5.30), $G_-(x,\bar{\lambda}_j)$ can be expressed as

$$G_-(x,\bar{\lambda}_j)=e^{\frac{i\bar{\lambda}_j x}{2}}T_+^{(1)}(x,\bar{\lambda}_j)(1,-\bar{\gamma}_j(x)), \tag{1.45}$$

where the column-matrix $T_+^{(1)}(x,\bar\lambda_j)$ is multiplied by the row-matrix $(1, -\bar\gamma_j(x))$.

To give a geometric interpretation of these representations, let $N_j^{(+)}(x)$ be the one-dimensional subspace in $\mathbb{C}^2$ spanned by $\begin{pmatrix}1\\ -\gamma_j(x)\end{pmatrix}$ and let $N_j^{(-)}(x)$ be its orthogonal complement spanned by $\begin{pmatrix}\bar\gamma_j(x)\\ 1\end{pmatrix}$. Then (1.43) and (1.45) amount to

$$N_j^{(+)}(x)=\operatorname{Im} G_+(x,\lambda_j),\qquad N_j^{(-)}(x)=\operatorname{Ker} G_-(x,\bar\lambda_j), \tag{1.46}$$

$j=1,\ldots,n$. The dependence of $N_j^{(\pm)}(x)$ on x is given by

$$N_j^{(+)}(x)=E(x,\lambda_j)\,N_j^{(+)},\qquad N_j^{(-)}(x)=E(x,\bar\lambda_j)\,N_j^{(-)}, \tag{1.47}$$

where the subspaces $N_j^{(+)}$ and $N_j^{(-)}$ are spanned by $\begin{pmatrix}1\\ -\gamma_j\end{pmatrix}$ and $\begin{pmatrix}1\\ \bar\gamma_j\end{pmatrix}$, respectively.

The properties of $G(\lambda)$, $G(x,\lambda)$ and $G_\pm(x,\lambda)$ listed above result from the analysis of the auxiliary linear problem (1.2) for $\psi(x)$, $\bar\psi(x)$ in $L_1(-\infty,\infty)$ carried out in Chapter I. Now we let them be the basic ingredients of the Riemann problem which consists in reconstructing $G_\pm(x,\lambda)$ (and hence $\psi(x)$, $\bar\psi(x)$, see (1.37)) from a given $G(\lambda)$.

More precisely, let the following data be given:

1) *a matrix $G(\lambda)$ from the ring $\mathfrak{R}^{(2\times 2)}$ subject to conditions 1–3;*

2) *if $\varkappa<0$, an n-tuple of pairwise distinct numbers λ_j, $\bar\lambda_j$, $\operatorname{Im}\lambda_j>0$, $j=1,\ldots,n$, and an n-tuple of non-zero numbers γ_j, $\bar\gamma_j$, $j=1,\ldots,n$.*

Consider the matrix $G(x,\lambda)$ given by (1.21) and, if $\varkappa<0$, the set of subspaces $N_j^{(\pm)}(x)$ given by (1.47).

The Riemann problem is to find, for each x, the matrices $G_\pm(x,\lambda)$ in $\mathfrak{R}_\pm^{(2\times 2)}$ with $c_\pm=1$ satisfying

$$G(x,\lambda)=G_+(x,\lambda)\,G_-(x,\lambda). \tag{1.48}$$

If $\varkappa>0$, $G_\pm(x,\lambda)$ are supposed non-degenerate in their domains of analyticity. If $\varkappa<0$, they are supposed non-degenerate except for the points λ_j, $\bar\lambda_j$, respectively, where

$$\operatorname{Im} G_+(x,\lambda_j)=N_j^{(+)}(x),\qquad \operatorname{Ker} G_-(x,\bar\lambda_j)=N_j^{(-)}(x),\qquad j=1,\ldots,n. \tag{1.49}$$

In the following section we shall prove that the problem has a unique solution and analyze the properties of the solution $G_\pm(x,\lambda)$. We will show that within the chosen functional class the matrices $G_\pm(x,\lambda)$ are characterized by the aforementioned properties 1–5.

§ 2. The Rapidly Decreasing Case. Analysis of the Riemann Problem

Consider the Riemann problem

$$G(x,\lambda)=G_+(x,\lambda)\,G_-(x,\lambda) \tag{2.1}$$

stated at the end of § 1. Here we shall analyze the problem for the given $G(x,\lambda)$ and the unkown $G_\pm(x,\lambda)$ in the functional classes indicated above. We will prove the following.

1. *The Riemann problem* (2.1) *is uniquely solvable.*
2. *The matrices*

$$F_+(x,\lambda)=G_+^{-1}(x,\lambda)\,E(x,\lambda) \tag{2.2}$$

and

$$F_-(x,\lambda)=G_-(x,\lambda)\,E(x,\lambda) \tag{2.3}$$

satisfy the differential equation of the auxiliary linear problem

$$\frac{d}{dx}F_\pm(x,\lambda)=\left(\frac{\lambda}{2i}\sigma_3+U_0(x)\right)F_\pm(x,\lambda), \tag{2.4}$$

with $U_0(x)$ having the form

$$U_0(x)=\sqrt{\varkappa}\begin{pmatrix}0 & \bar{\psi}(x)\\ \psi(x) & 0\end{pmatrix}, \tag{2.5}$$

where $\psi(x)$, $\bar{\psi}(x)$ are in $L_1(-\infty,\infty)$.

3. *For real λ, $G_\pm(x,\lambda)$ have the asymptotic behaviour* (1.39)–(1.42) *as $|x|\to\infty$, where $b(\lambda)$ enters into the definition* (1.22) *of $G(\lambda)$ and $a(\lambda)$ is given by*

$$a(\lambda)=\prod_{j=1}^{n}\frac{\lambda-\lambda_j}{\lambda-\bar{\lambda}_j}\exp\left\{\frac{1}{2\pi i}\int_{-\infty}^{\infty}\frac{\log(1+\varepsilon|b(\mu)|^2)}{\mu-\lambda-i0}\,d\mu\right\}. \tag{2.6}$$

Note that if $\varepsilon=1$, there are no Blaschke factors in (2.6).

4. *The reduced monodromy matrix of the auxiliary linear problem (2.4) has the form*

$$T(\lambda)=\begin{pmatrix} a(\lambda) & \varepsilon\bar{b}(\lambda) \\ b(\lambda) & \bar{a}(\lambda) \end{pmatrix}. \tag{2.7}$$

If $\varepsilon=-1$, *the discrete spectrum coincides with the set* λ_j, $\bar{\lambda}_j$, $j=1,\dots,n$, *and the transition coefficients for the discrete spectrum are* γ_j, $\bar{\gamma}_j$, $j=1,\dots,n$.

Let us now turn to the proof of these assertions.

1. *The unique solvability of the Riemann problem.*

Here the simplest way is to exploit the general theory of Gohberg and Krein who considered the Riemann problem

$$G(\lambda)=G_+(\lambda)G_-(\lambda) \tag{2.8}$$

for a non-degenerate $n\times n$ matrix $G(\lambda)$ from the ring $\mathfrak{R}^{(n\times n)}$, defined on the whole real line and normalized to I as $|\lambda|\to\infty$. The theorem we need reads that *if* $G(\lambda)+G^*(\lambda)$ *is positive definite then (2.8) has a unique solution in the class of matrices* $G_+(\lambda)$ *and* $G_-(\lambda)$ *belonging to the rings* $\mathfrak{R}_\pm^{(n\times n)}$, *non-degenerate in their domains of analyticity and normalized to I as* $|\lambda|\to\infty$.

The analysis is based on the reduction of the Riemann problem to a Wiener-Hopf equation which can be carried out as follows. Rewrite (2.8) as

$$G_-(\lambda)=G_+^{-1}(\lambda)G(\lambda). \tag{2.9}$$

Since $G_+(\lambda)$ is non-degenerate, Wiener's theorem implies that $G_+^{-1}(\lambda)$ can be expressed as

$$G_+^{-1}(\lambda)=I+\int_0^\infty \Omega_+(s)e^{i\lambda s}\,ds, \tag{2.10}$$

where $\Omega_+(s)$ is in $L_1^{(n\times n)}(0,\infty)$. The Fourier transform applied to (2.9) then makes the Riemann problem equivalent to the *Wiener-Hopf equation*

$$\Omega_+(s)+\Phi(s)+\int_0^\infty \Omega_+(s')\Phi(s-s')\,ds'=0, \quad s\geqslant 0, \tag{2.11}$$

where

$$G(\lambda)=I+\int_{-\infty}^\infty \Phi(s)e^{i\lambda s}\,ds. \tag{2.12}$$

The analysis of equation (2.11) is what the Gohberg-Krein theory is mainly concerned with.

The theorem cited above proves the unique solvability of (2.1) in the *regular case*, i.e. when $G_\pm(x,\lambda)$ are non-degenerate in their domains of analyticity, or, equivalently, when there are no discrete eigenvalues. Indeed, $G(x,\lambda)$ has the form

$$G(x,\lambda)=\begin{pmatrix} 1 & \varepsilon\bar{b}(\lambda)e^{-i\lambda x} \\ -b(\lambda)e^{i\lambda x} & 1 \end{pmatrix} \tag{2.13}$$

so that for $\varepsilon=1$

$$\frac{G(x,\lambda)+G^*(x,\lambda)}{2}=I. \tag{2.14}$$

For $\varepsilon=-1$, $G(x,\lambda)$ is Hermitian positive-definite by virtue of the condition (A_1) from § I.6,

$$|b(\lambda)|<1. \tag{2.15}$$

By the uniqueness theorem for the Riemann problem (2.1), the involution (1.24) for $G(x,\lambda)$ extends to $G_\pm(x,\lambda)$

$$G^*_\pm(x,\lambda)=\tau G_\mp(x,\bar{\lambda})\tau \tag{2.16}$$

with $\tau=\sigma_3$ if $\varepsilon=1$ and $\tau=I$ if $\varepsilon=-1$. In particular, if $\varepsilon=-1$, $G_-(x,\lambda)$ is the Hermitian conjugate of $G_+(x,\lambda)$,

$$G^*_+(x,\lambda)=G_-(x,\bar{\lambda}). \tag{2.17}$$

Let us now consider the Riemann problem with zeros, i.e. the general problem (2.1) with the given λ_j, $\bar{\lambda}_j$, $\operatorname{Im}\lambda_j>0$; γ_j, $\bar{\gamma}_j$, $j=1,\dots,n$, and the relations (1.49). We shall assume from the very beginning that $\varkappa<0$.

To start with, let there be only one pair of zeros, λ_0, $\bar{\lambda}_0$, $\operatorname{Im}\lambda_0>0$, with the corresponding γ_0, $\bar{\gamma}_0$, and let $N_0^{(\pm)}(x)$ be the associated pair of orthogonal subspaces (see (1.47)). To simplify the notation we shall also leave out the dependence on the parameter x. We look for solutions $G_\pm(\lambda)$ in the form

$$G_+(\lambda)=\tilde{G}_+(\lambda)B(\lambda), \qquad G_-(\lambda)=B^{-1}(\lambda)\tilde{G}_-(\lambda), \tag{2.18}$$

where $\tilde{G}_\pm(\lambda)$ are solutions of the regular Riemann problem. *We shall find a matrix factor $B(\lambda)$ such that*

a) *$B(\lambda)$ is analytic in the upper half-plane and $B^{-1}(\lambda)$ is analytic in the lower half-plane*;

b) $\lim_{|\lambda|\to\infty} B(\lambda)=I$; (2.19)

c) $\det B(\lambda) \neq 0$ *for* $\operatorname{Im}\lambda \geqslant 0$ *and* $\det B^{-1}(\lambda) \neq 0$ *for* $\operatorname{Im}\lambda \leqslant 0$ *except at* $\lambda = \lambda_0$ *or* $\bar{\lambda}_0$ *respectively, where*

$$\operatorname{Im} B(\lambda_0) = \tilde{G}_+^{-1}(\lambda_0) N_0^{(+)} = \tilde{N}_0^{(+)}, \tag{2.20}$$

$$\operatorname{Ker} B^{-1}(\bar{\lambda}_0) = \tilde{G}_-(\bar{\lambda}_0) N_0^{(-)} = \tilde{N}_0^{(-)}. \tag{2.21}$$

In addition, by virtue of the involution (2.17) the subspaces $\tilde{N}_0^{(\pm)}$ are orthogonal.

These requirements unambiguously identify $B(\lambda)$ with the *matrix Blaschke-Potapov factor*

$$B(\lambda) = I + \frac{\bar{\lambda}_0 - \lambda_0}{\lambda - \bar{\lambda}_0} P, \quad B^{-1}(\lambda) = I + \frac{\lambda_0 - \bar{\lambda}_0}{\lambda - \lambda_0} P, \tag{2.22}$$

where the orthogonal projection operator P is determined by

$$\operatorname{Im}(I - P) = \tilde{N}_0^{(+)}, \quad \operatorname{Ker}(I - P) = \tilde{N}_0^{(-)} \tag{2.23}$$

and has the form

$$P = \frac{1}{1 + |\beta|^2} \begin{pmatrix} |\beta|^2 & \bar{\beta} \\ \beta & 1 \end{pmatrix}. \tag{2.24}$$

Here

$$\beta = \frac{\tilde{G}_+^{(11)}(\lambda_0)\gamma_0 + \tilde{G}_+^{(21)}(\lambda_0)}{\tilde{G}_+^{(12)}(\lambda_0)\gamma_0 + \tilde{G}_+^{(22)}(\lambda_0)}, \tag{2.25}$$

with the obvious notation for the matrix elements of $\tilde{G}_+(\lambda_0)$.

The Blaschke-Potapov factor defined by (2.22) satisfies the generalized unitarity condition

$$B^*(\lambda) = B^{-1}(\bar{\lambda}). \tag{2.26}$$

Restoring the dependence on x we find that $P(x)$ has the form (2.24) with

$$\beta(x) = \frac{\tilde{G}_+^{(11)}(x, \lambda_0)\gamma_0(x) + \tilde{G}_+^{(21)}(x, \lambda_0)}{\tilde{G}_+^{(12)}(x, \lambda_0)\gamma_0(x) + \tilde{G}_+^{(22)}(x, \lambda_0)}, \tag{2.27}$$

and $\gamma_0(x) = \gamma_0 e^{i\lambda_0 x}$. If the denominator in the expression for $\beta(x)$ vanishes for some x, then (2.24) still makes sense and P turns into $\frac{1}{2}(I + \sigma_3)$.

In the general case, given the zeros λ_j, $\bar{\lambda}_j$ and the subspaces $N_j^{(\pm)}$, $j = 1, \ldots, n$, the Riemann problem can be solved in a similar way. The factor

$B(\lambda)$ should be replaced by the *ordered product of the Blaschke-Potapov factors*

$$\Pi(\lambda)=B_1(\lambda)\ldots B_n(\lambda)=\prod_{j=1}^{\widehat{n}}\left(I+\frac{\bar{\lambda}_j-\lambda_j}{\lambda-\bar{\lambda}_j}P_j\right) \tag{2.28}$$

where the orthogonal projection operators P_j are determined by the subspaces $N_j^{(\pm)}$. They are most simply constructed step by step. Suppose the unitary factors $B_1(\lambda), \ldots, B_{k-1}(\lambda)$ are found. Then P_k is determined by

$$\operatorname{Im}(I-P_k)=B_{k-1}^{-1}(\lambda_k)\ldots B_1^{-1}(\lambda_k)\tilde{G}_+^{-1}(\lambda_k)N_k^{(+)}=\tilde{N}_k^{(+)} \tag{2.29}$$

and

$$\operatorname{Ker}(I-P_k)=B_{k-1}^{-1}(\bar{\lambda}_k)\ldots B_1^{-1}(\bar{\lambda}_k)\tilde{G}_-(\bar{\lambda}_k)N_k^{(-)}=\tilde{N}_k^{(-)}. \tag{2.30}$$

Let us now show the *uniqueness of the solution of (2.1)*. Suppose there are two solutions $G_\pm(x,\lambda)$ and $G'_\pm(x,\lambda)$. Suppressing once again the dependence on x, from (2.1) we have

$$G_+'^{-1}(\lambda)G_+(\lambda)=G'_-(\lambda)G_-^{-1}(\lambda) \tag{2.31}$$

for real λ. The left hand side is analytic in the upper half-plane except at $\lambda=\lambda_j$, and the right hand side is analytic in the lower half-plane except at $\lambda=\bar{\lambda}_j$, $j=1,\ldots,n$. For large $|\lambda|$ both sides of (2.31) are normalized to I. If there are no actual singularities at these points, then by the Liouville theorem both sides are equal to I identically, whence the uniqueness.

To prove the regularity, consider, for definiteness, the left hand side of (2.31). In the neighbourhood of $\lambda=\lambda_j$ the matrices $G_+(\lambda)$ and $G_+'^{-1}(\lambda)$ have the expansions

$$G_+(\lambda)=A+O(|\lambda-\lambda_j|),\quad G_+'^{-1}(\lambda)=\frac{B}{\lambda-\lambda_j}+O(1), \tag{2.32}$$

with

$$AB=BA=0. \tag{2.33}$$

Since

$$\operatorname{Im}A=\operatorname{Im}G_+(\lambda_j)=N_j^{(+)}, \tag{2.34}$$

it follows from (2.33) that the subspace $N_j^{(+)}$ is contained in $\operatorname{Ker}B$. Both spaces being one-dimensional, they coincide with each other:

$$N_j^{(+)}=\operatorname{Ker}B. \tag{2.35}$$

There is a similar expansion for $G'_+(\lambda)$ and $G'^{-1}_+(\lambda)$; as before,

$$\operatorname{Im} A' = N_j^{(+)} = \operatorname{Ker} B'. \tag{2.36}$$

Now it is clear that the residue of $G'^{-1}_+(\lambda)\,G_+(\lambda)$ at $\lambda=\lambda_j$ equals $B'A$ and therefore vanishes.

The same argument works for the right hand side of (2.31) proving that both sides of this equation are regular in the whole plane.

The uniqueness theorem shows in particular that the involution (2.17) also extends to the general case of the Riemann problem with zeros.

2. *Derivation of the differential equation.*

Let us consider the matrices $F_\pm(x,\lambda)$ (see (2.2), (2.3)). Obviously, they satisfy

$$F_-(x,\lambda) = F_+(x,\lambda)\,G(\lambda). \tag{2.37}$$

The matrix $F_+(x,\lambda)$ is analytic and non-degenerate in the upper half-plane except for simple poles at $\lambda=\lambda_j, j=1,\dots,n$; $F_-(x,\lambda)$ is analytic in the lower half-plane and has simple zeros at $\lambda=\bar{\lambda}_j, j=1,\dots,n$. In addition, $F_\pm(x,\lambda)$ satisfy

$$\operatorname{Im} F_+^{-1}(x,\lambda_j) = N_j^{(+)}, \quad \operatorname{Ker} F_-(x,\bar{\lambda}_j) = N_j^{(-)}, \quad j=1,\dots,n, \tag{2.38}$$

where the subspaces $N_j^{(\pm)}$ are determined by the γ_j, $\bar{\gamma}_j$ (see § 1) and do not depend on x.

Later it will be shown that $F_\pm(x,\lambda)$ are absolutely continuous functions of x. So, differentiating (2.37) with respect to x gives

$$\frac{dF_-}{dx}(x,\lambda) = \frac{dF_+}{dx}(x,\lambda)\,G(\lambda) = \frac{dF_+}{dx}(x,\lambda)\,F_+^{-1}(x,\lambda)\,F_-(x,\lambda), \tag{2.39}$$

or

$$\frac{dF_+}{dx}(x,\lambda)\,F_+^{-1}(x,\lambda) = \frac{dF_-}{dx}(x,\lambda)\,F_-^{-1}(x,\lambda). \tag{2.40}$$

Both left and right hand sides of (2.40) may be analytically continued into their respective half-planes in spite of the fact that $F_+(x,\lambda)$ is singular at $\lambda=\lambda_j$ and $F_-^{-1}(x,\lambda)$ is singular at $\lambda=\bar{\lambda}_j, j=1,\dots,n$.

The proof is similar to that of the uniqueness theorem. We set

$$F_+(x,\lambda) = \frac{A(x)}{\lambda-\lambda_j} + O(1) \tag{2.41}$$

and

$$F_+^{-1}(x,\lambda) = B(x) + O(|\lambda - \lambda_j|). \tag{2.42}$$

We have

$$\operatorname{Im} B(x) = N_j^{(+)} \tag{2.43}$$

with the obvious identities

$$A(x)B(x) = B(x)A(x) = 0. \tag{2.44}$$

Since $N_j^{(+)}$ does not depend on x it follows that $\operatorname{Ker}\frac{dA}{dx}(x)$ contains $N_j^{(+)}$, so that by (2.43) and (2.44) the residue of $\frac{dF_+}{dx}(x,\lambda)F_+^{-1}(x,\lambda)$ at $\lambda=\lambda_j$ equals $\frac{dA}{dx}(x)B(x)$ and therefore vanishes. Hence the left hand side of (2.40) is non-singular. The same argument applies to the right hand side of (2.40).

Thus, for each x, $\frac{dF_+}{dx}(x,\lambda)F_+^{-1}(x,\lambda)$ *is an entire function of* λ. Let us analyze its asymptotic behaviour as $|\lambda|\to\infty$.

In the lower half λ-plane we shall make use of the representation

$$F_-(x,\lambda) = \left(I + \int_0^\infty \Phi_-(x,s)e^{-i\lambda s}\,ds\right)E(x,\lambda) \tag{2.45}$$

(recall that $G_-(x,\lambda)$ belongs to the ring $\mathfrak{R}_-^{(2\times 2)}$). Assume for the moment that $\Phi_-(x,s)$ is absolutely continuous in x and s and that $\frac{\partial\Phi_-}{\partial x}$, $\frac{\partial\Phi_-}{\partial s}$, $\frac{\partial^2\Phi_-}{\partial x\,\partial s}$ as functions of s belong to $L_1^{(2\times 2)}(0,\infty)$. Then, for $\operatorname{Im}\lambda\leqslant 0$, $F_-(x,\lambda)$ has the asymptotic behaviour

$$F_-(x,\lambda) = \left(I + \frac{\Phi_-(x,0)}{i\lambda} + o\left(\frac{1}{|\lambda|}\right)\right)E(x,\lambda), \tag{2.46}$$

as $|\lambda|\to\infty$, which allows differentiation with respect to x. It follows that, as $|\lambda|\to\infty$ in the lower half-plane,

$$\frac{dF_-}{dx}(x,\lambda)F_-^{-1}(x,\lambda) = \frac{\lambda\sigma_3}{2i} + \frac{1}{2}[\sigma_3, \Phi_-(x,0)] + o(1). \tag{2.47}$$

Similarly, the representation

$$F_+^{-1}(x,\lambda)=E^{-1}(x,\lambda)\left(I+\int_0^\infty \Phi_+(x,s)e^{i\lambda s}\,ds\right) \tag{2.48}$$

implies for $|\lambda|\to\infty$, $\operatorname{Im}\lambda\geqslant 0$,

$$\begin{aligned}\frac{dF_+}{dx}(x,\lambda)F_+^{-1}(x,\lambda)&=-F_+(x,\lambda)\frac{dF_+^{-1}}{dx}(x,\lambda)\\&=\frac{\lambda\sigma_3}{2i}+\frac{1}{2}[\sigma_3,\Phi_+(x,0)]+o(1).\end{aligned} \tag{2.49}$$

By the Liouville theorem we then obtain

$$\frac{dF_+}{dx}(x,\lambda)F_+^{-1}(x,\lambda)=\frac{dF_-}{dx}(x,\lambda)F_-^{-1}(x,\lambda)=\frac{\lambda\sigma_3}{2i}+U_0(x), \tag{2.50}$$

where

$$U_0(x)=\tfrac{1}{2}[\sigma_3,\Phi_+(x,0)]=\tfrac{1}{2}[\sigma_3,\Phi_-(x,0)]. \tag{2.51}$$

We thus conclude that the auxiliary linear equation (2.4) is satisfied. The matrix $U_0(x)$ is off-diagonal and satisfies

$$U_0^*(x)=\tau U_0(x)\tau, \tag{2.52}$$

which follows from (2.16) reexpressed in terms of $F_\pm(x,\lambda)$. Hence it can be written as $U_0(x)=\begin{pmatrix}0 & \varepsilon\bar{\varphi}(x)\\ \varphi(x) & 0\end{pmatrix}$ and coincides with the matrix (2.5) upon introducing a parameter $\varkappa$: $\varphi(x)=\sqrt{\varkappa}\,\psi(x)$. The parameter $\varkappa$ here is somewhat artificial; it is needed for (2.5) to coincide with (I.2.4) literally.

Let us return to the hypothesis that the kernels $\Phi_\pm(x,s)$ are differentiable. In general, our assumptions on $b(\lambda)$ do not imply this property, and the validity of the differential equation (2.4) will be demonstrated in the next subsection by means of a closure procedure.

We point out that the above derivation of (2.4) did not use any special features of the matrix $G(x,\lambda)$ except the unique solvability of the Riemann problem, the involution and the explicit form of the x-dependence. So, there is a fairly general relationship between the Riemann problem and the differential equation (2.4), which is, besides, local in x. All these requirements on $G(x,\lambda)$ will be used in our analysis of the properties of $G_\pm(x,\lambda)$ and $U_0(x)$ as functions of x.

3. *The asymptotic behaviour of* $G_\pm(x,\lambda)$ *as* $|x|\to\infty$.

First we shall consider the regular case of the Riemann problem. We shall exploit the Wiener-Hopf equation (2.11)

$$\Omega_+(x,s)+\Phi(x,s)+\int_0^\infty \Omega_+(x,s')\Phi(x,s-s')\,ds'=0, \tag{2.53}$$

$s \geqslant 0$, with the x-dependence indicated explicitly. The matrix kernel $\Phi(x,s)$ is given by (1.30)

$$\Phi(x,s)=\begin{pmatrix} 0 & \varepsilon\bar{\beta}(-s-x) \\ -\beta(s-x) & 0 \end{pmatrix}, \tag{2.54}$$

where

$$\beta(s)=\frac{1}{2\pi}\int_{-\infty}^\infty b(\lambda)e^{-i\lambda s}\,d\lambda, \tag{2.55}$$

and $\Omega_+(x,s)$ is determined from

$$G_+^{-1}(x,\lambda)=I+\int_0^\infty \Omega_+(x,s)e^{i\lambda s}\,ds. \tag{2.56}$$

The matrix $\Phi_-(x,s)$ involved in the representation

$$G_-(x,\lambda)=I+\int_0^\infty \Phi_-(x,s)e^{-i\lambda s}\,ds \tag{2.57}$$

has the following expression in terms of the solution $\Omega_+(x,s)$ of the Wiener-Hopf equation:

$$\Phi_-(x,s)=\Phi(x,-s)+\int_0^\infty \Omega_+(x,s')\Phi(x,-s-s')\,ds'. \tag{2.58}$$

To emphasize that the variable x is merely a parameter in the Wiener-Hopf equation we introduce the following notation for the matrix elements of $\Omega_+(x,s)$:

$$\Omega_+(x,s)=\begin{pmatrix} A_x(s) & B_x(s) \\ C_x(s) & D_x(s) \end{pmatrix}. \tag{2.59}$$

By using (2.54), the matrix equation (2.53) can be written as

$$A_x(s)=\int_0^\infty \beta(s-x-s')B_x(s')\,ds', \tag{2.60}$$

$$B_x(s) = -\varepsilon\bar{\beta}(-s-x) - \varepsilon \int_0^\infty k_x(s,s')\,B_x(s')\,ds' \tag{2.61}$$

and

$$D_x(s) = -\varepsilon \int_0^\infty \bar{\beta}(-s-x+s')\,C_x(s')\,ds', \tag{2.62}$$

$$C_x(s) = \beta(s-x) - \varepsilon \int_0^\infty l_x(s,s')\,C_x(s')\,ds', \tag{2.63}$$

where

$$k_x(s,s') = \int_{-x}^\infty \bar{\beta}(u-s)\,\beta(u-s')\,du, \tag{2.64}$$

$$l_x(s,s') = \int_x^\infty \beta(s-u)\,\bar{\beta}(s'-u)\,du. \tag{2.65}$$

Clearly, equations (2.61) and (2.63) together with (2.60) and (2.62) are equivalent to the initial Wiener-Hopf equation (2.53). Their solvability is an immediate consequence of the aforementioned theorem of Gohberg and Krein. Still, the dependence of $\Omega_+(x,s)$ on the parameter x remains to be analyzed because it governs the behaviour of $\psi(x)$, $\bar{\psi}(x)$ in x, as is shown by the first formula in (2.51),

$$\psi(x) = \frac{1}{\sqrt{\varkappa}}\,C_x(s)|_{s=0}, \qquad \bar{\psi}(x) = -\frac{1}{\sqrt{\varkappa}}\,B_x(s)|_{s=0}. \tag{2.66}$$

To study the behaviour of the solutions of (2.61) *and* (2.63) *as functions of x we proceed as follows.* The equations involve the integral operators $\mathbf{K}_x$ and $\mathbf{L}_x$, with the kernels $k_x(s,s')$ and $l_x(s,s')$, respectively, bounded on the space $L_1(0,\infty)$ and continuous in x in the sense of operator norm. To estimate the norm of the occurring integral operators it will be enough to use an obvious estimate

$$\|A\| \leqslant \max_{0 \leqslant s' < \infty} \int_0^\infty |A(s,s')|\,ds. \tag{2.67}$$

The Gohberg-Krein theory gives that for each x the operators $\mathbf{I}+\varepsilon\mathbf{K}_x$ and $\mathbf{I}+\varepsilon\mathbf{L}_x$ have inverses in $L_1(0,\infty)$. It follows that $\Omega_+(x,s)$ as an element of $L_1^{(2\times 2)}(0,\infty)$ depends continuously on x, so that for each λ the matrices $G_\pm(x,\lambda)$ are continuous in x.

Let us show that the norms of $(\mathbf{I}+\varepsilon\,\mathbf{K}_x)^{-1}$, $(\mathbf{I}+\varepsilon\mathbf{L}_x)^{-1}$, *in* $L_1(0,\infty)$ *are uniformly bounded in* x, $-\infty<x<\infty$. To this end we will prove that the operators $\mathbf{K}_x$ and $\mathbf{L}_x$ have limits, $\mathbf{K}_\pm$ and $\mathbf{L}_\pm$, as $x\to\pm\infty$, where convergence is taken in the operator norm, and that the inverses $(\mathbf{I}+\varepsilon\mathbf{K}_\pm)^{-1}$ and $(\mathbf{I}+\varepsilon\mathbf{L}_\pm)^{-1}$ exist and are bounded operators.

Consider, for definiteness, the operator $\mathbf{K}_x$ and write it as

$$\mathbf{K}_x=\mathbf{K}+\mathbf{R}_x, \tag{2.68}$$

where $\mathbf{K}$ is an integral operator with the kernel $k(s-s')$,

$$k(s)=\int_{-\infty}^{\infty}\beta(u+s)\bar{\beta}(u)\,du, \tag{2.69}$$

and the kernel $r_x(s,s')$ of the operator $\mathbf{R}_x$ is

$$r_x(s,s')=-\int_{-\infty}^{-x}\bar{\beta}(u-s)\beta(u-s')\,du. \tag{2.70}$$

The norm of $\mathbf{R}_x$ is bounded by

$$\begin{aligned}\|\mathbf{R}_x\| &\leqslant \max_{0\leqslant s'<\infty}\int_0^{\infty}\int_{-\infty}^{-x}|\beta(u-s)\beta(u-s')|\,du\,ds\\ &\leqslant \max_{0\leqslant s'<\infty}\int_{-\infty}^{-x}\int_{-\infty}^{u}|\beta(s)\beta(u-s')|\,ds\,du\leqslant\left(\int_{-\infty}^{-x}|\beta(u)|\,du\right)^2.\end{aligned} \tag{2.71}$$

Hence the norm vanishes as $x\to+\infty$.

The operator $\mathbf{I}+\varepsilon\,\mathbf{K}$ has an inverse, because the inversion problem reduces to a scalar Wiener-Hopf equation equivalent to the Riemann problem for the function

$$1+\varepsilon\int_{-\infty}^{\infty}k(s)e^{i\lambda s}\,ds=1+\varepsilon|b(\lambda)|^2=a_+(\lambda)a_-(\lambda). \tag{2.72}$$

Its unique solvability is obvious when $\varepsilon=1$ and follows from the condition (A_1) (see (2.15)) when $\varepsilon=-1$. The solution $a_+(\lambda)$ is given by (2.6) with the Blaschke factors left out, and $a_-(\lambda)=\bar{a}_+(\bar{\lambda})$.

Thus we have shown that the norm of $(\mathbf{I}+\varepsilon\,\mathbf{K}_x)^{-1}$ is uniformly bounded in x in the neighbourhood of $+\infty$.

Now we consider the neighbourhood of $-\infty$. One should not be misled by (2.64) into thinking that $\mathbf{K}_x$ vanishes as $x\to-\infty$. This is made clear by introducing a new function

$$f_x(s) = B_x(s-x), \quad s \geqslant x, \tag{2.73}$$

so that (2.61) becomes

$$f_x(s) = -\varepsilon\bar{\beta}(-s) - \varepsilon \int_x^\infty q(s, s') f_x(s') \, ds'. \tag{2.74}$$

The kernel $q(s, s')$ does not depend on x and has the form

$$q(s, s') = \int_0^\infty \bar{\beta}(u-s)\beta(u-s') \, du. \tag{2.75}$$

The shift takes $L_1(0, \infty)$ into $L_1(x, \infty)$ and $\mathbf{K}_x$ into the operator $\mathbf{Q}_x$ on $L_1(x, \infty)$ with the kernel $q(s, s')$ where $s, s' \geqslant x$. Now, $L_1(x, \infty)$ embeds naturally into $L_1(-\infty, \infty)$; let $\mathbf{Q}_x$ also denote the operator on $L_1(-\infty, \infty)$ with the kernel

$$q_x(s, s') = \theta(s-x)\theta(s'-x) q(s, s'), \tag{2.76}$$

where $\theta(s) = 1$ for $s > 0$ and $\theta(s) = 0$ for $s < 0$. As $x \to -\infty$, the family $\mathbf{Q}_x$ has a limit $\mathbf{Q}$ with the kernel $q(s, s')$ where convergence is taken in the norm of $L_1(-\infty, \infty)$.

To prove that $\mathbf{I} + \varepsilon\mathbf{Q}$ has an inverse we will show that the equation

$$f(s) = g(s) - \varepsilon \int_0^\infty q(s, s') f(s') \, ds' \tag{2.77}$$

is uniquely solvable in $L_1(-\infty, \infty)$. Let

$$F(\lambda) = \int_{-\infty}^\infty f(s) e^{i\lambda s} \, ds, \quad G(\lambda) = \int_{-\infty}^\infty g(s) e^{i\lambda s} \, ds \tag{2.78}$$

and apply the Fourier transform to (2.77). Then

$$F(\lambda) = G(\lambda) - \varepsilon\bar{b}(\lambda) \Pi_+(b(\lambda) F(\lambda)), \tag{2.79}$$

where the projection operator Π_+ is defined as follows: if

$$\zeta(\lambda) = \int_{-\infty}^\infty \xi(s) e^{i\lambda s} \, ds, \tag{2.80}$$

then

$$(\Pi_+\zeta)(\lambda)=\int_0^\infty \xi(s)e^{i\lambda s}\,ds\,. \tag{2.81}$$

Equation (2.79) has a unique solution which can be written in closed form

$$F(\lambda)=G(\lambda)-\varepsilon\frac{\bar{b}(\lambda)}{a_+(\lambda)}\Pi_+\left(\frac{b(\lambda)}{a_-(\lambda)}G(\lambda)\right), \tag{2.82}$$

with $a_\pm(\lambda)$ as defined in (2.72).

Indeed, introducing a new function $\Phi(\lambda)$ by

$$F(\lambda)-G(\lambda)=-\varepsilon\bar{b}(\lambda)\Phi(\lambda), \tag{2.83}$$

from (2.79) we have the equation

$$\Phi(\lambda)=\Pi_+(b(\lambda)G(\lambda)-\varepsilon|b(\lambda)|^2\Phi(\lambda)), \tag{2.84}$$

which shows, in particular, that $\Phi(\lambda)$ belongs to the ring $\mathfrak{R}_+$. Using the factorization (2.72) we can write (2.84) in the form

$$\Pi_+(b(\lambda)G(\lambda)-a_+(\lambda)a_-(\lambda)\Phi(\lambda))=0, \tag{2.85}$$

and easily get an expression for $\Phi(\lambda)$,

$$\Phi(\lambda)=\frac{1}{a_+(\lambda)}\Pi_+\left(\frac{b(\lambda)}{a_-(\lambda)}G(\lambda)\right), \tag{2.86}$$

which implies (2.82).

Thus we have shown that $\mathbf{I}+\varepsilon\mathbf{K}_x$ has a bounded inverse in the neighbourhood of $-\infty$. This completes our proof of solvability, uniformly in x, of the integral equation (2.61).

Equation (2.63) is analyzed in a similar way. To prove that $(\mathbf{I}+\varepsilon\mathbf{L}_x)^{-1}$ is uniformly bounded in the neighbourhood of $-\infty$, a representation of the type (2.68) should be used; the neighbourhood of $+\infty$ should be treated by the method just described.

Let us apply the above results for studying the asymptotic behaviour of the solutions of the Riemann problem as $|x|\to\infty$. For half of the matrix elements of $G_\mp^{-1}(x,\lambda)$ the asymptotics are trivial.

In fact, the $L_1(0,\infty)$-norms of the inhomogeneous terms in (2.61) and (2.63) are

$$\int_0^\infty |\beta(-s-x)|\,ds = \int_{-\infty}^{-x} |\beta(s)|\,ds \tag{2.87}$$

and

$$\int_0^\infty |\beta(s-x)|\,ds = \int_{-x}^{\infty} |\beta(s)|\,ds, \tag{2.88}$$

and vanish as $x \to +\infty$ and $x \to -\infty$, respectively. Therefore

$$\|B_x\| \to 0, \quad \|A_x\| \to 0 \tag{2.89}$$

as $x \to +\infty$ and

$$\|C_x\| \to 0, \quad \|D_x\| \to 0 \tag{2.90}$$

as $x \to -\infty$. So, for every λ with $\operatorname{Im}\lambda \geqslant 0$, the asymptotic behaviour of the first row of $G_{\mp}^{-1}(x,\lambda)$, as $x \to +\infty$, and of the second row, as $x \to -\infty$, is given by (1.39) and (1.41).

Next, consider the first row of $G_{\mp}^{-1}(x,\lambda)$, as $x \to -\infty$. Putting $B_x(s) = f_x(s+x)$ we have $f_x(s) \to f(s)$, as $x \to -\infty$, in the norm of $L_1(-\infty, \infty)$. Since $f(s)$ satisfies (2.77) with $g(s) = -\varepsilon\bar{\beta}(-s)$, we find for real λ, as $x \to -\infty$,

$$\begin{aligned}
(G_{\mp}^{-1}(x,\lambda))_{12} &= \int_0^\infty B_x(s)\,e^{i\lambda s}\,ds \\
&= e^{-i\lambda x} \int_{-\infty}^{\infty} f(s)\,e^{i\lambda s}\,ds + o(1) \\
&= -\frac{\varepsilon\bar{b}(\lambda)}{a_+(\lambda)}\,e^{-i\lambda x}\left(a_+(\lambda) - \Pi_+\left(\frac{\varepsilon|b(\lambda)|^2}{a_-(\lambda)}\right)\right) + o(1) \\
&= -\frac{\varepsilon\bar{b}(\lambda)}{a_+(\lambda)}\,e^{-i\lambda x} \\
&\quad \times \left(a_+(\lambda) + \Pi_+\left((1 - a_+(\lambda)) + \left(\frac{1}{a_-(\lambda)} - 1\right)\right)\right) + o(1) \\
&= -\frac{\varepsilon\bar{b}(\lambda)}{a_+(\lambda)}\,e^{-i\lambda x} + o(1).
\end{aligned} \tag{2.91}$$

Moreover, from (2.60) it follows that, as $x \to -\infty$,

$$
\begin{aligned}
(G_{+}^{-1}(x,\lambda))_{11} &= 1+\int_{0}^{\infty} A_x(s)\,e^{i\lambda s}\,ds \\
&= 1+\int_{0}^{\infty}\int_{0}^{\infty} \beta(s-s'-x) f_x(x+s')\,e^{i\lambda s}\,ds'\,ds \\
&= 1+\int_{0}^{\infty}\int_{-\infty}^{\infty} \beta(s-s') f(s')\,e^{i\lambda s}\,ds'\,ds+o(1) \\
&= 1-\Pi_{+}\left(\frac{\varepsilon|b(\lambda)|^2}{a_{+}(\lambda)}\right)+o(1) \\
&= 1+\Pi_{+}\left(\left(\frac{1}{a_{+}(\lambda)}-1\right)+(1-a_{-}(\lambda))\right)+o(1) \\
&= \frac{1}{a_{+}(\lambda)}+o(1).
\end{aligned}
\tag{2.92}
$$

In a similar way, as $x\to+\infty$,

$$
(G_{+}^{-1}(x,\lambda))_{21} = \frac{b(\lambda)}{a_{+}(\lambda)}\,e^{i\lambda x}+o(1) \tag{2.93}
$$

and

$$
(G_{+}^{-1}(x,\lambda))_{22} = \frac{1}{a_{+}(\lambda)}+o(1). \tag{2.94}
$$

So, identifying $a(\lambda)$ with $a_{+}(\lambda)$, we have reproduced the asymptotic formulae (1.39) and (1.41).

The asymptotic behaviour of $G_{-}(x,\lambda)$ follows from (2.1), (2.13), and the asymptotic behaviour of $G_{+}^{-1}(x,\lambda)$ just established. It is given by (1.40) and (1.42) upon identifying $\bar{a}(\lambda)$ with $a_{-}(\lambda)$.

Thus we deduce that $G_{\pm}(x,\lambda)$ are composed of the Jost solutions, $T_{\pm}(x,\lambda)$, of the auxiliary linear problem (2.4) according to (1.5), (1.6) and (1.8), (1.19). The matrix

$$
T(\lambda)=\begin{pmatrix} a(\lambda) & \varepsilon\bar{b}(\lambda) \\ b(\lambda) & \bar{a}(\lambda)\end{pmatrix} \tag{2.95}
$$

plays the role of the reduced monodromy matrix for $T_{\pm}(x,\lambda)$.

Let us now proceed, for $\varkappa<0$, to the general case of the Riemann problem with zeros. For simplicity we shall again consider a single pair of zeros $\lambda_0, \bar{\lambda}_0$, $\operatorname{Im}\lambda_0>0$, and of subspaces $N_0^{(\pm)}(x)$ of the form (1.47). The formulae

(2.18), (2.22), (2.24) and (2.27) solving the Riemann problem with zeros involve the solution $\tilde{G}_+(x,\lambda)$ of the regular Riemann problem at $\lambda=\lambda_0$. Therefore, we also need the asymptotic behaviour of $\tilde{G}_+(x,\lambda)$, as $|x|\to\infty$, for complex λ in the upper half-plane. An inspection of the preceding argument confirms that all the formulae remain valid for such λ with the exception of (2.91) and (2.93) in which $\frac{\bar{b}(\lambda)}{a_+(\lambda)}$ and $\frac{b(\lambda)}{a_+(\lambda)}$ must be replaced by 0. Indeed, consider, for instance, the limit of the matrix element $(G_+^{-1}(x,\lambda))_{12}$ as $x\to-\infty$. One has

$$\int_x^\infty f_x(s)e^{i\lambda(s-x)}dx=\int_x^\infty f(s)e^{i\lambda(s-x)}dx+o(1). \tag{2.96}$$

Let

$$g(x)=\int_x^\infty f(s)e^{i\lambda(s-x)}dx=\int_0^\infty f(s+x)e^{i\lambda s}ds. \tag{2.97}$$

We will show that for $\operatorname{Im}\lambda>0$ the function $g(x)$ vanishes, as $x\to-\infty$. The second identity in (2.97) implies that for such λ, $g(x)$ is a convolution of two functions in $L_1(-\infty,\infty)$, so that it is itself in $L_1(-\infty,\infty)$. On the other hand, the first identity in (2.97) shows that $g(x)$ is absolutely continuous and

$$\frac{dg(x)}{dx}=-i\lambda g(x)-f(x), \tag{2.98}$$

so that its derivative also lies in $L_1(-\infty,\infty)$. Since $g(x)$ vanishes as $x\to+\infty$, it follows that it also vanishes as $x\to-\infty$.

Thus, for $\operatorname{Im}\lambda>0$ we have the asymptotic formulae

$$\tilde{G}_+(x,\lambda)=\begin{pmatrix}1 & 0\\ 0 & a_+(\lambda)\end{pmatrix}+o(1) \tag{2.99}$$

as $x\to+\infty$ and

$$\tilde{G}_+(x,\lambda)=\begin{pmatrix}a_+(\lambda) & 0\\ 0 & 1\end{pmatrix}+o(1) \tag{2.100}$$

as $x\to-\infty$.

Now consider the projection operator $P(x)$ entering into the definition of the Blaschke-Potapov factor (2.22). The asymptotic formulae (2.99), (2.100) combined with (2.27) imply that

$$\lim_{x\to+\infty} \beta(x,\lambda_0)=0, \qquad \lim_{x\to-\infty} \beta(x,\lambda_0)=\infty. \tag{2.101}$$

Therefore for $P(x)$ we have

$$\lim_{x\to+\infty} P(x)=\begin{pmatrix} 0 & 0 \\ 0 & 1 \end{pmatrix}, \qquad \lim_{x\to-\infty} P(x)=\begin{pmatrix} 1 & 0 \\ 0 & 0 \end{pmatrix}, \tag{2.102}$$

whence

$$\lim_{x\to+\infty} B(\lambda)=\begin{pmatrix} 1 & 0 \\ 0 & \dfrac{\lambda-\lambda_0}{\lambda-\bar{\lambda}_0} \end{pmatrix}, \qquad \lim_{x\to-\infty} B(\lambda)=\begin{pmatrix} \dfrac{\lambda-\lambda_0}{\lambda-\bar{\lambda}_0} & 0 \\ 0 & 1 \end{pmatrix}. \tag{2.103}$$

Thus the asymptotic behaviour, as $|x|\to\infty$, of the solutions $G_\pm(x,\lambda)$ of the Riemann problem with zeros is given by (1.39)–(1.42) with $a(\lambda)$ replaced by $\dfrac{\lambda-\lambda_0}{\lambda-\bar{\lambda}_0}a_+(\lambda)$ and $\bar{a}(\lambda)$ by $\dfrac{\lambda-\bar{\lambda}_0}{\lambda-\lambda_0}a_-(\lambda)$. Moreover, it can easily be verified that $G_+(x,\lambda_0)$ is composed, according to (1.5) and (1.19), of the columns of the Jost solutions, $T_\pm(x,\lambda_0)$, which are proportional to each other and decay exponentially as $|x|\to\infty$.

Consequently, λ_0 is an eigenvalue of the auxiliary linear problem (2.4) and γ_0 plays the role of the associated transition coefficient for the discrete spectrum.

The case of several pairs of zeros λ_j, $\bar{\lambda}_j$, $\operatorname{Im}\lambda_j>0$, and subspaces $N_j^{(\pm)}(x)$, $j=1,\dots,n$, is treated in a similar way. The Blaschke-Potapov factors involved in $\Pi(\lambda)$ (see (2.28)) become diagonal as $|x|\to\infty$; the asymptotic behaviour of $G_\pm(x,\lambda)$ is given by (1.39)–(1.42) with $a(\lambda)$ given by (2.6). This completes the proofs for Subsection 4.

The proof in Subsection 2 has shown that both the regular Riemann problem and the one with zeros are related to equation (2.4) with coefficients $\tilde{U}_0(x)$ and $U_0(x)$ of the form (2.5), respectively. Comparing (2.18), (2.28), (2.46) and (2.51) leads to the following relationship between $U_0(x)$ and $\tilde{U}_0(x)$:

$$U_0(x)=\tilde{U}_0(x)+\Delta_0(x), \tag{2.104}$$

with

$$\Delta_0(x)=\tfrac{1}{2}[\sigma_3,\pi(x)], \tag{2.105}$$

where the matrix $\pi(x)$ is determined by the asymptotic behaviour of $\Pi(x,\lambda)$ as $|\lambda|\to\infty$,

$$\Pi(x,\lambda)=I+\frac{1}{\lambda}i\pi(x)+O\left(\frac{1}{|\lambda|^2}\right). \tag{2.106}$$

It has the form

$$\pi(x)=\frac{1}{i}\sum_{j=1}^{n}(\bar{\lambda}_j-\lambda_j)P_j(x). \tag{2.107}$$

Here the $P_j(x)$ are the orthogonal projection operators entering into (2.28).

The above information about the asymptotic behaviour of $\tilde{G}_+(x,\lambda)$, as $|x|\to\infty$, together with formulae such as (2.27) show that $\Delta_0(x)$ is absolutely integrable in the neighbourhood of $\pm\infty$. Besides, $\pi(x)$ and hence $\Delta_0(x)$ are continuous in x. If then follows that $\Delta_0(x)$ belongs to $L_1^{(2\times 2)}(-\infty,\infty)$. We shall need this fact later.

Now we shall prove that $\psi(x)$, $\bar{\psi}(x)$ *are in* $L_1(-\infty,\infty)$. First we shall discuss the regular Riemann problem and show that $B_x(s)$ and $C_x(s)$ as functions of x for each $s\geqslant 0$ are elements of $L_1(-\infty,\infty)$ depending continuously on s.

Let us prove this, say, for $B_x(s)$. We shall interprete (2.61) as an equation in the space of functions of two variables, $f(x,s)$, absolutely integrable in x on the whole line and continuous in s for $s\geqslant 0$ in the sense indicated above.

In other words, the space in question is just the tensor product $L_1(-\infty,\infty)\otimes C[0,\infty)$ where $C[0,\infty)$ is the space of bounded continuous functions on $[0,\infty)$. The norm in $L_1(-\infty,\infty)\otimes C[0,\infty)$ is given by

$$\|f\|=\max_{0\leqslant s<\infty}\int_{-\infty}^{\infty}|f(x,s)|\,dx. \tag{2.108}$$

Clearly, the inhomogeneous term in (2.61) lies in this space. It is easily verified that the operator $\mathbf{K}_x$ with the kernel $k_x(s,s')$ is bounded on $L_1(-\infty,\infty)\otimes C[0,\infty)$. In fact, since $\beta(x)$ is in $L_1(-\infty,\infty)$, we have

$$|k_x(s,s')|\leqslant\int_{-\infty}^{\infty}|\beta(u-s)\beta(u-s')|\,du=\tilde{k}(s-s'), \tag{2.109}$$

where

$$K=\int_{-\infty}^{\infty}|k(s)|\,ds<\infty. \tag{2.110}$$

It follows that

$$\|\mathbf{K}_x f\|\leqslant K\cdot\|f\|, \tag{2.111}$$

where $f(x, s)$ is any element of $L_1(-\infty, \infty) \otimes C[0, \infty)$. Now, by using the representation (2.64) for $k_x(s, s')$ it is easily shown that $\int\limits_{-\infty}^{\infty} (\mathbf{K}_x f)(x, s)\, dx$ is a continuous function of s.

This allows to prove that $\mathbf{I} + \varepsilon \mathbf{K}_x$ has a unique inverse on $L_1(-\infty, \infty) \otimes C[0, \infty)$. Indeed, the results obtained may be interpreted as the existence of $(\mathbf{I} + \varepsilon \mathbf{K}_x)^{-1}$ on the space $C(-\infty, \infty) \otimes L_1(0, \infty)$ with the natural norm

$$\|g\| = \max_{-\infty < x < \infty} \int\limits_0^{\infty} |g(x, s)|\, ds. \tag{2.112}$$

The space $L_1(-\infty, \infty) \otimes C[0, \infty)$ is "almost" the dual of the latter and, according to (2.64), $\mathbf{K}_x$ is a formally self-adjoint operator. Therefore $(\mathbf{I} + \varepsilon \mathbf{K}_x)^{-1}$ exists and is bounded on $L_1(-\infty, \infty) \otimes C[0, \infty)$ as well. More accurately, we could reproduce the existence proof for $(\mathbf{I} + \varepsilon \mathbf{K}_x)^{-1}$ based on the Gohberg-Krein theory in the space $C[0, \infty)$ instead of $L_1(0, \infty)$.

By virtue of (2.66) we then conclude that in the regular case of the Riemann problem the functions $\psi(x)$, $\bar{\psi}(x)$ belong to $L_1(-\infty, \infty)$. For the case of the Riemann problem with zeros one should use (2.104) and the aforementioned fact that $\Delta_0(x)$ is absolutely integrable.

Let us now prove that $F_\pm(x, \lambda)$ are absolutely continuous in x, which will justify the derivation of (2.4) *in Subsection 2.*

Suppose for the moment that $\beta(x)$ has two derivatives belonging to $L_1(-\infty, \infty)$. Then it is easily seen that $\Omega_+(x, s)$ is a solution to (2.53) differentiable with respect to x and s and $\dfrac{\partial \Omega_+}{\partial x}, \dfrac{\partial \Omega_+}{\partial s}, \dfrac{\partial^2 \Omega_+}{\partial x\, \partial s}$ are in $L_1^{(2\times 2)}(-\infty, \infty)$ as functions of s. As was shown in Subsection 2, this implies that the matrices $F_\pm(x, \lambda)$ are absolutely continuous in x and satisfy the differential equation (2.4). To treat the general case it is sufficient to approximate $\beta(x)$ by functions $\beta_n(x)$ in $L_1(-\infty, \infty)$ with the properties indicated above. Then the associated $U_0^{(n)}(x)$ will converge to $U_0(x)$ as $n \to \infty$ in the norm of $L_1^{(2\times 2)}(-\infty, \infty)$, while $F_\pm^{(n)}(x, \lambda)$ will satisfy the differential equation (2.4),

$$\frac{dF_\pm^{(n)}}{dx}(x, \lambda) = \left(\frac{\lambda}{2i}\sigma_3 + U_0^{(n)}(x)\right) F_\pm^{(n)}(x, \lambda). \tag{2.113}$$

For fixed λ these converge to $F_\pm(x, \lambda)$ in the norm of $C^{(2\times 2)}(-\infty, \infty)$. Since differentiation is a closed operator, $F_\pm(x, \lambda)$ are absolutely continuous and satisfy the auxiliary linear equation (2.4).

So we have proved all the assertions indexed 1–4 at the beginning of this section. Here we shall add several comments.

1. Everything was discussed in a most general setting assuming $\psi(x)$, $\bar{\psi}(x)$ are in $L_1(-\infty, \infty)$ and $b(\lambda)$ is the Fourier transform of an absolutely integrable function. The Riemann problem and the relationship between $b(\lambda)$ and $\psi(x)$, $\bar{\psi}(x)$ can also be investigated in other functional classes. In particular, the simplest case here is when $b(\lambda)$ is of Schwartz type. Then $\psi(x)$, $\bar{\psi}(x)$ prove to be of Schwartz type, too.

2. The functions $b(\lambda)$, $\bar{b}(\lambda)$ and the set $\lambda_j, \bar{\lambda}_j; \gamma_j, \bar{\gamma}_j, j=1, \ldots, n$, are independent input data for the Riemann problem. Therefore we may consider the situation when $b(\lambda)$, $\bar{b}(\lambda)$ vanish identically, i.e. $G(\lambda)=I$. In this case the determination of the parameters of the Blaschke-Potapov matrix factors in (2.28) reduces to solving a system of linear algebraic equations which will be written in closed form and solved in § 5. The associated auxiliary linear problem (2.4) is then called *reflectionless* because one of the transition coefficients, $b(\lambda)$, vanishes and the other, $a(\lambda)$, is a product of elementary Blaschke factors. The functions $\psi(x)$, $\bar{\psi}(x)$ in this case provide pure soliton solutions of the NS equation which will be discussed at length in § 5.

This completes our analysis of the Riemann problem in the rapidly decreasing case. In the next section we shall consider its implications for the NS model.

§ 3. Application of the Inverse Scattering Problem to the NS Model

The investigation of the Riemann problem in § 2 allows us to solve the inverse problem in the rapidly decreasing case, i.e. to give an explicit procedure for inverting the mapping

$$\mathscr{F}: (\psi(x), \bar{\psi}(x)) \rightarrow (b(\lambda), \bar{b}(\lambda); \lambda_j, \bar{\lambda}_j, \gamma_j, \bar{\gamma}_j, j=1, \ldots, n) \tag{3.1}$$

from $\psi(x)$, $\bar{\psi}(x)$ to the transition coefficients and discrete spectrum of the auxiliary linear problem

$$\frac{dF}{dx} = U(x, \lambda) F. \tag{3.2}$$

In fact, the results of §§ I.5–I.6 show that $\mathscr{F}$ maps the functions $\psi(x)$, $\bar{\psi}(x)$ belonging to $L_1(-\infty, \infty)$ into the functions $b(\lambda)$, $\bar{b}(\lambda)$ belonging to $\mathfrak{R}_0$, the ring of Fourier transforms of functions in $L_1(-\infty, \infty)$ (see § I.6). The discrete spectrum comes into play only if $\varkappa < 0$, and then the condition (A) from § I.6 is assumed; it implies that $b(\lambda)$ is subject to an additional constraint,

$$|b(\lambda)| < 1 \tag{3.3}$$

for all λ, and the λ_j are pairwise distinct and $\operatorname{Im}\lambda_j > 0$. Also, none of the γ_j, $j=1, \ldots, n$, vanishes. Within these classes the mapping $\mathscr{F}$ is one-to-one.

Indeed, the initial data of the Riemann problem (see the end of § 1) are parametrized by the data on the right hand side of (3.1). The analysis of the Riemann problem in § 2 shows that $\psi(x)$, $\bar{\psi}(x)$ defined by (2.66) give rise to these data as transition coefficients and discrete eigenvalues of the auxiliary linear problem. Technically, the inversion of $\mathscr{F}$ is based on the matrix Wiener-Hopf equation with the x-dependence of a special form as described in § 2.

The mapping $\mathscr{F}$ has an inverse in other functional classes as well. For instance, $\mathscr{F}$ may be restricted to the phase space $\mathscr{M}_0$ of the NS model consisting of functions $(\psi(x), \bar{\psi}(x))$ in Schwartz space. Then $b(\lambda)$, $\bar{b}(\lambda)$ are also in Schwartz space and $\mathscr{F}$ is one-to-one. In Chapter III we shall see that both $\mathscr{F}$ and $\mathscr{F}^{-1}$ are differentiable mappings in these classes.

Let us now use this information for a complete description of the NS dynamics in the rapidly decreasing case. From § I.7 we know that if a complex-valued function $\psi(x, t)$ satisfies the initial-value problem

$$i\frac{\partial \psi}{\partial t} = -\frac{\partial^2 \psi}{\partial x^2} + 2\varkappa |\psi|^2 \psi, \tag{3.4}$$

$$\psi(x, t)|_{t=0} = \psi(x), \tag{3.5}$$

then the dynamics of the transition coefficients and discrete spectrum for the auxiliary linear problem with the potential

$$U(x, t, \lambda) = \frac{\lambda\sigma_3}{2i} + \sqrt{\varkappa}\,(\bar{\psi}(x, t)\sigma_+ + \psi(x, t)\sigma_-) \tag{3.6}$$

is given by

$$\begin{aligned} &b(\lambda, t) = e^{-i\lambda^2 t} b(\lambda); \quad \lambda_j(t) = \lambda_j, \\ &\gamma_j(t) = e^{-i\lambda_j^2 t}\gamma_j, \quad j=1, \ldots, n. \end{aligned} \tag{3.7}$$

Here $b(\lambda)$, λ_j and γ_j result from the initial data, $\psi(x)$, of (3.4)–(3.5) by applying the mapping $\mathscr{F}$.

Now we will prove the converse, i.e. given (3.7), the function $\psi(x, t)$ resulting from the data in (3.7) via $\mathscr{F}^{-1}$ satisfies the NS equation. We shall assume $b(\lambda)$ to be a Schwartz function because the dynamics (3.7) leaves the Schwartz space invariant.

For the proof consider the Riemann problem (2.1)

$$G(x,t,\lambda)=G_+(x,t,\lambda)\,G_-(x,t,\lambda) \tag{3.8}$$

where the dependence on t is also taken into account. By (3.7) we have

$$G(x,t,\lambda)=E^{-1}(t,\lambda^2)\,G(x,\lambda)\,E(t,\lambda^2), \tag{3.9}$$

where $E(t,\lambda^2)=\exp\left\{\frac{\lambda^2 t}{2i}\sigma_3\right\}$ has already occurred more then once. For $\varkappa<0$ there are additional relations due to the presence of zeros,

$$\operatorname{Im} G_+(x,t,\lambda_j)=N_j^{(+)}(x,t), \qquad \operatorname{Ker} G_-(x,t,\bar{\lambda}_j)=N_j^{(-)}(x,t), \tag{3.10}$$

where

$$\begin{gathered} N_j^{(+)}(x,t)=E^{-1}(t,\lambda_j^2)\,N_j^{(+)}(x), \qquad N_j^{(-)}(x,t)=E^{-1}(t,\bar{\lambda}_j^2)\,N_j^{(-)}(x), \\ j=1,\ldots,n. \end{gathered} \tag{3.11}$$

By the results of § 2, the Riemann problem (3.9)–(3.10) is uniquely solvable in the Schwartz class for each x and t.

Let $F_\pm(x,t,\lambda)$ be defined by

$$F_+(x,t,\lambda)=G_+^{-1}(x,t,\lambda)\,E(x,\lambda)\,E^{-1}(t,\lambda^2), \tag{3.12}$$

$$F_-(x,t,\lambda)=G_-(x,t,\lambda)\,E(x,\lambda)\,E^{-1}(t,\lambda^2). \tag{3.13}$$

In § 2 we have shown that, for t fixed, $F_\pm(x,t,\lambda)$ satisfy the differential equation (3.2) of the auxiliary linear problem. Let us show that they also satisfy a differential equation with respect to t for a fixed x. The Wiener-Hopf equation (2.53) depending on t implies that they are differentiable with respect to t.

To derive the desired equation in t (as in § 2 for the equation in x) we rewrite (3.9) in the form

$$F_-(x,t,\lambda)=F_+(x,t,\lambda)\,G(\lambda), \tag{3.14}$$

which gives

$$\frac{\partial}{\partial t}F_-(x,t,\lambda)\,F_-^{-1}(x,t,\lambda)=\frac{\partial}{\partial t}F_+(x,t,\lambda)\,F_+^{-1}(x,t,\lambda). \tag{3.15}$$

By virtue of (3.11), the subspaces $\operatorname{Im} F_+^{-1}(x,t,\lambda_j)$ and $\operatorname{Ker} F_-(x,t,\bar{\lambda}_j)$ depend neither on x nor on t. The same argument as in § 2, Subsection 2, shows that the functions $\frac{\partial F_\pm}{\partial t}(x,t,\lambda)\,F_\pm^{-1}(x,t,\lambda)$ are non-singular in their respective half-planes and hence, by (3.15), give rise to an entire function of λ. To analyze it we will use, following § 2, the integral representation

$$F_-(x,t,\lambda)=\left(I+\int_0^\infty \Phi_-(x,t,s)e^{-i\lambda s}\,ds\right)E(x,\lambda)E^{-1}(t,\lambda^2) \tag{3.16}$$

with the resulting asymptotic behaviour

$$F_-(x,t,\lambda)=\left(I+\frac{\Phi_-(x,t,0)}{i\lambda}-\frac{1}{\lambda^2}\frac{\partial\Phi_-}{\partial s}(x,t,0)+O\left(\frac{1}{|\lambda|^3}\right)\right)$$
$$\times E(x,\lambda)E^{-1}(t,\lambda^2), \tag{3.17}$$

as $|\lambda|\to\infty$, $\operatorname{Im}\lambda\leqslant 0$. Differentiation with respect to t gives

$$\frac{\partial F_-}{\partial t}(x,t,\lambda)F_-^{-1}(x,t,\lambda)=V_-(x,t,\lambda)+O\left(\frac{1}{|\lambda|}\right), \tag{3.18}$$

where

$$V_-(x,t,\lambda)=\lambda^2 V_2+\lambda V_1+V_0, \tag{3.19}$$

with

$$V_2=\frac{i\sigma_3}{2},\qquad V_1(x,t)=\frac{1}{2}[\Phi_-(x,t,0),\sigma_3]=-U_0(x,t) \tag{3.20}$$

(see (2.51)) and

$$V_0(x,t)=\frac{i}{2}\left[\sigma_3,\frac{\partial\Phi_-}{\partial s}(x,t,0)\right]+\frac{i}{2}[\Phi_-(x,t,0),\sigma_3]\Phi_-(x,t,0). \tag{3.21}$$

We shall now express $V_0(x,t)$ in terms of $U_0(x,t)$. Notice that the differential equation (3.2) yields an infinite sequence of identities involving $\Phi_-(x,t,s)$ and its derivatives with respect to x and s for $s=0$. In fact, successive integration by parts in (3.16) and differentiation with respect to x give the following asymptotic expansion:

$$\frac{\partial F_-}{\partial x}(x,t,\lambda)F_-^{-1}(x,t,\lambda)=\frac{\lambda\sigma_3}{2i}+U_0(x)+\sum_{n=1}^{\infty}\frac{F_n(x,t)}{(i\lambda)^n}+O(|\lambda|^{-\infty}), \tag{3.22}$$

as $|\lambda|\to\infty$, $\operatorname{Im}\lambda\leqslant 0$ (cf. § 2). In particular, we have

$$F_1(x,t)=\frac{1}{2}\left(\left[\sigma_3,\frac{\partial\Phi_-}{\partial s}(x,t,0)\right]\right.$$

$$\left.+[\Phi_-(x,t,0),\sigma_3]\Phi_-(x,t,0)+2\frac{\partial\Phi_-}{\partial x}(x,t,0)\right). \tag{3.23}$$

On the other hand, from (3.2) it follows that the coefficients $F_n(x,t)$ vanish

$$F_n(x,t)=0, \qquad n=1,2,\ldots \tag{3.24}$$

The first of these identities implies that $V_0(x,t)$ can be expressed as

$$V_0(x,t)=-i\frac{\partial\Phi_-}{\partial x}(x,t,0). \tag{3.25}$$

By virtue of (3.20), the off-diagonal part of $\frac{\partial\Phi_-}{\partial x}(x,t,0)$ coincides with $-\frac{\partial U_0(x,t)}{\partial x}\sigma_3$. To find the diagonal part, consider (3.24) for $n=1$ and split off its diagonal part. From (3.20) and (3.23) we deduce that it equals $-\sigma_3 U_0^2(x,t)$, so finally we obtain

$$V_0(x,t)=i\sigma_3 U_0^2(x,t)+i\frac{\partial U_0(x,t)}{\partial x}\sigma_3. \tag{3.26}$$

By comparing (3.20) and (3.26) with (I.2.7) we see that $V_-(x,t,\lambda)$ coincides with the matrix $V(x,t,\lambda)$ from § I.2

$$V_-(x,t,\lambda)=V(x,t,\lambda). \tag{3.27}$$

In a similar manner, we find the asymptotic behaviour of $F_+(x,t,\lambda)$ as $|\lambda|\to\infty$, $\operatorname{Im}\lambda\geqslant 0$,

$$\frac{\partial F_+}{\partial t}(x,t,\lambda)F_+^{-1}(x,t,\lambda)=V_+(x,t,\lambda)+O\left(\frac{1}{|\lambda|}\right), \tag{3.28}$$

where $V_+(x,t,\lambda)$ has the form (3.19). The Liouville theorem then shows that $V_+(x,t,\lambda)$ coincides with $V_-(x,t,\lambda)$

$$V_+(x,t,\lambda)=V_-(x,t,\lambda)=V(x,t,\lambda), \tag{3.29}$$

and $F_\pm(x,t,\lambda)$ satisfy the required differential equation

$$\frac{\partial F_\pm}{\partial t}(x,t,\lambda)=V(x,t,\lambda)F_\pm(x,t,\lambda). \tag{3.30}$$

Combined with the differential equation (3.2) this implies that the connection $(U(x,t,\lambda), V(x,t,\lambda))$ satisfies the zero curvature condition (I.2.10)

$$\frac{\partial U}{\partial t}-\frac{\partial V}{\partial x}+[U,V]=0. \tag{3.31}$$

Thus, starting with the Riemann problem (3.9)–(3.10) we have defined a connection $(U(x,t,\lambda), V(x,t,\lambda))$ *of the form* (I.2.3)–(I.2.8) *satisfying the zero curvature condition*. It follows that $\psi(x,t)$ is indeed a solution of the NS equation. The above discussion also proves the global *unique solvability of the initial-value problem (3.4)–(3.5) for the NS model in the Schwartz class* (if $\varkappa<0$ the condition (A) is assumed as well).

It is now clear that $\mathscr{F}$ *is a nonlinear change of variables linearizing the NS equation.*

The method for solving the initial value problem for the NS equation can be presented as a commutative diagram

$$\begin{array}{ccc} (\psi(x),\bar\psi(x)) & \xrightarrow{\quad\mathscr{F}\quad} & (b(\lambda),\bar b(\lambda);\lambda_j,\bar\lambda_j,\gamma_j,\bar\gamma_j) \\ \Big\downarrow \tau_1 & & \Big\downarrow \tau_2 \\ (\psi(x,t),\bar\psi(x,t)) & \xleftarrow{\quad\mathscr{F}^{-1}\quad} & (b(\lambda,t),\bar b(\lambda,t);\lambda_j,\bar\lambda_j,\gamma_j(t),\bar\gamma_j(t)) \end{array} \tag{3.32}$$

Here τ_1 is the t-displacement according to (3.4) and τ_2 is the t-displacement given explicitly by (3.7).

It is instructive to look at the linear approximation for $\mathscr{F}$ as $\varkappa\to 0$, when the NS equation goes into the linear Schrödinger equation

$$i\frac{\partial\psi}{\partial t}=-\frac{\partial^2\psi}{\partial x^2}. \tag{3.33}$$

For that purpose consider the asymptotic behaviour of the transition coefficients $a(\lambda)$ and $b(\lambda)$ as $\varkappa\to 0$.

The integral equation (I.5.37) for the Jost solution $T_-(x,\lambda)$ yields, as $\varkappa\to 0$,

$$T_-(x,\lambda)=E(x,\lambda)+\int_{-\infty}^{x}E(x-y)\,U_0(y)\,E(y,\lambda)\,dy+O(|\varkappa|). \tag{3.34}$$

Letting $x \to +\infty$ in this formula gives

$$a(\lambda) = 1 + O(|\varkappa|), \qquad b(\lambda) = \sqrt{\varkappa} \int_{-\infty}^{\infty} \psi(x) e^{-i\lambda x} dx + O(|\varkappa|). \tag{3.35}$$

So the discrete spectrum disappears *and $\mathscr{F}$ turns into the Fourier transform.* The time dynamics of $b(\lambda)$ given by (3.7) is obviously the same as that of the Fourier transform of $\psi(x, t)$ subject to (3.33).

In the general case, $\varkappa \neq 0$, this argument allows us to interprete $\mathscr{F}$ *as a nonlinear analogue of the Fourier transform*. The scheme for integrating the NS equation via the inverse scattering method (diagram (3.32)) *becomes a nonlinear analogue of the Fourier method.*

§ 4. Relationship Between the Riemann Problem Method and the Gelfand-Levitan-Marchenko Integral Equations Formulation

This section is of technical nature. It discusses an alternative, more traditional method for solving the inverse problem, based on the Gelfand-Levitan-Marchenko equation, and establishes its relation to the Riemann problem.

In contrast with the latter method, based on the typical factorization problem for matrix-valued functions (2.1), the Gelfand-Levitan-Marchenko method employs a *special conjugation problem for vector-valued analytic functions* suggested by the relation (1.3) for the Jost solutions,

$$T_-(x, \lambda) = T_+(x, \lambda)\, T(\lambda). \tag{4.1}$$

Recall that $T(\lambda)$ has the form

$$T(\lambda) = \begin{pmatrix} a(\lambda) & \varepsilon \bar{b}(\lambda) \\ b(\lambda) & \bar{a}(\lambda) \end{pmatrix} \tag{4.2}$$

with $\varepsilon = \operatorname{sign} \varkappa$.

To formulate the problem let us write down the relation (4.1) for the first column, $T_-^{(1)}(x, \lambda)$, of $T_-(x, \lambda)$ in the form

$$\frac{1}{a(\lambda)} T_-^{(1)}(x, \lambda) = T_+^{(1)}(x, \lambda) + r(\lambda)\, T_+^{(2)}(x, \lambda), \tag{4.3}$$

where

$$r(\lambda) = \frac{b(\lambda)}{a(\lambda)}. \tag{4.4}$$

The left hand side of (4.3) has an analytic continuation into the upper half λ-plane with the exception of $\lambda = \lambda_j$, $j = 1, \ldots, n$, where it has simple poles. By (I.6.20),

$$T_-^{(1)}(x, \lambda_j) = \gamma_j T_+^{(2)}(x, \lambda_j), \tag{4.5}$$

we obtain

$$\operatorname{res} \frac{1}{a(\lambda)} T_-^{(1)}(x, \lambda)|_{\lambda = \lambda_j} = c_j T_+^{(2)}(x, \lambda_j), \tag{4.6}$$

with

$$c_j = \frac{\gamma_j}{\dot{a}(\lambda_j)}, \quad j = 1, \ldots, n, \tag{4.7}$$

the dot indicating the derivative with respect to λ. Then, asymptotically as $|\lambda| \to \infty$,

$$\frac{1}{a(\lambda)} T_-^{(1)}(x, \lambda) e^{\frac{i\lambda x}{2}} = \begin{pmatrix} 1 \\ 0 \end{pmatrix} + o(1). \tag{4.8}$$

The first term, $T_+^{(1)}(x, \lambda)$, on the right hand side of (4.3) is analytic in the lower half-plane and has there the asymptotic behaviour

$$T_+^{(1)}(x, \lambda) e^{\frac{i\lambda x}{2}} = \begin{pmatrix} 1 \\ 0 \end{pmatrix} + o(1), \tag{4.9}$$

as $|\lambda| \to \infty$. The column $T_+^{(2)}(x, \lambda)$ in the second term is analytic in the upper half-plane and, asymptotically as $|\lambda| \to \infty$,

$$T_+^{(2)}(x, \lambda) e^{-\frac{i\lambda x}{2}} = \begin{pmatrix} 0 \\ 1 \end{pmatrix} + o(1). \tag{4.10}$$

The columns $T_+^{(1)}(x, \lambda)$ and $T_+^{(2)}(x, \lambda)$ are related to each other by the involution (I.5.30),

$$\bar{T}_+^{(1)}(x, \bar{\lambda}) = \tilde{\sigma} T_+^{(2)}(x, \lambda), \tag{4.11}$$

where $\tilde{\sigma} = \sigma_1$ for $\varkappa > 0$ and $\tilde{\sigma} = i\sigma_2$ for $\varkappa < 0$.

Equation (4.3) combined with (4.6) and (4.8)–(4.11) constitutes the conjugation problem referred to above. Given a function $r(\lambda)$ on the real line and parameters λ_j, c_j, $j=1,\dots,n$, it allows to recover the columns $T_+^{(1)}(x,\lambda)$, $T_+^{(2)}(x,\lambda)$ and $\frac{1}{a(\lambda)}T_-^{(1)}(x,\lambda)$ with the necessary analyticity properties.

Like the Riemann problem, this conjugation problem can be reduced to a system of integral equations. To derive it consider the representations

$$T_+^{(1)}(x,\lambda)=\begin{pmatrix}1\\0\end{pmatrix}e^{-\frac{i\lambda x}{2}}+\int_x^\infty \Gamma_+(x,y)\begin{pmatrix}1\\0\end{pmatrix}e^{-\frac{i\lambda y}{2}}dy, \tag{4.12}$$

$$T_+^{(2)}(x,\lambda)=\begin{pmatrix}0\\1\end{pmatrix}e^{\frac{i\lambda x}{2}}+\int_x^\infty \Gamma_+(x,y)\begin{pmatrix}0\\1\end{pmatrix}e^{\frac{i\lambda y}{2}}dy \tag{4.13}$$

and insert them into (4.3). Substracting $\begin{pmatrix}1\\0\end{pmatrix}e^{-\frac{i\lambda x}{2}}$ from both sides of the resulting equation and performing Fourier transform with respect to λ we obtain

$$\Gamma_+(x,y)\begin{pmatrix}1\\0\end{pmatrix}+\omega(x+y)\begin{pmatrix}0\\1\end{pmatrix}+\int_x^\infty \Gamma_+(x,s)\begin{pmatrix}0\\1\end{pmatrix}\omega(s+y)\,ds=0 \tag{4.14}$$

for $y\geqslant x$, where

$$\omega(x)=\frac{1}{4\pi}\int_{-\infty}^{\infty} r(\lambda)e^{\frac{i\lambda x}{2}}d\lambda+\frac{1}{2i}\sum_{j=1}^{n}c_j e^{\frac{i\lambda_j x}{2}}. \tag{4.15}$$

By using the involution (I.5.18)

$$\bar{\Gamma}_+(x,y)=\sigma\Gamma_+(x,y)\sigma, \tag{4.16}$$

equation (4.14) can be written in matrix form,

$$\Gamma_+(x,y)+\Omega(x+y)+\int_x^\infty \Gamma_+(x,s)\Omega(s+y)\,ds=0 \tag{4.17}$$

for $y\geqslant x$ with

$$\Omega(x)=\omega(x)\sigma_-+\varepsilon\bar{\omega}(x)\sigma_+. \tag{4.18}$$

The integral equation (4.17) for the unknown matrix $\Gamma_+(x,y)$ is called the *Gelfand-Levitan-Marchenko equation from the right.*

In a similar manner, (4.1) gives an equation for $T_+^{(2)}(x,\lambda)$,

$$\frac{1}{a(\lambda)} T_+^{(2)}(x,\lambda) = \tilde{r}(\lambda) T_-^{(1)}(x,\lambda) + T_-^{(2)}(x,\lambda), \tag{4.19}$$

with

$$\tilde{r}(\lambda) = -\varepsilon \frac{\bar{b}(\lambda)}{a(\lambda)}, \tag{4.20}$$

which leads to the *Gelfand-Levitan-Marchenko equation from the left*

$$\Gamma_-(x,y) + \tilde{\Omega}(x+y) + \int_{-\infty}^{x} \Gamma_-(x,s)\tilde{\Omega}(s+y)\,ds = 0 \tag{4.21}$$

for $x \geqslant y$. Here

$$\tilde{\Omega}(x) = \varepsilon \bar{\tilde{\omega}}(x)\sigma_- + \tilde{\omega}(x)\sigma_+, \tag{4.22}$$

$$\tilde{\omega}(x) = \frac{1}{4\pi} \int_{-\infty}^{\infty} \tilde{r}(\lambda) e^{-\frac{i\lambda x}{2}} d\lambda + \frac{1}{2i} \sum_{j=1}^{n} \tilde{c}_j e^{-\frac{i\lambda_j x}{2}}, \tag{4.23}$$

and

$$\tilde{c}_j = \frac{1}{\gamma_j \dot{a}(\lambda_j)}, \quad j = 1, \ldots, n. \tag{4.24}$$

The kernel $\Gamma_-(x,y)$ occurs in the integral representation (I.5.10)

$$T_-(x,\lambda) = E(x,\lambda) + \int_{-\infty}^{x} \Gamma_-(x,y) E(y,\lambda)\,dy. \tag{4.25}$$

The analytical tools used for studying the Wiener-Hopf equation apply to (4.17) and (4.21) as well. A characteristic difference is that now we are dealing with compact integral operators.

In the general case, let $\psi(x)$, $\bar{\psi}(x)$ be absolutely integrable over the whole line. Since $a(\lambda) \neq 0$ for λ real, Wiener's theorem implies that the function

$$F(x) = \int_{-\infty}^{\infty} r(\lambda) e^{\frac{i\lambda x}{2}} d\lambda \tag{4.26}$$

is absolutely integrable on the whole line. The contribution into $\omega(x)$ from the discrete spectrum is rapidly decaying as $t \to +\infty$ (see (4.15)). Therefore the well-known theorem of functional analysis yields that the integral operator $\mathbf{\Omega}_x$ on $L_1^{(2\times 2)}(x,\infty)$ defined by

$$\boldsymbol{\Omega}_x f(s) = \int_x^{\infty} f(s')\Omega(s+s')\,ds' \tag{4.27}$$

is compact and its norm vanishes as $x \to +\infty$. In a similar way,

$$\tilde{\boldsymbol{\Omega}}_x f(s) = \int_{-\infty}^{x} f(s')\tilde{\Omega}(s+s')\,ds' \tag{4.28}$$

is a compact operator on $L_1^{(2\times 2)}(-\infty, x)$ with the norm vanishing as $x \to -\infty$.

The method for solving the inverse problem through the Gelfand-Levitan-Marchenko integral equations (4.17) and (4.21) is based on the following fact.

Suppose we are given functions $r(\lambda)$, $\tilde{r}(\lambda)$ in the ring $\mathfrak{R}_0$ and, if $\varepsilon = -1$, a set of pairwise distinct λ_j, $\operatorname{Im}\lambda_j > 0$, and of c_j, $\tilde{c}_j$, $j = 1, \ldots, n$, subject to:

1. *For all real λ*

$$|r(\lambda)| = |\tilde{r}(\lambda)| < 1 \tag{4.29}$$

for $\varepsilon = 1$ and

$$|r(\lambda)| = |\tilde{r}(\lambda)| < \infty \tag{4.30}$$

for $\varepsilon = -1$.

2. *The consistency relations*

$$\frac{\bar{r}(\lambda)}{\tilde{r}(\lambda)} = -\varepsilon\,\frac{a(\lambda)}{\bar{a}(\lambda)}, \quad c_j \tilde{c}_j = \frac{1}{\dot{a}^2(\lambda_j)}, \tag{4.31}$$

hold, $j = 1, \ldots, n$, where $a(\lambda)$ is given by (cf. (I.6.22) and (I.6.23))

$$a(\lambda) = \prod_{j=1}^{n} \frac{\lambda - \lambda_j}{\lambda - \bar{\lambda}_j} \exp\left\{\frac{1}{2\pi i}\int_{-\infty}^{\infty} \frac{\log(1 - \varepsilon|r(\mu)|^2)}{\lambda - \mu + i0}\,d\mu\right\}. \tag{4.32}$$

Given these data, let $\Omega(x)$ and $\tilde{\Omega}(x)$ be defined by (4.15), (4.18) and (4.22), (4.23), (4.24), respectively.

Then we claim that

1. *For each x the Gelfand-Levitan-Marchenko integral equations (4.17) and (4.21) have unique solutions, $\Gamma_{\pm}(x, y)$, in $L_1^{(2\times 2)}(x, \infty)$ or $L_1^{(2\times 2)}(-\infty, x)$, respectively.*

2. *The matrices $T_{\pm}(x, \lambda)$ resulting from these solutions via (4.12), (4.13) and (4.25) satisfy the involution* (cf. (I.5.19))

$$\bar{T}_{\pm}(x, \lambda) = \sigma T_{\pm}(x, \lambda)\sigma \tag{4.33}$$

and the differential equations

$$\frac{d}{dx}T_{\pm}(x,\lambda)=\left(\frac{\lambda\sigma_3}{2i}+U_0^{(\pm)}(x)\right)T_{\pm}(x,\lambda), \tag{4.34}$$

where $U_0^{(\pm)}(x)$ *are given by* (cf. (I.5.32) and (I.5.33))

$$U_0^{(\pm)}(x)=\pm(\sigma_3\Gamma_{\pm}(x,x)\sigma_3-\Gamma_{\pm}(x,x)) \tag{4.35}$$

and are absolutely integrable near $\pm\infty$, *respectively.*

3. *The consistency relation*

$$U_0^{(+)}(x)=U_0^{(-)}(x)=U_0(x) \tag{4.36}$$

holds, so that $U_0(x)$ *is in* $L_1^{(2\times 2)}(-\infty,\infty)$ *and has the form (2.5).*

4. *The transition coefficients for the continuous spectrum of the auxiliary linear problem (3.2) with the potential* $U_0(x)$ *coincide with* $a(\lambda)$ *and* $b(\lambda)=a(\lambda)r(\lambda)$, *whereas the discrete spectrum consists of the* $\lambda_j,\bar{\lambda}_j$ *with the transition coefficients* $\gamma_j,\bar{\gamma}_j$ *where* $\gamma_j=\dot{a}(\lambda_j)c_j$, $j=1,\ldots,n$.

Notice that if the Gelfand-Levitan-Marchenko method is used for solving the inverse problem, both equations from the right (4.17) and from the left (4.21) are to be considered. The former serves to determine the behaviour of $U_0^{(+)}(x)$ near $+\infty$ whilst the latter that of $U_0^{(-)}(x)$ near $-\infty$. The fact that $U_0^{(+)}(x)$ coincides with $U_0^{(-)}(x)$ needs a special demonstration.

Instead of giving an independent proof of the assertions 1–4 we will show how to derive the Gelfand-Levitan-Marchenko equations from the Wiener-Hopf equation discussed in § 2. In particular, this will imply the above assertions. For simplicity we shall concentrate on the regular Riemann problem where no discrete spectrum is present.

Recall that the Wiener-Hopf equation has the form (see § 2)

$$\Omega_+(x,s)+\Phi(x,s)+\int_0^\infty \Omega_+(x,s')\Phi(x,s-s')\,ds'=0 \tag{4.37}$$

for $s\geqslant 0$ with

$$\Phi(x,s)=\begin{pmatrix} 0 & \varepsilon\bar{\beta}(-s-x) \\ -\beta(s-x) & 0 \end{pmatrix} \tag{4.38}$$

and

$$\beta(x)=\frac{1}{2\pi}\int_{-\infty}^{\infty} b(\lambda)e^{-i\lambda x}\,d\lambda. \tag{4.39}$$

Putting

$$G_{+}^{-1}(x,\lambda)=I+\int_0^\infty \Omega_{+}(x,s)e^{i\lambda s}ds \tag{4.40}$$

we find that $G_{+}^{-1}(x,\lambda)$ is composed of the columns of the Jost solutions,

$$G_{+}^{-1}(x,\lambda)=\frac{1}{a(\lambda)}(T_{-}^{(1)}(x,\lambda),\, T_{+}^{(2)}(x,\lambda))\,E^{-1}(x,\lambda), \tag{4.41}$$

of the auxiliary linear problem (3.2) with the potential

$$U_0(x)=\tfrac{1}{2}[\sigma_3,\Omega_{+}(x,0)]. \tag{4.42}$$

The remaining columns of the Jost solutions make up a matrix $G_{-}(x,\lambda)$,

$$G_{-}(x,\lambda)=(T_{+}^{(1)}(x,\lambda),\, T_{-}^{(2)}(x,\lambda))\,E^{-1}(x,\lambda), \tag{4.43}$$

which can be represented as

$$G_{-}(x,\lambda)=I+\int_0^\infty \Phi_{-}(x,s)e^{-i\lambda s}ds, \tag{4.44}$$

with

$$\Phi_{-}(x,s)=\Phi(x,-s)+\int_0^\infty \Omega_{+}(x,s')\Phi(x,-s-s')ds'. \tag{4.45}$$

Recall that, according to the general Gohberg-Krein theory, the Wiener-Hopf system (4.37) is Fredholm, i.e. the operator $\mathbf{I}+\mathbf{\Phi}$ on the space $L_1^{(2\times 2)}(0,\infty)$ involved in (4.37) can be written as

$$\mathbf{I}+\mathbf{\Phi}=\mathbf{A}+\mathbf{K}, \tag{4.46}$$

where $\mathbf{A}$ has a bounded inverse and $\mathbf{K}$ is compact. The equation

$$f+\mathbf{\Phi}f=g \tag{4.47}$$

becomes

$$f+\mathbf{A}^{-1}\mathbf{K}f=\mathbf{A}^{-1}g, \tag{4.48}$$

with $\mathbf{A}^{-1}\mathbf{K}$ compact. The transformation of (4.47) into (4.48) is sometimes called *regularization.*

It will be shown that the Gelfand-Levitan-Marchenko equations are obtained from the Wiener-Hopf equation through a particular regularization.

Denote the matrix elements of $\Omega_{+}(x,s)$ by

$$\Omega_+(x,s)=\begin{pmatrix} A_x(s) & B_x(s) \\ C_x(s) & D_x(s) \end{pmatrix}. \tag{4.49}$$

The specific off-diagonal form of the matrix kernel $\Phi(x,s)$ allows us to reduce (4.37) to two independent integral equations for $B_x(s)$ and $C_x(s)$ (see § 2)

$$B_x(s)+\varepsilon\bar{\beta}(-s-x)+\varepsilon\int_0^\infty k_x(s,s')B_x(s')\,ds'=0 \tag{4.50}$$

and

$$C_x(s)-\beta(s-x)+\varepsilon\int_0^\infty l_x(s,s')C_x(s')\,ds'=0, \tag{4.51}$$

with

$$k_x(s,s')=\int_{-x}^\infty \bar{\beta}(u-s)\beta(u-s')\,du, \tag{4.52}$$

$$l_x(s,s')=\int_x^\infty \beta(s-u)\bar{\beta}(s'-u)\,du. \tag{4.53}$$

Consider, for definiteness, (4.50). Denoting $\beta_x(s)=\beta(-x-s)$ we regard $B_x(s)$ and the free term, $\varepsilon\bar{\beta}_x(s)$, as elements of $L_1(0,\infty)$ and write (4.50) as

$$(\mathbf{I}+\varepsilon\mathbf{K}_x)B_x+\varepsilon\bar{\beta}_x=0, \tag{4.54}$$

where $\mathbf{K}_x$ is an integral operator with the kernel $k_x(s,s')$. We have (see § 2)

$$\mathbf{I}+\varepsilon\mathbf{K}_x=\mathbf{I}+\varepsilon\mathbf{K}+\mathbf{R}_x, \tag{4.55}$$

where $\mathbf{K}$ is an integral operator with the kernel $k(s-s')$,

$$k(s)=\int_{-\infty}^\infty \beta(u+s)\bar{\beta}(u)\,du, \tag{4.56}$$

and the kernel $r_x(s,s')$ of the operator $\mathbf{R}_x$ is

$$r_x(s,s')=-\varepsilon\int_{-\infty}^{-x} \beta(u-s')\bar{\beta}(u-s)\,du. \tag{4.57}$$

The operator $\mathbf{I}+\varepsilon\mathbf{K}$ has a unique inverse since its inversion problem amounts to the scalar Riemann problem for the function

$$1+\varepsilon \int_{-\infty}^{\infty} k(s) e^{i\lambda s} ds = 1+\varepsilon|b(\lambda)|^2 = a_+(\lambda) a_-(\lambda). \tag{4.58}$$

Here $a_+(\lambda)=a(\lambda)$ and $a_-(\lambda)=\bar{a}(\bar{\lambda})$ with $a(\lambda)$ given by (2.6) but without the Blaschke factors (see § 2). Now $\mathbf{R}_x$ is a compact operator on $L_1(0, \infty)$. To see this it is enough to verify that the functions $h(s)=\mathbf{R}_x f(s)$ are equicontinuous in the mean and $\int_A^\infty |h(s)|\,ds$ is small for large A uniformly with respect to $f(s)$ from a bounded set in $L_1(0, \infty)$. There are elementary estimates

$$\begin{aligned}
&\int_0^\infty |h(s+\delta)-h(s)|\,ds \\
&= \int_0^\infty \left| \int_0^\infty \int_{-\infty}^{-x} (\bar{\beta}(u-s-\delta)-\bar{\beta}(u-s))\beta(u-s')f(s')\,du\,ds' \right| ds \\
&\leqslant \int_{-\infty}^{\infty} |\beta(u)|\,du \int_0^\infty |f(s')|\,ds' \int_{-\infty}^{\infty} |\beta(s+\delta)-\beta(s)|\,ds
\end{aligned} \tag{4.59}$$

and

$$\begin{aligned}
\int_A^\infty |h(s)|\,ds &= \int_A^\infty \left| \int_0^\infty \int_{-\infty}^{-x} \bar{\beta}(u-s)\beta(u-s')f(s')\,du\,ds' \right| ds \\
&\leqslant \int_{-\infty}^{\infty} |\beta(u)|\,du \int_0^\infty |f(s')|\,ds' \int_{-\infty}^{-x-A} |\beta(s)|\,ds,
\end{aligned} \tag{4.60}$$

which imply compactness.

A regularization of (4.54) is given by

$$B_x + (\mathbf{I}+\varepsilon\mathbf{K})^{-1}(\mathbf{R}_x B_x + \varepsilon\bar{\beta}_x) = 0. \tag{4.61}$$

We will show that (4.61) actually coincides with the Gelfand-Levitan-Marchenko equation.

To find an explicit expression for $(\mathbf{I}+\varepsilon\mathbf{K})^{-1}$ we use a standard method for solving the scalar Wiener-Hopf equation

$$(\mathbf{I}+\varepsilon\mathbf{K})f(s)=g(s), \qquad s\geqslant 0. \tag{4.62}$$

Extend the inhomogeneous term $g(x)$ to be zero for $s\leqslant 0$ and take the Fourier transform,

$$\hat{f}(\lambda)=\int_{-\infty}^{\infty} f(s)e^{i\lambda s}\,ds, \qquad \hat{g}(\lambda)=\int_0^\infty g(s)e^{i\lambda s}\,ds. \tag{4.63}$$

Equation (4.62) becomes

$$\hat{f}(\lambda)+\varepsilon|b(\lambda)|^2\hat{f}_+(\lambda)=\hat{g}(\lambda) \tag{4.64}$$

or

$$(1+\varepsilon|b(\lambda)|^2)\hat{f}_+(\lambda)=\hat{g}(\lambda)-\hat{f}_-(\lambda), \tag{4.65}$$

where

$$\hat{f}_\pm(\lambda)=\int_0^\infty f(\pm s)e^{\pm i\lambda s}ds,\quad \hat{f}(\lambda)=\hat{f}_+(\lambda)+\hat{f}_-(\lambda). \tag{4.66}$$

By virtue of (4.58) this gives

$$\hat{f}_+(\lambda)=\frac{1}{a(\lambda)}\Pi_+\left(\frac{\hat{g}(\lambda)}{\bar{a}(\lambda)}\right), \tag{4.67}$$

where the projection operator Π_+ introduced in § 2 is determined by

$$\Pi_+\hat{f}_+=\hat{f}_+,\qquad \Pi_+\hat{f}_-=0. \tag{4.68}$$

Remark that by this property of Π_+, an arbitrary absolutely integrable extention of $g(s)$ to the half-line $s\leqslant 0$ may be chosen (not necessarily the zero one).

Finally, the solution of (4.62) has the form

$$f(s)=\frac{1}{2\pi}\int_{-\infty}^\infty \hat{f}_+(\lambda)e^{-i\lambda s}d\lambda,\qquad s\geqslant 0, \tag{4.69}$$

with $\hat{f}_+(\lambda)$ given by (4.67). This leads to an explicit expression for $(\mathbf{I}+\varepsilon\mathbf{K})^{-1}$.

The formulae obtained will serve to modify (4.61). Introducing the Fourier transform of $B_x(s)$,

$$B_x(s)=\frac{1}{2\pi}\int_{-\infty}^\infty B_+(\lambda)e^{-i\lambda s}d\lambda, \tag{4.70}$$

we write (4.61) as

$$a(\lambda)B_+(\lambda)+\Pi_+\left(\varepsilon\frac{\bar{b}(\lambda)}{\bar{a}(\lambda)}e^{-i\lambda x}+\frac{1}{\bar{a}(\lambda)}\int_{-\infty}^\infty\int_0^\infty r_x(s,s')B_x(s')e^{i\lambda s}ds'ds\right)=0, \tag{4.71}$$

where the natural extension of $\bar{\beta}_x(s)$ and $r_x(s,s')$ over the negative s is taken, which is possible according to the above remark. By using (4.57) the last term in (4.71) is easily reduced to

$$\begin{aligned}
&\int_{-\infty}^{\infty}\int_{0}^{\infty} r_x(s,s')B_x(s')e^{i\lambda s}\,ds'\,ds \\
&= -\varepsilon\bar{b}(\lambda)e^{-i\lambda x}\int_{0}^{\infty}\int_{-\infty}^{0}\beta(u-x-s)B_x(s)e^{i\lambda u}\,du\,ds \\
&= -\varepsilon\bar{b}(\lambda)e^{-i\lambda x}\Pi_-(b(\lambda)B_+(\lambda)e^{i\lambda x}), \qquad (4.72)
\end{aligned}$$

where Π_- is a projection operator complementary to Π_+,

$$\Pi_- = \mathbf{I} - \Pi_+. \qquad (4.73)$$

Then (4.71) becomes

$$\begin{aligned}
a(\lambda)B_+(\lambda) + \varepsilon\Pi_+(\bar{r}(\lambda)e^{-i\lambda x} - \bar{r}(\lambda)e^{-i\lambda x} \\
\times\Pi_-(a(\lambda)B_+(\lambda)r(\lambda)e^{i\lambda x})) = 0, \qquad (4.74)
\end{aligned}$$

with $r(\lambda)$ defined by (4.4). Letting

$$\varphi_x(s) = \frac{1}{2\pi}\int_{-\infty}^{\infty} a(\lambda)B_+(\lambda)e^{-i\lambda s}\,d\lambda \qquad (4.75)$$

and taking the inverse Fourier transform we find from (4.74)

$$\varphi_x(s) + \varepsilon\bar{w}(-x-s) - \varepsilon\int_{0}^{\infty}\int_{-\infty}^{-x}\bar{w}(u-s)w(u-s')\cdot\varphi_x(s')\,du\,ds' = 0, \qquad s>0, \qquad (4.76)$$

where

$$w(x) = \frac{1}{2\pi}\int_{-\infty}^{\infty} r(\lambda)e^{-i\lambda x}\,d\lambda. \qquad (4.77)$$

Let

$$\varphi_x(s) = 2\varphi(x, x+2s) \qquad (4.78)$$

and recall that $w(x) = 2\omega(-2x)$, so that finally (4.76) becomes

$$\varphi(x,y) + \varepsilon\bar{\omega}(x+y) - \varepsilon\int_{x}^{\infty}\int_{x}^{\infty}\varphi(x,z)\omega(z+z')\cdot\bar{\omega}(z'+y)\,dz'\,dz = 0, \qquad y \geqslant x. \qquad (4.79)$$

This is the same equation as the one which results from the Gelfand-Levitan-Marchenko system (4.17) for the first row of $\Gamma_+(x,y)$ when $(\Gamma_+(x,y))_{11}$ is reduced out.

We also point out that the identification of $\varphi(x,y)$ with the matrix element $\varepsilon\bar{\beta}_+(x,y)=(\Gamma_+(x,y))_{12}$ (see (I.5.20)) follows directly from comparing (4.13), (4.40), (4.41), (4.49), (4.70) and (4.75).

To derive an equation relating the matrix element $(\Gamma_+(x,y))_{11}=\alpha_+(x,y)$ with $\bar{\beta}_+(x,y)$ (see (I.5.20)) we consider the expression (4.45) for $\Phi_-(x,s)$ in terms of $\Omega_+(x,s)$. Denote $\alpha_x(s)=(\Phi_-(x,s))_{11}$; then (4.45) yields

$$\alpha_x(s)=-\int_0^\infty \beta(-s-s'-x)B_x(s')\,ds', \tag{4.80}$$

for $s\geqslant 0$, whence immediately

$$\alpha_x(s)=-\int_0^\infty \varphi_x(s')w(-s-s'-x)\,ds'. \tag{4.81}$$

Letting

$$\alpha_x(s)=2\alpha_+(x,x+2s) \tag{4.82}$$

and using (4.81) we have

$$\alpha_+(x,y)+\varepsilon\int_x^\infty \bar{\beta}_+(x,z)\omega(z+y)\,dz=0, \tag{4.83}$$

for $y\geqslant x$. This equation coincides with the one relating the matrix elements of the first row of $\Gamma_+(x,y)$. Note that the identification of $\frac{1}{2}\alpha_x\left(\frac{y-x}{2}\right)$ with the matrix element $(\Gamma_+(x,y))_{11}$ is also a direct consequence of (4.12), (4.43) and (4.44).

Thus we have shown that (4.79), (4.83) together with the involution (4.16) are equivalent to (4.17), the Gelfand-Levitan-Marchenko equation from the right.

In a similar manner, (4.51) leads to (4.21), the Gelfand-Levitan-Marchenko equation from the left.

So we have seen how the Wiener-Hopf equation turns into the Gelfand-Levitan-Marchenko equations upon a special regularization. We conclude with several comments comparing the two approaches to the inverse problem.

1. The Riemann problem starts from a single independent function, $b(\lambda)$, whose Fourier transform behaves as $\psi(x)$ for all x. On the other hand,

the Gelfand-Levitan-Marchenko approach deals with two dependent functions, $\omega(x)$ and $\tilde{\omega}(x)$, behaving as $\psi(x)$ near $+\infty$ or $-\infty$, respectively.

2. The initial data $b(\lambda)$, λ_j, γ_j, $j=1,\dots,n$, are mutually independent whereas in the Gelfand-Levitan-Marchenko approach the eigenvalues λ_j of the discrete spectrum cannot vary without varying at least one of the functions $r(\lambda)$ or $\tilde{r}(\lambda)$.

3. In contrast to the Wiener-Hopf equation the Gelfand-Levitan-Marchenko integral equations are associated with compact operators.

4. The derivation of the auxiliary linear differential equation in the Riemann problem framework proceeds in a most simple way and is local in x. In the Gelfand-Levitan-Marchenko approach locality is lost and one is faced with an additional problem of identifying $U_0^{(+)}(x)$ with $U_0^{(-)}(x)$.

This concludes our comparison of the two approaches to solving the inverse problem. In the next section an explicit solution of the inverse problem will be presented in an important special case when $b(\lambda)$ vanishes. This will give us solitons for the NS model.

§ 5. The Rapidly Decreasing Case. Soliton Solutions

In this section we shall discuss an important special case of the inverse problem *when it can be solved in closed form*. Namely, we shall consider the case when

$$b(\lambda)=0 \tag{5.1}$$

for all λ, so that the Riemann factorization problem trivially reduces to determining the matrix Blaschke-Potapov factors. The latter problem makes sense only for $\varkappa<0$, which will be assumed in what follows.

As indicated in § 1, the ratio of the transition coefficients, $\dfrac{b(\lambda)}{a(\lambda)}$, plays the role of reflection coefficient in the scattering theory for the auxiliary linear problem. Therefore, in the case (5.1) the problem itself and everything related with it is referred to as *reflectionless*.

Now we proceed to the inverse problem. We begin with a single pair of zeros λ_0, $\bar{\lambda}_0$, $\operatorname{Im}\lambda_0>0$, and coefficients γ_0, $\bar{\gamma}_0$, $\gamma_0\neq 0$. The solution of the Riemann problem is

$$G_+(x,\lambda)=B(x,\lambda), \quad G_-(x,\lambda)=B^{-1}(x,\lambda), \tag{5.2}$$

(see (2.18)) where $B(x,\lambda)$ is the matrix Blaschke-Potapov factor,

$$B(x,\lambda)=I+\frac{\bar{\lambda}_0-\lambda_0}{\lambda-\bar{\lambda}_0}P(x), \tag{5.3}$$

and $P(x)$ is an orthogonal projection operator,

$$P(x)=\frac{1}{1+|\gamma_0(x)|^2}\begin{pmatrix}|\gamma_0(x)|^2 & \bar{\gamma}_0(x)\\ \gamma_0(x) & 1\end{pmatrix}. \tag{5.4}$$

Here $\gamma_0(x)=\gamma_0 e^{i\lambda_0 x}$ (see (2.22), (2.24) and (2.27)). The matrix $U_0(x)$ of the form (2.5),

$$U_0(x)=\sqrt{\varkappa}\begin{pmatrix}0 & \bar{\psi}(x)\\ \psi(x) & 0\end{pmatrix}, \tag{5.5}$$

occurring in the auxiliary linear problem (2.4) is given by (2.105),

$$U_0(x)=\tfrac{1}{2}[\sigma_3,\pi(x)], \tag{5.6}$$

where $\pi(x)$ is a coefficient in the expansion

$$B(x,\lambda)=I+\frac{i\pi(x)}{\lambda}+O\left(\frac{1}{|\lambda|^2}\right) \tag{5.7}$$

and has the form (see (2.106), (2.107))

$$\pi(x)=\frac{\bar{\lambda}_0-\lambda_0}{i}P(x). \tag{5.8}$$

As a result, there is the following simple expression for $\psi(x)$:

$$\psi(x)=\frac{2\,\mathrm{Im}\,\lambda_0}{\sqrt{\varkappa}}\cdot\frac{\gamma_0(x)}{1+|\gamma_0(x)|^2}. \tag{5.9}$$

Formula (5.9) provides the simplest example of a *reflectionless function* $\psi(x)$. It is specified by two arbitrary complex numbers λ_0, γ_0 subject to $\mathrm{Im}\,\lambda_0>0$, $\gamma_0\neq 0$, is infinitely differentiable and decays exponentially as $|x|\to\infty$.

Next we shall discuss the evolution of these initial data under the NS equation. The formulae (3.7) for the dynamics of the transition coefficients imply that (5.1) remains invariant and

$$\gamma_0(x,t)=e^{-i\lambda_0^2 t}\gamma_0(x). \tag{5.10}$$

It follows that $\psi(x, t)$, which is a solution of the NS equation, remains reflectionless and, as before, is given by a formula such as (5.9),

$$\psi(x, t) = \frac{2\,\mathrm{Im}\,\lambda_0}{\sqrt{\varkappa}} \cdot \frac{\gamma_0(x, t)}{1+|\gamma_0(x, t)|^2}\,. \tag{5.11}$$

With the definition

$$\begin{gathered} A = \frac{\mathrm{Im}\,\lambda_0}{\sqrt{|\varkappa|}}, \qquad u = 2\,\mathrm{Im}\,\lambda_0, \qquad v = 2\,\mathrm{Re}\,\lambda_0, \\ x_0 = \frac{1}{\mathrm{Im}\,\lambda_0} \log|\gamma_0|, \qquad \varphi_0 = \arg\gamma_0, \end{gathered} \tag{5.12}$$

(5.11) may be written as

$$\psi(x, t) = A \cdot \frac{\exp\left\{i\left(\varphi_0 + \frac{vx}{2} + \frac{(u^2-v^2)}{4}t - \frac{\pi}{2}\right)\right\}}{\operatorname{ch}\left\{\frac{u}{2}(x - vt - x_0)\right\}}\,. \tag{5.13}$$

The expression (5.13) shows that $\psi(x, t)$ *is a smooth function localized along the line*

$$x(t) = x_0 + vt \tag{5.14}$$

so that its center moves with constant velocity v. *Besides, the solution oscillates both in space and time with frequencies* $\frac{v}{2}$ *and* $\frac{u^2-v^2}{4}$, *respectively. The parameters* A, x_0 *and* φ_0 *play the role of amplitude, initial center and initial phase, respectively.*

We see that the NS solution (5.13) is a solitary wave with the following properties:

1. *Propagation does not change its shape.*

2. *It has finite energy and, moreover, all the integrals of the motion are finite.*

Following the established tradition, solutions with the above properties will be called *solitons in the broad sense of the word.* In the physics literature the term "soliton" sometimes refers to a general particle-like solution, i.e. a localized solution of finite energy.

So, the function $\psi(x, t)$ *given by (5.13) will be called a soliton solution of the NS equation in the rapidly decreasing case. It describes the free motion of a soliton.*

Notice that the existence of solitons is due to the nonlinear term in the NS equation: there are no solitons in the linear limit $\varkappa \to 0$. In fact, consider the solution $\psi(x,t)$ as $\varkappa \to 0$. For the limit to be finite it is necessary that $\gamma_0 = \sqrt{\varkappa}\,\tilde{\gamma}_0$, and then

$$\psi_0(x,t) = \lim_{\varkappa \to 0} \psi(x,t) = c_0 e^{i\lambda_0 x - i\lambda_0^2 t}, \tag{5.15}$$

with $c_0 = 2\tilde{\gamma}_0 \operatorname{Im} \lambda_0$. Obviously, $\psi_0(x,t)$ satisfies the linear Schrödinger equation and has property 1, but both its energy and momentum are infinite. Furthermore, the general solution $\tilde{\psi}(x,t)$ of the linear Schrödinger equation is given by the Fourier integral

$$\tilde{\psi}(x,t) = \int_{-\infty}^{\infty} e^{i\lambda x - i\lambda^2 t} \varphi(\lambda)\, d\lambda. \tag{5.16}$$

In order that its energy and momentum be finite, $\varphi(\lambda)$ must be sufficiently smooth and decaying. However, the method of stationary phase yields for such $\varphi(\lambda)$ that $\tilde{\psi}(x,t)$ decays as $\frac{1}{\sqrt{|t|}}$ along every direction $x - vt$, as $|t| \to \infty$. Therefore, property 1 is violated here. We may thus affirm that *soliton is an essentially nonlinear phenomenon*.

Let us now discuss the general reflectionless case when there are n pairs of zeros, $\lambda_j, \bar{\lambda}_j$, $\operatorname{Im} \lambda_j > 0$, and parameters $\gamma_j, \bar{\gamma}_j$, $j = 1, \ldots, n$. The solution of the associated Riemann problem is

$$G_+(x,\lambda) = \Pi(x,\lambda), \qquad G_-(x,\lambda) = \Pi^{-1}(x,\lambda), \tag{5.17}$$

where $\Pi(x,\lambda)$ is an ordered product of matrix Blaschke-Potapov factors,

$$\Pi(x,\lambda) = \prod_{j=1}^{\overset{\frown}{n}} B_j(x,\lambda), \tag{5.18}$$

$$B_j(x,\lambda) = I + \frac{\bar{\lambda}_j - \lambda_j}{\lambda - \bar{\lambda}_j} P_j(x) \tag{5.19}$$

(see (2.28)). The $P_j(x)$ are the orthogonal projection operators uniquely determined by the $\gamma_j, \bar{\gamma}_j$ via the relations (1.49) for the subspaces $\operatorname{Im} \Pi(x,\lambda_j)$ and $\operatorname{Ker} \Pi^{-1}(x,\bar{\lambda}_j)$, $j = 1, \ldots, n$. Using the generalized unitarity property

$$\Pi^*(x,\lambda) = \Pi^{-1}(x,\bar{\lambda}), \tag{5.20}$$

we can write (1.49) as

$$\Pi^{-1}(x,\bar{\lambda}_j)\,\xi_j=\Pi^*(x,\lambda_j)\,\xi_j=0\,, \tag{5.21}$$

where

$$\xi_j(x)=\begin{pmatrix}\bar{\gamma}_j(x)\\ 1\end{pmatrix} \tag{5.22}$$

is a column-vector with $\gamma_j(x)=e^{i\lambda_j x}\gamma_j$, $j=1,\ldots,n$.

As was explained in § 2, (5.21) allows to find the projection operators $P_1(x),\ldots,P_n(x)$ recursively. As before, $U_0(x)$ is given by (5.6) where $\pi(x)$ is a coefficient in the asymptotic expansion

$$\Pi(x,\lambda)=I+\frac{i\pi(x)}{\lambda}+O\left(\frac{1}{|\lambda|^2}\right). \tag{5.23}$$

More explicitly,

$$\pi(x)=\frac{1}{i}\sum_{j=1}^{n}(\bar{\lambda}_j-\lambda_j)\,P_j(x)\,. \tag{5.24}$$

Now we shall describe an alternative method for determining $\Pi(x,\lambda)$. It consists in resolving $\Pi^{-1}(x,\lambda)$ into partial fractions

$$\Pi^{-1}(x,\lambda)=I+\sum_{j=1}^{n}\frac{A_j(x)}{\lambda-\lambda_j} \tag{5.25}$$

and finding the matrix coefficients $A_j(x)$. Equations (5.20) and (5.21) show that $A_j(x)$ can be represented as

$$A_j(x)=z_j(x)\,\xi_j^*(x)\,, \tag{5.26}$$

where $z_j(x)=\begin{pmatrix}p_j(x)\\ q_j(x)\end{pmatrix}$ and $\xi_j^*(x)=(\gamma_j(x),1)$ is a row-vector conjugate to the column-vector $\xi_j(x)$, $j=1,\ldots,n$. It follows in particular that all the $A_j(x)$ are rank one matrices.

For the proof let us consider the following expansions in the neighbourhood of $\lambda=\lambda_j$ (cf. § 2):

$$\Pi^{-1}(x,\lambda)=\frac{A_j(x)}{\lambda-\lambda_j}+O(1) \tag{5.27}$$

and

$$\Pi(x,\lambda)=B_j(x)+O(|\lambda-\lambda_j|)\,, \tag{5.28}$$

so that

$$A_j(x)\,B_j(x)=B_j(x)A_j(x)=0. \tag{5.29}$$

Together with (1.49),

$$\operatorname{Im} B_j(x)=\left\{\begin{pmatrix}1\\ -\gamma_j(x)\end{pmatrix}\right\}, \tag{5.30}$$

this shows that $A_j(x)$ has rank one and can be represented as in (5.26).

Equations (5.21) allow us to determine the unknown vectors $z_j(x)$ involved in (5.26). Indeed, substituting (5.26) into (5.25) and using (5.21) we get a system of linear algebraic equations

$$\xi_j(x)+\sum_{k=1}^{n}\frac{\xi_k^*(x)\,\xi_j(x)}{\lambda_j-\bar{\lambda}_k}\,z_k(x)=0,\quad j=1,\dots,n. \tag{5.31}$$

The inner products $\xi_k^*\,\xi_j$ are

$$\xi_k^*(x)\,\xi_j(x)=1+\gamma_k(x)\,\bar{\gamma}_j(x) \tag{5.32}$$

so that (5.31) breaks up into two disjoint systems for the first and second vector components of the $z_j(x)$. In particular, for the first components $p_j(x)$ we have

$$\sum_{k=1}^{n} M_{jk}(x)p_k(x)=-\bar{\gamma}_j(x),\quad j=1,\dots,n, \tag{5.33}$$

where

$$M_{jk}(x)=\frac{1+\bar{\gamma}_j(x)\gamma_k(x)}{\bar{\lambda}_j-\lambda_k},\quad j,k=1,\dots,n. \tag{5.34}$$

The matrix $\pi(x)$ can be expressed in terms of the $A_j(x)$ as

$$\pi(x)=i\sum_{j=1}^{n}A_j(x), \tag{5.35}$$

whence by virtue of (5.6), (5.26) and (5.35) we have

$$\psi(x)=\frac{i}{\sqrt{\varkappa}}\sum_{j=1}^{n}\bar{p}_j(x). \tag{5.36}$$

Let $M(x)$ be the $n\times n$ matrix with matrix elements $\bar{M}_{jk}(x)$ and let

$$M_1(x) = \begin{pmatrix} & M(x) & & \vdots\, \gamma_1(x) \\ & & & \vdots\, \gamma_n(x) \\ \cdots\cdots & \cdots\cdots & \cdots\cdots & \vdots \\ 1 & & 1 & 0 \end{pmatrix}. \tag{5.37}$$

Then Cramer's rule gives the required expression for $\psi(x)$:

$$\psi(x) = \frac{i}{\sqrt{\varkappa}} \frac{\det M_1(x)}{\det M(x)}. \tag{5.38}$$

We have derived general closed-form expressions for reflectionless functions $\psi(x)$, $\bar{\psi}(x)$. They depend on $2n$ complex parameters λ_j, γ_j satisfying $\operatorname{Im}\lambda_j > 0$, $\gamma_j \neq 0$ and such that the λ_j are pairwise distinct; $\psi(x)$, $\bar{\psi}(x)$ are Schwartz functions and, moreover, decay exponentially as $|x| \to \infty$.

In fact, $\psi(x)$, $\bar{\psi}(x)$ are smooth (and, in particular, $M(x)$ is non-degenerate) because the projection operators $P_j(x)$ are non-singular for all x, which is easily verified by recursion. To show their exponential decay observe that $\gamma_j(x) = O(e^{-\operatorname{Im}\lambda_j x})$ as $x \to +\infty$ so that the $\xi_j(x)$ approach the constant vector $\begin{pmatrix} 0 \\ 1 \end{pmatrix}$ with the same order. It then follows that

$$B_j(x, \lambda) = \begin{pmatrix} 1 & 0 \\ 0 & \dfrac{\lambda - \lambda_j}{\lambda - \bar{\lambda}_j} \end{pmatrix} + O(e^{-ax}), \tag{5.39}$$

with $a = \min\limits_{j=1,\ldots,n} \{\operatorname{Im}\lambda_j\}$ so that $\psi(x) = O(e^{-ax})$ as $x \to +\infty$. The estimate $\psi(x) = O(e^{ax})$ as $x \to -\infty$ follows from a relation similar to the previous one,

$$B_j(x, \lambda) = \begin{pmatrix} \dfrac{\lambda - \lambda_j}{\lambda - \bar{\lambda}_j} & 0 \\ 0 & 1 \end{pmatrix} + O(e^{ax}), \tag{5.40}$$

which follows from the fact that $\gamma_j(x) = O(e^{-\operatorname{Im}\lambda_j x})$ as $x \to -\infty$ and hence the vectors $\dfrac{1}{\gamma_j(x)} \xi_j(x)$ approach $\begin{pmatrix} 1 \\ 0 \end{pmatrix}$ exponentially.

Now let a reflectionless $\psi(x)$ of the form (5.38) be taken as an initial value for the NS equation. The solution $\psi(x, t)$ is given by (5.38) upon replacing $\gamma_j(x)$ by $\gamma_j(x, t)$ as prescribed by (3.7):

$$\gamma_j(x, t) = e^{-i\lambda_j^2 t} \gamma_j(x), \quad j = 1, \ldots, n. \tag{5.41}$$

The solution remains reflectionless. *We will show that it corresponds to n interacting solitons. Namely, we will show that generically $\psi(x,t)$ can be expressed as a sum of one-soliton solutions*

$$\psi(x,t)=\sum_{j=1}^{n}\psi_j^{(\pm)}(x,t)+O(e^{-ac|t|}). \tag{5.42}$$

Here $\psi_j^{(\pm)}(x,t)$, $j=1,\dots,n$, are solitons with parameters A_j, v_j, $x_{0j}^{(\pm)}$ and $\varphi_{0j}^{(\pm)}$ defined by (5.12) from the following data λ_j and $\gamma_j^{(\pm)}$:

$$\gamma_j^{(+)}=\gamma_j\prod_{v_k<v_j}\frac{\lambda_j-\bar{\lambda}_k}{\lambda_j-\lambda_k}\prod_{v_k>v_j}\frac{\lambda_j-\lambda_k}{\lambda_j-\bar{\lambda}_k}, \tag{5.43}$$

$$\gamma_j^{(-)}=\gamma_j\prod_{v_k<v_j}\frac{\lambda_j-\lambda_k}{\lambda_j-\bar{\lambda}_k}\prod_{v_k>v_j}\frac{\lambda_j-\bar{\lambda}_k}{\lambda_j-\lambda_k}, \tag{5.44}$$

with $c=\min\limits_{j\neq k}|v_j-v_k|$. "Generically" means here that all velocities v_j are distinct.

For the proof of (5.42)-(5.44) it is sufficient to show that $\psi(x,t)$ approaches the one-soliton solution $\psi_j^{(\pm)}(x,t)$ along the trajectory C_j of a particular soliton,

$$x-v_jt=\text{const}, \tag{5.45}$$

and decays exponentially in all other directions as $t\to\pm\infty$.

This can be derived by examining the explicit formula (5.38). Instead of doing so we shall outline a more simple and elegant method based on a direct analysis of $\Pi(x,t,\lambda)$.

Observe that each of the $\gamma_j(x,t)$ either decays or grows exponentially, as $t\to\pm\infty$, along any direction except its own trajectory C_j. In fact, (5.41) yields

$$\gamma_j(x,t)|_{x-vt=c_0}=e^{-\frac{u_j}{2}(c_0+(v-v_j)t)}\,e^{\frac{i}{4}(2v_jc_0+(u_j^2+2vv_j-v_j^2)t)}\gamma_j,\quad j=1,\dots,n. \tag{5.46}$$

For the corresponding vectors $\xi_k(x,t)$ this gives, along C_j,

$$\lim_{t\to+\infty}\xi_k(x,t)=\begin{pmatrix}0\\1\end{pmatrix},\qquad \lim_{t\to-\infty}\frac{\xi_k(x,t)}{\gamma_k(x,t)}=\begin{pmatrix}1\\0\end{pmatrix}, \tag{5.47}$$

if $v_k<v_j$ and

$$\lim_{t\to+\infty} \frac{\xi_k(x,t)}{\gamma_k(x,t)} = \begin{pmatrix}1\\0\end{pmatrix}, \qquad \lim_{t\to-\infty} \xi_k(x,t) = \begin{pmatrix}0\\1\end{pmatrix}, \tag{5.48}$$

if $v_k > v_j$.

To determine the asymptotic behaviour of $\Pi(x,t,\lambda)$ along C_j it is convenient to use, in contrast with (5.18), the ordered product

$$\Pi(x,t,\lambda) = \prod_{\substack{k=1\\k\neq j}}^{\widehat{n}} \tilde{B}_k(x,t,\lambda)\tilde{B}_j(x,t,\lambda), \tag{5.49}$$

where the Blaschke-Potapov factor $\tilde{B}_j(x,t,\lambda)$ associated with a pair of zeros $\lambda_j, \bar{\lambda}_j$ is shifted to the extreme right. Here each of the $\tilde{B}_k(x,t,\lambda)$ for $k\neq j$ has asymptotically the form

$$\begin{pmatrix}1 & 0\\ 0 & \dfrac{\lambda-\lambda_k}{\lambda-\bar{\lambda}_k}\end{pmatrix} \quad \text{or} \quad \begin{pmatrix}\dfrac{\lambda-\lambda_k}{\lambda-\bar{\lambda}_k} & 0\\ 0 & 1\end{pmatrix}$$

depending on whether the asymptotic value of $\xi_k(x,t)$ is proportional to $\begin{pmatrix}0\\1\end{pmatrix}$ or $\begin{pmatrix}1\\0\end{pmatrix}$, respectively. Indeed, computing successively the asymptotic factors $\tilde{B}_k^{(\pm)}(\lambda)$, $k\neq j$, $k=1,\dots,n$, we obtain diagonal projection operators and hence diagonal matrices $\tilde{B}_k^{(\pm)}(\lambda)$ which leave invariant the subspaces spanned by $\begin{pmatrix}0\\1\end{pmatrix}$ and $\begin{pmatrix}1\\0\end{pmatrix}$.

Therefore equations (5.21) reduce asymptotically to a single equation for $\tilde{B}_j^{(\pm)}(\lambda)$, the asymptotic value of $\tilde{B}_j(x,t,\lambda)$ along C_j as $t\to\pm\infty$,

$$\tilde{B}_j^{(\pm)*}(\lambda_j)\,\xi_j^{(\pm)} = 0, \tag{5.50}$$

where

$$\xi_j^{(\pm)} = \begin{pmatrix}\tilde{\gamma}_j^{(\pm)}\\ 1\end{pmatrix}, \tag{5.51}$$

and the $\gamma_j^{(\pm)}$ are given by (5.43), (5.44). To complete the proof of (5.42) it is enough to observe that this argument also shows that $\psi(x,t)$ decays exponentially along every direction other than the trajectories C_j, $j=1,\dots,n$, as $t\to\pm\infty$.

The assertion proved above has a natural interpretation: *the solution $\psi(x,t)$ in (5.42) describes the process of interaction of n solitons moving freely* *and going apart from each other for large positive or negative times.* For that reason $\psi(x, t)$ is called an *n-soliton solution.*

The formulae obtained also allow a clear interpretation in terms of general scattering theory. In contrast with what happens in the linear theory, the one-soliton solution (5.13) is associated with a soliton particle, rather than with a wave train. A soliton is characterized by its velocity v, the position of its center of inertia $x(t)$ and the internal motion parameters A, φ_0. As $t \to \pm\infty$, the n-soliton solution $\psi(x, t)$ describes the free motion of n solitons with parameters $(v_j, x_{0j}^{(\pm)}, A_j, \varphi_{0j}^{(\pm)})$ given by (5.43), (5.44) and (5.12). It is convenient to relable the solitons according to their velocities, so that $\infty > v_1 > \ldots > v_n > -\infty$. Then, as $t \to -\infty$, their centers of inertia are separated by large intervals of order $a c |t|$ where $c = \min_{j \neq k} |v_j - v_k|$, and the quickest soliton is located on the left of all the others.

In this way, the asymptotic state associated with the n-soliton solution, as $t \to -\infty$, reproduces the motion of n solitons separated in space and coming together in the course of time. At finite times the picture of spatially separated solitons breaks down and the n-soliton solution describes the interaction of solitons. Finally, as $t \to +\infty$, the separated solitons reappear, the quickest being on the right of all the others. So it has interacted with all of them at finite times. A similar conclusion follows for all other solitons. In particular, the distance between solitons increases with time.

The above picture is typical of scattering theory which deals with asymptotic states described in terms of free particles. Scattering may only change the parameters of the particles and, possibly, their total number.

Here we encounter a very specific scattering process. Namely, the number of particles, their velocities and half of internal motion parameters – the amplitudes – remain invariant. The only effect of scattering is a variation in the centers of inertia and phases of internal motion. From (5.43), (5.44) and (5.12) one can deduce a relationship between the parameters of the asymptotic motion:

$$x_{0j}^{(+)} = x_{0j}^{(-)} + \Delta x_{0j}, \qquad \varphi_{0j}^{(+)} = \varphi_{0j}^{(-)} + \Delta\varphi_{0j}, \tag{5.52}$$

where

$$\Delta x_{0j} = \frac{2}{\operatorname{Im}\lambda_j}\left(\sum_{k=j+1}^{n} \log\left|\frac{\lambda_j - \bar{\lambda}_k}{\lambda_j - \lambda_k}\right| - \sum_{k=1}^{j-1} \log\left|\frac{\lambda_j - \bar{\lambda}_k}{\lambda_j - \lambda_k}\right|\right) \tag{5.53}$$

and

$$\Delta\varphi_{0j} = 2\left(\sum_{k=j+1}^{n} \arg\left(\frac{\lambda_j - \bar{\lambda}_k}{\lambda_j - \lambda_k}\right) - \sum_{k=1}^{j-1} \arg\left(\frac{\lambda_j - \bar{\lambda}_k}{\lambda_j - \lambda_k}\right)\right) \pmod{2\pi}. \tag{5.54}$$

It is characteristic that the increments in the coordinates x_{0j} and phases φ_{0j} are additively expressed through the two-particle increments

$$\begin{aligned}\Delta x_{01} &= \frac{2}{\operatorname{Im}\lambda_1}\log\left|\frac{\lambda_1-\bar{\lambda}_2}{\lambda_1-\lambda_2}\right|,\\ \Delta x_{02} &= -\frac{2}{\operatorname{Im}\lambda_2}\log\left|\frac{\bar{\lambda}_1-\lambda_2}{\lambda_1-\lambda_2}\right|,\\ \Delta\varphi_{01} &= 2\arg\frac{\lambda_1-\bar{\lambda}_2}{\lambda_1-\lambda_2},\\ \Delta\varphi_{02} &= -2\arg\frac{\bar{\lambda}_1-\lambda_2}{\lambda_1-\lambda_2}\end{aligned} \tag{5.55}$$

when $v_1>v_2$, with the interchange $1\leftrightarrow 2$ when $v_2>v_1$. The sum is taken over all two-particle interactions of the given soliton with the others. This specific scattering property, when n-particle scattering reduces to that of two particles, is commonly called *factorization*.

The factorization of scattering is sometimes included in the definition of the soliton along with properties 1–2. In this case one usually refers to the soliton in the narrow sense. However, in this book we shall only deal with solitons in the narrow sense and so simply call them solitons. In the next chapter we shall interprete the process of solitons scattering from the Hamiltonian standpoint.

To conclude this section we point out that a generic situation, when all velocities v_j are distinct, is essential for interpreting the n-soliton solution in terms of scattering theory. However, the solution itself obviously makes sense even if two or more velocities coincide. In this case solitons with the same velocity do not go apart but produce a bound state. In particular, a two-soliton solution with $v_1=v_2=0$ is periodic in time with frequency $(\operatorname{Im}\lambda_1)^2-(\operatorname{Im}\lambda_2)^2$.

This observation also refers to the constraints on the initial parameters λ_j, γ_j of the n-soliton solution. The algebraic formula (5.38) allows some of the λ_j to coincide and even reach the real axis, and some of the γ_j to vanish. Then the resulting $\psi(x,t)$ may vanish or go out of the Schwartz class (for instance, by developing a singularity); nevertheless, it will satisfy the NS equation by virtue of the algebraic nature of (5.38). As we shall see in the following chapter, these singular solutions are immaterial for the Hamiltonian interpretation of the NS model.

§ 6. Solution of the Inverse Problem in the Case of Finite Density. The Riemann Problem Method

In this section we begin the solution of the inverse problem under the finite density boundary conditions. The problem is to recover the functions $\psi(x)$, $\bar{\psi}(x)$ from the transition coefficients $a_\varrho(\lambda)$, $b_\varrho(\lambda)$ and characteristics of the discrete spectrum λ_j and γ_j. We shall restrict our attention to the case when $\psi(x)$, $\bar{\psi}(x)$ take their boundary values, as $x \to \pm\infty$, in the Schwartz sense.

As in the rapidly decreasing case, there are two approaches to solving the inverse problem, one based on the Riemann problem and the other on the Gelfand-Levitan-Marchenko formalism.

This section outlines the first one based on the matrix Riemann problem. Its natural formulation involves the Riemann surface Γ of the function $k(\lambda)=\sqrt{\lambda^2-\omega^2}$ with a contour $\mathscr{R}_\omega$ consisting of points (λ, ε) where $\varepsilon = \pm 1$ and λ lies in $\mathbb{R}_\omega$ (i.e. λ is real and $|\lambda| \geqslant \omega$ - see § I.8). The contour $\mathscr{R}_\omega$ devides Γ into two pieces - the sheets $\Gamma_\pm$.

As in the rapidly decreasing case, we start with a formula relating the Jost solutions,

$$T_-(x,\lambda) = T_+(x,\lambda)\, T_\varrho(\lambda) \tag{6.1}$$

(see (I.8.43)) where λ is in $\mathbb{R}_\omega$ and $T_\varrho(\lambda)$ is the reduced monodromy matrix,

$$T_\varrho(\lambda) = \begin{pmatrix} a_\varrho(\lambda) & \overline{b_\varrho}(\lambda) \\ b_\varrho(\lambda) & \overline{a_\varrho}(\lambda) \end{pmatrix}. \tag{6.2}$$

The involutions (I.8.41)-(I.8.42) show that, for λ in $\mathbb{R}_\omega$,

$$T_\pm(x,\lambda - i0) = \frac{i(\lambda+k)}{\omega}\, \sigma_1 \overline{T_\pm}(x,\lambda+i0)\sigma_3 \tag{6.3}$$

and

$$T_\varrho(\lambda - i0) = \sigma_3 \overline{T_\varrho}(\lambda + i0)\sigma_3. \tag{6.4}$$

In analogy with § 1 we set

$$S_+(x,\lambda) = (T_-^{(1)}(x,\lambda),\, T_+^{(2)}(x,\lambda)) \tag{6.5}$$

and

$$S_-(x,\lambda) = (T_+^{(1)}(x,\lambda),\, T_-^{(2)}(x,\lambda)). \tag{6.6}$$

These matrices may be analytically continued to the sheets $\Gamma_\pm$, respectively, (see § I.8) and for λ in $\mathscr{R}_\omega$ satisfy

$$S_-(x,\lambda)=S_+(x,\lambda)S_\varrho(\lambda), \tag{6.7}$$

where

$$S_\varrho(\lambda)=\frac{1}{a_\varrho(\lambda)}\begin{pmatrix}1 & \overline{b_\varrho}(\lambda)\\ -b_\varrho(\lambda) & 1\end{pmatrix}. \tag{6.8}$$

In the scattering theory for the auxiliary linear problem, $S_\varrho(\lambda)$ plays the role of *scattering matrix* and $\dfrac{1}{a_\varrho(\lambda)}$ and $\dfrac{b_\varrho(\lambda)}{a_\varrho(\lambda)}$ are interpreted as *transmission and reflection coefficients*, respectively.

In terms of $S_\pm(x,\lambda)$ the asymptotic behaviour (I.8.28)–(I.8.31), as $|\lambda|\to\infty$, becomes

$$S_+(x,\lambda)E^{-1}(x,k(\lambda))=S(\theta)\left(I+O\left(\frac{1}{|\lambda|}\right)\right), \tag{6.9}$$

where λ is in Γ_+ and $\operatorname{Im}\lambda>0$,

$$S_+(x,\lambda)E^{-1}(x,k(\lambda))=\frac{2\lambda}{\omega}\sigma_2 S^{-1}(\theta)\left(I+O\left(\frac{1}{|\lambda|}\right)\right), \tag{6.10}$$

where λ is in Γ_+ and $\operatorname{Im}\lambda<0$; also,

$$S_-(x,\lambda)E^{-1}(x,k(\lambda))=e^{-\frac{i\theta}{2}}S(\theta)\left(I+O\left(\frac{1}{|\lambda|}\right)\right), \tag{6.11}$$

where λ is in Γ_- and $\operatorname{Im}\lambda<0$,

$$S_-(x,\lambda)E^{-1}(x,k(\lambda))=\frac{2\lambda}{\omega}\sigma_2 e^{\frac{i\theta}{2}}S^{-1}(\theta)\left(I+O\left(\frac{1}{|\lambda|}\right)\right), \tag{6.12}$$

where λ is in Γ_- and $\operatorname{Im}\lambda>0$. Remind that $E(x,k)=\exp\left\{\dfrac{kx\sigma_3}{2i}\right\}$ with the definition

$$S(\theta)=\begin{pmatrix}1 & 0\\ 0 & e^{\frac{i\theta}{2}}\end{pmatrix}. \tag{6.13}$$

Now consider the matrices

$$G_+(x,\lambda)=a_\varrho(\lambda)\,G(\theta)\,E(x,k(\lambda))\,S_+^{-1}(x,\lambda) \tag{6.14}$$

and

$$G_-(x,\lambda)=S_-(x,\lambda)\,E^{-1}(x,k(\lambda))\,G^{-1}(\theta), \tag{6.15}$$

where $G(\theta)=e^{-\frac{i\theta}{2}}S(\theta)$. These have an analytic continuation to the respective sheets $\Gamma_\pm$, possibly with the exception of the branch points $\lambda=\pm\omega$ (see below), and provide a solution to the Riemann problem,

$$G_+(x,\lambda)\,G_-(x,\lambda)=G_\varrho(x,\lambda), \tag{6.16}$$

where

$$G_\varrho(x,\lambda)=G(\theta)\,E(x,k(\lambda))\,G_\varrho(\lambda)\,E^{-1}(x,k(\lambda))\,G^{-1}(\theta)$$
$$=\begin{pmatrix} 1 & e^{-\frac{i\theta}{2}-ikx}\,\overline{b_\varrho}(\lambda) \\ -e^{\frac{i\theta}{2}+ikx}\,b_\varrho(\lambda) & 1 \end{pmatrix} \tag{6.17}$$

and

$$G_\varrho(\lambda)=\begin{pmatrix} 1 & \overline{b_\varrho}(\lambda) \\ -b_\varrho(\lambda) & 1 \end{pmatrix} \tag{6.18}$$

(cf. § 1).

The matrices $G_\pm(x,\lambda)$ are non-degenerate on the sheets $\Gamma_\pm$ except at $\lambda_j^{(\pm)}=(\lambda_j,\pm)$, $j=1,\ldots,n$. More precisely, we have

$$\det G_+(x,\lambda)=\frac{e^{-\frac{i\theta}{2}}\omega^2}{2k(\lambda-k)}\,a_\varrho(\lambda) \tag{6.19}$$

for λ in Γ_+ and

$$\det G_-(x,\lambda)=\frac{2e^{\frac{i\theta}{2}}k(\lambda-k)}{\omega^2}\,a_\varrho^*(\lambda) \tag{6.20}$$

for λ in Γ_-, where $a_\varrho^*(\lambda)$ is the analytic continuation of $\overline{a_\varrho}(\lambda)$ to the sheet Γ_-. The functions $a_\varrho(\lambda)$ and $a_\varrho^*(\lambda)$ vanish precisely at $\lambda=\lambda_j^{(\pm)}$, $j=1,\ldots,n$. In addition (see (I.8.36), (I.9.22) and (6.14)–(6.15)) we have

$$\operatorname{Im} G_+(x,\lambda_j^{(+)})=\tilde{N}_j^{(+)}(x) \tag{6.21}$$

and

$$\operatorname{Ker} G_-(x, \lambda_j^{(-)}) = \tilde{N}_j^{(-)}(x), \tag{6.22}$$

where $\tilde{N}_j^{(+)}(x)$ and $\tilde{N}_j^{(-)}(x)$ are one-dimensional subspaces in $\mathbb{C}^2$ spanned by

$$\begin{pmatrix} 1 \\ -e^{\frac{i\theta}{2}+ik_jx}\gamma_j \end{pmatrix} \quad \text{and} \quad \begin{pmatrix} e^{-\frac{i\theta}{2}+ik_jx}\gamma_j \\ 1 \end{pmatrix},$$

respectively, $\gamma_j = -\bar{\gamma}_j$ are the transition coefficients for the discrete spectrum, and $k_j = i\sqrt{\omega^2 - \lambda_j^2}$, $j = 1, \ldots, n$.

Thus, equation (6.16) describes a Riemann problem with zeros on the surface Γ. Let us continue the list of the properties of its ingredients $G_\varrho(x, \lambda)$ and $G_\pm(x, \lambda)$.

We begin with $G_\varrho(x, \lambda)$ and the λ_j, γ_j, $j = 1, \ldots, n$.

1) *$G_\varrho(x, \lambda)$ has the form (6.17) with an integral representation for $b_\varrho(\lambda)$,*

$$b_\varrho(\lambda) = \frac{1}{k} \int_{-\infty}^{\infty} \beta_\varrho^{(1)}(x) e^{ikx}\, dx + \frac{\lambda}{k} \int_{-\infty}^{\infty} \beta_\varrho^{(2)}(x) e^{ikx}\, dx, \tag{6.23}$$

for λ in $\mathscr{R}_\omega$, where $\beta_\varrho^{(1,2)}(x)$ are real-valued Schwartz functions. (This representation results from (I.9.15) through integration by parts.)

In particular, it follows that $G_\varrho(x, \lambda)$ has the asymptotic behaviour

$$G_\varrho(x, \lambda) = I + O\left(\frac{1}{|\lambda|}\right) \tag{6.24}$$

as $|\lambda| \to \infty$ and satisfies the involution

$$G_\varrho(x, \lambda - i0) = \sigma_3 Q^{-1}(\theta)\, \overline{G_\varrho}(x, \lambda + i0)\, Q(\theta)\, \sigma_3 . \tag{6.25}$$

2) *The pairwise distinct real numbers λ_j lie in the gap $-\omega < \lambda_j < \omega$, and the $\gamma_j \neq 0$ are pure imaginary, $j = 1, \ldots, n$.*

The next three properties characterize the relationship between $b_\varrho(\lambda)$ and λ_j, γ_j (see § I.9).

3) *The condition (θ).*

4) *The condition for the determination of signs.*

5) *The relationship*

$$\operatorname{sign} i\gamma_j = \operatorname{sign} \frac{da_\varrho}{d\lambda}(\lambda_j), \tag{6.26}$$

where $a_\varrho(\lambda)$ is recovered from $b_\varrho(\lambda)$, θ and $\lambda_1, \ldots, \lambda_n$ according to

$$a_\varrho(\lambda)=e^{\frac{i\theta}{2}}\prod_{j=1}^{n}\frac{\lambda+k-\lambda_j-k_j}{\lambda+k-\lambda_j+k_j}$$

$$\times\exp\left\{\frac{1}{2\pi i}\int_{\mathbb{R}_\omega}\frac{\log(1+|b_\varrho(\mu)|^2)}{k(\mu)}\left(1+\frac{k}{\mu-\lambda}\right)d\mu\right\}. \qquad (6.27)$$

The properties of $G_\pm(x,\lambda)$, besides the degeneracy property mentioned above, are as follows.

1) *The asymptotic behaviour as* $|\lambda|\to\infty$:

$$G_\pm(x,\lambda)=I+O\left(\frac{1}{|\lambda|}\right), \qquad (6.28)$$

where respectively λ lies in $\Gamma_\pm$ and $\pm\operatorname{Im}\lambda>0$; when $\pm\operatorname{Im}\lambda<0$ we have

$$G_+(x,\lambda)=\frac{\omega}{2\lambda}G^2(\theta)\sigma_2\left(I+O\left(\frac{1}{|\lambda|}\right)\right) \qquad (6.29)$$

and

$$G_-(x,\lambda)=\frac{2\lambda}{\omega}\sigma_2 G^{-2}(\theta)\left(I+O\left(\frac{1}{|\lambda|}\right)\right). \qquad (6.30)$$

2) *The involution property for* λ *in* $\mathbb{R}_\omega$:

$$G_+(x,\lambda-i0)=\frac{\omega}{i(\lambda+k)}\sigma_3 G^2(\theta)\overline{G_+}(x,\lambda+i0)\sigma_1 \qquad (6.31)$$

and

$$G_-(x,\lambda-i0)=\frac{i(\lambda+k)}{\omega}\sigma_1\overline{G_-}(x,\lambda+i0)G^{-2}(\theta)\sigma_3, \qquad (6.32)$$

consistent with the asymptotic formulae (6.28)-(6.30).

3) *The asymptotic behaviour as* $|x|\to\infty$ *for* λ *in* $\mathbb{R}_\omega$:

$$G_\mp^{-1}(x,\lambda)=Q^{-1}(\theta)E_\varrho(\lambda)\begin{pmatrix} e^{\frac{i\theta}{2}} & 0 \\ \frac{b_\varrho(\lambda)}{a_\varrho(\lambda)}e^{\frac{i\theta}{2}+ikx} & \frac{1}{a_\varrho(\lambda)}\end{pmatrix}+o(1), \qquad (6.33)$$

$$G_-(x,\lambda) = Q^{-1}(\theta)E_\varrho(\lambda)\begin{pmatrix} e^{\frac{i\theta}{2}} & \overline{b_\varrho}(\lambda)e^{-ikx} \\ 0 & \bar{a}_\varrho(\lambda) \end{pmatrix} + o(1) \tag{6.34}$$

as $x \to +\infty$, and

$$G_+^{-1}(x,\lambda) = E_\varrho(\lambda)\begin{pmatrix} \frac{e^{\frac{i\theta}{2}}}{a_\varrho(\lambda)} & -\frac{\bar{b}_\varrho(\lambda)}{a_\varrho(\lambda)}e^{-ikx} \\ 0 & 1 \end{pmatrix} + o(1), \tag{6.35}$$

$$G_-(x,\lambda) = E_\varrho(\lambda)\begin{pmatrix} \bar{a}_\varrho(\lambda)e^{\frac{i\theta}{2}} & 0 \\ -b_\varrho(\lambda)e^{\frac{i\theta}{2}+ikx} & 1 \end{pmatrix} + o(1) \tag{6.36}$$

as $x \to -\infty$. Here $E_\varrho(\lambda) = E_\varrho(x,\lambda)|_{x=0}$ and $Q(\theta) = \exp\frac{i\theta\sigma_3}{2}$.

4) *The behaviour at the branch points* $\lambda = \pm\omega$.

The exact form of this property depends on the behaviour of $b_\varrho(\lambda)$ at $\lambda = \pm\omega$. First consider the case $\lambda = \omega$. There are two possibilities.

a) *A virtual level,*

$$|b_\varrho(\omega)| < \infty\,. \tag{6.37}$$

In this case $G_-(x,\omega)$ and $G_+^{-1}(x,\omega)$ are degenerate, so that

$$\operatorname{Ker} G_+^{-1}(x,\omega) = N_\omega^{(+)}, \qquad \operatorname{Ker} G_-(x,\omega) = N_\omega^{(-)}, \tag{6.38}$$

where $N_\omega^{(+)}$ and $N_\omega^{(-)}$ are one-dimensional subspaces in $\mathbb{C}^2$ spanned by $\begin{pmatrix} 1 \\ -c_+e^{\frac{i\theta}{2}} \end{pmatrix}$ and $\begin{pmatrix} c_+e^{-\frac{i\theta}{2}} \\ 1 \end{pmatrix}$, respectively, and

$$c_+ = b_\varrho(\omega) + ia_\varrho(\omega) \tag{6.39}$$

(see (I.9.9)). By virtue of (I.9.13) we have in addition $c_+ = -\bar{c}_+$.

b) *The generic case*

$$b_\varrho(\lambda) = \frac{b_+}{k} + O(1), \quad b_+ \neq 0 \tag{6.40}$$

near $\lambda = \omega$. Here $G_-(x, \omega)$ is non-degenerate and $G_+(x, \lambda)$ can be expressed as

$$G_+(x, \lambda) = \frac{G_+(x)}{k} + O(1) \tag{6.41}$$

near $\lambda = \omega$, with a non-degenerate $G_+(x)$ (see (I.9.11) and (6.19)-(6.20)).

The case $\lambda = -\omega$ is examined in a similar manner. If $\lambda = -\omega$ is a virtual level, the constant c_+ (6.39) is replaced by

$$c_- = b_\varrho(-\omega) - i a_\varrho(-\omega). \tag{6.42}$$

The formulation of the Riemann problem and the properties of $G_\varrho(x, \lambda)$ and $G_\pm(x, \lambda)$ look more complicated then those in the rapidly decreasing case. This is due primarily to the nature of the continuous spectrum of the auxiliary linear problem, especially to the existence of a gap in the spectrum - the interval $-\omega < \lambda < \omega$.

The properties of $G_\varrho(x, \lambda)$ and $G_\pm(x, \lambda)$ listed above were actually established in §§ I.8-I.9 during the investigation of the auxiliary linear problem. Let us now turn to solving the inverse problem. The solution is based on the matrix Riemann problem with zeros

$$G_\varrho(x, \lambda) = G_+(x, \lambda) G_-(x, \lambda), \tag{6.43}$$

where the matrix $G_\varrho(x, \lambda)$, the zeros λ_j, the constants γ_j and the parameter θ, $0 \leq \theta < 2\pi$, play the role of initial data, and $G_\pm(x, \lambda)$ give the solution of the Riemann problem. Here the variable x stands for a parameter.

The data $G_\varrho(x, \lambda)$, λ_j and γ_j are supposed to satisfy conditions 1)-5). It is required that the solution $G_\pm(x, \lambda)$ is analytic on $\Gamma_\pm$ except, possibly, at the branch points, satisfies the degeneracy conditions (6.21)-(6.22) and has properties 1)-2) and 4).

Then we claim the following.

I. *The Riemann problem has a unique solution.*

II. *The matrices $S_\pm(x, \lambda)$ constructed from $G_\pm(x, \lambda)$ according to (6.14)-(6.15) satisfy the differential equation of the auxiliary linear problem*

$$\frac{dS_\pm(x, \lambda)}{dx} = \left(\frac{\lambda \sigma_3}{2i} + U_0(x)\right) S_\pm(x, \lambda), \tag{6.44}$$

with

$$U_0(x) = \sqrt{\varkappa}\,(\bar{\psi}(x)\sigma_+ + \psi(x)\sigma_-). \tag{6.45}$$

III. *The solution $G_\pm(x,\lambda)$ has the asymptotic behaviour, as $x \to \pm\infty$, prescribed by property 3).*

IV. *The functions $\psi(x)$, $\bar{\psi}(x)$ satisfy the finite density boundary conditions*

$$\lim_{x\to-\infty} \psi(x) = \varrho, \qquad \lim_{x\to+\infty} \psi(x) = \varrho e^{i\theta}, \tag{6.46}$$

with $\varrho = \dfrac{\omega}{2\sqrt{\varkappa}}$ and the boundary values taken in the Schwartz sense.

V. *The functions $a_\varrho(\lambda)$ and $b_\varrho(\lambda)$, with $a_\varrho(\lambda)$ given by (6.27), are the transition coefficients for the continuous spectrum of the auxiliary linear problem (6.44), and $S_\pm(x,\lambda)$ are composed of the corresponding Jost solutions according to (6.5)–(6.6). The discrete spectrum of the auxiliary linear problem consists of the n-tuple $\lambda_1,\dots,\lambda_n$, and $\gamma_1,\dots,\gamma_n$ are the associated transition coefficients.*

The proof of these facts can be given along the lines of § 2. Since the Riemann surface Γ has genus 0, it is convenient to use the uniformization variable z as in § I.9,

$$\lambda(z) = \frac{1}{2}\left(z + \frac{\omega^2}{z}\right), \qquad k(z) = \frac{1}{2}\left(z - \frac{\omega^2}{z}\right), \tag{6.47}$$

so that $\mathscr{R}_\omega$ is mapped onto the real axis of the complex z-plane. The sheets $\Gamma_\pm$ are mapped onto the upper and lower half planes, respectively; a neighbourhood of $\lambda = \infty$ on $\Gamma_\pm$ with $\pm\operatorname{Im}\lambda > 0$ is mapped onto a neighbourhood of $z = \infty$, and a neighbourhood of $\lambda = \infty$ on $\Gamma_\pm$ with $\pm\operatorname{Im}\lambda < 0$ onto a neighbourhood of $z = 0$. The involution $\lambda - i0 \to \lambda + i0$ on $\mathscr{R}_\omega$ goes into the involution $z \to \dfrac{\omega^2}{z}$ on the real axis.

From (I.8.13)–(I.8.14) and (I.9.14) we derive integral representations for $G_\pm(x,z) = G_\pm(x,\lambda(z))$,

$$G_+(x,z) = \frac{z^2}{z^2-\omega^2}\left(I - \frac{\omega}{z}G^2(\theta)\sigma_2 + \int_0^\infty \Phi_+^{(1)}(x,s)e(s,z)\,ds + \frac{1}{z}\int_0^\infty \Phi_+^{(2)}(x,s)e(s,z)\,ds\right) \tag{6.48}$$

and

$$G_-(x,z) = I + \frac{\omega}{z}\sigma_2 G^{-2}(\theta)$$

$$+ \int_0^\infty \Phi_-^{(1)}(x,s)e(s,z)\,ds + \frac{1}{z}\int_0^\infty \Phi_-^{(2)}(x,s)e(s,z)\,ds, \tag{6.49}$$

where

$$e(s,z) = e^{2isk(z)} = e^{is\left(z-\frac{\omega^2}{z}\right)}. \tag{6.50}$$

These integral representations generalize those given by (1.36) in the rapidly decreasing case. They incorporate both the asymptotic behaviour (6.28)–(6.30) and the singularities of $G_+(x,z)$ at $z=\pm\omega$ (see condition 4)).

The integral representations (6.48)–(6.49) provide a basis for proving the assertions I–V. They can be used to derive a system of integral equations – an analogue of the Wiener-Hopf equation in the rapidly decreasing case, equivalent to the original Riemann problem. However, there arise several technical difficulties. First, various types of behaviour of $b_\varrho(\lambda)$ at the branch points $\lambda=\pm\omega$ should be examined separately; there are four cases altogether. Second, by properties 3)–5), the discrete spectrum data λ_j, γ_j are not independent of the continuous spectrum data $b_\varrho(\lambda)$; in particular, for the case of virtual level they appear in the behaviour of $G_+^{-1}(x,\lambda)$ at $\lambda=\pm\omega$ (see condition 4)). Hence the solution of the Riemann problem with zeros cannot be expressed as a product of Blaschke-Potapov factors and a solution of the regular Riemann problem with the same continuous spectrum data (cf. § 2).

Therefore, a detailed analysis of the Riemann problem (6.43) along the lines of § 2 would be rather cumbersome and is not so instructive as to be presented here. Instead, the next section will discuss more thouroughly a different approach to solving the inverse problem based on the Gelfand-Levitan-Marchenko formalism, which will also provide a proof for the assertions I–V.

To conclude this section we note that, as in § 3, the Riemann problem method allows to show that *if the data $b_\varrho(\lambda)$, λ_j, γ_j depend on t according to (I.10.7),*

$$b_\varrho(\lambda,t)=e^{-i\lambda k t}b_\varrho(\lambda,0), \quad \gamma_j(t)=e^{-i\lambda_j k_j t}\gamma_j(0),$$
$$\lambda_j(t)=\lambda_j(0)\,; \quad j=1,\dots,n, \tag{6.51}$$

then the resulting $\psi(x,t)$ satisfies the NS equation under the finite density boundary conditions.

For that purpose the dependence on t should be inserted into the Riemann problem (6.43)

$$G_\varrho(x,t,\lambda)=G_+(x,t,\lambda)\,G_-(x,t,\lambda), \tag{6.52}$$

where

$$G_\varrho(x,t,\lambda)=E^{-1}(t,\lambda k(\lambda))\,G_\varrho(x,\lambda)\,E(t,\lambda k(\lambda)). \tag{6.53}$$

From (6.52)–(6.53) it follows that

$$\frac{\partial F_+}{\partial t}(x,t,\lambda)\,F_+^{-1}(x,t,\lambda)=\frac{\partial F_-}{\partial t}(x,t,\lambda)\,F_-^{-1}(x,t,\lambda), \tag{6.54}$$

where

$$F_\pm(x,t,\lambda)=S_\pm(x,t,\lambda)\,E^{-1}(t,\lambda k(\lambda)). \tag{6.55}$$

Then, in complete analogy with § 3, in addition to the equation in x,

$$\frac{\partial F_\pm}{\partial x}=U(x,t,\lambda)\,F_\pm, \tag{6.56}$$

we also have an equation in t,

$$\frac{\partial F_\pm}{\partial t}=V_\varrho(x,t,\lambda)\,F_\pm, \tag{6.57}$$

where $V_\varrho(x,t,\lambda)$ is the same as in § I.2.

Thus, in the case of finite density as well, the Riemann problem method comes down to the zero curvature condition, so that the function $\psi(x,t)$ obtained via the inverse problem satisfies the NS equation.

§ 7. Solution of the Inverse Problem in the Case of Finite Density. The Gelfand-Levitan-Marchenko Formulation

Here we outline another method for solving the inverse problem. In contrast to the previous one based on the Riemann problem of analytic factorization for matrix-valued functions, this approach exploits a special conjugation problem for vector-valued analytic functions, motivated by the relation (6.1) for the Jost solutions.

In terms of the variable z, (6.1) is

$$T_-(x,z)=T_+(x,z)\,T_\varrho(z), \tag{7.1}$$

where $\operatorname{Im} z=0$ and

$$T_{\pm}(x,z)=T_{\pm}(x,\lambda(z)), \qquad T_{\varrho}(z)=T_{\varrho}(\lambda(z)). \tag{7.2}$$

The involutions (6.3)–(6.4) take the form

$$T_{\pm}\left(x,\frac{\omega^2}{z}\right)=\frac{iz}{\omega}\sigma_1\overline{T_{\pm}}(x,z)\sigma_3, \tag{7.3}$$

$$T_{\varrho}\left(\frac{\omega^2}{z}\right)=\sigma_3\overline{T_{\varrho}}(z)\sigma_3. \tag{7.4}$$

To state the required conjugation problem, consider (7.1) for the first column $T_-^{(1)}(x,z)$ of $T_-(x,z)$ and rewrite it as

$$\frac{1}{a_{\varrho}(z)}T_-^{(1)}(x,z)=T_+^{(1)}(x,z)+r_{\varrho}(z)\,T_+^{(2)}(x,z), \tag{7.5}$$

where we have denoted (cf. § 4)

$$r_{\varrho}(z)=\frac{b_{\varrho}(z)}{a_{\varrho}(z)}. \tag{7.6}$$

The vector function $F_1(x,z)=\dfrac{1}{a_{\varrho}(z)}T_-^{(1)}(x,z)$ on the right hand side of (7.5) may be analytically continued into the upper half-plane of the variable z with the exception of $z=z_j=\lambda_j+i\sqrt{\omega^2-\lambda_j^2}, j=1,\dots,n$ where it has simple poles, and $z=0$ where it has an essential singularity. By virtue of (I.9.22),

$$T_-^{(1)}(x,z_j)=\gamma_j\,T_+^{(2)}(x,z_j), \tag{7.7}$$

we find

$$\operatorname{res} F_1(x,z)|_{z=z_j}=c_j\,T_+^{(2)}(x,z_j), \tag{7.8}$$

where

$$c_j=\frac{\gamma_j}{\dot{a}_{\varrho}(z_j)}, \qquad j=1,\dots,n, \tag{7.9}$$

a dot indicating differentiation with respect to z.

In the neighbourhood of $z=0$, for $\operatorname{Im} z\geqslant 0$, (I.8.28), (I.8.33) and (I.9.5) yield asymptotically

$$F_1(x,z)e\left(\frac{x}{4},z\right)=\begin{pmatrix}0\\ \frac{i\omega}{z}e^{\frac{i\theta}{2}}\end{pmatrix}+O(1), \tag{7.10}$$

where the notation $e(x,z)$ was introduced in § 6. From (I.8.28), (I.8.32) and (I.9.4) we have, asymptotically as $|z|\to\infty$, $\operatorname{Im} z\geqslant 0$,

$$F_1(x,z)e\left(\frac{x}{4},z\right)=\begin{pmatrix}e^{-\frac{i\theta}{2}}\\ 0\end{pmatrix}+O\left(\frac{1}{|z|}\right). \tag{7.11}$$

Now consider the right hand side of (7.5). The first term, $T_+^{(1)}(x,z)$, may be analytically continued into the lower half-plane of the variable z with the exception of $z=0$ where it has an essential singularity. From (I.8.30), (I.8.32)–(I.8.33) and (I.9.4)–(I.9.5) we find the asymptotic formulae

$$T_+^{(1)}(x,z)e\left(\frac{x}{4},z\right)=\begin{pmatrix}0\\ \frac{i\omega}{z}e^{\frac{i\theta}{2}}\end{pmatrix}+O(1), \tag{7.12}$$

as $z\to 0$, and

$$T_+^{(1)}(x,z)e\left(\frac{x}{4},z\right)=\begin{pmatrix}e^{-\frac{i\theta}{2}}\\ 0\end{pmatrix}+O\left(\frac{1}{|z|}\right), \tag{7.13}$$

as $|z|\to\infty$.

The column $T_+^{(2)}(x,z)$ in the second term may be analytically continued into the upper half-plane and is related to $T_+^{(1)}(x,z)$ by the involution (I.8.36),

$$T_+^{(2)}(x,z)=\sigma_1\overline{T_+^{(1)}(x,\bar z)}. \tag{7.14}$$

The relation (7.5) combined with the analyticity properties stated above, relations (7.8), (7.14) and the asymptotic formulae (7.10)–(7.13) is what constitutes the required special conjugation problem. The data prescribed are the function $r_\varrho(z)$ defined on the real line and the parameters z_j, c_j; $j=1,\ldots,n$. The data $r_\varrho(z)$ and z_j, c_j are not all independent. They are subject to the following conditions resulting from § I.9.

1) *The function $r_\varrho(z)$ lies in Schwartz space and together with all its derivatives vanishes at $z=0$.*

This is a consequence of a similar property of $b_\varrho(z)$ (see the integral representation (6.23)).

2) *The involution property*

$$r_\varrho\left(\frac{\omega^2}{z}\right) = -\bar{r}_\varrho(z) \tag{7.15}$$

(see (7.4)).

3) *The inequality*

$$|r_\varrho(z)| \leqslant 1 \tag{7.16}$$

holds, with equality possibly attained only at $z=\pm\omega$, in which case

$$r_\varrho(\pm\omega) = \mp i \tag{7.17}$$

so that we are in a generic situation.

This follows from the normalization relation

$$|r_\varrho(z)|^2 = 1 - \frac{1}{|a_\varrho(z)|^2} = \frac{|b_\varrho(z)|^2}{1+|b_\varrho(z)|^2} \tag{7.18}$$

together with (I.9.11) which holds when $|a_\varrho(\pm\omega)| = \infty$.

4) *The condition* (θ)

$$e^{i\theta} = \prod_{j=1}^{n} \frac{\bar{z}_j}{z_j} \exp\left\{\frac{1}{\pi i} \int_{-\infty}^{\infty} \frac{\log(1-|r_\varrho(z)|^2)}{z}\, dz\right\}. \tag{7.19}$$

This is a variant of (I.9.44) since

$$1 + |b_\varrho(z)|^2 = \frac{1}{1-|r_\varrho(z)|^2}\,. \tag{7.20}$$

Here $|z_j| = \omega$ because the real numbers λ_j lie in the gap $(-\omega, \omega)$.

5) *The positivity condition: the quantities $m_j = -\dfrac{c_j}{z_j}$, $j=1, \ldots, n$, are positive real numbers.*

This follows from

$$c_j = \frac{\gamma_j}{\left.\dfrac{da_\varrho(z)}{dz}\right|_{z=z_j}} = \frac{\gamma_j}{\dfrac{da_\varrho(\lambda)}{d\lambda}} \left.\frac{dz(\lambda)}{d\lambda}\right|_{\lambda=\lambda_j}, \tag{7.21}$$

$$\frac{d\lambda(z)}{dz}=\frac{1}{2}\left(1-\frac{\omega^2}{z^2}\right) \tag{7.22}$$

together with relation (6.26)

$$\operatorname{sign} i\gamma_j=\operatorname{sign}\frac{da_\varrho}{d\lambda}(\lambda_j), \quad j=1,\dots,n \tag{7.23}$$

(see § I.9).

The formulation of the conjugation problem in the case of finite density (as well as that of the Riemann problem in § 6) looks more complicated than in the rapidly decreasing case in § 4. Yet, unlike the Riemann problem, it can be investigated in quite a similar way as in the rapidly decreasing case by reducing it to a system of integral equations.

To derive this system we shall make use of the integral representations (I.8.13)–(I.8.14) for the Jost solutions,

$$T_+(x,z)=Q^{-1}(\theta)E_\varrho(x,z)+\int_x^\infty \Gamma_+(x,y)Q^{-1}(\theta)E_\varrho(y,z)\,dy \tag{7.24}$$

and

$$T_-(x,z)=E_\varrho(x,z)+\int_{-\infty}^x \Gamma_-(x,y)E_\varrho(y,z)\,dy, \tag{7.25}$$

where

$$E_\varrho(x,z)=E_\varrho(x,\lambda(z))=\begin{pmatrix}1 & -\dfrac{i\omega}{z}\\ \dfrac{i\omega}{z} & 1\end{pmatrix} e^{-\frac{ix}{4}\left(z-\frac{\omega^2}{z}\right)\sigma_3} \tag{7.26}$$

(see (I.8.9)). Insert these expressions into (7.5), substract from both sides the first column $\mathscr{E}_\varrho(x,z)$ of $Q^{-1}(\theta)E_\varrho(x,z)$,

$$\mathscr{E}_\varrho(x,z)=\begin{pmatrix} e^{-\frac{i\theta}{2}}\\ \dfrac{i\omega}{z}e^{\frac{i\theta}{2}}\end{pmatrix} e\left(-\frac{x}{4},z\right), \tag{7.27}$$

multiply both sides of the resulting equation by $e\left(\frac{y}{4}, z\right)$, $y \geqslant x$, and integrate over z from $-\infty$ to ∞ (compare with the manipulations in § 4). Let us evaluate the integrals which occur.

First consider the left hand side; denote it **L**. From (7.10)–(7.11) it follows that the vector function $(F_1(x,z)-\mathscr{E}_\varrho(x,z))e\left(\frac{x}{4}, z\right)$ is regular at $z=0$ and of order $O\left(\frac{1}{|z|}\right)$ as $|z|\to\infty$. So using (7.8) and the Jordan lemma we deduce that

$$\mathbf{L}=2\pi i\sum_{j=1}^{n} c_j T_+^{(2)}(x, z_j)\, e\left(\frac{y}{4}, z_j\right). \tag{7.28}$$

Now consider the right hand side; denote it **R**. Here we encounter the integrals $\int_{-\infty}^{\infty} e(x,z)\,dz$ and $\int_{-\infty}^{\infty} e(x,z)\frac{dz}{z}$ taken in the sense of generalized functions. We have

$$\begin{aligned}\int_{-\infty}^{\infty} e(x,z)\,dz &= \int_0^{\infty} e^{i\left(z-\frac{\omega^2}{z}\right)x}\,dz+\int_{-\infty}^{0} e^{i\left(z-\frac{\omega^2}{z}\right)x}\,dz\\ &= \int_0^{\infty} e^{i\left(z-\frac{\omega^2}{z}\right)x}\left(1+\frac{\omega^2}{z^2}\right)dz=\int_{-\infty}^{\infty} e^{ipx}\,dp=2\pi\delta(x),\end{aligned} \tag{7.29}$$

where the second integral in the first equality was modified by a change of variables $z\to-\frac{\omega^2}{z}$. In a similar manner we find

$$\int_{-\infty}^{\infty} e(x,z)\frac{dz}{z}=0,\qquad \int_{-\infty}^{\infty} e(x,z)\frac{dz}{z^2}=\frac{2\pi}{\omega^2}\delta(x). \tag{7.30}$$

These formulae lead to

$$\mathbf{R}=8\pi e^{-\frac{i\theta}{2}}\left(\Gamma_+(x,y)\begin{pmatrix}1\\0\end{pmatrix}+\begin{pmatrix}\xi(x+y)\\ \tilde{\eta}(x+y)\end{pmatrix}+\int_x^{\infty}\Gamma_+(x,s)\begin{pmatrix}\tilde{\xi}(s+y)\\ \tilde{\eta}(s+y)\end{pmatrix}ds\right), \tag{7.31}$$

where

$$\tilde{\xi}(x) = \frac{\omega}{8\pi i} \int_{-\infty}^{\infty} r_\varrho(z) e\left(\frac{x}{4}, z\right) \frac{dz}{z}, \tag{7.32}$$

$$\tilde{\eta}(x) = \frac{e^{i\theta}}{8\pi} \int_{-\infty}^{\infty} r_\varrho(z) e\left(\frac{x}{4}, z\right) dz. \tag{7.33}$$

Now by using the equation $\mathbf{L} = \mathbf{R}$ and the representation (7.24) we obtain

$$\Gamma_+(x, y) \begin{pmatrix} 1 \\ 0 \end{pmatrix} + \begin{pmatrix} \xi(x+y) \\ \eta(x+y) \end{pmatrix} + \int_x^{\infty} \Gamma_+(x, s) \begin{pmatrix} \xi(s+y) \\ \eta(s+y) \end{pmatrix} ds = 0 \tag{7.34}$$

for $y \geqslant x$, with

$$\begin{pmatrix} \xi(x) \\ \eta(x) \end{pmatrix} = \begin{pmatrix} \tilde{\xi}(x) \\ \tilde{\eta}(x) \end{pmatrix} + \frac{1}{4i} \sum_{j=1}^{n} c_j \begin{pmatrix} \dfrac{\omega}{iz_j} \\ e^{i\theta} \end{pmatrix} e\left(\frac{x}{4}, z_j\right). \tag{7.35}$$

Recalling the involutions (I.8.26)-(I.8.27)

$$\overline{\Gamma_\pm}(x, y) = \sigma_1 \Gamma_\pm(x, y) \sigma_1, \tag{7.36}$$

we can write (7.34) in matrix form

$$\Gamma_+(x, y) + \Omega(x+y) + \int_x^{\infty} \Gamma_+(x, s) \Omega(s+y) ds = 0 \tag{7.37}$$

for $y \geqslant x$, with

$$\Omega(x) = \begin{pmatrix} \xi(x) & \bar{\eta}(x) \\ \eta(x) & \xi(x) \end{pmatrix}. \tag{7.38}$$

Here we have used that, by (7.15) and the positivity condition, $\xi(x)$ is a real-valued function.

Equation (7.37) is an integral equation for the matrix $\Gamma_+(x, y)$; it is called the *Gelfand-Levitan-Marchenko equation from the right*. Notice that it has the same structure as the Gelfand-Levitan-Marchenko equation in the rapidly decreasing case in § 4. However, $\Omega(x)$ is no more an off-diagonal matrix but has a diagonal part proportional to the unit matrix.

The matrix $U_0(x)$ involved in the auxiliary linear problem,

$$\frac{dT_\pm(x,\lambda)}{dx} = \left(\frac{\lambda\sigma_3}{2i} + U_0(x)\right) T_\pm(x,\lambda), \tag{7.39}$$

is expressed in terms of $\Gamma_\pm(x,y)$ according to (I.8.17)

$$U_0(x) = U_+ + \sigma_3 \Gamma_+(x,x)\sigma_3 - \Gamma_+(x,x), \tag{7.40}$$

with

$$U_+ = Q^{-1}(\theta)\, U_- Q(\theta), \qquad U_- = \frac{\omega}{2}\sigma_1 . \tag{7.41}$$

In a similar manner, (7.1) yields

$$\frac{1}{a_\varrho(z)} T_+^{(2)}(x,z) = \tilde{r}_\varrho(z)\, T_-^{(1)}(x,z) + T_-^{(2)}(x,z), \tag{7.42}$$

with

$$\tilde{r}_\varrho(z) = -\frac{\overline{b_\varrho}(z)}{a_\varrho(z)} . \tag{7.43}$$

Interpreting this equation as an appropriate conjugation problem gives the *Gelfand-Levitan-Marchenko equation from the left*

$$\Gamma_-(x,y) + \tilde{\Omega}(x+y) + \int_{-\infty}^{x} \Gamma_-(x,s)\tilde{\Omega}(s+y)\,ds = 0, \qquad y \leqslant x. \tag{7.44}$$

The kernel $\Omega(x)$ has the form

$$\tilde{\Omega}(x) = \begin{pmatrix} \tilde{\xi}(x) & \tilde{\eta}(x) \\ \bar{\tilde{\eta}}(x) & \bar{\tilde{\xi}}(x) \end{pmatrix}, \tag{7.45}$$

where

$$\tilde{\xi}(x) = \frac{i\omega}{8\pi} \int_{-\infty}^{\infty} \tilde{r}_\varrho(z)\, e\left(-\frac{x}{4}, z\right) \frac{dz}{z} + \frac{\omega}{4} \sum_{j=1}^{n} \frac{\tilde{c}_j}{z_j} e\left(-\frac{x}{4}, z_j\right) \tag{7.46}$$

and

$$\tilde{\eta}(x) = \frac{1}{8\pi} \int_{-\infty}^{\infty} \tilde{r}_\varrho(z)\, e\left(-\frac{x}{4}, z\right) dz + \frac{1}{4i} \sum_{j=1}^{n} \tilde{c}_j\, e\left(-\frac{x}{4}, z_j\right), \tag{7.47}$$

with

$$\tilde{c}_j = \frac{1}{\gamma_j \dot{a}_\varrho(z_j)}, \quad j=1,\ldots,n. \tag{7.48}$$

Relation (7.45) takes account of the fact that, by (7.15) and the positivity condition, $\tilde{\xi}(x)$ is real-valued.

The matrix $U_0(x)$ can be expressed in terms of $\Gamma_-(x,y)$ according to (I.8.17):

$$U_0(x) = U_- + \Gamma_-(x,x) - \sigma_3 \Gamma_-(x,x)\sigma_3. \tag{7.49}$$

The integral equations (7.37) and (7.44) provide a basis for solving the inverse problem in the case of finite density by the Gelfand-Levitan-Marchenko method. Suppose we are given functions $r_\varrho(z)$, $\tilde{r}_\varrho(z)$ and a set of numbers z_j, c_j, $\tilde{c}_j$, and θ, $0 \leqslant \theta < 2\pi$, subject to the following conditions.

1. *The set* $\{r_\varrho(z), z_j, c_j; j=1,\ldots,n\}$ *satisfies conditions 1)–5).*
2. *The functions* $r_\varrho(z)$ *and* $\tilde{r}_\varrho(z)$ *are related by*

$$\frac{\tilde{r}_\varrho(z)}{\bar{\tilde{r}}_\varrho(z)} = -\frac{\bar{a}_\varrho(z)}{a_\varrho(z)}, \tag{7.50}$$

with

$$a_\varrho(z) = e^{\frac{i\theta}{2}} \prod_{j=1}^{n} \frac{z-z_j}{z-\bar{z}_j} \exp\left\{\frac{1}{2\pi i}\int_{-\infty}^{\infty} \frac{\log(1-|r_\varrho(s)|^2)}{z-s+i0}\, ds\right\}. \tag{7.51}$$

3. *The coefficients* c_j *and* $\tilde{c}_j$ *satisfy*

$$c_j \tilde{c}_j = \frac{1}{\dot{a}_\varrho^2(z_j)}, \quad j=1,\ldots,n. \tag{7.52}$$

Then we claim the following.

I. *The Gelfand-Levitan-Marchenko equations (7.37) and (7.44) have unique solutions in* $L_1^{(2\times 2)}(x,\infty)$ *and* $L_1^{(2\times 2)}(-\infty,x)$. *Their matrix solutions* $\Gamma_\pm(x,y)$ *are of Schwartz type as* $x,y \to \pm\infty$, *respectively.*

II. *The matrices* $T_\pm(x,z)$ *derived from* $\Gamma_\pm(x,y)$ *according to (7.24)–(7.25) satisfy the differential equations*

$$\frac{dT_\pm(x,z)}{dx} = \left(\frac{\lambda(z)\sigma_3}{2i} + U_0^{(\pm)}(x)\right) T_\pm(x,z), \tag{7.53}$$

where $U_0^{(+)}(x)$ *and* $U_0^{(-)}(x)$ *are given by the right hand sides of (7.40) and (7.49), respectively.*

III. *The matrices* $U_0^{(\pm)}(x)$ *have the form*

$$U_0^{(\pm)}(x) = \sqrt{\varkappa}\begin{pmatrix} 0 & \bar{\psi}_\pm(x) \\ \psi_\pm(x) & 0 \end{pmatrix}, \tag{7.54}$$

where $\psi_\pm(x)$ behave asymptotically as

$$\lim_{x\to-\infty} \psi_-(x) = \varrho, \quad \lim_{x\to+\infty} \psi_+(x) = e^{i\theta}\varrho, \quad \varrho = \frac{\omega}{2\sqrt{\varkappa}}, \tag{7.55}$$

the asymptotic values being taken in the Schwartz sense.

IV. *The relation*

$$U_0^{(+)}(x) = U_0^{(-)}(x) = U_0(x) \tag{7.56}$$

holds, so that $U_0(x)$ satisfies the finite density boundary conditions.

V. *The transition coefficients for the continuous spectrum of the auxiliary linear problem with potential matrix $U_0(x)$ are $a_\varrho(\lambda) = a_\varrho(z(\lambda))$ and $b_\varrho(\lambda) = a_\varrho(\lambda) r_\varrho(z(\lambda))$; the discrete spectrum consists of the eigenvalues λ_j, $-\omega < \lambda_j < \omega$ with transition coefficients $\gamma_j = c_j \dot{a}_\varrho(z_j)$, $j = 1, \ldots, n$.*

Now we shall give the proofs.

I. *The unique solvability of the Gelfand-Levitan-Marchenko equations.*

Consider, for definiteness, equation (7.44) and write it in operator form,

$$(\mathbf{I} + \mathbf{\Omega}_x)\Gamma_x = -\Omega_x, \tag{7.57}$$

where $\Gamma_x(y) = \Gamma_-(x, y)$ and $\Omega_x(y) = \tilde{\Omega}(x+y)$ belong to $L_1^{(2\times 2)}(-\infty, x)$, and $\mathbf{\Omega}_x$ is an integral operator with the kernel $\tilde{\Omega}(s+y)$,

$$(\mathbf{\Omega}_x f)(y) = \int_{-\infty}^{x} f(s)\tilde{\Omega}(s+y)\,ds. \tag{7.58}$$

The variable x plays the role of a parameter. In order not to overload our notation we have omitted the symbols $-$ and $\sim$ in the entries of (7.57).

The kernel $\tilde{\Omega}(s)$ is a Schwartz function for $s \to -\infty$ so that $\mathbf{\Omega}_x$ is a compact operator on $L_1^{(2\times 2)}(-\infty, x)$ whose norm vanishes as $x \to -\infty$ (cf. § 4). Therefore, for (7.57) to have a unique solution it is sufficient to show that the homogeneous equation

$$f + \mathbf{\Omega}_x f = 0 \tag{7.59}$$

has only a trivial solution in $L_1^{(2\times 2)}(-\infty, x)$.

To show this let us first consider this equation in the Hilbert space $L_2^{(2\times 2)}(-\infty, x)$ of square integrable 2×2 matrix functions with the inner product

$$\langle f, g\rangle = \int_{-\infty}^{x} \operatorname{tr} f(s) g^*(s)\, ds, \tag{7.60}$$

the asterisk indicating Hermitian conjugation. The operator $\mathbf{\Omega}_x$ in this space is defined by the same expression (7.58) and is compact. By examining the kernel $\tilde{\Omega}(s)$ it is easily seen that a solution of (7.59) which belongs to $L_1^{(2\times 2)}(-\infty, x)$ also belongs to $L_2^{(2\times 2)}(-\infty, x)$. So it is enough to show that (7.59) has no nontrivial solutions in $L_2^{(2\times 2)}(-\infty, x)$. Actually, we shall prove a stronger result that the operator $\mathbf{I}+\mathbf{\Omega}_x$ is positive definite on $L_2^{(2\times 2)}(-\infty, x)$.

First suppose there is no discrete spectrum. Then $\mathbf{\Omega}_x$ can be regarded as a restriction of an operator $\mathbf{\Omega}$ on $L_2^{(2\times 2)}(-\infty, \infty)$ defined by

$$(\mathbf{\Omega} f)(s) = \int_{-\infty}^{\infty} f(s')\tilde{\Omega}(s+s')\, ds'. \tag{7.61}$$

More precisely, $L_2^{(2\times 2)}(-\infty, x)$ is embedded into $L_2^{(2\times 2)}(-\infty, \infty)$ in such a way that its elements, $f(s)$, are extended to be zero for $s \geqslant x$. We will show that $\mathbf{I}+\mathbf{\Omega}$ is positive definite, hence so is $\mathbf{I}+\mathbf{\Omega}_x$.

The representations (7.45)-(7.47) for the kernel $\tilde{\Omega}(x)$ may be written as

$$\tilde{\Omega}(x) = \frac{1}{8\pi} \int_{-\infty}^{\infty} E_\varrho(x, z) R(z)\, dz, \tag{7.62}$$

with

$$R(z) = \begin{pmatrix} 0 & \tilde{r}_\varrho(z) \\ \bar{\tilde{r}}_\varrho(z) & 0 \end{pmatrix}. \tag{7.63}$$

The matrix $E_\varrho(x, z)$ in the integrand satisfies the relations

$$\frac{1}{8\pi} \int_{\mathbb{R}_\omega} E_\varrho(x, z) E_\varrho^*(y, z)\, dz = \delta(x-y) I, \tag{7.64}$$

$$\frac{1}{8\pi} \int_{-\infty}^{\infty} E_\varrho^*(x, z) E_\varrho(x, z')\, dx = \delta(z-z') I, \tag{7.65}$$

with z and z' in $\mathbb{R}_\omega$ (i.e., $|z|, |z'| \geqslant \omega$). They have the meaning of *completeness and orthogonality relations* for eigenfunctions of the differential operator $\mathscr{L}_- = i\sigma_3 \dfrac{d}{dx} + \dfrac{\omega}{2}\sigma_2$ which governs the asymptotic behaviour of the differential operator $\mathscr{L}$ of the auxiliary linear problem, as $x \to -\infty$ (see § I.9).

The proof of (7.64) makes use of (7.29)-(7.30) and of the involution $z \to \frac{\omega^2}{z}$ which maps $\mathbb{R}_\omega$ onto the gap $-\omega \leqslant z \leqslant \omega$. Equation (7.65) follows from the usual representation for the δ-function as an integral over exponentials and the change of variable formula

$$\delta(\varphi(z)) = \sum_l \frac{1}{\left|\frac{d\varphi}{dz}(z_l)\right|} \delta(z - z_l), \tag{7.66}$$

where $\varphi(z_l) = 0$. It is essential here that z and z' are in $\mathbb{R}_\omega$. If z is in $\mathbb{R}_\omega$ and z' in the gap $(-\omega, \omega)$, then instead of (7.65) we have the relation

$$\frac{1}{8\pi} \int_{-\infty}^{\infty} E_\varrho^*(x, z) E_\varrho(x, z') dx = \frac{\omega}{z} \delta\left(z' - \frac{\omega^2}{z}\right) \sigma_2 . \tag{7.67}$$

We may interprete (7.64) as a requirement that the operator $\mathbf{E}_\varrho$ from $L_2^{(2\times 2)}(-\infty, \infty)$ to $L_2^{(2\times 2)}(\mathbb{R}_\omega)$ defined by

$$(\mathbf{E}_\varrho f)(z) = \hat{f}(z) = \frac{1}{\sqrt{8\pi}} \int_{-\infty}^{\infty} f(x) E_\varrho(x, z) dx \tag{7.68}$$

is isometric. The adjoint operator $\mathbf{E}_\varrho^*$ is given by

$$(\mathbf{E}_\varrho^* \hat{f})(x) = f(x) = \frac{1}{\sqrt{8\pi}} \int_{\mathbb{R}_\omega} \hat{f}(z) E_\varrho^*(x, z) dz \tag{7.69}$$

and, by (7.65) is also isometric, so that

$$\mathbf{E}_\varrho^* \mathbf{E}_\varrho = \mathbf{I}, \qquad \mathbf{E}_\varrho \mathbf{E}_\varrho^* = \mathbf{I}. \tag{7.70}$$

Now we will show that the operator

$$\hat{\mathbf{\Omega}} = \mathbf{E}_\varrho \mathbf{\Omega} \mathbf{E}_\varrho^* \tag{7.71}$$

conjugate to $\mathbf{\Omega}$ and acting on $L_2^{(2\times 2)}(\mathbb{R}_\omega)$ is a multiplication operator by the matrix-function

$$(\hat{\mathbf{\Omega}} \hat{f})(z) = \hat{f}(z) R(z). \tag{7.72}$$

For that purpose consider the kernel $\hat{\Omega}(z, z')$ of $\hat{\mathbf{\Omega}}$ as a generalized function

$$\hat{\Omega}(z, z') = \frac{1}{(8\pi)^2} \int_{-\infty}^{\infty} \int_{-\infty}^{\infty} \int_{-\infty}^{\infty} E_\varrho^*(y, z') E_\varrho(x+y, z'') \times R(z'') E_\varrho(x, z)\, dx\, dy\, dz'', \tag{7.73}$$

where the representation (7.62) for $\tilde{\Omega}(x)$ is used. The variables x and y in $E_\varrho(x+y, z'')$ separate,

$$E_\varrho(x+y, z'') = E_\varrho(y, z'') E\left(x, \frac{1}{2}\left(z'' - \frac{\omega^2}{z''}\right)\right), \tag{7.74}$$

and the integral over y is evaluated by using (7.65) and (7.67). This gives a δ-function which reduces out integration over z'', so that

$$\hat{\Omega}(z, z') = \frac{1}{8\pi} \int_{-\infty}^{\infty} \left[E\left(x, \frac{1}{2}\left(z' - \frac{\omega^2}{z'}\right)\right) R(z') + \frac{\omega}{z'} \sigma_2 E\left(x, -\frac{1}{2}\left(z' - \frac{\omega^2}{z'}\right)\right) R\left(\frac{\omega^2}{z'}\right) \right] E_\varrho(x, z)\, dx. \tag{7.75}$$

Next we employ the involution (see (7.15))

$$R\left(\frac{\omega^2}{z}\right) = \sigma_2 R(z) \sigma_2 \tag{7.76}$$

to carry the diagonal matrix $E(x, \cdot)$ to the right across the off-diagonal matrices $R(\cdot)$ and σ_2. As a result we find

$$\begin{aligned} \hat{\Omega}(z, z') &= \frac{1}{8\pi} \int_{-\infty}^{\infty} R(z') \left[E\left(x, -\frac{1}{2}\left(z' - \frac{\omega^2}{z'}\right)\right) + \frac{\omega}{z'} \sigma_2 E\left(x, \frac{1}{2}\left(z' - \frac{\omega^2}{z'}\right)\right) \right] E_\varrho(x, z)\, dx \\ &= \frac{1}{8\pi} \int_{-\infty}^{\infty} R(z') E_\varrho^*(x, z') E_\varrho(x, z)\, dx = \delta(z - z') R(z), \end{aligned} \tag{7.77}$$

which proves (7.72).

The positive definiteness of $\mathbf{I} + \mathbf{\Omega}$ and hence of $\mathbf{I} + \mathbf{\Omega}_x$ follows from that of the matrix

$$I + R(z) = \begin{pmatrix} 1 & \tilde{r}_\varrho(z) \\ \bar{\tilde{r}}_\varrho(z) & 1 \end{pmatrix}, \tag{7.78}$$

which is guaranteed by

$$|\tilde{r}_{\varrho}(z)|<1 \tag{7.79}$$

for $|z|>\omega$ (see property 3) and (7.50)).

Let us now consider the general case when the discrete spectrum is also present. To prove that $\mathbf{I}+\mathbf{\Omega}_x$ is positive definite we shall split it into terms corresponding to the continuous and discrete spectra, respectively:

$$\mathbf{I}+\mathbf{\Omega}_x=\mathbf{I}+\mathbf{\Omega}_x^{(c)}+\mathbf{\Omega}_x^{(d)}. \tag{7.80}$$

Here $\mathbf{\Omega}_x^{(c)}$ and $\mathbf{\Omega}_x^{(d)}$ are integral operators on $L_2^{(2\times 2)}(-\infty,x)$ with the kernels $\tilde{\Omega}^{(c)}(s+s')$ and $\tilde{\Omega}^{(d)}(s+s')$, respectively, where

$$\tilde{\Omega}^{(c)}(s)=\frac{1}{8\pi}\int_{-\infty}^{\infty} E_{\varrho}(s,z)\,R(z)\,dz, \tag{7.81}$$

$$\tilde{\Omega}^{(d)}(s)=\sum_{j=1}^{n} C_j\, e\left(-\frac{s}{4},z_j\right), \tag{7.82}$$

and the Hermitian 2×2 matrices C_j have the form

$$C_j=\frac{\tilde{m}_j}{4}\begin{pmatrix}\omega & -iz_j\\ i\bar{z}_j & \omega\end{pmatrix} \tag{7.83}$$

with

$$\tilde{m}_j=\frac{\tilde{c}_j}{z_j},\quad j=1,\ldots,n. \tag{7.84}$$

The decomposition (7.80) follows from (7.45)-(7.47) and

$$e(s,z_j)=\overline{e(s,z_j)}, \tag{7.85}$$

which is obvious since $|z_j|=\omega$.

We have already proved that $\mathbf{I}+\mathbf{\Omega}_x^{(c)}$ is positive definite on $L_2^{(2\times 2)}(-\infty,x)$. It is therefore sufficient to show that $\mathbf{\Omega}_x^{(d)}$ is non-negative. This in turn follows from the fact that the matrices C_j are non-negative.

In fact, in this case, for any $f(s)$ in $L_2^{(2\times 2)}(-\infty,x)$ we have

$$\langle\mathbf{\Omega}_x^{(d)}f,f\rangle=\sum_{j=1}^{n}\int_{-\infty}^{x}\int_{-\infty}^{x}\operatorname{tr} f(s)\,C_j f^*(s')\,e\left(-\frac{s}{4},z_j\right)\overline{e\left(-\frac{s'}{4},z_j\right)}\,ds\,ds'\geqslant 0. \tag{7.86}$$

The simplest way to verify this estimate is to reduce each of the C_j to diagonal form. Clearly, each term in (7.86) will then be non-negative.

Let us now show that all the C_j are non-negative. Since $|z_j|=\omega$, they are degenerate, hence it is sufficient to show that

$$\tilde{m}_j>0, \quad j=1,\ldots,n. \tag{7.87}$$

These inequalities result from

$$m_j\tilde{m}_j=-\frac{1}{z_j^2\dot{a}_\varrho^2(z_j)} \tag{7.88}$$

and the estimate

$$\frac{1}{z_j^2\dot{a}_\varrho^2(z_j)}<0, \quad j=1,\ldots,n, \tag{7.89}$$

which is a consequence of the positivity condition, the condition (θ), and formulae (7.15), (7.51).

This completes the proof of the fact that $\mathbf{I}+\mathbf{\Omega}_x$ is positive definite, and hence (7.44) has a unique solution.

To conclude, we point out that by virtue of the uniqueness theorem and the involution property of $\tilde{\Omega}(x)$,

$$\bar{\tilde{\Omega}}(x)=\sigma_1\tilde{\Omega}(x)\sigma_1, \tag{7.90}$$

the solution $\Gamma(x,y)$ has the same property,

$$\bar{\Gamma}(x,y)=\sigma_1\Gamma(x,y)\sigma_1. \tag{7.91}$$

Equation (7.37) is examined is a similar manner.

II. *Derivation of the differential equations for* $T_\pm(x,z)$.

It is enough to show that $\Gamma_\pm(x,z)$ satisfy partial differential equations

$$\frac{\partial}{\partial x}\Gamma_\pm(x,y)+\sigma_3\frac{\partial}{\partial y}\Gamma_\pm(x,y)\sigma_3-U_0^{(\pm)}(x)\Gamma_\pm(x,y)+\sigma_3\Gamma_\pm(x,y)\sigma_3U_\pm=0, \tag{7.92}$$

where

$$U_0^{(\pm)}(x)=U_\pm\mp(\Gamma_\pm(x,x)-\sigma_3\Gamma_\pm(x,x)\sigma_3) \tag{7.93}$$

(see (I.8.15)–(I.8.17)).

Indeed, these equations were derived in § I.8 when analyzing the auxiliary linear problem. It was indicated there that they are equivalent to the differential equations (7.53) for the matrices $T_\pm(x,z)$ obtained from $\Gamma_\pm(x,y)$ according to (7.24)–(7.26).

For definiteness, consider $\Gamma_-(x, y)$. Differentiate (7.44) with respect to x and y and add up the results after multiplying the second one by σ_3 from both sides. Then

$$\frac{\partial}{\partial x}\Gamma(x,y)+\sigma_3\frac{\partial}{\partial y}\Gamma(x,y)\sigma_3+\Omega'(x+y)+\sigma_3\Omega'(x+y)\sigma_3+\Gamma(x,x)\Omega(x+y)$$

$$+\int_{-\infty}^{x}\left(\frac{\partial}{\partial x}\Gamma(x,s)\Omega(s+y)+\sigma_3\Gamma(x,s)\Omega'(s+y)\sigma_3\right)ds=0, \qquad (7.94)$$

where the prime indicates the derivative with respect to the argument and for notational simplicity we have omitted the symbols $-$ and $\sim$ in the entries.

The equation involves the matrix $\sigma_3\Omega'(x)\sigma_3+\Omega'(x)$ proportional to the unit matrix. From (7.46) by using the involution (7.15) and the positivity condition we find

$$\frac{d\xi(x)}{dx}=\frac{\omega}{4}(\eta(x)+\bar{\eta}(x)), \qquad (7.95)$$

so that (7.45) yields

$$\Omega'(x)+\sigma_3\Omega'(x)\sigma_3=\frac{\omega}{2}(\sigma_1\Omega(x)-\sigma_3\Omega(x)\sigma_3\sigma_1)$$

$$=U_-\Omega(x)-\sigma_3\Omega(x)\sigma_3 U_-, \qquad (7.96)$$

where account was taken of $U_-=\frac{\omega}{2}\sigma_1$. By using this equation, the last term in the integrand in (7.94) can be transformed into

$$\int_{-\infty}^{x}\sigma_3\Gamma(x,s)\Omega'(s+y)\sigma_3\,ds=-\int_{-\infty}^{x}\sigma_3\Gamma(x,s)\sigma_3\Omega'(s+y)\,ds$$

$$+\int_{-\infty}^{x}\sigma_3\Gamma(x,s)\sigma_3(\Omega'(s+y)+\sigma_3\Omega'(s+y)\sigma_3)\,ds$$

$$=-\sigma_3\Gamma(x,x)\sigma_3\Omega(x+y)+\int_{-\infty}^{x}\left[\sigma_3\frac{\partial\Gamma}{\partial s}(x,s)\sigma_3\Omega(s+y)\right.$$

$$\left.+\sigma_3\Gamma(x,s)\sigma_3(U_-\Omega(s+y)-\sigma_3\Omega(s+y)\sigma_3 U_-)\right]ds, \qquad (7.97)$$

where we have integrated by parts. Using (7.93) and (7.96)–(7.97) we can write (7.94) as

$$\frac{\partial}{\partial x}\Gamma(x,y)+\sigma_3\frac{\partial}{\partial y}\Gamma(x,y)\sigma_3+U_0^{(-)}(x)\Omega(x+y)-\sigma_3\Omega(x+y)\sigma_3 U_-$$
$$+\int\limits_{-\infty}^{x}\left[\left(\frac{\partial}{\partial x}\Gamma(x,s)+\sigma_3\frac{\partial}{\partial s}\Gamma(x,s)\sigma_3\right)\Omega(s+y)\right.$$
$$\left.+\sigma_3\Gamma(x,s)\sigma_3(U_-\Omega(s+y)-\sigma_3\Omega(s+y)\sigma_3 U_-)\right]ds=0. \tag{7.98}$$

Let the terms $U_0^{(-)}(x)\Omega(x+y)$ and $\sigma_3\Omega(x+y)\sigma_3 U_-$ on the left hand side of this equation be modified by replacing $\Omega(x+y)$ by the right hand side of the Gelfand-Levitan-Marchenko equation written in the form

$$\Omega(x+y)=-\Gamma(x,y)-\int\limits_{-\infty}^{x}\Gamma(x,s)\Omega(s+y)\,ds. \tag{7.99}$$

With the definition

$$\Phi(x,y)=\frac{\partial}{\partial x}\Gamma(x,y)+\sigma_3\frac{\partial}{\partial y}\Gamma(x,y)\sigma_3-U_0^{(-)}(x)\Gamma(x,y)+\sigma_3\Gamma(x,y)\sigma_3 U_-, \tag{7.100}$$

we rewrite (7.98) as

$$\Phi(x,y)+\int\limits_{-\infty}^{x}\Phi(x,s)\Omega(s+y)\,ds=0, \tag{7.101}$$

which means that $\Phi(x,y)$ as a function of y satisfies the homogeneous equation (7.59). By virtue of the uniqueness theorem,

$$\Phi(x,y)=0 \tag{7.102}$$

for all $x, y, y\leqslant x$; this shows the validity of (7.92).

The equation for $\Gamma_+(x,y)$ can be proved in a similar way.

III. *The behaviour of* $U_0^{(\pm)}(x)$ *as* $x\to\pm\infty$.

By (7.40) and (7.49), the matrices $U_0^{(\pm)}(x)$ are off-diagonal, and the involution (7.91) assures that they have the special form (7.54).

The study of the asymptotic behaviour of the non-zero entries of $U_0^{(\pm)}(x)$ is based on the following argument. The norm of the operators $\mathbf{\Omega}_x$ and $\tilde{\mathbf{\Omega}}_x$ vanishes as $x\to+\infty$ or $x\to-\infty$, respectively, so that the integral equations (7.37) and (7.44) may be solved by successive approximations. Each iteration gives a function of Schwartz type for $x\to\pm\infty$, and this property is also shared by the solutions $\Gamma_\pm(x,y)$. In particular, $\Gamma_\pm(x,x)$ are of Schwartz type for $x\to\pm\infty$. This implies the required behaviour of $\psi_\pm(x)$ as $x\to\pm\infty$.

We also note that these properties of $\Gamma_{\pm}(x, y)$ yield the following asymptotic behaviour of $T_{\pm}(x, z)$ for real z:

$$T_{-}(x, z) = E_{\varrho}(x, z) + o(1) \tag{7.103}$$

as $x \to -\infty$ and

$$T_{+}(x, z) = Q^{-1}(\theta) E_{\varrho}(x, z) + o(1) \tag{7.104}$$

as $x \to +\infty$.

IV. *The consistency relation* $U_0^{(+)}(x) = U_0^{(-)}(x)$.

For the proof it is enough to show that the matrices $T_{+}(x, z)$ and $T_{-}(x, z)$ are linearly dependent, i.e., differ by a right matrix factor independent of x. Indeed in this case both $T_{+}(x, z)$, $T_{-}(x, z)$ satisfy the same differential equation, and hence $U_0^{(+)}(x)$ and $U_0^{(-)}(x)$ coincide as coefficients in (7.53).

We will show that (7.1) holds for real z with $T_{\varrho}(z)$ of the form

$$T_{\varrho}(z) = \begin{pmatrix} a_{\varrho}(z) & \overline{b_{\varrho}}(z) \\ b_{\varrho}(z) & \overline{a_{\varrho}}(z) \end{pmatrix}, \tag{7.105}$$

where $a_{\varrho}(z)$ is given by (7.51) and

$$b_{\varrho}(z) = a_{\varrho}(z) r_{\varrho}(z). \tag{7.106}$$

For the proof observe that the unique solvability of the integral equations (7.37) and (7.44) established earlier is equivalent to the existence and uniqueness theorem for two special conjugation problems

$$F_1(x, z) = T_{+}^{(1)}(x, z) + r_{\varrho}(z) T_{+}^{(2)}(x, z) \tag{7.107}$$

and

$$F_2(x, z) = \tilde{r}_{\varrho}(z) T_{-}^{(1)}(x, z) + T_{-}^{(2)}(x, z), \tag{7.108}$$

whose exact formulation was given above. The data $\{r_{\varrho}(z), z_j, c_j\}$ and $\{\tilde{r}_{\varrho}(z), z_j, \tilde{c}_j\}$ of the two problems are interrelated by conditions 1–3. Starting from these relations we will show that

$$F_1(x, z) = \frac{1}{a_{\varrho}(z)} T_{-}^{(1)}(x, z), \quad F_2(x, z) = \frac{1}{a_{\varrho}(z)} T_{+}^{(2)}(x, z), \tag{7.109}$$

which is equivalent to the required formula (7.1).

For the proof multiply (7.107) by $\overline{r_{\varrho}}(z)$

$$\overline{r_{\varrho}}(z) F_1(x, z) = \overline{r_{\varrho}}(z) T_{+}^{(1)}(x, z) + |r_{\varrho}(z)|^2 T_{+}^{(2)}(x, z), \tag{7.110}$$

and, using the involution (7.14), write it down as

$$\sigma_1 \overline{F_1}(x,z) = \overline{r_\varrho}(z)\, T_+^{(1)}(x,z) + T_+^{(2)}(x,z). \tag{7.111}$$

By substracting (7.110) and (7.111) and using (7.18) we obtain

$$\overline{r_\varrho}(z)\, F_1(x,z) - \sigma_1 \overline{F_1}(x,z) = -\frac{1}{|a_\varrho(z)|^2}\, T_+^{(2)}(x,z). \tag{7.112}$$

Now we exploit condition 2 given by (7.50) to transform the last equation into

$$\frac{1}{a_\varrho(z)}\, T_+^{(2)}(x,z) = \tilde{r}_\varrho(z)\, a_\varrho(z)\, F_1(x,z) + \sigma_1 \overline{a_\varrho}(z)\, \overline{F_1}(x,z). \tag{7.113}$$

Thus we have converted the conjugation problem (7.107) into a conjugation problem of the type (7.108). Next we are going to apply the uniqueness theorem. For that it is enough to show that the vector functions $a_\varrho(z)\, F_1(x,z)$ and $\frac{1}{a_\varrho(z)}\, T_2^{(+)}(x,z)$ fit into the setting of the conjugation problem (7.108).

First consider the column $a_\varrho(z)\, F_1(x,z)$. It has an analytic continuation into the upper half z-plane with the same asymptotic behaviour as for the column $T_-^{(1)}(x,z)$, as $z \to 0$ or $|z| \to \infty$. In a generic situation we have

$$a_\varrho(z) = \frac{a_\pm}{z \mp \omega} + O(1), \qquad a_\pm \neq 0 \tag{7.114}$$

and $a_\varrho(z)\, F_1(x,z)$ is regular at $z = \pm\omega$. In fact, using $r_\varrho(\pm\omega) = \mp i$ (see condition 3)) and the relation

$$T_+^{(1)}(x, \pm\omega) = \pm i\, T_+^{(2)}(x, \pm\omega) \tag{7.115}$$

we find

$$F_1(x,z) = O(|z \mp \omega|) \tag{7.116}$$

in the neighbourhood of $z = \pm\omega$. In turn, (7.115) is a consequence of a similar property of the columns of $Q^{-1}(\theta)\, E_\varrho(x, \pm\omega)$ by virtue of the differential equation (7.53) and the asymptotic formula (7.104).

Now, the column $\frac{1}{a_\varrho(z)}\, T_+^{(2)}(x,z)$, as well as $F_2(x,z)$, has an analytic continuation into the upper half-plane with the exception of $z = z_j$, $j = 1, \ldots, n$, where it has simple poles. By condition 3 (see (7.52) and (7.8)) we have

$$
\begin{aligned}
\operatorname{res} \frac{1}{a_\varrho(z)} T_+^{(2)}(x, z)\big|_{z=z_j} &= \frac{1}{\dot{a}_\varrho(z_j)} T_+^{(2)}(x, z_j) \\
&= \frac{1}{c_j \dot{a}_\varrho(z_j)} \operatorname{res} F_1(x, z)\big|_{z=z_j} \\
&= \tilde{c}_j (a_\varrho(z) F_1(x, z))\big|_{z=z_j}, \qquad j=1, \dots, n. \qquad (7.117)
\end{aligned}
$$

So the columns $\frac{1}{a_\varrho(z)} T_+^{(2)}(x, z)$ and $a_\varrho(z) F_1(x, z)$ are subject to the same conditions as imposed on the columns $F_2(x, z)$ and $T_-^{(1)}(x, z)$ in the conjugation problem (7.108). Therefore these columns coincide so that equations (7.109) hold.

V. *Transition coefficients and discrete spectrum.*

The results of Subsections I–IV imply that $T_\pm(x, z)$ are the Jost solutions for the auxiliary linear problem with potential matrix $U_0(x) = U_0^{(+)}(x) = U_0^{(-)}(x)$. The functions $a_\varrho(\lambda)$ and $b_\varrho(\lambda)$ play the role of transition coefficients for the continuous spectrum; the quantities λ_j are the discrete spectrum eigenvalues with transition coefficients $\gamma_j, j=1, \dots, n$.

This completes our general analysis of the inverse problem in the finite density case via the Gelfand-Levitan-Marchenko method. The results obtained may serve as a proof of the assertions I–V of the Riemann problem method in the previous section. Namely, the data $\{b_\varrho(\lambda), \bar{b}_\varrho(\lambda); \lambda_j, \gamma_j\}$ of the Riemann problem satisfying conditions 1)–5) of § 6 give rise to the data $\{r_\varrho(\lambda), \tilde{r}_\varrho(\lambda); z_j, c_j, \tilde{c}_j\}$ of the Gelfand-Levitan-Marchenko method satisfying conditions 1–3 of this section. Therefore, the results obtained for the two special conjugation problems yield the assertions I–V of § 6.

The next section will deal with an important special case of the inverse problem where the coefficient $b_\varrho(z)$ (and hence $r_\varrho(z)$, $\tilde{r}_\varrho(z)$) vanishes identically. The corresponding Gelfand-Levitan-Marchenko equations reduce to a system of linear algebraic equations and can be solved in closed form. This will give us soliton solutions for the NS model under the finite density boundary conditions.

§ 8. Soliton Solutions in the Case of Finite Density

As in the rapidly decreasing case, soliton solutions are associated with a *reflectionless linear problem*, that is, with such $\psi(x)$, $\bar{\psi}(x)$ that the corresponding $b_\varrho(\lambda)$ vanishes identically.

In that case the restrictions on the initial data simplify considerably. Namely, the set $\{\lambda_j, c_j, \tilde{c}_j; j=1, \dots, n\}$ must satisfy the following conditions.

1. *The numbers* λ_j *lie in the gap* $-\omega<\lambda_j<\omega$ *and are pairwise distinct.*

2. *The quantities* $m_j=\frac{c_j}{z_j}$ *with*

$$z_j=\lambda_j+i\sqrt{\omega^2-\lambda_j^2}, \quad |z_j|=\omega, \tag{8.1}$$

are real and satisfy

$$m_j<0, \quad j=1,\ldots,n. \tag{8.2}$$

3. *The condition* (θ) *holds,*

$$e^{i\theta}=\prod_{j=1}^{n}\frac{\bar{z}_j}{z_j}. \tag{8.3}$$

4. *The relation*

$$c_j\tilde{c}_j=\frac{1}{\dot{a}_\varrho^2(z_j)}; \quad j=1,\ldots,n, \tag{8.4}$$

holds, where

$$a_\varrho(z)=e^{\frac{i\theta}{2}}\prod_{j=1}^{n}\frac{z-z_j}{z-\bar{z}_j}, \tag{8.5}$$

and a dot indicates differentiation with respect to z.

In terms of $\tilde{m}_j=\frac{\tilde{c}_j}{z_j}$, (8.2) becomes $\tilde{m}_j>0, j=1,\ldots,n$.

Notice that in the reflectionless situation the function $a_\varrho(z)$ is regular at $z=\pm\omega$, so that these points are virtual levels.

The set $\{\lambda_j, c_j, \tilde{c}_j\}$ is related to the original data $\{\lambda_j, \gamma_j\}$ of the Riemann problem by

$$\gamma_j=c_j\dot{a}_\varrho(z_j), \quad j=1,\ldots,n. \tag{8.6}$$

By virtue of (8.2)-(8.3) and (8.5) the quantities γ_j are pure imaginary and

$$\operatorname{sign} i\gamma_j=\operatorname{sign}\frac{z_j\dot{a}_\varrho(z_j)}{i}=\varepsilon_j, \quad j=1,\ldots,n. \tag{8.7}$$

Now we turn to the inverse problem; to begin with, let $n=1$. By (8.3), the eigenvalue λ_1 is expressed explicitly through θ as

$$\lambda_1=-\omega\cos\frac{\theta}{2}. \tag{8.8}$$

Indeed, the condition (θ) gives

$$z_1 = -\omega e^{-\frac{i\theta}{2}}, \quad 0 \leqslant \theta < 2\pi, \tag{8.9}$$

so that (8.8) follows from (8.1).

The Gelfand-Levitan-Marchenko equation from the left is

$$\Gamma_-(x,y) + \tilde{\Omega}(x+y) + \int_{-\infty}^{x} \Gamma_-(x,s)\tilde{\Omega}(s+y)\,ds = 0 \tag{8.10}$$

for $y \leqslant x$ with $\tilde{\Omega}(x)$ given by (7.45)–(7.47). In our case it has the form

$$\tilde{\Omega}(x) = M_1 N_1^\tau e^{\frac{\nu_1 x}{2}}. \tag{8.11}$$

Here

$$\nu_1 = \frac{1}{i} k_1 = \sqrt{\omega^2 - \lambda_1^2} > 0, \tag{8.12}$$

and the columns M_1 and N_1 have the form

$$M_1 = \frac{\sqrt{\bar{m}_1}}{2}\begin{pmatrix}\omega \\ i\bar{z}_1\end{pmatrix}, \quad N_1 = \frac{\sqrt{\bar{m}_1}}{2}\begin{pmatrix}1 \\ \dfrac{z_1}{i\omega}\end{pmatrix}, \tag{8.13}$$

where the numeric value of the square root is taken. For notational uniformity with the subsequent formulae for $n > 1$, we write z_1 rather than θ.

So the kernel, $\tilde{\Omega}(x+y)$, of the integral equation (8.10) is one-dimensional, and the equation may be solved in closed form.

Representing $\Gamma_-(x,y)$ as

$$\Gamma_-(x,y) = f_1(x) N_1^\tau e^{\frac{\nu_1 y}{2}}, \tag{8.14}$$

we find a linear algebraic equation for the column $f_1(x)$

$$f_1(x) + M_1 e^{\frac{\nu_1 x}{2}} + A(x) f_1(x) = 0, \tag{8.15}$$

where $A(x)$ is given by

$$A(x) = N_1^\tau M_1 \int_{-\infty}^{x} e^{\nu_1 s}\,ds = \frac{\bar{m}_1 \omega}{2\nu_1} e^{\nu_1 x}. \tag{8.16}$$

This yields an expression for $f_1(x)$

$$f_1(x) = -\frac{e^{\frac{\nu_1 x}{2}}}{1+A(x)} M_1. \tag{8.17}$$

Evaluating $\psi(x)$ through the general formulae (7.49) and (7.54), we find

$$\psi(x) = \varrho \frac{i\gamma_1 + e^{i\theta} e^{\nu_1 x}}{i\gamma_1 + e^{\nu_1 x}}, \tag{8.18}$$

where we used (8.6) and the connection between z_1 and θ (8.9). Remind that $i\gamma_1 > 0$ by (8.7), so that the denominator in (8.18) does not vanish, and hence $\psi(x)$ is regular on the whole real line. The finite density boundary conditions

$$\lim_{x\to-\infty} \psi(x) = \varrho, \qquad \lim_{x\to+\infty} \psi(x) = \varrho e^{i\theta} \tag{8.19}$$

are satisfied with the exponential order $O(e^{-\nu_1|x|})$.

The solution $\psi(x, t)$ of the NS equation evolving from the initial data $\psi(x)$ is given by the general formulae (6.51) upon replacing γ_1 by $\gamma_1(t)$ in (8.18),

$$\gamma_1(t) = e^{\lambda_1 \nu_1 t} \gamma_1. \tag{8.20}$$

It can be expressed as

$$\psi(x, t) = \psi_\theta(x - vt, x_0) = \varrho \frac{1 + e^{i\theta} \exp\{\nu_1(x - vt - x_0)\}}{1 + \exp\{\nu_1(x - vt - x_0)\}}, \tag{8.21}$$

with

$$v = \lambda_1 = -\omega \cos\frac{\theta}{2}, \qquad x_0 = \frac{1}{\nu_1} \log i\gamma_1. \tag{8.22}$$

From (8.21) it is clear that $\psi(x, t)$ *describes a wave propagating with velocity v. Since*

$$|\psi(x, t)|^2 = \varrho^2 - \frac{\nu_1^2}{\omega^2 \operatorname{ch}^2\left\{\frac{\nu_1}{2}(x - vt - x_0)\right\}}, \tag{8.23}$$

it follows that the wave is localized near $x = x_0 + vt$.

By construction, the solution has finite energy; moreover, all other integrals of the motion are finite. By the definition in § 5, $\psi(x, t)$ *is a soliton in a wider* *sense for the NS model in the finite density case. Later we shall see that the scattering of these solitons is factorizable, so that $\psi(x, t)$ is a soliton in the usual sense.*

As opposed to the rapidly decreasing solitons which are parametrized by four real parameters, our soliton $\psi(x, t)$ depends on two parameters: the velocity v and the coordinate x_0 of its center of inertia at time $t=0$. The velocity of the soliton cannot be arbitrary but satisfies $|v|<\omega$. The parameter $v_1=\sqrt{\omega^2-v^2}$ characterizing the amplitude of the soliton vanishes as $|v|\to\omega$ $(\theta\to 0)$. From the point of view of physics, $\psi(x, t)$ describes a solitary wave propagating over a condensate of constant density; there is a natural bound for its velocity.

Now consider the general case of arbitrary n. As before, the kernel $\tilde{\Omega}(x+y)$ of the integral equation (8.10) is degenerate and can be written as

$$\tilde{\Omega}(x+y)=\sum_{j=1}^{n} M_j N_j^{\tau} e^{\frac{\nu_j(x+y)}{2}}, \tag{8.24}$$

with

$$\nu_j=\operatorname{Im} z_j=\sqrt{\omega^2-\lambda_j^2} \tag{8.25}$$

and

$$M_j=\frac{\sqrt{\tilde{m}_j}}{2}\begin{pmatrix}\omega\\ i\bar{z}_j\end{pmatrix},\qquad N_j=\frac{\sqrt{\tilde{m}_j}}{2}\begin{pmatrix}1\\ \frac{z_j}{i\omega}\end{pmatrix} \tag{8.26}$$

where $\sqrt{\tilde{m}_j}>0, j=1,\ldots,n$.

We ask for a solution of (8.10) of the form

$$\Gamma_-(x, y)=\sum_{j=1}^{n} f_j(x) N_j^{\tau} e^{\frac{\nu_j y}{2}}. \tag{8.27}$$

This leads to a system of linear algebraic equations for the columns $f_j(x)$,

$$f_j(x)+M_j e^{\frac{\nu_j x}{2}}+\sum_{l=1}^{n} A_{jl}(x) f_l(x)=0, \tag{8.28}$$

where the functions $A_{jl}(x)$ are given by

$$A_{jl}(x)=2 N_l^{\tau} M_j \frac{e^{\frac{1}{2}(\nu_j+\nu_l)x}}{\nu_j+\nu_l}=\omega\frac{\sqrt{\tilde{m}_j\tilde{m}_l}(z_j+z_l)}{2 z_j(\nu_j+\nu_l)} e^{\frac{1}{2}(\nu_j+\nu_l)x}. \tag{8.29}$$

The last expression simplifies by using $\nu_j = \dfrac{z_j - \bar{z}_j}{2i}$ and $|z_j| = \omega$. As a result we have

$$A_{jl}(x) = \frac{i\omega\sqrt{\tilde{m}_j \tilde{m}_l}}{z_j - \bar{z}_l}\, e^{\frac{1}{2}(\nu_j + \nu_l)x}\,. \tag{8.30}$$

We emphasize that (8.28) splits into two systems for the first and the second components of the $f_j(x)$, respectively. The second components, $p_j(x)$, satisfy the following linear algebraic system:

$$p_j(x) + \frac{i\bar{z}_j\sqrt{\tilde{m}_j}}{2}\, e^{\frac{\nu_j x}{2}} + \sum_{l=1}^{n} A_{jl}(x)\, p_l(x) = 0\,, \quad j = 1, \ldots, n\,. \tag{8.31}$$

The function $\psi(x)$ is expressed in terms of the $p_j(x)$ as

$$\psi(x) = \frac{2\varrho}{\omega} \sum_{j=1}^{n} \sqrt{\tilde{m}_j}\, p_j(x)\, e^{\frac{\nu_j x}{2}} + \varrho\,, \tag{8.32}$$

which by Cramer's rule can be written as

$$\psi(x) = \varrho\, \frac{\det(I + A_1(x))}{\det(I + A(x))} + \varrho\,. \tag{8.33}$$

Here $A(x)$ is a $n \times n$ matrix with elements $A_{jl}(x)$, and $A_1(x)$ has the form

$$A_1(x) = \left(\begin{array}{c:c} A(x) & \begin{matrix} e_1(x) \\ \vdots \\ e_n(x) \end{matrix} \\ \hdashline d_1(x) \ldots d_n(x) & -1 \end{array}\right), \tag{8.34}$$

where

$$d_j(x) = \sqrt{\tilde{m}_j}\, e^{\frac{\nu_j x}{2}}\,, \qquad e_j(x) = \frac{i\omega}{z_j}\, d_j(x)\,, \qquad j = 1, \ldots, n\,. \tag{8.35}$$

In (8.33)–(8.35) we have a final expression for reflectionless functions $\psi(x)$, $\bar{\psi}(x)$ under the finite density boundary conditions.

Remark that the smoothness of $\psi(x)$, i.e. the non-degeneracy of $I + A(x)$, as well as the validity of the finite density boundary conditions (8.19), are a consequence of the general assertions I–V established in the preceding section. However, they may be verified directly starting from (8.34)–(8.35). The limiting values (8.19) are now approached faster than in the Schwartz sense, namely, with the exponential order $O(e^{-\nu|x|})$, where $\nu = \min_{j=1,\dots,n} \{\nu_j\}$. This will easily follow from the discussion below.

The NS solution $\psi(x,t)$ with the inital value $\psi(x)$ of the form (8.33)–(8.35) is given by the same formulae upon replacing $\tilde{m}_j$ by $\tilde{m}_j(t)$,

$$\tilde{m}_j(t) = e^{-\lambda_j \nu_j t} \tilde{m}_j, \quad j = 1, \dots, n. \tag{8.36}$$

Let us verify that this solution describes the *n-soliton scattering process*.

With no loss of generality we may assume the parameters $\lambda_1, \dots, \lambda_n$ of $\psi(x,t)$ to be ordered, $\lambda_1 > \lambda_2 > \dots > \lambda_n$. Then, as $t \to \pm\infty$, $\psi(x,t)$ is expressed as a sum of one-soliton solutions,

$$\begin{aligned}\psi(x,t) = \psi_1^{(-)}(x,t) &+ e^{i\theta_1}(\psi_2^{(-)}(x,t) - \varrho) \\ &+ \dots + e^{i(\theta_1 + \dots + \theta_{n-1})}(\psi_n^{(-)}(x,t) - \varrho) + O(e^{-\nu c|t|})\end{aligned} \tag{8.37}$$

as $t \to -\infty$, and

$$\begin{aligned}\psi(x,t) = \psi_n^{(+)}(x,t) &+ e^{i\theta_n}(\psi_{n-1}^{(+)}(x,t) - \varrho) \\ &+ \dots + e^{i(\theta_n + \dots + \theta_2)}(\psi_1^{(+)}(x,t) - \varrho) + O(e^{-\nu c|t|})\end{aligned} \tag{8.38}$$

as $t \to +\infty$, with $c = \min_{j \neq l} \{|v_j - v_l|\}$.

The $\psi_j^{(\pm)}(x,t)$ are here solitons with parameters θ_j, v_j, $x_{0j}^{(\pm)}$:

$$\psi_j^{(\pm)}(x,t) = \psi_{\theta_j}(x - v_j t + x_{0j}^{(\pm)}), \tag{8.39}$$

where

$$e^{i\theta_j} = \frac{\bar{z}_j}{z_j}, \quad 0 \leqslant \theta_j < 2\pi, \quad v_j = \lambda_j = -\omega \cos\frac{\theta_j}{2} \tag{8.40}$$

and

$$\begin{aligned}x_{0j}^{(-)} = x_{0j} &- \frac{1}{2\nu_j} \sum_{l=j+1}^{n} \log \frac{(v_l - v_j)^2 + (\nu_l + \nu_j)^2}{(v_l - v_j)^2 + (\nu_l - \nu_j)^2} \\ &+ \frac{1}{2\nu_j} \sum_{l=1}^{j-1} \log \frac{(v_l - v_j)^2 + (v_l + \nu_j)^2}{(v_l - v_j)^2 + (\nu_l - \nu_j)^2},\end{aligned} \tag{8.41}$$

$$x_{0j}^{(+)} = x_{0j} + \frac{1}{2\nu_j} \sum_{l=j+1}^{n} \log \frac{(v_l - v_j)^2 + (\nu_l + \nu_j)^2}{(v_l - v_j)^2 + (\nu_l - \nu_j)^2}$$

$$- \frac{1}{2\nu_j} \sum_{l=1}^{j-1} \log \frac{(v_l - v_j)^2 + (\nu_l + \nu_j)^2}{(v_l - v_j)^2 + (\nu_l - \nu_j)^2}, \tag{8.42}$$

with

$$x_{0j} = \frac{1}{\nu_j} \log i\varepsilon_j \gamma_j = \frac{1}{\nu_j} \log |\gamma_j|, \quad j = 1, \ldots, n. \tag{8.43}$$

Here ε_j is the sign of the real parameter $i\gamma_j$ uniquely determined by (8.7).

The proof of these statements will be based on the explicit formulae (8.33)–(8.35). Consider, for definiteness, the case $t \to -\infty$. It is sufficient to show that along the trajectory C_j of a particular soliton

$$x - v_j t = \text{const} \tag{8.44}$$

the solution $\psi(x,t)$ approaches the one-soliton solution $e^{i(\theta_1 + \ldots + \theta_{j-1})} \psi_j^{(-)}(x,t)$ as $t \to -\infty$, whereas along a generic line $x - vt = \text{const}$ it takes asymptotic values ϱ, $\varrho e^{i(\theta_1 + \ldots + \theta_j)}$ or $\varrho e^{i\theta}$, $\theta \equiv \theta_1 + \ldots + \theta_n \pmod{2\pi}$, when $v > v_1$, $v_j > v > v_{j+1}$ or $v_n > v$, respectively. These limiting values are approached with the exponential order $O(e^{-\nu c|t|})$.

Let us proceed to the proof of these statements. Write the matrix elements of $A(x,t)$ in the form

$$A_{jl}(x,t) = \frac{i\omega}{z_j - \bar{z}_l} e^{\zeta_j(x,t) + \zeta_l(x,t)}, \tag{8.45}$$

where

$$\zeta_j(x,t) = \frac{\nu_j}{2}(x - v_j t) + \frac{1}{2} \log \tilde{m}_j; \quad j = 1, \ldots, n, \tag{8.46}$$

and modify in turn the numerator and the denominator of (8.33).

We start with the denominator. There is an obvious relation

$$\det(I + A(x,t))$$

$$= 1 + \sum_{l=1}^{n} (i\omega)^l \cdot \sum_{1 \leq j_1 < \ldots < j_l \leq n} \Delta(j_1, \ldots, j_l) \exp 2(\zeta_{j_1}(x,t) + \ldots + \zeta_{j_l}(x,t)), \tag{8.47}$$

where $\Delta(j_1, \ldots, j_l)$ is the principal minor of order l of the $n \times n$ matrix

$$D = \begin{pmatrix} \frac{1}{z_1-\bar{z}_1} & \dots & \frac{1}{z_1-\bar{z}_n} \\ \vdots & & \vdots \\ \frac{1}{z_n-\bar{z}_1} & \dots & \frac{1}{z_n-\bar{z}_n} \end{pmatrix}, \tag{8.48}$$

composed of the rows and columns with indices $j_1, \dots, j_l$. To compute $\Delta(j_1, \dots, j_l)$ we use the well-known relation

$$\det \begin{pmatrix} \frac{1}{a_1+b_1} & \dots & \frac{1}{a_1+b_n} \\ \vdots & & \vdots \\ \frac{1}{a_n+b_1} & \dots & \frac{1}{a_n+b_n} \end{pmatrix} = \prod_{1 \le i < j \le n} (a_i - a_j)(b_i - b_j) \prod_{i,j=1}^{n} (a_i + b_j)^{-1}, \tag{8.49}$$

which gives

$$\Delta(j_1, \dots, j_l) = \prod_{p=1}^{n} (z_{j_p} - \bar{z}_{j_p})^{-1} \cdot \prod_{1 \le p < q \le l} \left| \frac{z_{j_p} - z_{j_q}}{z_{j_p} - \bar{z}_{j_q}} \right|^2. \tag{8.50}$$

For $x - v_j t = \text{const}$ we have

$$\lim_{t \to -\infty} \zeta_l(x, t) = -\infty; \quad l > j, \tag{8.51}$$

$$\lim_{t \to -\infty} \zeta_l(x, t) = +\infty; \quad l < j, \tag{8.52}$$

so that by (8.47) we find the asymptotic behaviour along C_j as $t \to -\infty$:

$$\begin{aligned} \det(I + A(x, t)) = {} & (i\omega)^{j-1} \exp 2(\zeta_1(x, t) + \dots + \zeta_{j-1}(x, t)) \\ & \times (i\omega \Delta(1, \dots, j) \exp 2\zeta_j(x, t) + \Delta(1, \dots, j-1) + O(e^{-\nu c|t|})). \end{aligned} \tag{8.53}$$

Next consider the numerator of (8.33). Similarly to (8.47) we have

$$\begin{aligned} \det(I + A_1(x, t)) = {} & \sum_{l=2}^{n+1} (i\omega)^{l-1} \cdot \sum_{1 \le j_1 < \dots < j_{l-1} \le n} \Delta_1(j_1, \dots, j_{l-1}) \exp 2(\zeta_{j_1}(x, t) \\ & + \dots + \zeta_{j_{l-1}}(x, t)), \end{aligned} \tag{8.54}$$

where $\Delta_1(j_1, \dots, j_{l-1})$ is the principal minor of order l of the matrix

$$D_1 = \begin{pmatrix} & & \vdots & \dfrac{1}{z_1} \\ & D & \vdots & \vdots \\ & & \vdots & \dfrac{1}{z_n} \\ \cdots & \cdots & \cdots & \\ 1 & \cdots\cdots & 1 & 0 \end{pmatrix}, \tag{8.55}$$

composed of the rows and columns with indices $j_1, \dots, j_{l-1}$ and $n+1$.

To compute $\Delta_1(j_1, \dots, j_{l-1})$ we take a side way. Let

$$D_1(z) = \begin{pmatrix} & & \vdots & \dfrac{1}{z_1} \\ & D & \vdots & \vdots \\ & & \vdots & \dfrac{1}{z_n} \\ \cdots & \cdots & \cdots & \\ \dfrac{1}{z-\bar{z}_1} & \cdots & \dfrac{1}{z-\bar{z}_n} & \dfrac{1}{z} \end{pmatrix}. \tag{8.56}$$

By the general formula (8.49) we have

$$\det D_1(z) = \frac{\det D}{z} \prod_{j=1}^{n} \frac{\bar{z}_j}{z_j} \cdot \frac{z - z_j}{z - \bar{z}_j}. \tag{8.57}$$

On the other hand, expanding the determinant of $D_1(z)$ along the last row and letting $z \to \infty$ gives

$$\lim_{z \to \infty} z \det D_1(z) = \det D + \det D_1. \tag{8.58}$$

By comparing these expressions we find

$$\det D_1 = \left(\prod_{j=1}^{n} \frac{\bar{z}_j}{z_j} - 1 \right) \det D. \tag{8.59}$$

The principal minors of D_1 are treated in a similar way. As a result, we get the final expression

$$\Delta_1(j_1, \dots, j_{l-1}) = \left(e^{i(\theta_{j_1} + \dots + \theta_{j_{l-1}})} - 1\right) \Delta(j_1, \dots, j_{l-1}), \tag{8.60}$$

where

$$e^{i\theta_j} = \frac{\bar{z}_j}{z_j}, \qquad j=1, \ldots, n. \tag{8.61}$$

Similarly to (8.53), we find the asymptotic behaviour along C_j, as $t \to -\infty$,

$$\det(I + A_1(x,t)) = (i\omega)^{j-1} \exp 2(\zeta_1(x,t) + \ldots + \zeta_{j-1}(x,t)) \\ \times (i\omega \Delta_1(1, \ldots, j) \exp 2\zeta_j(x,t) + \Delta_1(1, \ldots, j-1) + O(e^{-\nu c|t|})). \tag{8.62}$$

Now insert the asymptotic formulae (8.53) and (8.62) into the expression (8.33) for $\psi(x,t)$. Using (8.46), (8.50) and (8.61) we deduce the asymptotic behaviour along C_j,

$$\psi(x,t) = \varrho e^{i(\theta_1 + \ldots + \theta_{j-1})} \frac{1 + e^{i\theta_j} a_j^{(-)} e^{\nu_j(x - v_j t)}}{1 + a_j^{(-)} e^{\nu_j(x - v_j t)}} + O(e^{-\nu c|t|}), \tag{8.63}$$

where

$$a_j^{(-)} = i\omega \tilde{m}_j \frac{\Delta(1, \ldots, j)}{\Delta(1, \ldots, j-1)} = \frac{\omega \tilde{c}_j}{2\nu_j z_j} \prod_{l=1}^{j-1} \left| \frac{z_l - z_j}{z_l - \bar{z}_j} \right|^2. \tag{8.64}$$

Let us modify the last expression. Using (8.5)–(8.6) we obtain

$$\frac{\omega \tilde{c}_j}{2\nu_j z_j} = \frac{Z_j}{i\gamma_j}, \tag{8.65}$$

with

$$Z_j = \frac{i\omega}{2\nu_j z_j \dot{a}_\varrho(z_j)} = -\frac{\omega}{z_j} e^{-\frac{i\theta}{2}} \prod_{\substack{l=1 \\ l \neq j}}^{n} \frac{z_l - \bar{z}_j}{z_l - z_j}. \tag{8.66}$$

Observe that, by virtue of (8.7), Z_j is real and

$$\operatorname{sign} Z_j = \varepsilon_j. \tag{8.67}$$

Therefore

$$Z_j = \varepsilon_j |Z_j| = \varepsilon_j \prod_{\substack{l=1 \\ l \neq j}}^{n} \left| \frac{z_l - \bar{z}_j}{z_l - z_j} \right|, \tag{8.68}$$

so that (8.64) becomes

$$a_j^{(-)} = \frac{1}{i\varepsilon_j \gamma_j^{(-)}}, \tag{8.69}$$

where

$$\gamma_j^{(-)}=\gamma_j\prod_{l=j+1}^{n}\left|\frac{z_l-z_j}{z_l-\bar{z}_j}\right|\cdot\prod_{l=1}^{j-1}\left|\frac{z_l-\bar{z}_j}{z_l-z_j}\right|. \tag{8.70}$$

Insert the expression for $a_j^{(-)}$ into (8.63) and use the equation $z_l=v_l+i\nu_l$. Finally, we find the following asymptotic behaviour along the trajectory C_j, as $t\to-\infty$,

$$\psi(x,t)=e^{i(\theta_1+\ldots+\theta_{j-1})}\psi_{\theta_j}(x-v_jt+x_{0j}^{(-)})+O(e^{-\nu c|t|}). \tag{8.71}$$

The behaviour of $\psi(x,t)$ along generic paths is analyzed by similar means. Asymptotically as $t\to\infty$, we have

$$\psi(x,t)=e^{i(\theta_1+\ldots+\theta_j)}\varrho+O(e^{-\nu c|t|}) \tag{8.72}$$

in the region $v_j>\tilde{v}>v_{j+1}$. This completes the proof of (8.37).

The limit $t\to+\infty$ is examined in a similar way.

Notice that this reasoning also proves the aforementioned fact that $\psi(x,t)$ approaches its boundary values with the order $O(e^{-\nu|x|})$ as $x\to\pm\infty$.

Thus we have shown that the *solution* $\psi(x,t)$ *describes the n-soliton interaction. As* $t\to\pm\infty$, *the solitons become free and go far apart from one another. So, as in the rapidly decreasing case,* $\psi(x,t)$ *will be called an n-soliton solution.*

The peculiarity of the present situation is that $\psi(x,t)$ "decays" into solitons $\psi_j^{(\pm)}(x,t)$ with distinct phase values θ_j. These phases are tied up with the velocities v_j of the asymptotic solitons and are therefore distinct. One can say that only those solitons interact which have different phases. For that reason soliton dynamics looks more natural in the extended phase space $\mathscr{M}_\varrho=\bigcup_{0\leq\theta<2\pi}\mathscr{M}_{\varrho,\theta}$ already mentioned in § I.1. The relation

$$\theta\equiv\sum_{j=1}^{n}\theta_j(\operatorname{mod}2\pi) \tag{8.73}$$

may then be regarded as a conservation law.

As in the rapidly decreasing case, the formulae (8.37)–(8.42) allow for a natural interpretation in terms of scattering theory. Namely, an n-soliton solution describes the *scattering process of n solitons*. For $t\to\pm\infty$, we are dealing with the free motion of n solitons separated in space with parameters $(v_j,x_{0j}^{(\pm)})$. Here for $t\to-\infty$ the centers of inertia of the solitons, $x_{0j}^{(-)}+v_jt$, are ordered from left to right in decreasing order of velocities; for $t\to+\infty$, the spatial order of solitons is reversed.

Scattering only changes the parameters $x_{0j}^{(\pm)}$ – the center of inertia coordinates at $t=0$. Their relationship follows from (8.41)–(8.42),

$$x_{0j}^{(+)} = x_{0j}^{(-)} + \Delta x_{0j}, \tag{8.74}$$

where

$$\Delta x_{0j} = -\frac{1}{\nu_j} \sum_{l=1}^{j-1} \log \frac{(v_l - v_j)^2 + (\nu_l + \nu_j)^2}{(v_l - v_j)^2 + (\nu_l - \nu_j)^2}$$
$$+ \frac{1}{\nu_j} \sum_{l=j+1}^{n} \log \frac{(v_l - v_j)^2 + (\nu_l + \nu_j)^2}{(v_l - v_j)^2 + (\nu_l - \nu_j)^2}. \tag{8.75}$$

These formulae show that the scattering increments in the coordinates x_{0j} are expressed as a sum over two-particle shifts,

$$\Delta x_{01} = \frac{1}{\nu_1} \log \frac{(v_1 - v_2)^2 + (\nu_1 + \nu_2)^2}{(v_1 - v_2)^2 + (\nu_1 - \nu_2)^2},$$
$$\Delta x_{02} = -\frac{1}{\nu_2} \log \frac{(v_1 - v_2)^2 + (\nu_1 + \nu_2)^2}{(v_1 - v_2)^2 + (\nu_1 - \nu_2)^2} \tag{8.76}$$

for $v_1 > v_2$ with the interchange $1 \leftrightarrow 2$ for $v_1 < v_2$. So, as in the rapidly decreasing case, *scattering factorizes*.

Hamiltonian aspects of soliton scattering will be discussed in the following chapter.

§ 9. Notes and References

1. The regular Riemann problem of analytic factorization has been extensively studied in the mathematical literature; see the books by N. I. Muskhelishvili [Mu 1968] and N. P. Vekua [V 1970]. The principal method here is to reduce it to singular integral equations. The latter are examined in various functional classes, mostly in Hölder classes. For our purposes it is preferable to deal with the normed rings $\mathfrak{R}^{(n\times n)}$ and $\mathfrak{R}_{\pm}^{(n\times n)}$ where the Riemann problem reduces naturally to a Wiener-Hopf integral equation. This approach was developed by I. C. Gohberg and M. G. Krein [GK 1958] who proved the existence theorem for the Riemann problem used in this chapter. A passage from the scattering matrix $S(\lambda)$ to $G(\lambda) = a(\lambda)S(\lambda)$ performed in § 1 is an essential step for applying the theorem.

2. The Riemann problem method in soliton theory was introduced in [ZS 1979], which also contained a suitable formulation of the problem with

zeros. Since then the Riemann problem method has gained in popularity and in use (see, for instance, [M 1981]). In [ZM 1975], [S 1975], [S 1979] for the first time the Riemann problem was taken as a basis for solving the inverse problem for a first order matrix linear differential operator.

3. The matrix $B(\lambda)$ generalizing the scalar Blaschke factor was introduced in a general setting in [P 1955]. In order to reduce the Riemann problem with zeros to the regular one in § 2 we follow [ZS 1979] and multiply $B(\lambda)$ by $G_+(\lambda)$ from the right. This operation leaves invariant $G(\lambda)$, the matrix to be factorized. Other authors (see, for instance, [ZMNP 1980] and [M 1981]) usually multiply $B(\lambda)$ by $G_+(\lambda)$ from the left. Then $G(\lambda)$ is modified by a similarity transformation.

4. A simple derivation of the differential equation with respect to x in § 2 and of the differential equation with respect to t in § 3 is the major conceptual benefit of the Riemann problem method, based, in fact, on the Liouville theorem only. The idea to derive the zero curvature condition from the Riemann problem with the matrix $G(x, t, \lambda)$ depending explicitly on x and t was suggested by V. E. Zakharov and A. B. Shabat; a detailed exposition can be found in [ZS 1979]. Similar ideas were also present in [K 1977] where actually a special Riemann problem on an algebraic curve was studied. It then became clear that the form of the matrices $U(x, t, \lambda)$ and $V(x, t, \lambda)$ in the zero curvature condition is governed only by the principal parts of the factorizing matrices $F_\pm(x, t, \lambda)$ at their essentially singular points (cf. §§ 2-3). Again, extensive use is made of the Liouville theorem (see [JMU 1981], [JM 1981a], [JM 1981b], [I 1984]).

5. If $\beta(s)$ lies in $L_2(-\infty, \infty)$, the operators $\mathbf{K}_x$ and $\mathbf{L}_x$ introduced in § 2 are well defined and bounded on $L_2(0, \infty)$. By (2.64), for $f(s)$ in $L_2(0, \infty)$ we have

$$\begin{aligned}\langle \mathbf{K}_x f, f\rangle &= \int_0^\infty \int_0^\infty k_x(s, s')f(s')\overline{f(s)}\,ds\,ds' \\ &= \int_{-x}^\infty \left|\int_0^\infty \beta(u-s)f(s)\,ds\right|^2 du \geqslant 0, \end{aligned} \tag{9.1}$$

so that the operator $\mathbf{K}_x$ (and also $\mathbf{L}_x$) is positive and monotone in x. Moreover, $\mathbf{I}+\varepsilon\mathbf{K}_\pm$ and $\mathbf{I}+\varepsilon\mathbf{L}_\pm$ have bounded inverses both for $\varepsilon=1$, which is obvious, and for $\varepsilon=-1$ by the condition (A). So it is quite easy to prove that $\mathbf{I}+\varepsilon\mathbf{K}_x$ and $\mathbf{I}+\varepsilon\mathbf{L}_x$ have inverses on $L_2(0, \infty)$ uniformly in x.

However, under our general assumptions on $\beta(s)$, these operators are only defined on $L_1(0, \infty)$. Therefore we have to appeal to the Gohberg-Krein theory and give a more detailed analysis in Subsection 3 of § 2.

6. In the derivation of the asymptotic behaviour of $G_\pm(x, \lambda)$ as $|x|\to\infty$, we exploited the explicit dependence of $G(x, \lambda)$ on x (see (2.13)) to obtain explicit expressions (2.64)-(2.65) for the kernels $k_x(s, s')$ and $l_x(s, s')$. On the other hand, the matrix $G(x, t, \lambda)$ for the NS equation also depends explicitly

on t (see (3.9)). It is then natural to ask whether one can investigate in this way the behaviour of $G_{\pm}(x, t, \lambda)$ and derive the asymptotic behaviour of the NS solution $\psi(x, t)$ as $t \to \pm\infty$. Asymptotic expressions of this kind were first obtained in [ZM 1976] and proved rigorously in [N 1980].

This problem is more difficult than that in Subsection 3 of § 2; it was solved in [I 1981]. It was shown that along the lines $x - vt = \text{const}$, as $t \to \pm\infty$, the Riemann problem simplifies and reduces to one where the matrix to be factorized does not depend on λ. The latter problem can be solved in closed form in terms of special functions. There emerge some interesting connections with the so-called isomonodromic solutions, self-similar solutions and Painlevé-type equations. This vast theme is not treated here and we can only refer the reader to the original papers [ARS 1980a], [ARS 1980b], [FN 1980], [U 1980a], [U 1980b], [A 1981], [JMU 1981], [JM 1981a], [JM 1981b]. The role of isomonodromic solutions of the Riemann problem in soliton theory is discussed in [I 1985].

7. The formalism of the Gelfand-Levitan-Marchenko integral equations was developed by I. M. Gelfand and B. M. Levitan [GL 1951] and V. A. Marchenko [M 1955] who gave a complete solution of the inverse problem for the radial Schrödinger equation (Schrödinger operator on the half-line). An elementary exposition of these methods and connections with M. G. Krein's approach [K 1954], [K 1955] can be found in the review paper [F 1959]. The Schrödinger equation on the whole line (one-dimensional Schrödinger operator) was studied by I. Cay and G. Moses [CM 1956]. A complete mathematical treatment of the problem for potentials $u(x)$ satisfying

$$\int_{-\infty}^{\infty} (1+|x|)\,|u(x)|\,dx < \infty \tag{9.2}$$

was given in [F 1958], [F 1964]. This work showed for the first time the necessity to consider both of the Gelfand-Levitan-Marchenko equations and established the relationship between their solutions.

The inverse problem for the NS model in the rapidly decreasing case was solved via this method in [ZS 1971] for $\varepsilon = -1$ and in [T 1973] for $\varepsilon = 1$. We also note that the inverse problem for the radial Dirac operator with zero mass was solved in [GL 1966].

8. Soliton solutions for the NS model in the rapidly decreasing case were first found and analyzed in [ZS 1971].

9. In § 6 we have already mentioned some technical difficulties in the Riemann problem arising in the case of finite density due to boundary points of the continuous spectrum.

Conceptually similar complications also occur in the analysis of the Riemann problem for the one-dimensional Schrödinger operator. The role of the surface Γ is now played by the Riemann surface of the function $k = \sqrt{\lambda}$; the branch point $\lambda = 0$ may give rise to a virtual level.

In this connection it might be of interest to consider the general Riemann problem on an arbitrary Riemann surface and develope an analogue of the Gohberg-Krein theory.

10. The Gelfand-Levitan-Marchenko equations for the case of finite density derived in § 7 were obtained in [ZS 1973] (see also [Fr 1972], [GK 1978], [AK 1984]). Here we follow the approach first suggested in [F 1958], [F 1964] for the one-dimensional Schrödinger operator (see also the review paper [F 1974]). The proof of the existence and uniqueness theorem for the integral equations (7.37) and (7.44) relies on the fact that the associated operators are positive and is similar to the approach mentioned in our note 5.

The Gelfand-Levitan-Marchenko equations in the rapidly decreasing case can also be treated by the method of § 7, with some technical simplifications.

11. In § 7 we assumed, for simplicity, that the boundary values in the finite density case are approached in the sense of Schwartz. In § I.2 it was explained to what extent this assumption may be relaxed. The restrictions on $a_\varrho(\lambda)$ and $b_\varrho(\lambda)$ at $\lambda = \pm\omega$ are then as follows

$$k b_\varrho(\lambda) = b_\pm + o(1), \quad k a_\varrho(\lambda) = a_\pm + o(1) \tag{9.3}$$

as $k \to 0$. The above method for solving the inverse problem applies here, too.

For the one-dimensional Schrödinger operator a natural restriction on the potential $u(x)$ in the direct and inverse problems is given by (9.2). This is precisely the condition stated in [F 1958], [F 1964]. However, the behaviour of the transition coefficients on the boundary of the continuous spectrum was treated inaccurately. This was the reason for the criticism in [DT 1979] after which the impression was formed that $u(x)$ should be subject to a stronger restriction,

$$\int_{-\infty}^{\infty} (1+x^2)\,|u(x)|\,dx < \infty\,. \tag{9.4}$$

Nevertheless, as is shown in [M 1977] and [L 1979], if the behaviour of the transition coefficients at $k=0$ is made more precise, the method of [F 1958], [F 1964] remains valid under the single condition (9.2).

12. Soliton solutions in the finite density case were analyzed in [ZS 1973]. We note their method for studying the interaction of solitons. It is based on the assumption that a multisoliton solution $\psi(x,t)$ can be expressed, as $t \to \pm\infty$, as a sum of spatially separated solitons. For such $\psi(x,t)$, $\bar{\psi}(x,t)$ the auxiliary linear problem can be solved in closed form, so that the transition coefficients for the discrete spectrum, and hence the $x_{0j}^{(\pm)}$, are evaluated explicitly.

In application to the rapidly decreasing case this method is presented in the book [ZMNP 1980].

The method outlined in § 8 is based on the direct analysis of the explicit formulae (8.33)–(8.36) for a multisoliton solution. The expression for the determinant of (8.49) which plays an important role in the computations can be found in the problem-book [FS 1977].

13. Throughout the text we made a pedantic use of the notation $\psi(x)$, $\bar{\psi}(x)$ (also $b(\lambda)$, $\bar{b}(\lambda)$) despite the fact that the functions in these pairs are the complex conjugates of each other. We persist in this notation by analogy with complex coordinates $z=x+iy$ and $\bar{z}=x-iy$ of the real plane $\mathbb{R}^2$, which is particularly convenient in the Hamiltonian formalism. Moreover, this notation allows an easy extension to the more general case where $\psi(x)$ and $\bar{\psi}(x)$ are completely independent, so that the bar ceases to indicate complex conjugation. Instead of the NS equation we then obtain a system

$$i\frac{\partial\psi}{\partial t}=-\frac{\partial^2\psi}{\partial x^2}+2\varkappa\psi^2\bar{\psi}, \quad i\frac{\partial\bar{\psi}}{\partial t}=\frac{\partial^2\bar{\psi}}{\partial x^2}-2\varkappa\bar{\psi}^2\psi. \tag{9.5}$$

All the results of Chapter I including the zero curvature representation and the analysis of the mapping $\mathscr{F}$ remain essentially valid for this system. Of course, various involutions for the Jost solutions hold no longer so that, for instance, the reduced monodromy matrix in the rapidly decreasing case has the form

$$T(\lambda)=\begin{pmatrix} a(\lambda) & \varepsilon\bar{b}(\lambda) \\ b(\lambda) & \bar{a}(\lambda)\end{pmatrix}. \tag{9.6}$$

Here $\bar{a}(\lambda)$ and $\bar{b}(\lambda)$ are not the complex conjugates of $a(\lambda)$ and $b(\lambda)$, respectively. The same refers to the discrete spectrum λ_j, $\bar{\lambda}_j$ and its transition coefficients γ_j, $\bar{\gamma}_j$. In the case of finite density the boundary conditions become

$$\lim_{x\to\pm\infty}\psi(x)=\varrho_1^{(\pm)}, \qquad \lim_{x\to\pm\infty}\bar{\psi}(x)=\varrho_2^{(\pm)}, \tag{9.7}$$

where in general $|\varrho_1^{\pm}|\neq|\varrho_2^{\pm}|$; it is only assumed that $\varrho_1^{(-)}\varrho_2^{(-)}=\varrho_1^{(+)}\varrho_2^{(+)}$.

As regards the Poisson structures introduced in § I.1, we note that (I.1.18) and the related formulae should be interpreted in the formally-complex sense. So, for example, the Hamiltonian (see (I.1.24)) becomes

$$H=\int_{-\infty}^{\infty}\left(\frac{\partial\psi}{\partial x}\frac{\partial\bar{\psi}}{\partial x}+\varkappa\bar{\psi}^2\psi^2\right)dx \tag{9.8}$$

and is a complex-valued functional.

All the results of Chapter II can also be generalized with the exception of one important fact. The matrix

$$G(x,\lambda)=\begin{pmatrix} 1 & \varepsilon\bar{b}(\lambda)e^{-i\lambda x} \\ -b(\lambda)e^{i\lambda x} & 1 \end{pmatrix} \tag{9.9}$$

of the Riemann problem (2.1) does not satisfy the hypothesis of the theorem of Gohberg and Krein referred to above because $b(\lambda)$ and $\bar{b}(\lambda)$ are now completely independent. Here in order for the Riemann problem to have a solution it is to be assumed that for every x all partial indices of $G(x,\lambda)$ vanish. Thus there arise severe but rather implicit restrictions on the inverse problem data. Moreover, the class of these data need not be invariant under the evolution in t.

In a similar manner, the Gelfand-Levitan-Marchenko formulation has to be supplemented by the requirement that the corresponding integral equations have a solution, which is a restriction on the initial data.

The general case under the rapidly decreasing boundary conditions was studied in detail in [AKNS 1974]. For the general finite density case the reader is referred to [GK 1978], [AK 1981], [AK 1984].

Since physical applications involve primarily the ordinary NS equation, we concentrate here on the model admitting the involution of complex conjugation.

14. Besides the Riemann problem and the Gelfand-Levitan-Marchenko equations there are other schemes for constructing solutions to a wide class of nonlinear equations. We can cite, for instance, the methods of [FA 1981], [QC 1983], [M 1985]. We think, however, that these methods are not so natural mathematically. The problem of identifying the solutions that belong to a given functional class has been studied within these schemes in lesser detail than in the Riemann problem method or the Gelfand-Levitan-Marchenko formalism.

References

[A 1981] Ablowitz, M. J.: Remarks on nonlinear evolution equations and ordinary differential equations of Painlevé type. Physica D, 3D, 129–141 (1981)

[AK 1981] Asano, N., Kato, Y.: Non-self-adjoint Zakharov-Shabat operator with a potential of the finite asymptotic values. I. Direct spectral and scattering problems. J. Math. Phys. *22*, 2780–2793 (1981)

[AK 1984] Asano, N., Kato, Y.: Non-self-adjoint Zakharov-Shabat operator with a potential of the finite asymptotic values. II. Inverse Problem. J. Math. Phys. *25*, 570–588 (1984)

[AKNS 1974] Ablowitz, M. J., Kaup, D. J., Newell, A. C., Segur, H.: The inverse scattering transform – Fourier analysis for nonlinear problems. Stud. Appl. Math. *53*, 249–315 (1974)

[ARS 1980 a] Ablowitz, M. J., Ramani, A., Segur, H.: A connection between nonlinear evolution equations and ordinary differential equations of P-type. I. J. Math. Phys. *21*, 715–721 (1980)

[ARS 1980 b] Ablowitz, M. J., Ramani, A., Segur, H.: A connection between nonlinear evolution equations and ordinary differential equations of P-type. II. J. Math. Phys. *21*, 1006–1015 (1980)

[DT 1979] Deift, P., Trubowitz, E.: Inverse scattering on the line. Comm. Pure Appl. Math. *32*, 121–251 (1979)

[F 1958] Faddeev, L. D.: On the relation between *S*-matrix and potential for the one-dimensional Schrödinger operator. Dokl. Akad. Nauk SSSR *121*, 63–66 (1958) [Russian]

[F 1959] Faddeev, L. D.: The inverse problem of the quantum theory of scattering. Uspekhi Mat. Nauk *14* (4), 57–119 (1959) [Russian]; English transl. in J. Math. Phys. *4*, 72–104 (1962)

[F 1964] Faddeev, L. D.: Properties of the *S*-matrix of the one-dimensional Schrödinger equation. Trudy Mat. Inst. Steklov *73*, 314–336 (1964) [Russian]; English transl. in Transl. Amer. Math. Soc. Ser. *2*, 65, 139–166 (1967)

[F 1974] Faddeev, L. D.: The inverse problem of quantion scattering theory. II. Sovrem. Probl. Mat. 3. VINITI, Moscow, 93–181 (1974) [Russian]; English transl. in Soviet J. Math. *5* (1976)

[FA 1981] Fokas, A. S., Ablowitz, M. J.: Linearization of the Korteweg-de Vries and Painlevé II equations. Phys. Rev. Lett. *47*, 1096–1100 (1981)

[FN 1980] Flaschka, H., Newell, A. C.: Monodromy and spectrum preserving deformations. Comm. Math. Phys. *76*, 65–116 (1980)

[Fr 1972] Frolov, I. S.: Inverse scattering problem for the Dirac system on the whole line. Dokl. Akad. Nauk SSSR *207*, 44–47 (1972) [Russian]; English transl. in Sov. Math. Dokl. *13*, 1468–1472 (1972)

[FS 1977] Faddeev, D. K., Sominski, I. S.: Problems in Higher Algebra. Moscow, Nauka 1977 [Russian]

[GK 1958] Gohberg, I. C., Krein, M. G.: Systems of integral equations on the half-line with kernels depending on the difference of the arguments. Uspekhi Mat. Nauk *13* (2), 3–72 (1958) [Russian]

[GK 1978] Gerdjikov, V. C., Kulish, P. P.: Completely integrable Hamiltonian system associated with the non-self-adjoint Dirac operator. Bulg. J. Phys. *5*, 337–348 (1978) [Russian]

[GL 1951] Gelfand, I. M., Levitan, B. M.: On the determination of a differential equation from its spectral function. Izv. Akad. Nauk SSR, Ser. Math. *15*, 309–360 (1951) [Russian]; English transl. in Amer. Math. Soc. transl. Ser. 2, *1*, 253–304 (1955)

[GL 1966] Gasymov, M. G., Levitan, B. M.: The inverse problem for a Dirac system. Dokl. Akad. Nauk SSSR *167*, 967–970 (1966) [Russian]; English transl. in Soviet Phys. Doklady *7*, 495–499 (1966)

[I 1981] Its, A. R.: Asymptotics of solutions of the nonlinear Schrödinger equation and isomonodromic deformations of systems of linear differential equations. Dokl. Akad. Nauk SSSR *261*, 14–18 (1981) [Russian]; English transl. in Soviet Math. Dokl. *24*, 452–456 (1981)

[I 1984] Its, A. R.: The Liouville theorem and the inverse scattering method. In: Differential geometry. Lie groups and mechanics. VI. Zapiski Nauchn. Semin. LOMI *133*, 113–125 (1984) [Russian]

[I 1985] Its, A. R.: Isomonodromic solutions of the zero-curvature equation. Izv. Akad. Nauk SSSR Ser. Mat. *49*, 530–565 (1985) [Russian]

[JM 1981 a] Jimbo, M., Miwa, T.: Monodromy preserving deformation of linear ordinary differential equations with rational coefficients. II. Physica D, 2D, 407–448 (1981)

[JM 1981b] Jimbo, M., Miwa, T.: Monodromy preserving deformation of linear ordinary differential equations with rational coefficients. III. Physica D, 4D, 26–46 (1981)

[JMU 1981] Jimbo, M., Miwa, T., Ueno, K.: Monodromy preserving deformation of linear ordinary differential equations with rational coefficients. I. General theory and τ-function. Physica D, 2D, 306–352 (1981)

[K 1954] Krein, M. G.: On integral equations generating differential equations of second order. Dokl. Akad. Nauk SSSR *97*, 21–24 (1954) [Russian]

[K 1955] Krein, M. G.: On determination of the potential of a particle from its S-matrix. Dokl. Akad. Nauk SSSR *105*, 433–436 (1955) [Russian]

[K 1977] Krichever, I. M.: Methods of algebraic geometry in the theory of nonlinear equations. Uspekhi Mat. Nauk *32* (6), 183–208 (1977) [Russian]; English transl. in Russian Math. Surveys *32* (6), 185–213 (1977)

[KM 1955] Kay, I., Moses, H. E.: The determination of the scattering potential from the spectral measure function. I. Continuous spectrum. Nuovo Cimento *2*, 916–961 (1955)

[KM 1956] Kay, I., Moses, H. E.: The determination of the scattering potential from the spectral measure function. II. Point eigenvalues and proper eigenfunctions. Nuovo Cimento *3*, 66–84 (1956)

[KM 1956a] Kay, I., Moses, H. E.: The determination of the scattering potential from the spectral measure function. III. Calculation of the scattering potential from the scattering operator for the one dimensional Schrödinger equation. Nuovo Cimento *3*, 276–304 (1956)

[L 1979] Levitan, B. M.: Sufficient conditions for the solvability of the inverse problem of scattering theory on the whole line. Mat. Sb. *108*, 350–357 (1979) [Russian]

[M 1955] Marchenko, V. A.: On reconstruction of the potential energy from phases of the scattered waves. Dokl. Akad. Nauk SSSR *104*, 695–698 (1955) [Russian]

[M 1977] Marchenko, V. A.: Sturm-Liouville Operators and their Applications. Kiev, Naukova Dumká 1977 [Russian]

[M 1981] Mikhailov, A. V.: The reduction problem and the inverse scattering method. Physica D, 3D, 73–117 (1981)

[M 1985] Marchenko, V. A.: Nonlinear Equations and Operator Algebras. Kiev, Naukova Dumka 1985 [Russian]

[Mu 1968] Muskhelishvili, N. I.: Singular Integral Equations. Boundary-value Problems of Function Theory and their Applications in Mathematical Physics. Moscow, Nauka 1968 [Russian]; English transl. of an earlier Russian edition: Groningen-Holland, P. Noordhoff N.V. 1953

[N 1980] Novoksenov, V. Ju.: Asymptotic behaviour as $t\to\infty$ of the solution of the Cauchy problem for the nonlinear Schrödinger equation. Dokl. Akad. Nauk SSSR *251*, 799–802 (1980) [Russian]

[P 1955] Potapov, B. P.: Multiplicative structure of J-contractive matrix functions. Trudy Moskov. Mat. Obšč. *4*, 125–236 (1955) [Russian]

[QC 1983] Quispel, G. R. W., Capel, H. W.: The anisotropic Heisenberg spin chain and the nonlinear Schrödinger equation. Physica A, *117A*, 76–102 (1983)

[S 1975] Shabat, A. B.: The inverse scattering problem for a system of differential equations. Funk. Anal. Prilož. *9* (3), 75–78 (1975) [Russian]

[S 1979] Shabat, A. B.: An inverse scattering problem. Differencialnye Uravneniya *15*, 1824–1834 (1979) [Russian]; English transl. in Diff. Equations *15*, 1299–1307 (1980)

[T 1973] Takhtajan, L. A.: Hamiltonian equations connected with the Dirac equation. In: Differential geometry, Lie groups and mechanics. I. Zapiski Nauchn. Semin. LOMI *37*, 66–76 (1973) [Russian]; English transl. in J. Sov. Math. *8*, 219–228 (1977)

[U 1980a] Ueno, K.: Monodromy preserving deformation and its application to soliton theory. I. Proc. Japan Acad., ser. A, *56* (3), 103–108 (1980)

[U 1980b] Ueno, K.: Monodromy preserving deformation and its application to soliton theory. II. Proc. Japan Acad., ser. A, *56* (5), 210–215 (1980)

[V 1970] Vekua, N. P.: Systems of Singular Integral Equations and Boundary-value problems. Moscow, Nauka 1970 [Russian]

[ZM 1975] Zakharov, V. E., Manakov, S. V.: The theory of resonant interaction of wave packets in nonlinear media. Zh. Eksp. Teor. Fiz. *69*, 1654–1673 (1975) [Russian]; English transl. in Soviet Phys. JETP *42*, 842–850 (1976)

[ZM 1976] Zakharov, V. E., Manakov, S. V.: Asymptotic behaviour of nonlinear wave systems integrable by the inverse scattering method. Zh. Eksp. Teor. Fiz. *71*, 203–215 (1976) [Russian]; English transl. in Soviet Phys. JETP *44*, 106–112 (1976)

[ZMNP 1980] Zakharov, V. E., Manakov, S. V., Novikov, S. P., Pitaievski, L. P.: Theory of Solitons. The Inverse Problem Method. Moscow, Nauka 1980 [Russian]; English transl.: New York, Plenum 1984

[ZS 1971] Zakharov, V. E., Shabat, A. B.: Exact theory of two-dimensional self-focusing and one-dimensional self-modulation of waves in non-linear media. Zh. Exp. Teor. Fiz. *61*, 118–134 (1971) [Russian]; English transl. in Soviet Phys. JETP *34*, 62–69 (1972)

[ZS 1973] Zakharov, V. E., Shabat, A. B.: Interaction between solitons in a stable medium. Zh. Exp. Teor. Fiz. *64*, 1627–1639 (1973) [Russian]; English transl. in Sov. Phys. JETP *37*, 823–828 (1973)

[ZS 1979] Zakharov, V. E., Shabat, A. B.: Integration of the nonlinear equations of mathematical physics by the method of the inverse scattering problem. II. Funk. Anal. Prilož. *13* (3), 13–22 (1979) [Russian]; English transl. in Funct. Anal. Appl. *13*, 166–174 (1979)

Chapter III
The Hamiltonian Formulation

In this chapter we return to the Hamiltonian formulation of the NS model in order to discuss the basic transformation of the inverse scattering method

$$\mathscr{F}: (\psi(x), \bar{\psi}(x)) \rightarrow (b(\lambda), \bar{b}(\lambda); \lambda_j, \gamma_j)$$

from the Hamiltonian standpoint. We shall describe the Poisson structure on the scattering data of the auxiliary linear problem induced through $\mathscr{F}$ from the initial Poisson structure defined in Chapter I. Under the rapidly decreasing or finite density boundary conditions, the NS model proves to be a completely integrable system, with $\mathscr{F}$ defining a transformation to action-angle variables. In particular, we will show that the integrals of the motion introduced in Chapter I are in involution. In these terms scattering of solitons amounts to a simple canonical transformation.

This chapter introduces an important element of the inverse scattering method, the classical r-matrix, whose universal role will be fully revealed only in Part II. Here we shall see that the r-matrix is a useful tool for computing the Poisson brackets of transition coefficients. Moreover, it will be shown that the r-matrix representation of the Poisson brackets can replace the zero curvature representation.

§ 1. Fundamental Poisson Brackets and the r-Matrix

We will show how to compute the Poisson brackets of the entries of the transition matrix $T(x, y, \lambda)$. The formulae of this section will be used in §§ 5 and 6 to describe the Poisson structure on transition coefficients under the rapidly decreasing and finite density boundary conditions, respectively.

Most of the computations in this section are purely local. We suppose $\psi(x)$, $\bar{\psi}(x)$ are defined on the interval $-L<x<L$; we shall only deal with compactly supported functionals, i.e. functionals depending only on $\psi(x)$,

$\bar{\psi}(x)$ for x inside the interval. A precise definition of compactly supported functionals was given in § I.1.

Recall that the Poisson bracket of such functionals is given by

$$\{F, G\} = i \int_{-L}^{L} \left(\frac{\delta F}{\delta \psi(x)} \frac{\delta G}{\delta \bar{\psi}(x)} - \frac{\delta F}{\delta \bar{\psi}(x)} \frac{\delta G}{\delta \psi(x)} \right) dx, \tag{1.1}$$

where, due to compact support, integration actually goes over a smaller interval. Boundary conditions are irrelevant here. From now on, along with real-valued functionals we shall also consider complex-valued ones. The Poisson structure extends by linearity to these functionals, and their Poisson bracket has the same form (1.1).

Our nearest goal is to compute all the 16 Poisson brackets between the matrix elements of $T(x, y, \lambda)$ for different values of λ. The definition of $T(x, y, \lambda)$ and the superposition property (I.3.7) imply that, for $-L<y<x<L$, the matrix elements are functionals with compact support. To be able to treat all their Poisson brackets simultaneously, it is convenient to adopt the following notation.

Let A and B be two matrix functionals with compact support, i.e., 2×2 matrices whose elements are functionals with compact support. Let

$$\{A \underset{,}{\otimes} B\} = i \int_{-L}^{L} \left(\frac{\delta A}{\delta \psi(x)} \otimes \frac{\delta B}{\delta \bar{\psi}(x)} - \frac{\delta A}{\delta \bar{\psi}(x)} \otimes \frac{\delta B}{\delta \psi(x)} \right) dx, \tag{1.2}$$

where $\otimes$ on the right hand side indicates tensor product. So $\{A \underset{,}{\otimes} B\}$ is a 4×4 matrix composed of various Poisson brackets of the matrix elements of A and B. We shall use the natural convention for tensor product,

$$A \otimes B = \begin{pmatrix} A_{11} B & A_{12} B \\ A_{21} B & A_{22} B \end{pmatrix}, \tag{1.3}$$

or

$$(A \otimes B)_{jk,mn} = A_{jm} B_{kn}, \tag{1.4}$$

where $jk, mn = 11, 12, 21, 22$ so that

$$\{A \underset{,}{\otimes} B\}_{jk,mn} = \{A_{jm}, B_{kn}\}. \tag{1.5}$$

This notation will really prove convenient as will be seen more than once. In particular, the basic properties of the Poisson bracket take the form

$$\{A \underset{,}{\otimes} B\} = -P\{B \underset{,}{\otimes} A\} P \tag{1.6}$$

for skew-symmetry,

$$\{A \underset{,}{\otimes} BC\} = \{A \underset{,}{\otimes} B\}(I \otimes C) + (I \otimes B)\{A \underset{,}{\otimes} C\} \tag{1.7}$$

for the derivation property and

$$\{A \underset{,}{\otimes} \{B \underset{,}{\otimes} C\}\} + P_{13} P_{23} \{C \underset{,}{\otimes} \{A \underset{,}{\otimes} B\}\} P_{23} P_{13} + P_{13} P_{12} \{B \underset{,}{\otimes} \{C \underset{,}{\otimes} A\}\} P_{12} P_{13} = 0 \tag{1.8}$$

for the Jacobi identity.

Let us explain the notation employed. In (1.6) there appears a 4×4 matrix P which is the permutation matrix in $\mathbb{C}^2 \otimes \mathbb{C}^2$ defined by

$$P(\xi \otimes \eta) = \eta \otimes \xi \tag{1.9}$$

for any vectors ξ and η in $\mathbb{C}^2$. From (1.9) it follows that

$$P^2 = I, \quad P(A \otimes B) = (B \otimes A) P, \tag{1.10}$$

where A and B are any 2×2 matrices and I denotes the unit 4×4 matrix (we are not afraid of confusion because the context will always make clear on which space I acts). In terms of the Pauli matrices σ_a (see § I.2), P is expressed as

$$P = \frac{1}{2}\left(I + \sum_{a=1}^{3} \sigma_a \otimes \sigma_a\right) \tag{1.11}$$

and in the basis 11, 12, 21, 22 it has the form

$$P = \begin{pmatrix} 1 & 0 & 0 & 0 \\ 0 & 0 & 1 & 0 \\ 0 & 1 & 0 & 0 \\ 0 & 0 & 0 & 1 \end{pmatrix}. \tag{1.12}$$

The operation $\{\underset{,}{\otimes}\}$ in (1.8) is defined, according to (1.2), for matrices of arbitrary dimension, so that $\{A \underset{,}{\otimes} \{B \underset{,}{\otimes} C\}\}$ is a matrix in $\mathbb{C}^2 \otimes \mathbb{C}^2 \otimes \mathbb{C}^2$ and P_{12} (P_{13} and P_{23}, respectively) denotes a matrix which equals the unit matrix in the third (respectively the second and the first) space and coincides with P in the product of the remaining two spaces.

Obviously, the representation of the basic properties of the Poisson bracket in terms of the operation $\{\underset{,}{\otimes}\}$ allows for $n \times n$, not necessarily 2×2, matrices A, B, C; the $n^2 \times n^2$ matrix P is then defined by (1.9) as before, and satisfies (1.10).

Let us now compute the Poisson brackets. Consider $U(z,\lambda)$ as a compactly supported matrix-functional of $\psi(x)$, $\bar{\psi}(x)$, $-L<x<L$. Remind that

$$U(z,\lambda)=\frac{\lambda}{2i}\sigma_3+U_0(z)=\frac{\lambda}{2i}\sigma_3+\sqrt{\varkappa}\,(\bar{\psi}(z)\sigma_+ +\psi(z)\sigma_-), \qquad (1.13)$$

where σ_3, σ_+ and σ_- are the Pauli matrices (see § I.2). The basic Poisson brackets from § I.1

$$\{\psi(x),\psi(y)\}=\{\bar{\psi}(x),\bar{\psi}(y)\}=0, \quad \{\psi(x),\bar{\psi}(y)\}=i\delta(x-y), \qquad (1.14)$$

easily lead to the Poisson bracket matrix $\{U(x,\lambda)\underset{,}{\otimes} U(y,\mu)\}$:

$$\{U(x,\lambda)\underset{,}{\otimes} U(y,\mu)\}=i\varkappa(\sigma_-\otimes\sigma_+ -\sigma_+\otimes\sigma_-)\delta(x-y). \qquad (1.15)$$

Now observe that the matrix on the right hand side can be expressed as

$$\sigma_-\otimes\sigma_+ -\sigma_+\otimes\sigma_- =\tfrac{1}{2}[P,\sigma_3\otimes I]=-\tfrac{1}{2}[P,I\otimes\sigma_3]. \qquad (1.16)$$

To verify this, it is enough to use the expression (1.11) for P and the commutation relations for the Pauli matrices

$$[\sigma_+,\sigma_-]=\sigma_3, \quad [\sigma_3,\sigma_+]=2\sigma_+, \quad [\sigma_3,\sigma_-]=-2\sigma_-. \qquad (1.17)$$

Equation (1.16) makes it possible to rewrite the right hand side of (1.15) as an expression linear in $U(x,\lambda)$ and $U(y,\mu)$.

In fact, by virtue of (1.16) it can be expressed as a commutator $\frac{-\varkappa}{\lambda-\mu}\left[P,\frac{\lambda}{2i}\sigma_3\otimes I+\frac{\mu}{2i}I\otimes\sigma_3\right]\delta(x-y)$. By (1.10), P commutes with $U_0(x)\otimes I+I\otimes U_0(x)$. Hence we can write the Poisson bracket matrix as

$$\{U(x,\lambda)\underset{,}{\otimes} U(y,\mu)\}=[r(\lambda-\mu),U(x,\lambda)\otimes I+I\otimes U(x,\mu)]\delta(x-y), \qquad (1.18)$$

with

$$r(\lambda)=-\frac{\varkappa}{\lambda}P. \qquad (1.19)$$

At first sight this formula is nothing but a rather cumbersome reformulation of the basic Poisson brackets (1.14). Yet, as we shall see below, it represents a universal property of the matrices $U(x,\lambda)$ involved in the zero curvature representation for all the models to be considered. Moreover, this property underlies the integrability itself and has a natural Lie-algebraic interpretation. For that reason (1.18) will be called *the fundamental Poisson brackets*.

We will now show that *(1.18) immediately yields the Poisson brackets between the entries of the transition matrix in the form*

$$\{T(x,y,\lambda) \overset{\otimes}{,} T(x,y,\mu)\} = [r(\lambda-\mu), T(x,y,\lambda) \otimes T(x,y,\mu)], \tag{1.20}$$

for $-L<y<x<L$.

We shall outline two derivations of this formula. One is based on the definition of $T(x,y,\lambda)$ as a multiplicative integral by a passage to the limit (see § I.2), the other makes use of the differential equation (see § I.3).

We begin with the first one. Cut the interval (y,x) into N segments Δ_n, $n=1,\ldots,N$, so that their maximal length Δ vanishes as $N\to\infty$. Then, by (I.2.14)-(I.2.16) we have

$$T(x,y,\lambda) = \lim_{N\to\infty} T_N(\lambda), \tag{1.21}$$

where

$$T_N(\lambda) = \overset{\curvearrowleft}{\prod_{n=1}^{N}} L_n(\lambda) \tag{1.22}$$

and

$$L_n(\lambda) = I + \int_{\Delta_n} U(x,\lambda)\,dx. \tag{1.23}$$

Now, by virtue of (1.18)

$$\{L_n(\lambda) \overset{\otimes}{,} L_m(\mu)\} = 0 \tag{1.24}$$

for $n \neq m$.

Indeed the computation of $\{L_n \overset{\otimes}{,} L_m\}$ leads to a vanishing integral $\int_{\Delta_n}\int_{\Delta_m} \delta(x-y)\,dx\,dy$.

It is essential in this reasoning that the fundamental Poisson brackets (1.18) contain only the generalized function $\delta(x-y)$ but not its derivatives. This is an important property of the potential $U(x,\lambda)$ of the auxiliary linear problem; we shall call it *ultralocality*.

From (1.22), (1.24) and (1.7) we obtain

$$\{T_N(\lambda) \overset{\otimes}{,} T_N(\mu)\} = \sum_{n=1}^{N} (\tilde{T}_n(\lambda) \otimes \tilde{T}_n(\mu)) \{L_n(\lambda) \overset{\otimes}{,} L_n(\mu)\} (T_{n-1}(\lambda) \otimes T_{n-1}(\mu)), \tag{1.25}$$

where

$$T_n(\lambda) = \overset{\curvearrowleft}{\prod_{k=1}^{n}} L_k(\lambda), \quad \tilde{T}_n(\lambda) = \overset{\curvearrowleft}{\prod_{k=n+1}^{N}} L_k(\lambda). \tag{1.26}$$

Then the fundamental Poisson brackets yield

$$\{L_n(\lambda) \underset{,}{\otimes} L_n(\mu)\} = \{\int_{\Delta_n} U(x,\lambda)\,dx \underset{,}{\otimes} \int_{\Delta_n} U(x,\mu)\,dx\}$$
$$= [r(\lambda-\mu), \int_{\Delta_n} U(x,\lambda)\,dx \otimes I + I \otimes \int_{\Delta_n} U(x,\mu)\,dx], \tag{1.27}$$

whence we find

$$\{L_n(\lambda) \underset{,}{\otimes} L_n(\mu)\} = [r(\lambda-\mu), L_n(\lambda) \otimes L_n(\mu)] + O(\Delta^2). \tag{1.28}$$

Using that the commutator acts as derivation with respect to multiplication, we find from (1.28)

$$\{T_N(\lambda) \underset{,}{\otimes} T_N(\mu)\} = [r(\lambda-\mu), T_N(\lambda) \otimes T_N(\mu)] + O(N\Delta^2). \tag{1.29}$$

Letting $N \to \infty$ in this equation and using $N\Delta = O(1)$ we obtain (1.20).

For the second method let us consider $T(x, y, \lambda)$ as a matrix functional of the entries of $U(z, \lambda)$ for $-L < y \leqslant z \leqslant x < L$. By the differentiation rule for composite functions, (1.1) gives

$$\{T_{ab}(x,y,\lambda), T_{cd}(x,y,\mu)\}$$
$$= \int_y^x \int_y^x \frac{\delta T_{ab}(x,y,\lambda)}{\delta U_{jk}(z,\lambda)} \{U_{jk}(z,\lambda), U_{lm}(z',\mu)\} \frac{\delta T_{cd}(x,y,\mu)}{\delta U_{lm}(z',\mu)}\,dz\,dz', \tag{1.30}$$

where repeated indices j, k, l, m imply summation from 1 to 2. Here the variation of $T(x, y, \lambda)$ in the definition of $\dfrac{\delta T(x,y,\lambda)}{\delta U(z,\lambda)}$ is taken with respect to a general variation $\delta U(z, \lambda)$ which need not preserve the special form of $U(z, \lambda)$.

Now by varying the differential equation

$$\frac{\partial T}{\partial x}(x,y,\lambda) = U(x,\lambda)\,T(x,y,\lambda) \tag{1.31}$$

for the transition matrix (see § I.3) with the initial condition

$$T(x,y,\lambda)|_{x=y} = I, \tag{1.32}$$

we find the equation

$$\frac{\partial}{\partial x}\delta T(x,y,\lambda) = U(x,\lambda)\,\delta T(x,y,\lambda) + \delta U(x,\lambda)\,T(x,y,\lambda) \tag{1.33}$$

with the initial condition

$$\delta T(x, y, \lambda)|_{x=y} = 0. \tag{1.34}$$

It is immediately verified that the solution of (1.33)–(1.34) is given by

$$\delta T(x, y, \lambda) = \int_y^x T(x, z, \lambda)\, \delta U(z, \lambda)\, T(z, y, \lambda)\, dz, \tag{1.35}$$

whence it follows that

$$\frac{\delta T_{ab}(x, y, \lambda)}{\delta U_{jk}(z, \lambda)} = T_{aj}(x, z, \lambda)\, T_{kb}(z, y, \lambda). \tag{1.36}$$

Now insert this formula into (1.30). We then deduce a relation which we shall again write in invariant form

$$\begin{aligned} &\{T(x, y, \lambda) \overset{\otimes}{,} T(x, y, \mu)\} \\ &= \int_y^x \int_y^x (T(x, z, \lambda) \otimes T(x, z', \mu)) \{U(z, \lambda) \overset{\otimes}{,} U(z', \mu)\} \\ &\quad \times (T(z, y, \lambda) \otimes T(z', y, \mu))\, dz\, dz'. \end{aligned} \tag{1.37}$$

Using the fundamental Poisson brackets we find

$$\begin{aligned} \{T(x, y, \lambda) \overset{\otimes}{,} T(x, y, \mu)\} = \int_y^x &(T(x, z, \lambda) \otimes T(x, z, \mu)) \\ &\times [r(\lambda - \mu), U(z, \lambda) \otimes I + I \otimes U(z, \mu)] \\ &\times (T(z, y, \lambda) \otimes T(z, y, \mu))\, dz. \end{aligned} \tag{1.38}$$

The matrices $U(z, \lambda)$ and $U(z, \mu)$ in the commutator on the right hand side of (1.38) stand on the left or on the right of $T(z, y, \lambda)$, $T(z, y, \mu)$ or of $T(x, z, \lambda)$, $T(x, z, \mu)$, respectively. By the differential equation (1.31) and

$$\frac{\partial T}{\partial y}(x, y, \lambda) = -T(x, y, \lambda)\, U(y, \lambda) \tag{1.39}$$

we deduce that the integrand of (1.38) is a total derivative with respect to z of the product

$$(T(x, z, \lambda) \otimes T(x, z, \mu))\, r(\lambda - \mu)\, (T(z, y, \lambda) \otimes T(z, y, \mu)).$$

Integration with the initial condition (1.32) gives (1.20).

To conclude this section we make a few comments on the formulae obtained.

1. By no means any given matrix $r(\lambda)$ can play the role of the classical r-matrix. *For the fundamental Poisson brackets (1.18) to be consistent with the skew-symmetry (1.6) and the Jacobi identity (1.8) it is sufficient that*

$$r(-\lambda) = -Pr(\lambda)P \tag{1.40}$$

and

$$[r_{12}(\lambda-\mu), r_{13}(\lambda)+r_{23}(\mu)]+[r_{13}(\lambda), r_{23}(\mu)]=0, \tag{1.41}$$

respectively. Obviously, these relations hold for the matrix $r(\lambda)$ in (1.19). *Conversely, if (1.40)–(1.41) hold, then (1.18) defines a Poisson structure on the space of functionals of matrix elements of* $U(x,\lambda)$. In Part II we will show how to construct some other solutions of (1.40)–(1.41) and prove that each of them gives rise to an integrable Hamiltonian system.

2. The two methods for deriving (1.20) outlined above are quite general and do not depend on a particular form of $U(x,\lambda)$ and $r(\lambda)$. Precisely, we have shown that if $U(x,\lambda)$ satisfies (1.18) with some $r(\lambda)$, then the Poisson brackets between the matrix elements of the transition matrix $T(x,y,\lambda)$ satisfy (1.20). Also, the local formula (1.18) is an infinitesimal version of (1.20).

3. The right hand sides of (1.18) and (1.20) contain an apparent singularity at $\lambda=\mu$ since the denominator in (1.19) vanishes at $\lambda=\mu$. However, by virtue of (1.10), P commutes with both $U(x,\lambda)\otimes I+I\otimes U(x,\lambda)$ and $T(x,y,\lambda)\otimes T(x,y,\lambda)$ so that the numerator in (1.18) and (1.20) also vanishes at $\lambda=\mu$, and the singularity cancels out ("L'Hopital's rule").

4. For $-L<x<y<L$ it follows from (1.20) that

$$\{T(x,y,\lambda) \overset{\otimes}{,} T(x,y,\mu)\} = -[r(\lambda-\mu), T(x,y,\lambda)\otimes T(x,y,\mu)], \tag{1.42}$$

since (see § I.3)

$$T(y,x,\lambda)=T^{-1}(x,y,\lambda). \tag{1.43}$$

5. Relation (1.20) extends to transition matrices for two arbitrary intervals (y,x) and (y',x') contained in $(-L,L)$. To show this observe that, by ultralocality, the Poisson brackets of the matrix elements of $T(x,y,\lambda)$ and $T(x',y',\mu)$ vanish for disjoint intervals (y,x) and (y',x') as well as for intervals with only one point in common. Therefore the superposition property (I.3.7), the derivation property (1.7), and (1.20) imply

$$\{T(x,y,\lambda) \underset{,}{\otimes} T(x',y',\mu)\} = (T(x,x'',\lambda) \otimes T(x',x'',\mu)) \times [r(\lambda-\mu), T(x'',y'',\lambda) \otimes T(x'',y'',\mu)] \times (T(y'',y,\lambda) \otimes T(y'',y',\mu)), \qquad (1.44)$$

where (y'', x'') is the intersection of the intervals (y, x) and (y', x').

§ 2. Poisson Commutativity of the Motion Integrals in the Quasi-Periodic Case

As a first application of the formulae derived in the previous section, *we will prove that the local integrals of the motion I_n constructed in § I.4 are in involution*,

$$\{I_n, I_m\} = 0. \qquad (2.1)$$

Still, first of all we have to show that the I_n are admissible functionals on the phase space $\mathscr{M}_{L,\theta}$ (see § I.1). *We will show the admissibility of the generating functional* $F_L(\lambda)$ defined in § I.2 by

$$F_L(\lambda) = \operatorname{tr} T_L(\lambda) Q(\theta). \qquad (2.2)$$

We shall deal with the quasi-periodic boundary conditions

$$\psi(x+2L) = e^{i\theta}\psi(x), \quad \bar{\psi}(x+2L) = e^{-i\theta}\bar{\psi}(x) \qquad (2.3)$$

and fix the fundamental domain $-L \leqslant x \leqslant L$. Consider the transition matrix $T(x, y, \lambda)$ for $-L < y < x < L$. As noted in § 1, its matrix elements are functionals with compact support. Their series expansion in $\psi(z)$, $\bar{\psi}(z)$ of the type (I.1.7) results from the Volterra integral equation (I.3.26) for the transition matrix,

$$T(x,y,\lambda) = E(x-y,\lambda) + \int_y^x T(x,z,\lambda)\, U_0(z)\, E(z-y,\lambda)\, dz, \qquad (2.4)$$

with $E(z,\lambda) = \exp \dfrac{\lambda z \sigma_3}{2i}$. The iterations here converge absolutely and give the required series for the matrix elements of $T(x, y, \lambda)$.

Now compute the variational derivatives $\dfrac{\delta T(x,y,\lambda)}{\delta \psi(z)}, \dfrac{\delta T(x,y,\lambda)}{\delta \bar{\psi}(z)}$. To this end we use (1.35) where we set

$$\delta U(z,\lambda)=\sqrt{\varkappa}\,(\delta\bar{\psi}(z)\sigma_{+}+\delta\psi(z)\sigma_{-}). \tag{2.5}$$

As a result we find, for $y<z<x$,

$$\frac{\delta T(x,y,\lambda)}{\delta\psi(z)}=\sqrt{\varkappa}\,T(x,z,\lambda)\sigma_{-}T(z,y,\lambda) \tag{2.6}$$

and

$$\frac{\delta T(x,y,\lambda)}{\delta\bar{\psi}(z)}=\sqrt{\varkappa}\,T(x,z,\lambda)\sigma_{+}T(z,y,\lambda). \tag{2.7}$$

For z in the fundamental domain but outside of (y,x), these variational defivatives vanish.

Thus the variational derivatives are discontinuous functions. Hence the matrix elements of $T(x,y,\lambda)$ cannot be regarded as admissible functionals in the sense of § I.1.

To prove that $F_L(\lambda)$ is admissible let us take the limit of (2.6)-(2.7) as $x\to L$, $y\to -L$, and consider the variational derivatives $\dfrac{\delta T_L(\lambda)}{\delta\psi(z)}$, $\dfrac{\delta T_L(\lambda)}{\delta\bar{\psi}(z)}$ of the monodromy matrix $T_L(\lambda)=T(L,-L,\lambda)$. From (2.6)-(2.7) it follows that these variational derivatives are smooth functions of z in the interval $-L<z<L$. Taking the limit as $z\to L-0$ and $z\to -L+0$ we find

$$\left.\frac{\delta T_L(\lambda)}{\delta\psi(z)}\right|_{z=L}=\sqrt{\varkappa}\,\sigma_{-}T_L(\lambda), \tag{2.8}$$

$$\left.\frac{\delta T_L(\lambda)}{\delta\bar{\psi}(z)}\right|_{z=L}=\sqrt{\varkappa}\,\sigma_{+}T_L(\lambda) \tag{2.9}$$

and

$$\left.\frac{\delta T_L(\lambda)}{\delta\psi(z)}\right|_{z=-L}=\sqrt{\varkappa}\,T_L(\lambda)\sigma_{-}, \tag{2.10}$$

$$\left.\frac{\delta T_L(\lambda)}{\delta\bar{\psi}(z)}\right|_{z=-L}=\sqrt{\varkappa}\,T_L(\lambda)\sigma_{+}. \tag{2.11}$$

This shows that the matrix elements of the monodromy matrix are not admissible functionals either.

To continue proving that $F_L(\lambda)$ is admissible, multiply the above equations by $Q(\theta)$ from the right and evaluate the trace. Using the elementary formulae

$$Q(\theta)\sigma_+ Q^{-1}(\theta)=e^{i\theta}\sigma_+, \tag{2.12}$$

$$Q(\theta)\sigma_- Q^{-1}(\theta)=e^{-i\theta}\sigma_-, \tag{2.13}$$

we obtain the required quasi-periodicity conditions

$$\left.\frac{\delta F_L(\lambda)}{\delta\bar{\psi}(z)}\right|_{z=L}=e^{i\theta}\left.\frac{\delta F_L(\lambda)}{\delta\bar{\psi}(z)}\right|_{z=-L}, \tag{2.14}$$

$$\left.\frac{\delta F_L(\lambda)}{\delta\psi(z)}\right|_{z=L}=e^{-i\theta}\left.\frac{\delta F_L(\lambda)}{\delta\psi(z)}\right|_{z=-L}. \tag{2.15}$$

For the z-derivatives $\frac{d}{dz}\frac{\delta F_L(\lambda)}{\delta\psi(z)}$ and $\frac{d}{dz}\frac{\delta F_L(\lambda)}{\delta\bar{\psi}(z)}$ there are similar formulae,

$$\left.\frac{d}{dz}\frac{\delta F_L(\lambda)}{\delta\bar{\psi}(z)}\right|_{z=L}=e^{i\theta}\left.\frac{d}{dz}\frac{\delta F_L(\lambda)}{\delta\bar{\psi}(z)}\right|_{z=-L}, \tag{2.16}$$

$$\left.\frac{d}{dz}\frac{\delta F_L(\lambda)}{\delta\psi(z)}\right|_{z=L}=e^{-i\theta}\left.\frac{d}{dz}\frac{\delta F_L(\lambda)}{\delta\psi(z)}\right|_{z=-L}. \tag{2.17}$$

For the derivation one should use the equations

$$\frac{\partial}{\partial z}\frac{\delta T(x,y,\lambda)}{\delta\psi(z)}=\sqrt{\varkappa}\,T(x,z,\lambda)[\sigma_-,U(z,\lambda)]\,T(z,y,\lambda), \tag{2.18}$$

$$\frac{\partial}{\partial z}\frac{\delta T(x,y,\lambda)}{\delta\bar{\psi}(z)}=\sqrt{\varkappa}\,T(x,z,\lambda)[\sigma_+,U(z,\lambda)]\,T(z,y,\lambda) \tag{2.19}$$

for $y<z<x$ which follow from (1.31), (1.39) and (2.6)–(2.7), and repeat the above reasoning taking into account the quasi-periodicity conditions

$$U(L,\lambda)=Q^{-1}(\theta)\,U(-L,\lambda)\,Q(\theta). \tag{2.20}$$

Conditions (2.14)–(2.17) together with the smoothness of $\frac{\delta F_L(\lambda)}{\delta\psi(z)}$, $\frac{\delta F_L(\lambda)}{\delta\bar{\psi}(z)}$ for $-L<z<L$ make it possible to extend these functions to the whole real axis in a smooth quasi-periodic way,

$$\frac{\delta F_L(\lambda)}{\delta\bar{\psi}(z+2L)}=e^{i\theta}\frac{\delta F_L(\lambda)}{\delta\bar{\psi}(z)},\quad \frac{\delta F_L(\lambda)}{\delta\psi(z+2L)}=e^{-i\theta}\frac{\delta F_L(\lambda)}{\delta\psi(z)}. \tag{2.21}$$

If $\psi(x)$, $\bar{\psi}(x)$ are infinitely differentiable, then the variational derivatives $\frac{\delta F_L(\lambda)}{\delta \psi(z)}$, $\frac{\delta F_L(\lambda)}{\delta \bar{\psi}(z)}$ are also infinitely differentiable quasi-periodic functions on the whole axis.

Finally we point out that $F_L(\lambda)$ is a real-analytic functional. Its expansion of the type (I.1.7) results from the corresponding expansion for the functional $\operatorname{tr} T(x,y,\lambda) Q(\theta)$ in the limit as $x \to L$, $y \to -L$. This finishes proving that $F_L(\lambda)$ is an admissible functional.

Now we will show that the integrals of the motion generated by $F_L(\lambda)$ are in involution:

$$\{F_L(\lambda), F_L(\mu)\} = 0 . \tag{2.22}$$

For that purpose we recall (1.20)

$$\{T(x,y,\lambda) \otimes T(x,y,\mu)\} = [r(\lambda-\mu), T(x,y,\lambda) \otimes T(x,y,\mu)], \tag{2.23}$$

for $-L < y < x < L$, and multiply it by $Q(\theta) \otimes Q(\theta)$ from the right. The elementary property

$$[r(\lambda), Q \otimes Q] = 0 \tag{2.24}$$

shows that (2.22) remains also valid when $T(x,y,\lambda)$ and $T(x,y,\mu)$ are replaced by $T(x,y,\lambda) Q(\theta)$ and $T(x,y,\mu) Q(\theta)$, respectively. Now take the matrix trace in $\mathbb{C}^2 \otimes \mathbb{C}^2$ of the resulting equation and use

$$\operatorname{tr}(A \otimes B) = \operatorname{tr} A \cdot \operatorname{tr} B , \tag{2.25}$$

where tr on the right indicates the trace in $\mathbb{C}^2$. Since the trace of the commutator is zero, we have

$$\{\operatorname{tr} T(x,y,\lambda) Q(\theta), \operatorname{tr} T(x,y,\mu) Q(\theta)\} = 0 . \tag{2.26}$$

Here one can take the limit as $x \to L$, $y \to -L$; then (2.22) results.

So we have shown that the motion integrals for the model in question are in involution. Equation (2.1) follows from (2.22) by expanding

$$p_L(\lambda) = -\lambda L + \frac{\theta}{2} + \varkappa \sum_{n=1}^{\infty} \frac{I_n}{\lambda^n} + O(|\lambda|^{-\infty}) , \tag{2.27}$$

where

$$p_L(\lambda) = \arccos \tfrac{1}{2} F_L(\lambda) \tag{2.28}$$

(see § I.4).

To every local integral I_n on $\mathcal{M}_{L,\theta}$ there corresponds a Hamiltonian flow

$$\frac{\partial \psi}{\partial t} = \{I_n, \psi\}, \qquad \frac{\partial \bar{\psi}}{\partial t} = \{I_n, \bar{\psi}\}. \tag{2.29}$$

For $n=1, 2$ the flows have a simple physical interpretation (see § I.1); for $n=3$ we recover the NS equation. The corresponding equation of motion (especially for $n>3$) are commonly called ***the higher nonlinear Schrödinger equations (higher NS equations)***.

What is remarkable here is that in addition to the first two (trivial) flows there are infinitely many commuting flows. This may be viewed as a manifestation of the "hidden symmetry" of the NS model.

To avoid misunderstanding, let us point out that the family $\operatorname{tr} T(x, y, \lambda) Q$ with an arbitrary matrix Q also gives functionals in involution, but these are irrelevant to our model. First, their variational derivatives are not smooth, so that these functionals are inadmissible. Next, even if we agree to extend the class of admissible functionals, then the functionals $\operatorname{tr} T(x, y, \lambda) Q$ will not Poisson commute with the I_n and so will be of no use for showing the complete integrability of the NS model.

The existence of infinitely many integrals of the motion in involution suggests that our model may be completely integrable. In case the phase space has finite dimension $2n$, there is the Liouville-Arnold theorem saying that a Hamiltonian system is completely integrable if it has a set of n (half the dimension of the phase space) integrals of the motion in involution. If so, the phase space is foliated by n-dimensional submanifolds on which the motion is linear.

In our case the phase space is infinite-dimensional and integrability is not so elementary since there is no analogue of the Liouville-Arnold theorem. Naively, one can assert that the "number" of the motion integrals, I_n, contained in $p_L(\lambda)$ is "half the dimension" of the phase space and these integrals are functionally independent.

However, to implement these ideas in the quasi-periodic case and, especially, to construct the angle variables which linearize the motion it is necessary to employ the analysis on Riemann surfaces in the general case of infinite genus - the techniques beyond the scope of this book. Therefore, we shall not elaborate the quasi-periodic case any further, reserving it to outline some basic constructions connected with the r-matrix (see §§ 3-5). In contrast, in the rapidly decreasing case and for the finite density boundary conditions, we shall give a full treatment of complete integrability and exhibit the corresponding action-angle variables.

§ 3. Derivation of the Zero Curvature Representation from the Fundamental Poisson Brackets

Here we will show in what sense *the existence of the fundamental Poisson brackets can replace the zero curvature condition. Namely, given the r-matrix and a matrix $U(x,\lambda)$ we shall construct a sequence of matrices $V_n(x,\lambda)$ appearing in the zero curvature representation for the higher NS equations, which are the equations of motion generated by the integrals I_n.* As in § 2, we shall concentrate on the case of quasi-periodic boundary conditions.

Consider the generating equations of motion

$$\frac{\partial \psi}{\partial t} = \{p_L(\mu), \psi\}, \qquad \frac{\partial \bar{\psi}}{\partial t} = \{p_L(\mu), \bar{\psi}\} \tag{3.1}$$

for all the higher NS equations. Here

$$p_L(\mu) = \arccos \tfrac{1}{2} F_L(\mu), \tag{3.2}$$

and μ plays the role of a parameter. *Let us show that (3.1) is equivalent to the zero curvature condition*

$$\frac{\partial U}{\partial t}(x,\lambda) - \frac{\partial V}{\partial x}(x,\lambda,\mu) + [U(x,\lambda), V(x,\lambda,\mu)] = 0, \tag{3.3}$$

satisfied for all λ (for notational simplicity, the dependence on t is suppressed), and derive an explicit expression for $V(x,\lambda,\mu)$.

To begin with, we compute the Poisson bracket matrix $\{T(x,y,\mu) \underset{,}{\otimes} U(z,\lambda)\}$ for $-L<y<z<x<L$. In analogy with the second derivation of (1.20), we use (1.18), (1.36) and the obvious identity

$$\frac{\delta U_{ab}(z,\lambda)}{\delta U_{cd}(z',\lambda)} = \delta_{ac}\delta_{bd}\delta(z-z'), \tag{3.4}$$

where δ_{ab} is the Kronecker δ-symbol, to obtain

$$\begin{aligned}\{T(x,y,\mu) \underset{,}{\otimes} U(z,\lambda)\} = &(T(x,z,\mu)\otimes I)\\ &\times [r(\mu-\lambda), U(z,\mu)\otimes I + I \otimes U(z,\lambda)]\\ &\times (T(z,y,\mu)\otimes I).\end{aligned} \tag{3.5}$$

This relation involves a 4×4 matrix

$$M(z,x,y;\lambda,\mu)=(T(x,z,\mu)\otimes I)r(\mu-\lambda)(T(z,y,\mu)\otimes I), \tag{3.6}$$

which appears in (3.5) through the commutator with $I\otimes U(x,\lambda)$. To modify the other terms on the right hand side of (3.5) containing $U(z,\mu)\otimes I$ we make use of the differential equations (1.39) and (1.31) for $T(x,z,\mu)$ and $T(z,y,\mu)$, respectively. We find that these terms sum up to $\frac{\partial}{\partial z}M(z,x,y;\lambda,\mu)$. So, finally, we have

$$\{T(x,y,\mu)\underset{,}{\otimes} U(z,\lambda)\} = \frac{\partial}{\partial z}M(z,x,y;\lambda,\mu) + [M(z,x,y;\lambda,\mu), I\otimes U(z,\lambda)]. \tag{3.7}$$

In what follows, along with the usual matrix trace we shall use the operation tr_1, which is matrix trace in the first factor of the tensor product $\mathbb{C}^2\otimes\mathbb{C}^2$; it carries a matrix in $\mathbb{C}^2\otimes\mathbb{C}^2$ into a matrix in $\mathbb{C}^2$ and is defined by linearity by

$$\mathrm{tr}_1(A\otimes B)=\mathrm{tr}\,A\cdot B, \tag{3.8}$$

where A and B are matrices in $\mathbb{C}^2$. The operation tr_1 is characterized by

$$\mathrm{tr}_1(I\otimes A)X = A\cdot \mathrm{tr}_1 X, \tag{3.9}$$

$$\mathrm{tr}_1 X(I\otimes A) = \mathrm{tr}_1 X\cdot A, \tag{3.10}$$

$$\mathrm{tr}_1(A\otimes I)X = \mathrm{tr}_1 X(A\otimes I), \tag{3.11}$$

where A is a matrix in $\mathbb{C}^2$ and X is a matrix in $\mathbb{C}^2\otimes\mathbb{C}^2$.

Now multiply both sides of (3.7) by $Q(\theta)\otimes I$ from the right, take the trace tr_1 and let $x\to L$, $y\to -L$. Using (3.9)–(3.10) we find

$$\{F_L(\mu), U(x,\lambda)\} = \frac{\partial}{\partial x}\tilde{V}(x,\lambda,\mu)+[\tilde{V}(x,\lambda,\mu), U(x,\lambda)], \tag{3.12}$$

with

$$\tilde{V}(x,\lambda,\mu)=\mathrm{tr}_1(M(x,L,-L;\lambda,\mu)(Q(\theta)\otimes I)) \tag{3.13}$$

where the variable z is replaced by x, $-L\leqslant x\leqslant L$. The left hand side of (3.12) is a 2×2 matrix composed of the Poisson brackets between $F_L(\mu)$ and the matrix elements of $U(x,\lambda)$.

The expression for $\tilde{V}(x,\lambda,\mu)$ can be simplified by using the explicit form of the r-matrix (1.19). We have

$$\tilde{V}(x,\lambda,\mu)=\frac{\varkappa}{\lambda-\mu}\operatorname{tr}_1((T(L,x,\mu)\otimes I)P(T(x,-L,\mu)Q(\theta)\otimes I))$$

$$=\frac{\varkappa}{\lambda-\mu}\operatorname{tr}_1(P(T(x,-L,\mu)Q(\theta)T(L,x,\mu)\otimes I))$$

$$=\frac{\varkappa}{\lambda-\mu}\operatorname{tr}_1((I\otimes T(x,-L,\mu)Q(\theta)T(L,x,\mu))P)$$

$$=\frac{\varkappa}{\lambda-\mu}T(x,-L,\mu)Q(\theta)T(L,x,\mu)\cdot\operatorname{tr}_1 P, \tag{3.14}$$

where we have used (3.11), (1.10) and (3.9). Now the explicit form of P (1.11) shows that

$$\operatorname{tr}_1 P=I, \tag{3.15}$$

which yields an expression for $\tilde{V}(x,\lambda,\mu)$,

$$\tilde{V}(x,\lambda,\mu)=\frac{\varkappa}{\lambda-\mu}T(x,-L,\mu)Q(\theta)T(L,x,\mu). \tag{3.16}$$

Let us show that $\tilde{V}(x,\lambda,\mu)$ satisfies the quasi-periodicity condition

$$\tilde{V}(x+2L,\lambda,\mu)=Q^{-1}(\theta)\tilde{V}(x,\lambda,\mu)Q(\theta). \tag{3.17}$$

Following the discussion of the preceding section, we compare the matrices $\tilde{V}(x,\lambda,\mu)$ evaluated at $x=L$ and $x=-L$. We have

$$\tilde{V}(L,\lambda,\mu)=\frac{\varkappa}{\lambda-\mu}T_L(\mu)Q(\theta) \tag{3.18}$$

and

$$\tilde{V}(-L,\lambda,\mu)=\frac{\varkappa}{\lambda-\mu}Q(\theta)T_L(\mu), \tag{3.19}$$

so that

$$\tilde{V}(L,\lambda,\mu)=Q^{-1}(\theta)\tilde{V}(-L,\lambda,\mu)Q(\theta). \tag{3.20}$$

This equation together with the smoothness of the matrix elements of $\tilde{V}(x,\lambda,\mu)$ for $-L<x<L$ assures that the latter extends to the whole line $-\infty<x<\infty$ under the quasi-periodicity condition (3.17).

Thus we have seen that both left and right hand sides of (3.12) are well defined under the boundary conditions in question. As a corollary, all the commuting flows on the phase space $\mathcal{M}_{L,\theta}$ induced by the general equations of motion

$$\frac{\partial \psi}{\partial t} = \{F_L(\mu), \psi\}, \qquad \frac{\partial \bar{\psi}}{\partial t} = \{F_L(\mu), \bar{\psi}\}, \tag{3.21}$$

or equivalently

$$\frac{\partial U(x,\lambda)}{\partial t} = \{F_L(\mu), U(x,\lambda)\}, \tag{3.22}$$

are representable as the zero curvature condition (3.3) with the matrix $\tilde{V}(x,\lambda,\mu)$ given by (3.16). Notice that our derivation based on the fundamental Poisson brackets was quite general.

However, equations (3.22) are non-local. Let us now show how the above formulae lead to the zero curvature representation for the flows generated by the local integrals I_n, the higher NS Hamiltonians.

To this end we recall the representation (I.4.5) for the transition matrix

$$T(x,y,\mu) = (I + W(x,\mu))\, e^{Z(x,y,\mu)} (I + W(y,\mu))^{-1}, \tag{3.23}$$

where $W(x,\mu)$ and $Z(x,y,\mu)$ are an off-diagonal and a diagonal matrix, respectively. As noted in § I.4, this decomposition is of asymptotic nature for large real μ and holds with the order $O(|\mu|^{-\infty})$, so that the matrices $W(x,\mu)$ and $Z(x,y,\mu) + \frac{i\mu\sigma_3}{2}(x-y)$ are given by asymptotic Taylor series in powers of μ^{-1}. Therefore all the operations below are taken asymptotically and all the formulae are valid with the order $O(|\mu|^{-\infty})$ which will not be specially mentioned any more.

Now insert the decomposition (3.23) into (3.16). The quasi-periodicity condition

$$W(x+2L,\mu) = Q^{-1}(\theta)\, W(x,\mu)\, Q(\theta) \tag{3.24}$$

and the fact that $Z(x,y,\mu)$ is diagonal yield an expression for $\tilde{V}(x,\lambda,\mu)$:

$$\tilde{V}(x,\lambda,\mu) = \frac{\varkappa (I + W(x,\mu))\, e^{Z_L(\mu)} Q(\theta) (I + W(x,\mu))^{-1}}{\lambda - \mu}. \tag{3.25}$$

Next, notice that (3.23)–(3.24) give the decomposition

$$T_L(\mu)\, Q(\theta) = (I + W(L,\mu))\, e^{Z_L(\mu)} Q(\theta) (I + W(L,\mu))^{-1}. \tag{3.26}$$

The fact that $T_L(\mu)Q(\theta)$ is unimodular and the definition

$$\operatorname{tr} T_L(\mu)\,Q(\theta) = 2\cos p_L(\mu) \tag{3.27}$$

then yield

$$e^{Z_L(\mu)}Q(\theta) = \cos p_L(\mu)\,I + i\sin p_L(\mu)\,\sigma_3 . \tag{3.28}$$

Substituting this into (3.25) we find

$$\begin{aligned}\tilde V(x,\lambda,\mu) = {} & \frac{\varkappa}{\lambda-\mu}\cos p_L(\mu)\,I + \frac{i\varkappa}{\lambda-\mu}\sin p_L(\mu) \\ & \times (I+W(x,\mu))\,\sigma_3\,(I+W(x,\mu))^{-1} .\end{aligned} \tag{3.29}$$

Now observe that the first term in (3.29) does not depend on x and is proportional to the unit matrix. Hence it does not contribute to the right hand side of (3.12) and may be left out. Thus we obtain the required expression for $\tilde V(x,\lambda,\mu)$

$$\tilde V(x,\lambda,\mu) = -2\sin p_L(\mu)\,V(x,\lambda,\mu) , \tag{3.30}$$

where

$$V(x,\lambda,\mu) = \frac{\varkappa}{2i(\lambda-\mu)}(I+W(x,\mu))\,\sigma_3\,(I+W(x,\mu))^{-1} . \tag{3.31}$$

After these transformations let us go back to the problem of the zero curvature representation for (3.1) or, equivalently, for the equation

$$\frac{\partial U(x,\lambda)}{\partial t} = \{p_L(\mu),\, U(x,\lambda)\} . \tag{3.32}$$

The elementary relation

$$\frac{d}{dt}\arccos f(t) = -\frac{1}{\sqrt{1-f^2(t)}}\,\frac{df(t)}{dt} \tag{3.33}$$

together with (3.12), (3.30) yields

$$\{p_L(\mu),\, U(x,\lambda)\} = \frac{\partial V}{\partial x}(x,\lambda,\mu) + [V(x,\lambda,\mu),\, U(x,\lambda)] . \tag{3.34}$$

It follows that (3.32) admits a zero curvature representation understood asymptotically to the order $O(|\mu|^{-\infty})$.

The matrix $V(x, \lambda, \mu)$ is the generating function for the matrices $V_n(x, \lambda)$ appearing in the zero curvature representation for the higher NS equations. In fact, it was shown in § I.4 that $W(x, \mu)$ has the asymptotic expansion

$$W(x, \mu) = \sum_{n=1}^{\infty} \frac{W_n(x)}{\mu^n} + O(|\mu|^{-\infty}), \tag{3.35}$$

where the $W_n(x)$ are polynomials in $\psi(x)$, $\bar{\psi}(x)$ and their derivatives at x. Substituting this expansion into (3.31) and expanding $\frac{1}{\lambda-\mu}$ in inverse powers of μ we find

$$V(x, \lambda, \mu) = \varkappa \sum_{n=1}^{\infty} \frac{V_n(x, \lambda)}{\mu^n} + O(|\mu|^{-\infty}), \tag{3.36}$$

where the coefficients $V_n(x, \lambda)$ are computed explicitly. The first two are $V_1 = \frac{i}{2}\sigma_3$, $V_2(x, \lambda) = -U(x, \lambda)$, and $V_3(x, \lambda)$ coincides with $V(x, \lambda)$ from § I.2 (see (I.2.6)–(I.2.8)).

By comparing (3.35) with (2.27) we see that all the higher NS equations

$$\frac{\partial \psi}{\partial t} = \{I_n, \psi\}, \qquad \frac{\partial \bar{\psi}}{\partial t} = \{I_n, \bar{\psi}\}, \tag{3.37}$$

$n = 1, 2, \ldots,$ *admit a zero curvature representation with matrices $U(x, \lambda)$ and $V_n(x, \lambda)$.* The matrix elements of $V_n(x, \lambda)$ are polynomials in λ, $\psi(x)$, $\bar{\psi}(x)$ and their derivatives at x, whose degree with respect to λ is $n-1$. The corresponding zero curvature equation involves a polynomial in λ of degree n; the coefficients of $\lambda, \ldots, \lambda^n$ vanish identically whereas the vanishing of the constant term is equivalent to the n-th NS equation.

We have thus shown that the fundamental Poisson brackets provide an alternative to the zero curvature representation. To conclude this section, we shall make the following general comment.

The zero curvature condition which is fundamental to the inverse scattering method looks somewhat mystical and appears in § I.2 as nothing more than a remarkable computational observation. The Hamiltonian approach based on the concepts of the r-matrix and fundamental Poisson brackets provides a natural explanation for this observation. However, it may also be said that the intrinsic meaning of the r-matrix remains obscure. We hope, nevertheless, that by the end of Part II it will be made quite transparent in the light of general Lie-algebraic considerations.

§ 4. Integrals of the Motion in the Rapidly Decreasing Case and in the Case of Finite Density

In this section we return, from the Hamiltonian standpoint, to the integrals of the motion under the rapidly decreasing or finite density boundary conditions.

We shall begin with the rapidly decreasing case. As was shown in § I.7, the local integrals of the motion, I_n, result from the corresponding functionals for the quasi-periodic case with $\theta=0$ by a passage to the limit as $L\to\infty$. By construction, the integrals I_n correspond to observables on the phase space $\mathscr{M}_0$. Hence their Poisson brackets, too, are obtained by taking the limit as $L\to\infty$, so that, by the arguments of § 2, the integrals I_n are in involution

$$\{I_n, I_m\}=0. \tag{4.1}$$

Thus *the flows on the space* $\mathscr{M}_0$ *generated by the higher NS equations commute with each other, and the equations themselves admit a zero curvature representation with* $U(x,\lambda)$ *and* $V_n(x,\lambda)$ *given by the same formulae as in § 3.*

The role of the generating function for the motion integrals I_n is played by $\frac{1}{i}\log a(\lambda)$ where the transition coefficient $a(\lambda)$ is as defined in § I.5. The asymptotic expansion of $\frac{1}{i}\log a(\lambda)$ in powers of λ^{-1} results from the corresponding expansion of $p_L(\lambda)+\lambda L$ as $L\to\infty$ (see § I.7). One may then expect that the functionals $\frac{1}{i}\log a(\lambda)$ are in involution, as well as $p_L(\lambda)$. This will be proved in §§ 6–7 by showing that the functionals $a(\lambda)$ for $\operatorname{Im}\lambda>0$ correspond to observables on the phase space $\mathscr{M}_0$ and have vanishing Poisson brackets $\{a(\lambda), a(\mu)\}$ and $\{a(\lambda), \bar{a}(\mu)\}$.

Let us now consider the finite density boundary conditions. As was noted in § I.10, here the quasi-periodic functionals I_n have no limit as $L\to\infty$. To regularize them one uses the asymptotic expansions of $p_L(\lambda)+kL$ in powers of λ^{-1} or of k^{-1} where $k(\lambda)=\sqrt{\lambda^2-\omega^2}$ (see § I.8). It is then possible to take the limit as $L\to\infty$ termwise and obtain functionals on the phase space $\mathscr{M}_{\varrho,\theta}$. However, as was noted in § I.10, there may occur inadmissible functionals, i.e. functionals which have no associated observable on $\mathscr{M}_{\varrho,\theta}$. Also, it was noted that *admissible functionals result from the expansion of* $p_L(\lambda)+kL$ *in powers of* k^{-1}. *Here we shall prove this claim by using the expression for the Poisson brackets* $\{p_L(\mu), U(x,\lambda)\}$ *derived in the previous section.*

Recall the expansion

$$p_L(\lambda) = -kL + \frac{\theta}{2} + \varkappa \sum_{n=1}^{\infty} \frac{J_n}{k^n} + O(|k|^{-\infty}), \tag{4.2}$$

for λ in $\mathbb{R}_\omega$ (i.e. $|\lambda| \geq \omega$ and λ is real), as $|\lambda| \to \infty$. The functionals J_n have limits as $L \to \infty$,

$$J_{n,\varrho} = \lim_{L\to\infty} J_n, \tag{4.3}$$

in particular, $J_{1,\varrho} = N_\varrho$, $J_{2,\varrho} = P_\varrho$, $J_{3,\varrho} = H_\varrho$. The generating function for the integrals of the motion $J_{n,\varrho}$ is $\frac{1}{i}\log a_\varrho(\lambda) e^{-\frac{i\theta}{2}}$. Its asymptotic expansion for $|\lambda| \to \infty$, λ in $\mathbb{R}_\omega$, results from (4.2) by taking the limit as $L \to \infty$ (see §§ I.9–I.10)

$$\frac{1}{i}\log a_\varrho(\lambda) e^{-\frac{i\theta}{2}} = \varkappa \sum_{n=1}^{\infty} \frac{J_{n,\varrho}}{k^n} + O(|k|^{-\infty}). \tag{4.4}$$

We will show that, if $n > 1$, the variational derivatives $\frac{\delta J_{n,\varrho}}{\delta \psi(x)}$ and $\frac{\delta J_{n,\varrho}}{\delta \bar{\psi}(x)}$ vanish as $|x| \to \infty$, so that the $J_{n,\varrho}$, $n > 1$, in contrast to $J_{1,\varrho}$ (see § I.1), are admissible functionals on $\mathscr{M}_{\varrho,\theta}$.

For that purpose recall the basic formulae of the preceding section,

$$\{p_L(\mu), U(x,\lambda)\} = \frac{\partial V}{\partial x}(x,\lambda,\mu) + [V(x,\lambda,\mu), U(x,\lambda)] \tag{4.5}$$

and

$$V(x,\lambda,\mu) = \frac{\varkappa}{2i(\lambda-\mu)}(I + W(x,\mu))\sigma_3(I + W(x,\mu))^{-1}. \tag{4.6}$$

These equations and the subsequent ones should be considered asymptotically to the order $O(|\mu|^{-\infty})$ which will be assumed for the rest of the section. The off-diagonal matrix $W(x,\mu)$ appearing in (4.6) has an asymptotic expansion

$$W(x,\mu) = \sum_{n=1}^{\infty} \frac{W_n(x)}{\mu^n} + O(|\mu|^{-\infty}) \tag{4.7}$$

and satisfies the Riccati equation of § I.4

$$\frac{dW}{dx} + i\mu\sigma_3 W - U_0 + W U_0 W = 0, \tag{4.8}$$

with $U_0(x) = \sqrt{\varkappa}\,(\bar{\psi}(x)\sigma_+ + \psi(x)\sigma_-)$.

Using an explicit expression for $U(x,\lambda)$ given in § 1 we see that the left hand side of (4.5) has the form

$$\{p_L(\mu), U(x,\lambda)\} = i\sqrt{\varkappa}\left(\frac{\delta p_L(\mu)}{\delta\psi(x)}\sigma_+ - \frac{\delta p_L(\mu)}{\delta\bar{\psi}(x)}\sigma_-\right) \tag{4.9}$$

and contains, in the limit as $L\to\infty$, the relevant variational derivatives $\frac{\delta J_{n,\varrho}}{\delta\psi(x)}, \frac{\delta J_{n,\varrho}}{\delta\bar{\psi}(x)}$. By comparing (4.5), (4.9) and (4.6) it follows that the behaviour of these variational derivatives, as $x\to\pm\infty$, is governed by $V(x,\lambda,\mu)$ and in the long run by $W(x,\mu)$. Thus we come down to the problem of determining the limit of $W(x,\mu)$ as $x\to\pm\infty$,

$$W_\pm(\mu) = \lim_{x\to\pm\infty} W(x,\mu) \tag{4.10}$$

for μ in $\mathbb{R}_\omega$, under the finite density boundary conditions. In terms of $U_0(x)$, these boundary conditions are

$$\lim_{x\to\pm\infty} U_0(x) = U_\pm\,, \qquad U_+ = Q^{-1}(\theta)\,U_-\,Q(\theta), \tag{4.11}$$

and $U_- = \frac{\omega}{2}\sigma_1$ (see § I.8).

The existence of the limits (4.10) is an immediate consequence of the expression for the $W_n(x)$, § I.4. To evaluate them we take the limit of (4.8), as $x\to-\infty$. Denoting $W(\mu) = W_-(\mu)$ we have

$$\frac{\omega}{2} W\sigma_1 W + i\mu\sigma_3 W - \frac{\omega}{2}\sigma_1 = 0. \tag{4.12}$$

Introducing a diagonal matrix $X = W\sigma_1$ we can write the last equation as

$$\frac{\omega}{2} X^2 + i\mu\sigma_3 X - \frac{\omega}{2} I = 0, \tag{4.13}$$

or

$$\left(X + \frac{i\mu\sigma_3}{\omega}\right)^2 = -\frac{k^2(\mu)}{\omega^2} I. \tag{4.14}$$

Equation (4.14) for a diagonal matrix X has four solutions. However, only one of them is consistent with the asymptotic expansion (4.7) for μ in $\mathbb{R}_\omega$ since

$$k(\mu)=\mu-\sum_{n=1}^{\infty}\frac{p_n}{\mu^n}+O(|\mu|^{-\infty}) \tag{4.15}$$

for all such μ (see § I.10). The solution is

$$X=\frac{i(k-\mu)}{\omega}\sigma_3, \tag{4.16}$$

so that

$$W_-(\mu)=\frac{\mu-k}{\omega}\sigma_2. \tag{4.17}$$

For $W_+(\mu)$ we have, from (4.11),

$$W_+(\mu)=Q^{-1}(\theta)\,W_-(\mu)\,Q(\theta). \tag{4.18}$$

Inserting these expressions into (4.6) we see that the limits of $V(x,\lambda,\mu)$, as $x\to\pm\infty$, for μ in $\mathbb{R}_\omega$, are

$$\lim_{x\to\pm\infty} V(x,\lambda,\mu)=V_\pm(\lambda,\mu), \tag{4.19}$$

where

$$V_+(\lambda,\mu)=Q^{-1}(\theta)\,V_-(\lambda,\mu)\,Q(\theta) \tag{4.20}$$

and

$$V_-(\lambda,\mu)=\frac{\varkappa((1+\eta^2)\sigma_3+2i\eta\sigma_1)}{2i(\lambda-\mu)(1-\eta^2)} \tag{4.21}$$

with the definition $\eta=\dfrac{\mu-k(\mu)}{\omega}$. It follows that the limits of the right hand side of (4.5) as $x\to\pm\infty$ have the form

$$P_\pm=\lim_{x\to\pm\infty}[V(x,\lambda,\mu),U(x,\lambda)]=\left[V_\pm(\lambda,\mu),\frac{\lambda\sigma_3}{2i}+U_\pm\right], \tag{4.22}$$

where

$$P_+(\mu)=Q^{-1}(\theta)P_-(\mu)Q(\theta),\qquad P_-(\mu)=-\frac{\varkappa\omega\sigma_2}{2k(\mu)}. \tag{4.23}$$

So we have shown in our case that the limits

$$\lim_{x\to\pm\infty}\left\{\frac{1}{i}\log a_\varrho(\mu)e^{-\frac{i\theta}{2}}, U(x,\lambda)\right\}=P_\pm(\mu) \tag{4.24}$$

exist. Now the asymptotic expansion (4.4), the expression for the variational derivatives

$$\frac{\delta J_{1,\varrho}}{\delta\psi(x)}=\bar\psi(x),\qquad \frac{\delta J_{1,\varrho}}{\delta\bar\psi(x)}=\psi(x) \tag{4.25}$$

(see § I.1), and (4.23) combined with (4.24) allow us to conclude that

$$\lim_{x\to\pm\infty}\frac{\delta J_{n,\varrho}}{\delta\psi(x)}=\lim_{x\to\pm\infty}\frac{\delta J_{n,\varrho}}{\delta\bar\psi(x)}=0. \tag{4.26}$$

We have thus shown that the functionals $J_{n,\varrho}$, $n>1$, are admissible on $\mathscr{M}_{\varrho,\theta}$. Relation (4.3) proves that they are in involution,

$$\{J_{n,\varrho}, J_{m,\varrho}\}=0. \tag{4.27}$$

Hence, in the finite density case these are the functionals the *higher NS equations* are naturally associated with,

$$\frac{\partial\psi}{\partial t}=\{J_{n,\varrho},\psi\},\qquad \frac{\partial\bar\psi}{\partial t}=\{J_{n,\varrho},\bar\psi\}. \tag{4.28}$$

These equations admit a zero curvature representation with the matrices $U(x,\lambda)$ and $V_{n,\varrho}(x,\lambda)$, $n>1$, determined by the asymptotic expansion

$$V(x,\lambda,\mu)=\varkappa\sum_{n=1}^{\infty}\frac{V_{n,\varrho}(x,\lambda)}{k^n(\mu)}+O(|k(\mu)|^{-\infty}) \tag{4.29}$$

for μ in $\mathbb{R}_\omega$. It is obtained by reexpanding the asymptotic series (3.36) in powers of $k^{-1}(\mu)$ (cf. a similar operation in § I.10). In particular, $V_{2,\varrho}(x,\lambda)=V_2(x,\lambda)$ and $V_{3,\varrho}(x,\lambda)=V_\varrho(x,\lambda)$ with $V_\varrho(x,\lambda)$ defined in § I.2.

So this section gives further evidence of the utility of the notion of an r-matrix. Using the general zero curvature representation (3.3) derived in § 3 from the fundamental Poisson brackets we were able to analyze the local integrals of the motion in the case of finite density and select the admissible functionals. Another application of the basic formulae of § 3 can be found in the next section.

§ 5. The Λ-Operator and a Hierarchy of Poisson Structures

In § 3 we derived the generating function, $V(x,\lambda,\mu)$, for the zero curvature representation of the higher NS equations. Here we shall introduce compact notation for these equations and their local integrals of the motion I_n. The Hamiltonian interpretation of the corresponding formulae leads naturally to a family (hierarchy) of Poisson structures. The NS model and the higher NS equations prove to be Hamilton's equations with respect to each of these structures. The associated Hamiltonians are just all of the local integrals I_n. A Lie-algebraic interpretation of these results will be given in Part II.

As usual, we begin with the quasi-periodic case. Recall (see § 3) that $V(x,\lambda,\mu)$, the generating function for the $V_n(x,\lambda)$,

$$V(x,\lambda,\mu)=\varkappa \sum_{n=1}^{\infty} \frac{V_n(x,\lambda)}{\mu^n}, \tag{5.1}$$

can be expressed as

$$V(x,\lambda,\mu)=\frac{\varkappa}{2i(\lambda-\mu)} M(x,\mu), \tag{5.2}$$

where

$$M(x,\mu)=(I+W(x,\mu))\sigma_3(I+W(x,\mu))^{-1}=\sigma_3+\sum_{n=1}^{\infty}\frac{M_n(x)}{\mu^n}. \tag{5.3}$$

The matrices $M_n(x)$ depend only on the functions $\psi(x)$, $\bar\psi(x)$ and their derivatives at x and have zero trace.

These identities and those to follow should be understood asymptotically to the order $O(|\mu|^{-\infty})$, which will be assumed for the rest of the section.Λ-Operator and

By comparing (5.1)–(5.3) we find

$$V_n(x,\lambda)=\frac{i}{2}\left(\lambda^{n-1}\sigma_3+\sum_{k=0}^{n-2}\lambda^k M_{n-k-1}(x)\right), \tag{5.4}$$

so that the n-th NS equation is given by the coefficients $M_k(x)$, $k\leqslant n-1$. By using (5.3) these can be calculated from the asymptotic expansion of $W(x,\mu)$ obtained in § I.4 by means of the Riccati equation. *Here we shall outline a more direct method for computing the $M_n(x)$.*

Observe that $M(x,\lambda)$ satisfies the differential equation

$$\frac{dM}{dx}=[U(x,\lambda),M], \tag{5.5}$$

with

$$U(x,\lambda)=\frac{\lambda\sigma_3}{2i}+U_0(x) \tag{5.6}$$

and the quasi-periodicity condition

$$M(x+2L,\lambda)=Q^{-1}(\theta)M(x,\lambda)Q(\theta) \tag{5.7}$$

(see (3.24)).

In fact, $M(x,\lambda)$ may be written as

$$M(x,\lambda)=i\operatorname{ctg}p_L(\lambda)I+\frac{1}{i\sin p_L(\lambda)}T(x,-L,\lambda)Q(\theta)T(L,x,\lambda) \tag{5.8}$$

(see (3.16) and (3.29)), whence by using the differential equations (1.31) and (1.39) for the transition matrix with respect to the first and the second variables we get (5.5).

Now express $M(x,\lambda)$ as

$$M(x,\lambda)=M^{(d)}(x,\lambda)+M^{(od)}(x,\lambda), \tag{5.9}$$

where $M^{(d)}$ and $M^{(od)}$ indicate respectively diagonal and off-diagonal parts of M, and insert this into (5.5). Splitting the resulting equation into diagonal and off-diagonal parts gives

$$\frac{dM^{(d)}}{dx}=[U_0(x),M^{(od)}], \tag{5.10}$$

$$\frac{dM^{(od)}}{dx}=\frac{\lambda}{2i}[\sigma_3,M^{(od)}]+[U_0(x),M^{(d)}]. \tag{5.11}$$

Using (5.10), we can formally write $M^{(d)}(x,\lambda)$ in terms of $M^{(od)}(x,\lambda)$ as

$$M^{(d)}(x,\lambda)=d^{-1}([U_0(\cdot),M^{(od)}(\cdot,\lambda)])(x)+\sigma_3. \tag{5.12}$$

Substituting this into (5.11) we obtain an integro-differential equation for $M^{(od)}(x,\lambda)$

$$\frac{dM^{(od)}}{dx}+i\lambda\sigma_3 M^{(od)}-[U_0(x),d^{-1}([U_0,M^{(od)}])(x)]=2\,U_0(x)\sigma_3\,, \qquad (5.13)$$

whose derivation made use of the fact that the diagonal matrix σ_3 anticommutes with the off-diagonal matrices $M^{(od)}(x,\lambda)$ and $U_0(x)$.

Let the operator Λ on the space of off-diagonal matrices $F(x)$ be defined by

$$\Lambda F(x)=i\sigma_3\left(\frac{dF}{dx}(x)-[U_0(x),d^{-1}([U_0(\cdot),F(\cdot)])(x)]\right). \qquad (5.14)$$

Using Λ we can write (5.13) in the following compact form:

$$(\Lambda-\lambda)M^{(od)}(x,\lambda)=-2i\,U_0(x). \qquad (5.15)$$

We then have, formally,

$$M^{(od)}(x,\lambda)=-2i(\Lambda-\lambda)^{-1}U_0(x). \qquad (5.16)$$

Expanding $(\Lambda-\lambda)^{-1}$ in inverse powers of λ we find an explicit expression for the coefficients $M_n^{(od)}(x)$,

$$M_n^{(od)}(x)=2i\Lambda^{n-1}U_0(x), \qquad (5.17)$$

so that

$$M_n^{(od)}(x)=\Lambda M_{n-1}^{(od)}(x), \qquad n>1, \qquad (5.18)$$

and

$$M_1^{(od)}(x)=2i\,U_0(x). \qquad (5.19)$$

The matrices $M_n^{(d)}(x)$ are recovered from the $M_n^{(od)}(x)$ according to (5.12),

$$M_n^{(d)}(x)=d^{-1}([U_0(\cdot),M_n^{(od)}(\cdot)])(x). \qquad (5.20)$$

By virtue of (5.18), Λ is sometimes referred to as a *recursion operator*. We shall use the more expressive though colloquial name of *Λ-operator*.

The above derivation contains an ambiguity in the integration operator d^{-1}. However, by virtue of (5.10) we know in advance that this operator is evaluated on matrices which are total derivatives with respect to x. The matrices $M_n^{(d)}(x)$ whose derivative we have in mind are of the form

$$M_n^{(d)}(x)=f_n(x)\sigma_3, \tag{5.21}$$

where the periodic function $f_n(x)$ is a polynomial in $\psi(x)$, $\bar{\psi}(x)$ and their derivatives with respect to x, with no free term. By definition, d^{-1} gives the latter matrix.

In particular, d^{-1} is consistent with the following boundary conditions for $M^{(d)}(x,\lambda)$ and $M^{(od)}(x,\lambda)$ considered as functionals of $\psi(x)$, $\bar{\psi}(x)$:

$$M^{(od)}(x,\lambda)|_{\psi=\bar{\psi}=0}=0, \quad M^{(d)}(x,\lambda)|_{\psi=\bar{\psi}=0}=\sigma_3. \tag{5.22}$$

The free term σ_3 in (5.12) results from these conditions.

At first sight our definition is a tautology. Still, the relevant fact is that if $M_n^{(od)}(x)$ is given by (5.18)–(5.19) then $[U_0(x), M_n^{(od)}(x)]$ proves to be always a total derivative. In a different way, one can say that $\Lambda^n U_0(x)$ is well defined for all $n \geqslant 0$.

Let us now use the expression for $M^{(od)}(x,\lambda)$ in order to completely determine $M(x,\lambda)$. Writing (5.12) as

$$M^{(d)}(x,\lambda)=\sigma_3-2id^{-1}([U_0,(\Lambda-\lambda)^{-1}U_0])(x), \tag{5.23}$$

we find an explicit expression for $M(x,\lambda)$ in terms of the Λ-operator:

$$M(x,\lambda)=\sigma_3-2i(\Lambda-\lambda)^{-1}U_0(x)-2id^{-1}([U_0,(\Lambda-\lambda)^{-1}U_0])(x). \tag{5.24}$$

As a first corollary of the above results *we shall derive a compact form of the higher NS equations*. The zero curvature representation

$$\frac{\partial U(x,\lambda)}{\partial t}-\frac{\partial V_n}{\partial x}(x,\lambda)+[U(x,\lambda),V_n(x,\lambda)]=0 \tag{5.25}$$

together with (5.4), (5.6) imply that the n-th NS equation, which is the constant term in (5.25) with respect to λ, is written as

$$\frac{\partial U_0}{\partial t}-\frac{i}{2}\frac{\partial M_{n-1}^{(od)}}{\partial x}+\frac{i}{2}[U_0,M_{n-1}^{(d)}]=0. \tag{5.26}$$

By (5.17) and (5.20) this equation can be written as

$$\frac{\partial U_0}{\partial t}+\frac{\partial \Lambda^{n-2}U_0}{\partial x}-[U_0,d^{-1}([U_0,\Lambda^{n-2}U_0])]=0. \tag{5.27}$$

With the definition (5.14), this leads to the desired expression for the n-th NS equation,

$$\frac{\partial U_0(x)}{\partial t} = i\sigma_3 \Lambda^{n-1} U_0(x). \tag{5.28}$$

Let us compare (5.28) with the Hamiltonian form of the same equation

$$\frac{\partial U_0(x)}{\partial t} = \{I_n, U_0(x)\}. \tag{5.29}$$

The matrix on the right hand side is given by

$$\{I_n, U_0(x)\} = i\sqrt{\varkappa}\left(\frac{\delta I_n}{\delta \psi(x)}\sigma_+ - \frac{\delta I_n}{\delta \bar{\psi}(x)}\sigma_-\right). \tag{5.30}$$

Let the matrix $\operatorname{grad} I_n(x)$ be defined by

$$\operatorname{grad} I_n(x) = \frac{1}{\sqrt{\varkappa}}\left(\frac{\delta I_n}{\delta \psi(x)}\sigma_+ + \frac{\delta I_n}{\delta \bar{\psi}(x)}\sigma_-\right). \tag{5.31}$$

We shall explain later why this notation is natural from the Hamiltonian point of view. The equation of motion (5.29) becomes

$$\frac{\partial U_0}{\partial t} = i\varkappa\sigma_3 \operatorname{grad} I_n. \tag{5.32}$$

By comparing (5.28) with (5.32) *we find a compact expression for the gradients of the local integrals of the motion,*

$$\operatorname{grad} I_n(x) = \frac{1}{\varkappa}\Lambda^{n-1} U_0(x), \tag{5.33}$$

or the recursion relation

$$\operatorname{grad} I_n(x) = \Lambda \operatorname{grad} I_{n-1}(x), \qquad n > 1, \tag{5.34}$$

where

$$\operatorname{grad} I_1(x) = \frac{1}{\varkappa} U_0(x). \tag{5.35}$$

Next we will show that the Λ-operator also provides expressions for the local integrals I_n. For that purpose, instead of the generating function $p_L(\lambda) = \arccos(\frac{1}{2}\operatorname{tr} T_L(\lambda) Q(\theta))$ it will be convenient to deal with its derivative with respect to λ. We will show that

$$\frac{dp_L(\lambda)}{d\lambda} = -\frac{1}{4}\int_{-L}^{L} \operatorname{tr} M(x,\lambda)\sigma_3\,dx. \tag{5.36}$$

We shall start from the basic representation for the monodromy matrix

$$T_L(\lambda) = \widehat{\exp}\int_{-L}^{L} U(x,\lambda)\,dx \tag{5.37}$$

(see §§ I.2–I.3). Differentiation with respect to λ gives

$$\frac{dT_L(\lambda)}{d\lambda} = \int_{-L}^{L} T(L,x,\lambda)\frac{\partial U}{\partial\lambda}(x,\lambda)\,T(x,-L,\lambda)\,dx$$

$$= \frac{1}{2i}\int_{-L}^{L} T(L,x,\lambda)\sigma_3 T(x,-L,\lambda)\,dx, \tag{5.38}$$

so that

$$\frac{d}{d\lambda}\operatorname{tr} T_L(\lambda)Q(\theta) = \frac{1}{2i}\int_{-L}^{L} \operatorname{tr}(T(x,-L,\lambda)Q(\theta)T(L,x,\lambda)\sigma_3)\,dx$$

$$= \frac{\sin p_L(\lambda)}{2}\int_{-L}^{L} \operatorname{tr} M(x,\lambda)\sigma_3\,dx. \tag{5.39}$$

In deriving the last identity we made use of (5.8). Now, (5.36) follows from (5.39) by the differentiation rule for composite functions.

The right hand side of (5.36) has an explicit expression in terms of the Λ-operator via (5.23),

$$\int_{-L}^{L} \operatorname{tr} M(x,\lambda)\sigma_3\,dx = 4L + 4i\int_{-L}^{L} d^{-1}\operatorname{tr}(U_0\sigma_3(\Lambda-\lambda)^{-1}U_0)(x)\,dx, \tag{5.40}$$

while the left hand side may be taken, along with $p_L(\lambda)$, as a generating function for local integrals of the motion,

$$\frac{d}{d\lambda}p_L(\lambda) = -L - \varkappa\sum_{n=1}^{\infty}\frac{nI_n}{\lambda^{n+1}}. \tag{5.41}$$

Comparing (5.41) with the expansion of (5.40) in inverse powers of λ *we finally obtain*

$$I_n = \frac{1}{i\varkappa n} \int_{-L}^{L} d^{-1}(\operatorname{tr} U_0 \sigma_3 \Lambda^n U_0)(x)\,dx, \qquad n \geqslant 1. \tag{5.42}$$

The coefficient of λ^{-1} in the expansion for the right hand side of (5.40) vanishes because $U_0(x)\sigma_3 U_0(x)$ is traceless.

It might be of interest to observe that (5.42) *can be extended over the negative indices n producing a sequence of non-local integrals I_n, $n<0$.*

The starting point here is the Taylor series expansion of the entire function $p_L(\lambda)$

$$p_L(\lambda) = \varkappa \sum_{n=0}^{\infty} I_{-n} \lambda^n. \tag{5.43}$$

The expressions (5.12) and (5.15) for the matrix $M(x,\lambda)$ given by (5.8), as well as (5.36) relating $\frac{dp_L}{d\lambda}(\lambda)$ to $M(x,\lambda)$, are independent of the asymptotic expansion in powers of λ^{-1}. By definition, the operator d^{-1} transforms $[U_0(x), M^{(od)}(x,\lambda)]$ into $M^{(d)}(x,\lambda)-\sigma_3$, in accordance with the boundary conditions (5.22). It results that the Taylor series coefficients of $M^{(od)}(x,\lambda)$,

$$M^{(od)}(x,\lambda) = -\sum_{n=0}^{\infty} M_{-n}^{(od)}(x)\lambda^n \tag{5.44}$$

by virtue of (5.15) satisfy

$$\Lambda M_{-n}^{(od)}(x) = M_{-n+1}^{(od)}(x) \tag{5.45}$$

and

$$\Lambda M_0^{(od)}(x) = 2i\,U_0(x). \tag{5.46}$$

These equations are extensions of (5.18)–(5.19), so that

$$M_n^{(od)}(x) = 2i\Lambda^{n-1} U_0(x) \tag{5.47}$$

holds for all integers *n*.

The same reasoning as for positive *n* leads to (5.42) for negative *n*.

Formally, the integral $I_0 = \frac{1}{\varkappa} p_L(0)$ does not fit into the family (5.42). Nevertheless one can show that L'Hopital's rule applies to (5.42) with the result

$$I_0 = \frac{1}{i\varkappa} \int_{-L}^{L} d^{-1}(\operatorname{tr} U_0 \sigma_3 \log \Lambda\, U_0)(x)\, dx. \tag{5.48}$$

We will not present here the involved and uninteresting details of the computation.

Expressions such as (5.33) for the gradients of the motion integrals are also valid for all integers n. For the proof one should expand the matrix $V(x, \lambda, \mu)$ of the zero curvature representation (3.34) into Taylor series in μ

$$V(x, \lambda, \mu) = \varkappa \sum_{n=0}^{\infty} V_{-n}(x, \lambda) \mu^n \tag{5.49}$$

and find the coefficients by using (5.2) and (5.44).

Of course, computationally the above formulae are of small efficiency because Λ cannot be inverted in closed form. However, we shall see that they are interesting enough withing the Hamiltonian ideology. Namely, these formulae will allow us to define the aforementioned *hierarchy of Poisson structures*.

To begin with, we shall explain why it is natural, for a matrix such as (5.31), to use the notation grad which we now extend to any observable F by

$$\operatorname{grad} F(x) = \frac{1}{\sqrt{\varkappa}} \left(\frac{\delta F}{\delta \psi(x)} \sigma_+ + \frac{\delta F}{\delta \bar{\psi}(x)} \sigma_- \right). \tag{5.50}$$

With this notation the Poisson bracket of two observables, F and G, on the phase space $\mathscr{M}_{L,\theta}$ is

$$\begin{aligned} \{F, G\} &= \frac{\varkappa}{i} \int_{-L}^{L} \operatorname{tr}(\operatorname{grad} F(x) \sigma_3 \operatorname{grad} G(x))\, dx \\ &= \frac{i\varkappa}{2} \int_{-L}^{L} \operatorname{tr}(\sigma_3 [\operatorname{grad} F(x), \operatorname{grad} G(x)])\, dx \end{aligned} \tag{5.51}$$

and defines a skew form on gradients, as prescribed by the general formulae of Hamiltonian mechanics. The equations of motion for coordinates $\psi(x)$, $\bar{\psi}(x)$ on the phase space $\mathscr{M}_{L,\theta}$ parametrized by matrices $U_0(x)$ are

$$\frac{\partial U_0(x)}{\partial t} = \{F, U_0(x)\} = i\varkappa \sigma_3 \operatorname{grad} F(x), \tag{5.52}$$

where F acts as a Hamiltonian. Thus the matrix $i\varkappa\sigma_3$ plays the role of the Jacobi matrix in this parametrization.

Let us now consider the n-th NS equation

$$\frac{\partial U_0(x)}{\partial t} = \{I_n, U_0(x)\} = i\varkappa\sigma_3 \operatorname{grad} I_n(x). \tag{5.53}$$

By (5.34) it can be written as

$$\frac{\partial U_0(x)}{\partial t} = i\varkappa\sigma_3 \Lambda^m \operatorname{grad} I_{n-m}(x), \tag{5.54}$$

with m an arbitrary integer. This suggests that the operator $i\varkappa\sigma_3\Lambda^m$ *can also play the role of the Jacobi matrix. The associated hypothetical Poisson structure has the form*

$$\{F, G\}_m = \frac{\varkappa}{i} \int_{-L}^{L} \operatorname{tr}(\operatorname{grad} F(x)\sigma_3 \Lambda^m \operatorname{grad} G(x))\, dx, \tag{5.55}$$

so that equation (5.54) with respect to this Poisson structure is defined by the Hamiltonian I_{n-m}:

$$\frac{\partial U_0(x)}{\partial t} = \{I_{n-m}, U_0(x)\}_m. \tag{5.56}$$

Our basic Poisson structure corresponds to $m=0$.

To make the argument precise, one must verify that the bracket (5.55) is well defined and skew-symmetric and satisfies the Jacobi identity. Here we will check the first two properties. The Jacobi identity will be established in Part II as a consequence of general Lie-algebraic considerations.

We shall start with the bracket $\{F, G\}_1$. The observable G, which is an admissible functional on the phase space $\mathcal{M}_{L,\theta}$ has the gradient satisfying the quasi-periodicity condition

$$\operatorname{grad} G(x+2L) = Q^{-1}(\theta) \operatorname{grad} G(x)\, Q(\theta), \tag{5.57}$$

the same as the quasi-periodicity condition for $U_0(x)$. The definition of the Λ-operator acting on $\operatorname{grad} G(x)$ involves the diagonal matrix $[U_0(x), \operatorname{grad} G(x)]$ periodic in x of period $2L$, as shown by (5.57). If it has zero mean,

$$\int_{-L}^{L} [U_0(x), \operatorname{grad} G(x)]\, dx = 0, \tag{5.58}$$

then $d^{-1}([U_0, \operatorname{grad} G])(x)$ will also be periodic, whereas the off-diagonal matrix $\Lambda \operatorname{grad} G(x)$ will be quasi-periodic. Therefore the integrand in (5.55) is periodic so that the integral does not depend on the choice of the fundamental domain.

Still, the action of d^{-1} on diagonal traceless matrices with zero mean is defined only up to an additive constant $c\sigma_3$. We will show that if F also satisfies (5.58) then $\{F, G\}_1$ is independent of c and hence well defined.

The ambiguity in $\Lambda \operatorname{grad} G(x)$ amounts to $c\sigma_3[U_0(x), \sigma_3] = -2c\, U_0(x)$; its contribution into (5.55) is

$$\frac{c\varkappa}{i} \int_{-L}^{L} \operatorname{tr}(\operatorname{grad} F(x)[U_0(x), \sigma_3])\, dx = ic\varkappa \int_{-L}^{L} \operatorname{tr}([U_0(x), \operatorname{grad} F(x)]\sigma_3)\, dx = 0, \tag{5.59}$$

if F satisfies (5.58). In fact, since $[U_0(x), \operatorname{grad} G(x)]$ is diagonal and traceless, it is proportional to σ_3 so that (5.59) is equivalent to (5.58) upon replacing G by F.

Thus the Poisson bracket $\{F, G\}_1$ is indeed well defined for F and G satisfying (5.58). Its skew-symmetry follows from the formal skew-symmetry of the operator $i\sigma_3\Lambda$.

The admissibility condition (5.58) has a clear Hamiltonian interpretation. In fact, the relation

$$\int_{-L}^{L} \operatorname{tr}([U_0(x), \operatorname{grad} G(x)]\sigma_3) = 0, \tag{5.60}$$

which is equivalent to (5.58), may be written as

$$\{G, I_1\}_0 = 0, \tag{5.61}$$

if one uses the definition (5.51) of the Poisson bracket $\{,\}_0$ and (5.35). By (5.55) and (5.34) the latter formula may be written in terms of the Poisson bracket $\{,\}_1$

$$\{G, I_0\}_1 = 0. \tag{5.62}$$

From the Jacobi identity claimed above we deduce that if F and G satisfy (5.62) then $\{F, G\}_1$ satisfies (5.62), too. *So (5.61) is a consistent restriction specifying the algebra of observables on which the new Poisson bracket is well defined.*

Let us now consider the Poisson bracket $\{F, G\}_n$ for any positive integer n. By reproducing the preceding arguments we see that this Poisson structure is defined for observables satisfying

$$\{G, I_1\}_0 = \ldots = \{G, I_n\}_0 = 0. \tag{5.63}$$

These conditions may be written in the form

$$\{G, I_{-n+1}\}_n = \ldots = \{G, I_0\}_n = 0, \tag{5.64}$$

which allows to identity the algebra of observables associated with $\{,\}_n$ (up to the yet unverified Jacobi identity).

Similar arguments apply to define the Poisson brackets $\{,\}_n$ for negative n. We shall discuss in detail only the case of $n = -1$.

The formal expression

$$\{F, G\}_{-1} = \frac{\varkappa}{i} \int_{-L}^{L} \operatorname{tr}(\operatorname{grad} F(x) \sigma_3 \Lambda^{-1} \operatorname{grad} G(x))\, dx \tag{5.65}$$

makes sense if $\operatorname{grad} G(x)$ lies in the range $\operatorname{Im} \Lambda$ of Λ,

$$\operatorname{grad} G(x) = \Lambda H(x), \tag{5.66}$$

where the off-diagonal matrix $H(x)$ satisfies

$$\int_{-L}^{L} [U_0(x), H(x)]\, dx = 0, \tag{5.67}$$

that is, lies in the domain of definition of Λ. Condition (5.67) can be written as

$$\begin{aligned} 0 &= \int_{-L}^{L} \operatorname{tr}(H(x) \sigma_3 \operatorname{grad} I_1(x))\, dx = \int_{-L}^{L} \operatorname{tr}(H(x) \sigma_3 \Lambda \operatorname{grad} I_0(x))\, dx \\ &= \int_{-L}^{L} \operatorname{tr}(\operatorname{grad} G(x) \sigma_3 \operatorname{grad} I_0(x))\, dx. \end{aligned} \tag{5.68}$$

We thus find a necessary condition for (5.65) to be well defined:

$$\{G, I_0\}_0 = 0. \tag{5.69}$$

If the same holds for F, then (5.65) is independent of the ambiguity in (5.66) which amounts to an additive term $c\, U_0(x)$. Indeed,

$$c\int_{-L}^{L}\operatorname{tr}(\operatorname{grad}F(x)\sigma_3\Lambda^{-1}U_0(x))\,dx$$

$$=c\varkappa\int_{-L}^{L}\operatorname{tr}(\operatorname{grad}F(x)\sigma_3\Lambda^{-1}\operatorname{grad}I_1(x))\,dx$$

$$=c\varkappa\int_{-L}^{L}\operatorname{tr}(\operatorname{grad}F(x)\sigma_3\operatorname{grad}I_0(x))\,dx=0\,. \tag{5.70}$$

Finally, condition (5.69) is not only necessary but also sufficient for $\operatorname{grad}G(x)$ to lie in $\operatorname{Im}\Lambda$. In fact, it can be stated as the condition of being orthogonal to $\operatorname{Ker}\Lambda$, the kernel of the skew-symmetric operator Λ, considered modulo the ambiguity $c\,U_0(x)$ that occurs in the definition of Λ.

The admissibility condition (5.69) can be written in terms of $\{,\}_{-1}$ as

$$\{G,I_1\}_{-1}=0 \tag{5.71}$$

and serves to specify the algebra of observables in a consistent way.

In a similar manner, associated with the Poisson bracket $\{,\}_{-n}$, $n>1$, are the admissibility conditions

$$\{G,I_0\}_0=\ldots=\{G,I_{-n+1}\}_0=0, \tag{5.72}$$

or

$$\{G,I_n\}_{-n}=\ldots=\{G,I_1\}_{-n}=0\,. \tag{5.73}$$

These specify the algebra of observables associated with the $(-n)$-th Poisson structure.

We point out that the integrals I_n are admissible for the whole hierarchy of Poisson structures $\{,\}_m$ and are in involution with respect to each structure,

$$\{I_k,I_n\}_m=0, \tag{5.74}$$

with k, n and m arbitrary integers.

This completes our discussion of the Poisson structure hierarchy in the quasi-periodic case. To summarize, let us point out that the distinctive property of these structures is the possibility to write down the higher NS equations in the form (5.56).

Finally, let us discuss the limit $L\to\infty$. To avoid irrelevant technical details we shall concentrate on the rapidly decreasing case.

The operator d^{-1} involved in Λ can be defined on the whole space of rapidly decreasing functions $f(x)$ in such a way that it is formally skew-symmetric:

$$d^{-1}f(x)=\frac{1}{2}\left(\int_{-\infty}^{x} f(y)\,dy-\int_{x}^{\infty} f(y)\,dy\right). \tag{5.75}$$

Its range contains, apart from Schwartz functions, also functions with non-vanishing limiting values as $|x|\to\infty$. Nevertheless, Λ maps Schwartz matrices (matrices with Schwartz matrix elements) into matrices of the same kind since d^{-1} is followed by multiplication by the rapidly decreasing function $U_0(x)$.

With this definition of Λ the above formulae for the integrals I_n, their gradients, and the hierarchy of Poisson structures remain valid if the interval of integration $(-L, L)$ in (5.42), (5.48), (5.51), and (5.55) is replaced by the whole real line. In contrast to the quasi-periodic case, the definition of the Poisson brackets $\{,\}_n$ for $n>0$ does not require any truncation of the algebra of observables. However, the Poisson structure proves to be degenerate: there is an annihilator - the centre of the Poisson bracket $\{,\}_n$. The annihilator consists of such observables F that $\operatorname{grad} F(x)$ lies in $\operatorname{Ker}\Lambda^n$.

For the Poisson structures $\{,\}_n$ with $n<0$ the admissibility condition survives, but is much weaker than in the quasi-periodic case. Namely, for F to be admissible the function $\operatorname{grad} F(x)$ must lie in $\operatorname{Im}\Lambda^{-n}$.

These conditions take their simplest form in terms of the action-angle variables to be constructed in § 7.

§ 6. Poisson Brackets of Transition Coefficients in the Rapidly Decreasing Case

As was indicated at the end of § 2, the proof of complete integrability in the rapidly decreasing case proceeds by an explicit construction of canonical variables of action-angle type. In the following section these variables will be written down in terms of the transition coefficients for the continuous and discrete spectra defined in §§ I.5–I.6. Here we shall establish some auxiliary formulae by evaluating *the Poisson brackets of the transition coefficients*.

Remind that transition coefficients for the continuous spectrum come from the reduced monodromy matrix

$$T(\lambda)=\lim_{\substack{x\to\infty\\ y\to-\infty}} E(-x,\lambda)\,T(x,y,\lambda)\,E(y,\lambda)=\begin{pmatrix} a(\lambda) & \varepsilon\bar{b}(\lambda)\\ b(\lambda) & \bar{a}(\lambda)\end{pmatrix}, \tag{6.1}$$

with $\varepsilon=\operatorname{sign}\varkappa$, λ real and $E(x,\lambda)=\exp\left\{\frac{\lambda x}{2i}\sigma_3\right\}$. Transition coefficients for the discrete spectrum occur only if $\varepsilon=-1$ and are defined by

$$T_-^{(1)}(x,\lambda_j)=\gamma_j T_+^{(2)}(x,\lambda_j), \quad j=1,\dots,n, \tag{6.2}$$

where the λ_j are the zeros of $a(\lambda)$ in the upper half λ-plane. Here $T_-^{(1)}(x,\lambda)$ and $T_+^{(2)}(x,\lambda)$ are the first and the second columns of the Jost solutions $T_\pm(x,\lambda)$, respectively, defined, for real λ, by

$$T_\pm(x,\lambda)=\lim_{y\to\pm\infty} T(x,y,\lambda)E(y,\lambda). \tag{6.3}$$

It is these columns of $T_\pm(x,\lambda)$ that have analytic continuation into the upper half-plane (for the details see § I.5–I.6).

We shall start from the basic relation established in § 1,

$$\{T(x,y,\lambda) \underset{,}{\otimes} T(x,y,\mu)\}=[r(\lambda-\mu), T(x,y,\lambda)\otimes T(x,y,\mu)], \quad y<x, \tag{6.4}$$

and take successively the limits indicated in (6.1), (6.3). Our discussion here will proceed on a formal level; an interpretation in terms of the algebra of observables on the phase space $\mathcal{M}_0$ will be given in the next section.

We begin by computing the Poisson brackets of the Jost solutions $T_-(x,\lambda)$ for real λ. For that purpose multiply (6.4) by $E(y,\lambda)\otimes E(y,\mu)$ from the right and let $y\to-\infty$. The left hand side is then the required Poisson bracket matrix $\{T_-(x,\lambda)\underset{,}{\otimes} T_-(x,\mu)\}$. To evaluate the limit on the right hand side, we split the commutator in (6.4) into two summands. Remark that each of them will be singular at $\lambda=\mu$, so that we shall specify, for definiteness, the generalized function $\frac{1}{\lambda-\mu}$ to be taken as p.v. $\frac{1}{\lambda-\mu}$. Since (6.4) is non-singular at $\lambda=\mu$, the final result is independent of this specification.

The matrices $E(y,\lambda)$ and $E(y,\mu)$ in $r(\lambda-\mu)(T(x,y,\lambda)\otimes T(x,y,\mu))$ are multiplied from the left by the corresponding transition matrices and in the limit, as $y\to-\infty$, give $r(\lambda-\mu)(T_-(x,\lambda)\otimes T_-(x,\mu))$. However, this is not the case for $T(x,y,\lambda)\otimes T(x,y,\mu)r(\lambda-\mu)$, so we shall write this term as a product of $T(x,y,\lambda)E(y,\lambda)\otimes T(x,y,\mu)E(y,\mu)$, which converges to $T_-(x,\lambda)\otimes T_-(x,\mu)$ as $y\to-\infty$, and $(E(-y,\lambda)\otimes E(-y,\mu))r(\lambda-\mu)(E(y,\lambda)\otimes E(y,\mu))$. The explicit form (1.19) of $r(\lambda)$ and the permutation property (1.10) allow us to write the latter as $(E(y,\mu-\lambda)\otimes E(y,\lambda-\mu))r(\lambda-\mu)$. This matrix has a limit, as $y\to-\infty$, in the sense of generalized functions. To find it we use the well-known relation

$$\lim_{y\to-\infty} \text{p.v.} \frac{e^{\pm i\lambda y}}{\lambda}=\mp\pi i\delta(\lambda) \tag{6.5}$$

and the explicit form of $E(y,\lambda)$ and $r(\lambda)$. As a result, for the limiting matrix $r_-(\lambda-\mu)$,

$$r_-(\lambda-\mu)=\lim_{y\to-\infty} E(y,\mu-\lambda)\otimes E(y,\lambda-\mu)r(\lambda-\mu), \tag{6.6}$$

we obtain

$$r_-(\lambda)=-\varkappa\begin{pmatrix} \text{p.v.}\,\frac{1}{\lambda} & 0 & 0 & 0 \\ 0 & 0 & -\pi i\delta(\lambda) & 0 \\ 0 & \pi i\delta(\lambda) & 0 & 0 \\ 0 & 0 & 0 & \text{p.v.}\,\frac{1}{\lambda} \end{pmatrix}. \tag{6.7}$$

So we have the final relation

$$\{T_-(x,\lambda)\overset{\otimes}{,} T_-(x,\mu)\}=r(\lambda-\mu)\,T_-(x,\lambda)\otimes T_-(x,\mu) - T_-(x,\lambda)\otimes T_-(x,\mu)\,r_-(\lambda-\mu). \tag{6.8}$$

In a similar way we derive

$$\{T_+(x,\lambda)\overset{\otimes}{,} T_+(x,\mu)\}=T_+(x,\lambda)\otimes T_+(x,\mu)\,r_+(\lambda-\mu) - r(\lambda-\mu)\,T_+(x,\lambda)\otimes T_+(x,\mu), \tag{6.9}$$

where

$$r_+(\lambda-\mu)=\lim_{y\to+\infty} E(y,\mu-\lambda)\otimes E(y,\lambda-\mu)r(\lambda-\mu) \tag{6.10}$$

and $r_+(\lambda)$ differs from $r_-(\lambda)$ by replacing i with $-i$. Finally, the ultralocality property (see § 1) yields

$$\{T_-(x,\lambda)\overset{\otimes}{,} T_+(x,\mu)\}=0. \tag{6.11}$$

The Poisson brackets of the reduced monodromy matrix can be derived from the above formulae. Thus, multiplying (6.8) by $E(-x,\lambda)\otimes E(-x,\mu)$ from the left, letting $x\to+\infty$ and using (6.10) we find

$$\{T(\lambda)\overset{\otimes}{,} T(\mu)\}=r_+(\lambda-\mu)\,T(\lambda)\otimes T(\mu)-T(\lambda)\otimes T(\mu)\,r_-(\lambda-\mu). \tag{6.12}$$

This relation is of crucial importance, so we shall also write it down in terms of matrix elements. The explicit form of $T(\lambda)$ shows that the 16 relations in (6.12) follow from the 6 basic ones,

$$\{a(\lambda),a(\mu)\}=0, \tag{6.13}$$

$$\{a(\lambda), \bar{a}(\mu)\} = 0, \tag{6.14}$$

$$\{a(\lambda), b(\mu)\} = \frac{\varkappa}{\lambda - \mu + i0} a(\lambda) b(\mu), \tag{6.15}$$

$$\{a(\lambda), \bar{b}(\mu)\} = -\frac{\varkappa}{\lambda - \mu + i0} a(\lambda) \bar{b}(\mu), \tag{6.16}$$

$$\{b(\lambda), b(\mu)\} = 0, \tag{6.17}$$

$$\{b(\lambda), \bar{b}(\mu)\} = 2\pi i |\varkappa| \cdot |a(\lambda)|^2 \delta(\lambda - \mu). \tag{6.18}$$

The generalized function $\frac{1}{\lambda + i0}$ appears because p.v. $\frac{1}{\lambda}$ and $\pi i \delta(\lambda)$ are combined according to the Sochocki-Plemelj formula

$$\frac{1}{\lambda \pm i0} = \text{p.v.}\, \frac{1}{\lambda} \mp \pi i \delta(\lambda). \tag{6.19}$$

We emphasize that (6.13)–(6.18) are consistent with the analyticity of $a(\lambda)$ in the upper half-plane so that the first four relations can be analytically continued in λ.

Let us now go over to computing the Poisson brackets of the discrete spectrum data, $\lambda_j, \bar{\lambda}_j, \gamma_j, \bar{\gamma}_j; j = 1, \ldots, n$. From (6.13) and (6.14) it immediately follows that

$$\{a(\lambda), \lambda_j\} = \{a(\lambda), \bar{\lambda}_j\} = 0 \tag{6.20}$$

and

$$\{\lambda_j, \lambda_k\} = \{\lambda_j, \bar{\lambda}_k\} = 0, \quad j, k = 1, \ldots, n. \tag{6.21}$$

To compute the Poisson brackets $\{\lambda_j, b(\mu)\}$ we proceed as follows. Consider (6.15)

$$\{a(\lambda), b(\mu)\} = \frac{\varkappa}{\lambda - \mu} a(\lambda) b(\mu), \tag{6.22}$$

where it is assumed that $\operatorname{Im}\lambda > 0$, and

$$a(\lambda) = \prod_{j=1}^{n} \frac{\lambda - \lambda_j}{\lambda - \bar{\lambda}_j} \tilde{a}(\lambda). \tag{6.23}$$

The function $\tilde{a}(\lambda)$ is analytic in the upper half-plane and has no zeros, hence the function $\log\tilde{a}(\lambda)$ is analytic for $\operatorname{Im}\lambda>0$. Now insert (6.23) into (6.22) and rewrite it as

$$\{\log a(\lambda), b(\mu)\} = \{\log\tilde{a}(\lambda), b(\mu)\} + \sum_{j=1}^{n}\left(\frac{\{\bar{\lambda}_j, b(\mu)\}}{\lambda-\bar{\lambda}_j} - \frac{\{\lambda_j, b(\mu)\}}{\lambda-\lambda_j}\right)$$
$$= \frac{\varkappa}{\lambda-\mu}\, b(\mu). \tag{6.24}$$

Since the right hand side of this equation is analytic for $\operatorname{Im}\lambda>0$, it follows that the left hand side has no singularities at $\lambda=\lambda_j$ so that

$$\{b(\mu), \lambda_j\}=0. \tag{6.25}$$

In a similar manner, for the Poisson bracket $\{\bar{a}(\lambda), b(\mu)\}$ we find

$$\{b(\mu), \bar{\lambda}_j\}=0, \quad j=1,\ldots,n. \tag{6.26}$$

To compute the remaining Poisson brackets involving the coefficients γ_j, $\bar{\gamma}_j$ one has to exploit the more general relations (6.8), (6.9) and (6.11). Let us first consider the most interesting bracket $\{a(\lambda), \gamma_j\}$.

We adopt the following notation for the components of the columns, $T_-^{(1)}(x,\lambda)$ and $T_+^{(2)}(x,\lambda)$, of the Jost solutions:

$$T_-^{(1)}(x,\lambda)=\begin{pmatrix} f_-(x,\lambda) \\ g_-(x,\lambda)\end{pmatrix}, \quad T_+^{(2)}(x,\lambda)=\begin{pmatrix} f_+(x,\lambda) \\ g_+(x,\lambda)\end{pmatrix}, \tag{6.27}$$

which extend analytically into the upper half-plane. From

$$T(\lambda)=T_+^{-1}(x,\lambda)\,T_-(x,\lambda) \tag{6.28}$$

we have

$$a(\lambda)=f_-(x,\lambda)g_+(x,\lambda)-f_+(x,\lambda)g_-(x,\lambda) \tag{6.29}$$

(see §§ I.5–I.6). With this notation (6.2) becomes

$$\gamma_j=\left.\frac{f_-(x,\mu)}{f_+(x,\mu)}\right|_{\mu=\lambda_j}=\left.\frac{g_-(x,\mu)}{g_+(x,\mu)}\right|_{\mu=\lambda_j}. \tag{6.30}$$

Substitute these expressions into $\{a(\lambda), \gamma_j\}$ taking for γ_j, say, the first equality in (6.30). The Poisson brackets which occur can be calculated by using (6.8), (6.9) and (6.11) and have the form

$$\{f_\pm(x,\lambda), f_\pm(x,\mu)\} = 0, \tag{6.31}$$

$$\{f_-(x,\lambda), f_+(x,\mu)\} = 0, \tag{6.32}$$

$$\{g_\pm(x,\lambda), f_\mp(x,\mu)\} = 0, \tag{6.33}$$

$$\{g_\pm(x,\lambda), f_\pm(x,\mu)\} = \mp\frac{\varkappa}{\lambda-\mu}(g_\pm(x,\lambda)f_\pm(x,\mu) - g_\pm(x,\mu)f_\pm(x,\lambda)). \tag{6.34}$$

Clearly, these can be analytically continued in μ, so that one may set $\mu = \lambda_j$. Collecting the non-vanishing terms we find

$$\begin{aligned}\{a(\lambda), \gamma_j\} &= \frac{\varkappa}{\lambda-\lambda_j}\left(\frac{f_-(x,\lambda)f_-(x,\lambda_j)}{f_+^2(x,\lambda_j)}\right.\\ &\quad\times(g_+(x,\lambda)f_+(x,\lambda_j) - g_+(x,\lambda_j)f_+(x,\lambda))\\ &\quad\left.-\frac{f_+(x,\lambda)}{f_+(x,\lambda_j)}(g_-(x,\lambda)f_-(x,\lambda_j) - g_-(x,\lambda_j)f_-(x,\lambda))\right)\\ &= \frac{\varkappa a(\lambda)\gamma_j}{\lambda-\lambda_j} + \frac{\varkappa}{\lambda-\lambda_j}f_+(x,\lambda)f_-(x,\lambda)\\ &\quad\times\left(\frac{g_-(x,\lambda_j)}{f_+(x,\lambda_j)} - \frac{g_+(x,\lambda_j)f_-(x,\lambda_j)}{f_+^2(x,\lambda_j)}\right).\end{aligned} \tag{6.35}$$

The second term above vanishes by (6.30), hence finally we obtain

$$\{a(\lambda), \gamma_j\} = \frac{\varkappa}{\lambda-\lambda_j}a(\lambda)\gamma_j. \tag{6.36}$$

In a similar way we deduce

$$\{a(\lambda), \bar{\gamma}_j\} = -\frac{\varkappa}{\lambda-\bar{\lambda}_j}a(\lambda)\bar{\gamma}_j \tag{6.37}$$

and

$$\{b(\lambda), \gamma_j\} = \{b(\lambda), \bar{\gamma}_j\} = 0, \quad j = 1, \dots, n. \tag{6.38}$$

Notice that (6.15), (6.16) and (6.36), (6.37) agree with the relations

$$\gamma_j = b(\lambda_j), \quad \bar{\gamma}_j = b^*(\lambda_j), \tag{6.39}$$

which make sense only for compactly supported $\psi(x)$, $\bar{\psi}(x)$. In that case $b(\lambda)$ has an analytic continuation into the whole plane whereas $b^*(\lambda)$ is the analytic continuation of $\bar{b}(\lambda)$ off the real line according to $b^*(\lambda)=\bar{b}(\bar{\lambda})$ (see § I.6).

Next, proceeding as when deriving (6.25) from (6.36)–(6.37) we get the Poisson brackets

$$\{\gamma_j, \lambda_k\} = \varkappa \gamma_j \delta_{jk}, \tag{6.40}$$

$$\{\gamma_j, \bar{\lambda}_k\} = 0, \quad j,k=1,\dots,n. \tag{6.41}$$

The non-vanishing right hand side of (6.40) results from comparing the residues at $\lambda=\lambda_j$ in a relation of the type (6.24).

Finally, using (6.30) together with (6.8), (6.9), and (6.11) we obtain

$$\{\gamma_j, \gamma_k\} = \{\gamma_j, \bar{\gamma}_k\} = 0, \quad j,k=1,\dots,n. \tag{6.42}$$

This completes the computation of the Poisson brackets of the transition coefficients for the continuous and discrete spectra.

Now recall that the input data of the inverse problem which are sufficient for parametrizing the functions $\psi(x)$, $\bar{\psi}(x)$ are the transition coefficients $b(\lambda)$, $\bar{b}(\lambda)$; γ_j, $\bar{\gamma}_j$ and the discrete spectrum λ_j, $\bar{\lambda}_j$, $j=1,\dots,n$, of the auxiliary linear problem (see §§ II.1–II.2). The coefficient $a(\lambda)$ is uniquely determined by means of the dispersion relation

$$a(\lambda) = \prod_{j=1}^{n} \frac{\lambda-\lambda_j}{\lambda-\bar{\lambda}_j} \exp\left\{\frac{1}{2\pi i}\int_{-\infty}^{\infty} \frac{\log(1+\varepsilon|b(\mu)|^2)}{\mu-\lambda}\, d\mu\right\}, \tag{6.43}$$

with $\operatorname{Im}\lambda>0$; if $\varepsilon=1$, the product over the zeros does not appear (see § I.6).

The non-vanishing Poisson brackets of $b(\lambda), \bar{b}(\lambda)$; $\gamma_j, \bar{\gamma}_j$ *and* $\lambda_j, \bar{\lambda}_j$ *are*

$$\{b(\lambda), \bar{b}(\mu)\} = 2\pi i |\varkappa| \cdot (1+\varepsilon|b(\lambda)|^2)\,\delta(\lambda-\mu) \tag{6.44}$$

and

$$\{\gamma_j, \lambda_k\} = \varkappa \gamma_j \delta_{jk}\,; \quad j,k=1,\dots,n. \tag{6.45}$$

It is not hard to see that the Poisson brackets containing $a(\lambda)$ are consistent with (6.44)–(6.45) and with the dispersion relation (6.43).

The resultant formulae (6.44)–(6.45) have a remarkably simple form. In the next section we shall give their rigorous interpretation and derive explicit expressions for canonical variables of action-angle type.

§ 7. Action-Angle Variables in the Rapidly Decreasing Case

This section practically concludes our general discussion of the Hamiltonian approach to the NS model under the rapidly decreasing boundary conditions. *We will show here that this model is completely integrable.* For the proof we will construct explicit canonical variables of action-angle type.

Already in § I.7 it has been shown that the involutory integrals I_n are functionals of only "half" of the inverse problem data $\{b(\lambda), \bar{b}(\lambda); \lambda_j, \bar{\lambda}_j, \gamma_j, \bar{\gamma}_j; j=1, \ldots, n\}$. Indeed, the generating function, $\log a(\lambda)$, for the integrals of the motion depends, by virtue of (6.43), only on $|b(\lambda)|^2$ and on the set $\lambda_j, \bar{\lambda}_j$. For the "second half" it is natural to take $\arg b(\lambda)$ and the set $\gamma_j, \bar{\gamma}_j$; the Poisson brackets (6.44)–(6.45) confirm this point of view.

Let us analyze this in detail proceeding formally in the spirit of § 6; the necessary comments will be given at the end of the section. Let

$$\varphi(\lambda) = -\arg b(\lambda) \tag{7.1}$$

be defined for $|b(\lambda)| \neq 0$; it is understood that $0 \leqslant \varphi(\lambda) < 2\pi$. We will show that

$$\{\varphi(\lambda), \varphi(\mu)\} = 0. \tag{7.2}$$

In fact, from (6.17) and (6.44) we have

$$\begin{aligned}\{e^{2i\varphi(\lambda)}, e^{2i\varphi(\mu)}\} &= \left\{\frac{\bar{b}(\lambda)}{b(\lambda)}, \frac{\bar{b}(\mu)}{b(\mu)}\right\} \\ &= -\frac{\bar{b}(\mu)}{b(\lambda)b^2(\mu)}\{\bar{b}(\lambda), b(\mu)\} \\ &\quad -\frac{\bar{b}(\lambda)}{b^2(\lambda)b(\mu)}\{b(\lambda), \bar{b}(\mu)\} = 0.\end{aligned} \tag{7.3}$$

Let us now find a quantity canonically conjugate to $\varphi(\lambda)$, which is a function of $|b(\lambda)|^2$. For any function $f(|b(\lambda)|^2)$ we have

$$\{f(|b(\lambda)|^2), \varphi(\mu)\} = \frac{f'(|b(\lambda)|^2)}{2i}\left\{|b(\lambda)|^2, \log\frac{\bar{b}(\mu)}{b(\mu)}\right\}$$
$$= \frac{f'(|b(\lambda)|^2)}{2i}\frac{b(\mu)}{\bar{b}(\mu)}\left\{|b(\lambda)|^2, \frac{\bar{b}(\mu)}{b(\mu)}\right\}$$
$$= \frac{1}{2i}\frac{f'(|b(\lambda)|^2)b(\mu)}{\bar{b}(\mu)}$$
$$\times\left(\frac{\bar{b}(\lambda)}{b(\mu)}\{b(\lambda), \bar{b}(\mu)\} - \frac{b(\lambda)\bar{b}(\mu)}{b^2(\mu)}\{\bar{b}(\lambda), b(\mu)\}\right)$$
$$= 2\pi|\varkappa| f'(|b(\lambda)|^2)(1+\varepsilon|b(\lambda)|^2)\delta(\lambda-\mu), \tag{7.4}$$

the prime over f indicating differentiation. The coefficient of $\delta(\lambda-\mu)$ on the right hand side of (7.4) turns into 1 if

$$f(x) = \frac{1}{2\pi\varkappa}\log(1+\varepsilon x). \tag{7.5}$$

Hence it follows that the quantities

$$\varrho(\lambda) = \frac{1}{2\pi\varkappa}\log(1+\varepsilon|b(\lambda)|^2), \quad \varphi(\lambda) = -\arg b(\lambda), \quad -\infty<\lambda<\infty \tag{7.6}$$

are *canonically conjugate variables*, that is the only non-vanishing Poisson bracket is

$$\{\varrho(\lambda), \varphi(\mu)\} = \delta(\lambda-\mu). \tag{7.7}$$

Notice that $\varrho(\lambda)$ is non-negative for all λ. Indeed, this is obvious for $\varepsilon=1$, while for $\varepsilon=-1$ this follows from

$$|b(\lambda)|<1, \tag{7.8}$$

due to the condition (A) (see § I.6). The ambiguity in the definition of $\varphi(\lambda)$ can be avoided if instead of $\varrho(\lambda)$ and $\varphi(\lambda)$ one considers complex-valued functions

$$\Phi(\lambda) = \sqrt{\varrho(\lambda)}\, e^{-i\varphi(\lambda)}, \quad \bar{\Phi}(\lambda) = \sqrt{\varrho(\lambda)}\, e^{i\varphi(\lambda)}, \tag{7.9}$$

which are well defined for all λ and vanish if $b(\lambda)=0$. Just as $b(\lambda)$, the function $\Phi(\lambda)$ is of Schwartz type; if $\varkappa<0$, its smoothness is assured by (7.8). From (7.2) and (7.7) we deduce the Poisson brackets

$$\{\Phi(\lambda), \Phi(\mu)\} = \{\bar{\Phi}(\lambda), \bar{\Phi}(\mu)\} = 0, \quad \{\Phi(\lambda), \bar{\Phi}(\mu)\} = i\delta(\lambda-\mu), \tag{7.10}$$

which are analogous to the initial Poisson brackets (1.14) for $\psi(x)$ and $\bar{\psi}(x)$.

Next, as it follows from § 6, the continuous spectrum data $b(\lambda), \bar{b}(\lambda)$ are in involution with the discrete spectrum data $\lambda_j, \bar{\lambda}_j, \gamma_j, \bar{\gamma}_j$. For the latter, it is also easy to construct canonical variables. Namely, write (6.45) as

$$\left\{\log \gamma_j, \frac{1}{\varkappa} \lambda_k\right\} = \delta_{jk} \tag{7.11}$$

and recall equation (6.41)

$$\left\{\log \bar{\gamma}_j, \frac{1}{\varkappa} \lambda_k\right\} = 0, \quad j, k = 1, \ldots, n. \tag{7.12}$$

By separating real and imaginary parts in (7.11)–(7.12) we conclude that the variables

$$\begin{aligned} p_j &= -\frac{2}{\varkappa} \operatorname{Re} \lambda_j, \quad & q_j &= \log |\gamma_j|, \\ \varrho_j &= -\frac{2}{\varkappa} \operatorname{Im} \lambda_j, \quad & \varphi_j &= -\arg \gamma_j \end{aligned} \tag{7.13}$$

have the non-vanishing Poisson brackets of the form

$$\{p_j, q_k\} = \delta_{jk}, \quad \{\varrho_j, \varphi_k\} = \delta_{jk}; \tag{7.14}$$

$j, k = 1, \ldots, n$. The variables p_j and q_j range over the whole real line, whereas $\varrho_j > 0$ (recall that $\varkappa < 0$) and $0 \leq \varphi_j < 2\pi$.

To summarize, the complete system of inverse problem data consists of the real-valued functions $\varrho(\lambda), \varphi(\lambda)$ (or complex-valued functions $\Phi(\lambda), \bar{\Phi}(\lambda)$) and the set of real variables for the discrete spectrum, $p_j, q_j, \varrho_j, \varphi_j$, $j = 1, \ldots, n$, which come in canonically conjugate pairs. The generating function for the motion integrals depends only on the involutory variables $\varrho(\lambda)$, p_j and ϱ_j. In analogy with Hamiltonian mechanics for finitely many degrees of freedom it is natural to call them *action variables. In particular, the Hamiltonian of our model is a function of these variables only. The conjugate variables, $\varphi(\lambda)$, q_j and φ_j, are angle variables.* Of course, it should be remembered that q_j, in contrast to $\varphi(\lambda)$ and φ_j, ranges over the whole real axis rather than from 0 to 2π. *The transformation to the inverse scattering data,*

$\mathscr{F}: (\psi(x), \bar{\psi}(x)) \to (\varrho(\lambda), \varphi(\lambda); p_j, q_j, \varrho_j, \varphi_j; j=1, \ldots, n)$, thouroughly analyzed in Chapters I–II is an invertible canonical transformation.

These results constitute the principal statement for the NS model, establishing its complete integrability.

To conclude the formal part of this section we shall write down some other useful formulae. First, the invertibility of $\mathscr{F}$ implies that the symplectic form Ω in the new variables has the canonical form

$$\Omega = \int_{-\infty}^{\infty} d\varrho(\lambda) \wedge d\varphi(\lambda) d\lambda + \sum_{j=1}^{n} (dp_j \wedge dq_j + d\varrho_j \wedge d\varphi_j). \tag{7.15}$$

Next we shall express the local integrals I_n in terms of $\varrho(\lambda)$, p_j and ϱ_j. For that we write the trace identities of § I.7 in new notation,

$$I_k = \int_{-\infty}^{\infty} \lambda^{k-1} \varrho(\lambda) d\lambda + \frac{(-1)^k}{ik\varkappa} \left(\frac{\varkappa}{2}\right)^k \sum_{j=1}^{n} ((p_j - i\varrho_j)^k - (p_j + i\varrho_j)^k). \tag{7.16}$$

In particular, the charge N, momentum P and Hamiltonian H become

$$N = \int_{-\infty}^{\infty} \varrho(\lambda) d\lambda + \sum_{j=1}^{n} \varrho_j, \tag{7.17}$$

$$P = \int_{-\infty}^{\infty} \lambda \varrho(\lambda) d\lambda - \frac{\varkappa}{2} \sum_{j=1}^{n} \varrho_j p_j, \tag{7.18}$$

$$H = \int_{-\infty}^{\infty} \lambda^2 \varrho(\lambda) d\lambda + \frac{\varkappa^2}{4} \sum_{j=1}^{n} \left(\varrho_j p_j^2 - \frac{1}{3} \varrho_j^3\right). \tag{7.19}$$

The passage to the new variables completely trivializes the dynamics of the NS model and reproduces the results of § I.7. In fact from (7.7), (7.14), and (7.19) we have

$$\frac{\partial \varrho}{\partial t}(\lambda, t) = \frac{dp_j(t)}{dt} = \frac{d\varrho_j(t)}{dt} = 0 \tag{7.20}$$

and

$$\frac{\partial \varphi}{\partial t}(\lambda, t) = \{H, \varphi\} = \lambda^2, \tag{7.21}$$

$$\frac{dq_j}{dt} = \{H, q_j\} = \frac{\varkappa^2}{2} \varrho_j \cdot p_j, \tag{7.22}$$

$$\frac{d\varphi_j}{dt} = \{H, \varphi_j\} = \frac{\varkappa^2}{4} (p_j^2 - \varrho_j^2), \tag{7.23}$$

which is equivalent to the already familiar formulae (I.7.11):

$$b(\lambda, t) = e^{-i\lambda^2 t} b(\lambda, 0), \quad \gamma_j(t) = e^{-i\lambda_j^2 t} \gamma_j(0),$$
$$\lambda_j(t) = \lambda_j(0); \quad j = 1, \ldots, n. \tag{7.24}$$

Of course, *all the higher NS equations are also completely integrable* so that the evolution under the general equation

$$\frac{\partial \psi}{\partial t} = \{I, \psi\}, \quad \frac{\partial \bar{\psi}}{\partial t} = \{I, \bar{\psi}\}, \tag{7.25}$$

where

$$I = \sum_k \alpha_k I_k \tag{7.26}$$

with α_k real, is given by

$$b(\lambda, t) = e^{-iI(\lambda)t} b(\lambda, 0), \quad \gamma_j(t) = e^{-iI(\lambda_j)t} \gamma_j(0),$$
$$\lambda_j(t) = \lambda_j(0), \quad j = 1, \ldots, n; \tag{7.27}$$

and

$$I(\lambda) = \sum_k \alpha_k \lambda^{k-1}. \tag{7.28}$$

So far the discussion remained on the formal level, in particular it did not touch the problem of admissibility of the quantities $a(\lambda)$, $b(\lambda)$, λ_j and γ_j as functionals on the phase space $\mathcal{M}_0$. Now we shall give the necessary refinements starting with the continuous spectrum data.

Let us consider the behaviour of the variational derivatives of $a(\lambda)$ and $b(\lambda)$ with respect to $\psi(x)$, $\bar{\psi}(x)$ as $|x| \to \infty$. The corresponding formulae (2.6)–(2.7) were derived in § 2. Taking their limit as $L \to \infty$ and recalling the definition (6.1) of the reduced monodromy matrix we find

$$\frac{\delta T(\lambda)}{\delta \psi(x)} = \sqrt{\varkappa}\, T_+^{-1}(x, \lambda) \sigma_- T_-(x, \lambda) \tag{7.29}$$

and

$$\frac{\delta T(\lambda)}{\delta \bar{\psi}(x)} = \sqrt{\varkappa}\, T_+^{-1}(x, \lambda) \sigma_+ T_-(x, \lambda). \tag{7.30}$$

This implies, by using the notation of § 6 and the involution property (I.5.19), that

$$\frac{\delta a(\lambda)}{\delta \psi(x)} = -\sqrt{\varkappa}\, f_+(x,\lambda) f_-(x,\lambda), \tag{7.31}$$

$$\frac{\delta a(\lambda)}{\delta \bar{\psi}(x)} = \sqrt{\varkappa}\, g_+(x,\lambda) g_-(x,\lambda) \tag{7.32}$$

and

$$\frac{\delta b(\lambda)}{\delta \psi(x)} = \sqrt{\varkappa}\, \overline{g_+}(x,\lambda) f_-(x,\lambda), \tag{7.33}$$

$$\frac{\delta b(\lambda)}{\delta \bar{\psi}(x)} = -\varepsilon\sqrt{\varkappa}\, \overline{f_+}(x,\lambda) g_-(x,\lambda). \tag{7.34}$$

Here (7.31)-(7.32) can be analytically continued into the upper half of the λ-plane.

Thus the variational derivatives of the functionals $a(\lambda)$ and $b(\lambda)$ with respect to $\psi(x)$, $\bar{\psi}(x)$ are smooth functions of x. Let us consider their behaviour as $|x|\to\infty$. We begin with $a(\lambda)$ for real λ. There are the following asymptotic formulae (see § I.5):

$$e^{\frac{i\lambda x}{2}} f_-(x,\lambda) = 1 + o(1), \tag{7.35}$$

$$e^{\frac{i\lambda x}{2}} g_-(x,\lambda) = o(1), \tag{7.36}$$

$$e^{\frac{i\lambda x}{2}} f_+(x,\lambda) = -\varepsilon \bar{b}(\lambda) + o(1), \tag{7.37}$$

$$e^{-\frac{i\lambda x}{2}} g_+(x,\lambda) = a(\lambda) + o(1) \tag{7.38}$$

as $x\to-\infty$ and

$$e^{-\frac{i\lambda x}{2}} f_+(x,\lambda) = o(1), \tag{7.39}$$

$$e^{-\frac{i\lambda x}{2}} g_+(x,\lambda) = 1 + o(1), \tag{7.40}$$

$$e^{\frac{i\lambda x}{2}} f_-(x,\lambda) = a(\lambda) + o(1), \tag{7.41}$$

$$e^{-\frac{i\lambda x}{2}} g_-(x,\lambda) = b(\lambda) + o(1) \tag{7.42}$$

as $x \to +\infty$ where the limiting values are taken in the Schwartz sense. It then follows that

$$\frac{\delta a(\lambda)}{\delta \psi(x)} = \varepsilon \sqrt{\varkappa}\, e^{-i\lambda x}(\overline{b(\lambda)} + o(1)), \qquad \frac{\delta a(\lambda)}{\delta \bar{\psi}(x)} = o(1) \tag{7.43}$$

as $x \to -\infty$, and

$$\frac{\delta a(\lambda)}{\delta \psi(x)} = o(1), \qquad \frac{\delta a(\lambda)}{\delta \bar{\psi}(x)} = \sqrt{\varkappa}\, e^{i\lambda x}(b(\lambda) + o(1)) \tag{7.44}$$

as $x \to +\infty$. *Hence, for λ real, $a(\lambda)$ is an inadmissible functional.*

Nevertheless, for $\operatorname{Im}\lambda > 0$, *the functional $a(\lambda)$ is admissible.* In fact, as was shown in § II.2 (see (II.2.99)-(II.2.100)) for such λ the asymptotic formulae (7.35)-(7.36) and (7.38)-(7.41) remain valid while (7.37) and (7.42) are replaced by

$$e^{-\frac{i\lambda x}{2}} f_+(x,\lambda) = o(1) \quad \text{as} \quad x \to -\infty \tag{7.45}$$

and

$$e^{\frac{i\lambda x}{2}} g_-(x,\lambda) = o(1) \quad \text{as} \quad x \to +\infty\,. \tag{7.46}$$

Therefore, for $\operatorname{Im}\lambda > 0$, the right hand sides of (7.31)-(7.32) are rapidly decreasing as $|x| \to \infty$, so that the variational derivatives $\frac{\delta a(\lambda)}{\delta \psi(x)}$ and $\frac{\delta a(\lambda)}{\delta \bar{\psi}(x)}$ are Schwartz functions. The real analyticity of the functional $a(\lambda)$ (and that of $b(\lambda)$), i.e. its series expansion with respect to $\psi(x)$, $\bar{\psi}(x)$ of the form (I.1.7), results from a similar expansion for the corresponding matrix element of the monodromy matrix $T_L(\lambda)$ (see § 2) by taking the limit as $L \to \infty$.

In a similar manner, for the functional $b(\lambda)$ we have

$$\frac{\delta b(\lambda)}{\delta \psi(x)} = \begin{cases} \sqrt{\varkappa}\, e^{-i\lambda x}(a(\lambda)+o(1)), & \text{as } x \to +\infty \\ \sqrt{\varkappa}\, e^{-i\lambda x}(\bar{a}(\lambda)+o(1)), & \text{as } x \to -\infty \end{cases}$$

$$\frac{\delta b(\lambda)}{\delta \bar{\psi}(x)} = o(1), \qquad \text{as } |x| \to \infty \tag{7.47}$$

where the limiting values are taken in the Schwartz sense.

We have thus shown that if $\operatorname{Im}\lambda > 0$, then $a(\lambda)$ and $\bar{a}(\lambda)$ are admissible functionals on the phase space $\mathscr{M}_0$. As for $b(\lambda)$ and $\bar{b}(\lambda)$, by virtue of (7.47) these should be thought of as generalized functions of the variable λ with values in the algebra of admissible functionals; in other words, $\int_{-\infty}^{\infty} \varphi_1(\lambda) b(\lambda)\, d\lambda$ and $\int_{-\infty}^{\infty} \varphi_2(\lambda) \bar{b}(\lambda)\, d\lambda$ are admissible functionals for any Schwartz functions $\varphi_1(\lambda)$ and $\varphi_2(\lambda)$. Moreover, functionals defined by absolutely convergent series

$$F = c + \sum_{n,m=0}^{\infty} \int_{-\infty}^{\infty} \dots \int_{-\infty}^{\infty} c_{nm}(\mu_1, \dots, \mu_n | \nu_1, \dots, \nu_m)$$
$$\times b(\mu_1) \dots b(\mu_n) \bar{b}(\nu_1) \dots \bar{b}(\nu_m)\, d\mu_1 \dots d\mu_n\, d\nu_1 \dots d\nu_m, \tag{7.48}$$

are also admissible if the coefficients $c_{nm}(\mu_1, \dots, \mu_n | \nu_1, \dots, \nu_m)$, which are generalized functions, are such that the variational derivatives $\frac{\delta F}{\delta b(\lambda)}$ and $\frac{\delta F}{\delta \bar{b}(\lambda)}$ are Schwartz functions (note the analogy with the definition in § I.1). In fact, for such F

$$\frac{\delta F}{\delta \psi(x)} = \int_{-\infty}^{\infty} \left(\frac{\delta F}{\delta b(\lambda)} \frac{\delta b(\lambda)}{\delta \psi(x)} + \frac{\delta F}{\delta \bar{b}(\lambda)} \frac{\delta \bar{b}(\lambda)}{\delta \psi(x)} \right) d\lambda \tag{7.49}$$

and

$$\frac{\delta F}{\delta \bar{\psi}(x)} = \int_{-\infty}^{\infty} \left(\frac{\delta F}{\delta b(\lambda)} \frac{\delta b(\lambda)}{\delta \bar{\psi}(x)} + \frac{\delta F}{\delta \bar{b}(\lambda)} \frac{\delta \bar{b}(\lambda)}{\delta \bar{\psi}(x)} \right) d\lambda, \tag{7.50}$$

so that, by (7.47), the variational derivatives $\frac{\delta F}{\delta \psi(x)}, \frac{\delta F}{\delta \bar{\psi}(x)}$ are Schwartz functions. The expansion of F with respect to $\psi(x)$, $\bar{\psi}(x)$ of the form (I.1.7) results from inserting the expansion for $b(\lambda)$ into (7.48).

There is a similar interpretation, via generalized functions, for $\Phi(\lambda)$, $\bar{\Phi}(\lambda)$. Admissible functionals of the form (7.48) are then given by similar series in $\Phi(\lambda)$ and $\bar{\Phi}(\lambda)$.

If $\varkappa > 0$, the whole algebra of observables is practically exhausted by the functionals (7.48). Indeed, the inverse problem allows us to express $\psi(x)$ and $\bar{\psi}(x)$ for every x as functionals of $b(\lambda)$, $\bar{b}(\lambda)$ such as (7.48) where the series converges absolutely for $|b(\lambda)|$ small enough, uniformly in x. This follows from the fact that in this case the iteration series for the Wiener-Hopf equation (II.2.53) is absolutely convergent. The convergence problem for the expansions of $\psi(x)$, $\bar{\psi}(x)$ for arbitrary $b(\lambda)$ is beyond the scope of this book.

In the above sense, $\Phi(\lambda)$ and $\bar{\Phi}(\lambda)$ (and also $b(\lambda)$ and $\bar{b}(\lambda)$) are coordinates on the phase space just as $\psi(x)$ and $\bar{\psi}(x)$.

If $\varkappa > 0$, this gives a complete picture of the phase space $\mathcal{M}_0$. There are two sets of complex canonical coordinates on $\mathcal{M}_0$, $\psi(x)$, $\bar{\psi}(x)$ and $\Phi(\lambda)$, $\bar{\Phi}(\lambda)$, related by a nonlinear invertible canonical transformation, $\mathcal{F}$. We point out once again that the existence of an inverse to $\mathcal{F}$ was shown in the course of solving the inverse problem in Chapter II. As for differentiability, it is implied by the calculations of this section.

In the coordinates $\Phi(\lambda)$, $\bar{\Phi}(\lambda)$, the local integrals I_n, which are the principal observables in our model, take the simple form

$$I_n = \int_{-\infty}^{\infty} \lambda^{n-1} |\Phi(\lambda)|^2 \, d\lambda, \qquad n = 1, 2, \ldots, \tag{7.51}$$

and so become the moments of $|\Phi(\lambda)|^2$.

These expressions have a natural wave interpretation. The function $|\Phi(\lambda)|^2$ plays the role of the independent mode distribution function for the wave train $\psi(x, t)$ representing the NS solution, so that its first moments are the charge (number of particles), momentum and energy of the train. The mapping $\mathcal{F}$ is a transformation to independent modes of the NS model. The existence of such modes in our model is itself far from trivial and relies on its complete integrability. In the linear limit, as $\varkappa \to 0$, $\mathcal{F}$ goes into the Fourier transform,

$$\Phi(\lambda) = \frac{1}{\sqrt{2\pi}} \int_{-\infty}^{\infty} \psi(x) e^{-i\lambda x} \, dx, \tag{7.52}$$

which reduces the linear Schrödinger equation to independent modes (see also § II.3).

If $\varkappa < 0$, there are, along with $b(\lambda)$ and $\bar{b}(\lambda)$, the discrete spectrum data λ_j, $\bar{\lambda}_j$ and γ_j, $\bar{\gamma}_j$, $j = 1, \ldots, n$. *The admissibility of $a(\lambda)$ for* $\operatorname{Im} \lambda > 0$ *implies that of the functionals λ_j, $\bar{\lambda}_j$ on the phase space $\mathcal{M}_0$.* In fact, the differentiation rule for composite functionals applied to $a(\lambda_j) = 0$ gives

$$\delta a(\lambda)|_{\lambda=\lambda_j} + \dot{a}(\lambda_j)\delta\lambda_j = 0, \tag{7.53}$$

the dot indicating differentiation with respect to λ. By using (6.30) and the analytic continuation of (7.31)-(7.32) for $\operatorname{Im}\lambda > 0$ we have

$$\frac{\delta\lambda_j}{\delta\psi(x)} = \sqrt{\varkappa}\,\frac{\gamma_j}{\dot{a}(\lambda_j)} f_+^2(x,\lambda_j), \qquad \frac{\delta\lambda_j}{\delta\bar{\psi}(x)} = -\sqrt{\varkappa}\,\frac{\gamma_j}{\dot{a}(\lambda_j)} g_+^2(x,\lambda_j), \tag{7.54}$$

$j=1,\ldots,n$. From (6.30), (7.35)-(7.36) and (7.39)-(7.40) we conclude that the variational derivatives $\frac{\delta\lambda_j}{\delta\psi(x)}, \frac{\delta\lambda_j}{\delta\bar{\psi}(x)}$ are Schwartz functions. An alternate derivation of (7.54) in the lines of § 6 would take into account (6.23) and deal with the residues of the variational derivatives $\frac{\delta\log a(\lambda)}{\delta\psi(x)}$ and $\frac{\delta\log a(\lambda)}{\delta\bar{\psi}(x)}$ at $\lambda=\lambda_j$.

Let us now consider the functionals γ_j, $\bar{\gamma}_j$. To evaluate their variational derivatives we shall use (6.30) taking, say, the first identity

$$\gamma_j = \frac{f_-(z,\lambda_j)}{f_+(z,\lambda_j)}, \qquad j=1,\ldots,n. \tag{7.55}$$

This gives

$$\begin{aligned}\frac{\delta\gamma_j}{\delta\psi(x)} = \frac{1}{f_+(z,\lambda_j)}\frac{\delta f_-(z,\lambda_j)}{\delta\psi(x)} + \frac{\dot{f}_-(z,\lambda_j)}{f_+(z,\lambda_j)}\frac{\delta\lambda_j}{\delta\psi(x)} \\ - \frac{f_-(z,\lambda_j)}{f_+^2(z,\lambda_j)}\frac{\delta f_+(z,\lambda_j)}{\delta\psi(x)} - \frac{f_-(z,\lambda_j)\dot{f}_+(z,\lambda_j)}{f_+^2(z,\lambda_j)}\frac{\delta\lambda_j}{\delta\psi(x)}\end{aligned} \tag{7.56}$$

with an analogous expression for $\frac{\delta\gamma_j}{\delta\bar{\psi}(x)}$. The variational derivatives of $f_\pm(z,\lambda_j)$ which occur are calculated by using (2.6)-(2.7). More precisely, letting there $L\to\infty$ and recalling the definition (6.3) of the Jost solutions, we have for $z>x$

$$\frac{\delta T_-(z,\lambda)}{\delta\psi(x)} = \sqrt{\varkappa}\,T(z,x,\lambda)\sigma_-T_-(z,\lambda) \tag{7.57}$$

and for $z<x$

$$\frac{\delta T_+(z,\lambda)}{\delta\psi(x)} = -\sqrt{\varkappa}\,T(z,x,\lambda)\sigma_-T_+(z,\lambda). \tag{7.58}$$

For $z<x$ (or $z>x$) the variational derivative $\frac{\delta T_-(z,\lambda)}{\delta\psi(x)}$ $\left(\text{respectively } \frac{\delta T_+(z,\lambda)}{\delta\psi(x)}\right)$ vanishes. To obtain $\frac{\delta T_-(z,\lambda)}{\delta\bar{\psi}(x)}$ and $\frac{\delta T_+(z,\lambda)}{\delta\bar{\psi}(x)}$ one has to replace σ_- by σ_+.

We shall now exploit the fact that the left hand side of (7.56) does not depend on z. Hence we may let $z\to x\pm 0$ in the right hand side of (7.56), so that (7.57)-(7.58) will simplify considerably. Letting, for instance, $z=x+0$, we find

$$\frac{\delta f_-(x+0,\lambda)}{\delta\psi(x)}=\frac{\delta f_+(x+0,\lambda)}{\delta\psi(x)}=0, \tag{7.59}$$

and taking account of (7.54),

$$\frac{\delta\gamma_j}{\delta\psi(x)}=\sqrt{\varkappa}\,\frac{\gamma_j}{\dot a(\lambda_j)}\,(\dot f_-(x,\lambda_j)f_+(x,\lambda_j)-f_-(x,\lambda_j)\dot f_+(x,\lambda_j)). \tag{7.60}$$

In a similar manner we deduce

$$\frac{\delta\gamma_j}{\delta\bar{\psi}(x)}=\sqrt{\varkappa}\,\frac{\gamma_j}{\dot a(\lambda_j)}\,(\dot g_+(x,\lambda_j)g_-(x,\lambda_j)-g_+(x,\lambda_j)\dot g_-(x,\lambda_j)). \tag{7.61}$$

Observe that (7.60)-(7.61) are consistent with (7.33)-(7.34) and with the identities

$$\gamma_j=b(\lambda_j),\qquad \bar\gamma_j=b^*(\bar\lambda_j);\qquad j=1,\ldots,n, \tag{7.62}$$

which are meaningful only for compactly supported $\psi(x)$, $\bar\psi(x)$. From (7.60)-(7.61) it follows that the variational derivatives $\frac{\delta\gamma_j}{\delta\psi(x)}$ and $\frac{\delta\gamma_j}{\delta\bar\psi(x)}$ are smooth functions, and (7.35)-(7.36), (7.38)-(7.41) and (7.45)-(7.46) imply that they decrease rapidly as $|x|\to\infty$. So $\frac{\delta\gamma_j}{\delta\psi(x)}$ and $\frac{\delta\gamma_j}{\delta\bar\psi(x)}$ are Schwartz functions, *hence γ_j, $\bar\gamma_j$ are admissible functionals on the phase space $\mathscr{M}_0$.* This completes the justification of the formal computations in § 6 and at the beginning of this section.

We shall now describe the phase space $\mathscr{M}_0$ in the new coordinates for the case $\varkappa<0$. Remind that the mapping $\mathscr{F}$ has been defined and analyzed only on an open subset, $\tilde{\mathscr{M}}_0$, of $\mathscr{M}_0$, consisting of pairs of functions $\psi(x)$, $\bar\psi(x)$ subject to the condition (A) which amounts to (7.8) together with the requirement that the zeros λ_j are simple (see § I.6). As was noted above, the fact that $\Phi(\lambda)$ and $\bar{\Phi}(\lambda)$ are smooth is equivalent to (7.8). The properties of $\mathscr{F}$ established earlier show that $\tilde{\mathscr{M}}_0$ is a disjoint union,

$$\tilde{\mathscr{M}}_0 = \bigcup_{n=0}^{\infty} \mathfrak{M}_n, \tag{7.63}$$

where the component $\mathfrak{M}_n$ is the product of the phase space $\mathfrak{M}_0$ equipped with complex canonical coordinates $\Phi(\lambda)$, $\bar{\Phi}(\lambda)$, and a finite-dimensional phase space Γ_n which is $\mathbb{R}^{4n}$ with certain submanifolds deleted. Canonical coordinates in $\mathbb{R}^{4n}$ are p_j, q_j and $\varrho_j, \varphi_i, j=1, \ldots, n$, where $-\infty < p_j, q_j < \infty$ and $0 \leqslant \varrho_j < \infty$, $0 \leqslant \varphi_j < 2\pi$, and the deleted submanifolds are given by $\varrho_j = 0$ and $(p_j - p_k)^2 + (\varrho_j - \varrho_k)^2 = 0, j, k = 1, \ldots, n$. In terms of the λ_j these submanifolds are defined by $\operatorname{Im}\lambda_j = 0$ and $\lambda_j = \lambda_k$. In fact, the product structure $\mathfrak{M}_n = \mathfrak{M}_0 \times \Gamma_n$ is compatible with the Poisson structure since the coordinates of the continuous spectrum are in involution with those of the discrete spectrum, and the Poisson bracket for $\Phi(\lambda)$, $\bar{\Phi}(\lambda)$ (as well as for p_j, ϱ_j; q_j, φ_j) is obviously non-degenerate. In particular, the finite-dimensional space Γ_n may be thought of as a reduction of $\mathfrak{M}_n$ with respect to the constraints $\Phi(\lambda) = \bar{\Phi}(\lambda) = 0$.

The components $\mathfrak{M}_n$, as well as the phase spaces Γ_n separately, are invariant under the flows induced by the higher NS equations. Restricted to Γ_n, these flows describe soliton dynamics, to be detailed in the next section.

The algebra of admissible functionals on $\mathfrak{M}_n$ is generated by products of admissible functionals of the type (7.48) constructed from $\Phi(\lambda)$ and $\bar{\Phi}(\lambda)$, and by smooth functions on the phase space Γ_n. Specifically, admissible functionals F on $\mathfrak{M}_n$ are given by absolutely convergent series in $\Phi(\lambda)$, $\bar{\Phi}(\lambda)$ of the form (7.48) where the coefficient functions c_{nm} depend in a smooth way on additional variables $p_j, q_j, \varrho_j, \varphi_j; j=1, \ldots, n$. Here apart from the decay of $\dfrac{\delta F}{\delta \Phi(\lambda)}$ and $\dfrac{\delta F}{\delta \bar{\Phi}(\lambda)}$, as $|\lambda| \to \infty$, it is also required that the series obtained from (7.48) by multiple differentiation with respect to additional variables be absolutely convergent.

The extension of $\mathscr{F}$ to the whole phase space $\mathscr{M}_0$ (without the condition (A)), the definition and analysis of the appropriate new variables constitute a difficult problem of global analysis connected with "sewing together" the components $\mathfrak{M}_n$ when the zeros acquire multiplicity or reach the real axis. Its discussion is beyond the scope of our text. Fortunately enough, the problem is marginal for our main topic, soliton dynamics. So, for instance, if a zero comes to the real axis, its contribution to the integrals of the motion vanishes.

This completes our analysis of the canonical transformation $\mathscr{F}$.

To conclude this section we will show what the hierarchy of Poisson structures defined in § 5 looks like in the new coordinates. It turns out that the action of the operator Λ reduces to multiplication by the variable λ. More precisely, the non-vanishing Poisson brackets for the l-th structure $\{,\}_l$ in complex coordinates on $\mathfrak{M}_n$ are

$$\{\Phi(\lambda), \bar{\Phi}(\mu)\}_l = \lambda^l \delta(\lambda-\mu), \tag{7.64}$$

$$\{\log\gamma_j, \lambda_k\}_l = \varkappa \lambda_k^l \delta_{jk}\,; \quad j,k=1,\ldots,n. \tag{7.65}$$

In particular, for $l \geqslant 0$ the Poisson structure $\{,\}_l$ is well defined on the whole algebra of functionals of the form (7.48). The annihilator mentioned in § 5 is generated by the quantities of the form $\left.\dfrac{d^k\Phi(\lambda)}{d\lambda^k}\right|_{\lambda=0}$, $k=0,\ldots,l-1$. If $l<0$ there are additional restrictions on the admissible functionals: their variational derivatives with respect to $\Phi(\lambda)$ and $\bar{\Phi}(\lambda)$ must vanish at $\lambda=0$ together with all their derivatives with respect to λ up to order $|l|-1$. The Jacobi identity for the Poisson structures $\{,\}_l$ is a trivial consequence of (7.64)-(7.65).

These formulae are most easy to verify for the equations of motion. In fact, the $(l+1)$-th NS equation whose Hamiltonian is I_{l+1} can be written as

$$\frac{\partial\Phi(\lambda)}{\partial t} = \{I_1, \Phi(\lambda)\}_l, \tag{7.66}$$

$$\frac{d\gamma_j}{dt} = \{I_1, \gamma_j\}_l\,; \quad j=1,\ldots,n, \tag{7.67}$$

which is characteristic of the l-th Poisson structure (see § 5).

Of course, this reasoning lacks rigour. There is a rigourous derivation (not presented here) which determines how the initial Poisson structure transforms under the mapping $\mathscr{F}$.

§ 8. Soliton Dynamics from the Hamiltonian Point of View

Here soliton solutions of the NS model in the rapidly decreasing case will be discussed from the Hamiltonian point of view. As we saw in § 7, the phase space for the n-soliton system is a finite-dimensional subspace, Γ_n, in $\mathfrak{M}_n$ defined by $b(\lambda)=0$ for all λ. It is parametrized by canonical coordinates $-\infty < p_j, q_j < \infty$, $0<\varrho_j<\infty$, $0 \leqslant \varphi_j < 2\pi$, $j=1,\ldots,n$, or by complex coordinates $\lambda_j, \bar{\lambda}_j$; $\gamma_j, \bar{\gamma}_j$, $\operatorname{Im}\lambda_j>0$, $\lambda_j \neq \lambda_k$ and $\gamma_j \neq 0$, related to one another by

$$\lambda_j = -\frac{\varkappa}{2}(p_j + i\varrho_j), \quad \gamma_j = e^{q_j - i\varphi_j}, \quad j = 1, \ldots, n \tag{8.1}$$

(cf. (7.13)).

The flows of the higher NS equations restricted to Γ_n induce completely integrable systems describing soliton dynamics. A set of $2n$ integrals in involution is formed by the ϱ_j and p_j, variables of action type; in terms of their conjugate angle-type variables, φ_j and q_j, the motion is linear.

The Hamiltonians given by the local motion integrals, I_l, of the NS model are expressed through action-type variables as

$$I_l = \frac{1}{i\varkappa l} \sum_{j=1}^{n} (\bar{\lambda}_j^l - \lambda_j^l), \quad l = 1, 2, \ldots. \tag{8.2}$$

In particular, the number of particles N, the momentum P, and the energy H are

$$N = \sum_{j=1}^{n} \varrho_j, \tag{8.3}$$

$$P = -\frac{\varkappa}{2} \sum_{j=1}^{n} \varrho_j p_j, \tag{8.4}$$

$$H = \frac{\varkappa^2}{4} \sum_{j=1}^{n} \left(\varrho_j p_j^2 - \frac{1}{3}\varrho_j^3\right). \tag{8.5}$$

These expressions are sums over independent modes. Each mode is described by coordinates p, q, ϱ, φ; associated to a separate mode in the phase space Γ_1 is a particle-like solution $\psi(x, t)$ of the NS equation, the soliton

$$\psi(x, t) = \frac{A \exp\left\{-i\left(\varphi + \dfrac{\varkappa p x}{2} + \dfrac{\varkappa^2}{4}(p^2 - \varrho^2)t + \dfrac{\pi}{2}\right)\right\}}{\operatorname{ch}\left\{\dfrac{\varkappa\varrho}{2}(x + \varkappa p t - x_0)\right\}}, \tag{8.6}$$

with

$$A = \frac{\sqrt{|\varkappa|}}{2}\varrho, \quad x_0 = -\frac{2q}{\varkappa\varrho} \tag{8.7}$$

(see § II.5). The momentum P and the energy E of the soliton,

$$P=-\frac{\varkappa}{2}\varrho p, \quad E=\frac{\varkappa^2}{4}\left(\varrho p^2-\frac{1}{3}\varrho^3\right) \tag{8.8}$$

are related by the *dispersion law*

$$E=\frac{1}{\varrho}P^2-\frac{\varkappa^2}{12}\varrho^3. \tag{8.9}$$

The last relation is typical of non-relativistic classical mechanics and allows the soliton to be interpreted as a particle of mass $m=\frac{\varrho}{2}$. The quantity x_0 is canonically conjugate to the momentum P and plays the role of particle's center of inertia coordinate. This interpretation is in agreement with the location of the maximum of $|\psi(x,t)|^2$ at $x=x_0+vt$ where the velocity $v=-\varkappa p=\frac{1}{m}P$ is given by the usual non-relativistic formula.

The coordinates ϱ and φ describe internal degrees of freedom and determine the oscillating behaviour of $\psi(x,t)$. Their contribution to the energy is given by $-\frac{\varkappa^2}{12}\varrho^3$ which may be interpreted as the internal energy (rest energy) of the particle.

The phase space Γ_n describes an n-soliton interacting system. In fact, the general n-soliton solution derived in § II.5 is not a superposition of one-soliton solutions. It splits into the sum of separate solitons only asymptotically as $|t|\to\infty$, when the interacting solitons move sufficiently far apart from one another. More precisely, the n-soliton solution parametrized by $\{p_j, q_j, \varrho_j, \varphi_j;\ j=1,\ldots,n\}$ can be expressed, for $t\to\pm\infty$, as a sum of one-soliton solutions parametrized by $p_j^{(\pm)}, q_j^{(\pm)}, \varrho_j^{(\pm)}, \varphi_j^{(\pm)}$ where

$$p_j^{(+)}=p_j^{(-)}=p_j, \quad \varrho_j^{(+)}=\varrho_j^{(-)}=\varrho_j, \tag{8.10}$$

$$q_j^{(\pm)}=q_j\pm\Delta q_j, \quad \varphi_j^{(\pm)}=\varphi_j\pm\Delta\varphi_j \tag{8.11}$$

and

$$\Delta q_j=\sum_{k=j+1}^{n}\log\left|\frac{\lambda_j-\bar\lambda_k}{\lambda_j-\lambda_k}\right|-\sum_{k=1}^{j-1}\log\left|\frac{\lambda_j-\bar\lambda_k}{\lambda_j-\lambda_k}\right|, \tag{8.12}$$

$$\Delta\varphi_j=\sum_{k=1}^{j-1}\arg\frac{\lambda_j-\bar\lambda_k}{\lambda_j-\lambda_k}-\sum_{k=j+1}^{n}\arg\frac{\lambda_j-\bar\lambda_k}{\lambda_j-\lambda_k}, \tag{8.13}$$

$j=1,\ldots,n$. Here it is assumed that $p_1>p_2>\ldots>p_n$.

The transformations $W_\pm$,

$$W_\pm : \{p_j, q_j, \varrho_j, \varphi_j; j=1,\ldots,n\} \to \{p_j^{(\pm)}, q_j^{(\pm)}, \varrho_j^{(\pm)}, \varphi_j^{(\pm)}; j=1,\ldots,n\}, \tag{8.14}$$

defined by (8.10)–(8.13) are canonical, i.e. preserve the Poisson brackets. Since the shifts Δq_j and $\Delta\varphi_j$ depend only on the variables of action type (generalized momenta), we have but to verify the relations

$$\{q_j^{(\pm)}, q_k^{(\pm)}\} = \{\varphi_j^{(\pm)}, \varphi_k^{(\pm)}\} = 0, \tag{8.15}$$

which look like integrability conditions

$$\frac{\partial \Delta q_j}{\partial p_k} = \frac{\partial \Delta q_k}{\partial p_j}, \quad \frac{\partial \Delta\varphi_k}{\partial \varrho_j} = \frac{\partial \Delta\varphi_j}{\partial \varrho_k}; \quad j,k=1,\ldots,n, \tag{8.16}$$

and the proof is straightforward. The latter formulae yield the existence of a function $K_n(p_1,\ldots,p_n;\varrho_1,\ldots,\varrho_n)$ such that

$$\Delta q_j = \frac{\partial K_n}{\partial p_j}, \quad \Delta\varphi_j = \frac{\partial K_n}{\partial \varrho_j}, \quad j=1,\ldots,n. \tag{8.17}$$

The functions $\pm K_n$ *are generating functions for the canonical transformations* $W_\pm$ in the sense of Hamiltonian mechanics.

Notice that $W_\pm$ are a very special kind of canonical transformations: their generating functions $\pm K_n$ depend only on the generalized momenta so that the latter remain invariant.

This picture allows to interprete the interaction of solitons in terms of the associated particles. Before and after scattering, as $t \to \pm\infty$, these particles are free. Their momenta are

$$P_j^{(+)} = P_j^{(-)} = P_j, \quad P_j = -\frac{\varkappa}{2}\varrho_j p_j, \tag{8.18}$$

"internal momenta" are $\varrho_j^{(+)} = \varrho_j^{(-)} = \varrho_j$, and their center of inertia coordinates $x_j^{(\pm)}(t)$ and phases $\varphi_j^{(\pm)}(t)$ depend linearly on time,

$$x_j^{(\pm)}(t) = x_{0j}^{(\pm)} + \frac{2P_j t}{\varrho_j}, \quad \varphi_j^{(\pm)}(t) = \varphi_j^{(\pm)} + \left(\frac{P_j^2}{\varrho_j^2} - \frac{\varkappa^2}{4}\varrho_j^2\right) t, \tag{8.19}$$

with

$$x_{0j}^{(\pm)} = -\frac{2}{\varkappa\varrho_j} q_j^{(\pm)}, \quad j=1,\ldots,n. \tag{8.20}$$

The parameters of the asymptotic motion are $P_j, \varrho_j, x_{0j}^{(\pm)}, \varphi_j^{(\pm)}$. They differ from $p_j, \varrho_j, q_j^{(\pm)}, \varphi_j^{(\pm)}$ by a trivial canonical transformation M of scaling type given by (8.18), (8.20).

The scattering transformation S,

$$S: \{P_j, \varrho_j, x_{0j}^{(-)}, \varphi_j^{(-)}; j=1, \ldots, n\} \to \{P_j, \varrho_j, x_{0j}^{(+)}, \varphi_j^{(+)}; j=1, \ldots, n\} \tag{8.21}$$

is canonical and can be expressed as a superposition of the transformations introduced above,

$$S = M W_+ W_-^{-1} M^{-1}. \tag{8.22}$$

The generating function

$$S_n(P_1, \ldots, P_n; \varrho_1, \ldots, \varrho_n) = 2 K_n\left(-\frac{2P_1}{\varkappa \varrho_1}, \ldots, -\frac{2P_n}{\varkappa \varrho_n}, \varrho_1, \ldots, \varrho_n\right) \tag{8.23}$$

determines S as follows:

$$x_{0j}^{(+)} = x_{0j}^{(-)} + \frac{\partial S_n}{\partial P_j}, \quad \varphi_j^{(+)} = \varphi_j^{(-)} + \frac{\partial S_n}{\partial \varrho_j}, \quad j=1, \ldots, n. \tag{8.24}$$

Let us compute the generating function explicitly. It will be more convenient to deal with $K_n(p_1, \ldots, p_n; \varrho_1, \ldots, \varrho_n)$ and (8.17). The explicit form of Δq_j and $\Delta \varphi_j$ in (8.12)–(8.13) implies that K_n can be expressed as a sum of two-particle contributions:

$$K_n(p_1, \ldots, p_n; \varrho_1, \ldots, \varrho_n) = \sum_{1 \leq j < k \leq n} K_2(p_j, p_k; \varrho_j, \varrho_k). \tag{8.25}$$

The function K_2 is determined by the system of equations

$$\frac{\partial K_2}{\partial p_1} = -\frac{\partial K_2}{\partial p_2} = \frac{1}{2} \log \frac{(p_1-p_2)^2 + (\varrho_1+\varrho_2)^2}{(p_1-p_2)^2 + (\varrho_1-\varrho_2)^2}, \tag{8.26}$$

$$\frac{\partial K_2}{\partial \varrho_1} = \operatorname{arctg} \frac{\varrho_1-\varrho_2}{p_1-p_2} - \operatorname{arctg} \frac{\varrho_1-\varrho_2}{p_1-p_2}, \tag{8.27}$$

$$\frac{\partial K_2}{\partial \varrho_2} = -\operatorname{arctg} \frac{\varrho_1+\varrho_2}{p_1-p_2} - \operatorname{arctg} \frac{\varrho_1-\varrho_2}{p_1-p_2}. \tag{8.28}$$

This system is easily solved with the result

$$K_2(p_1, p_2; \varrho_1, \varrho_2) = \frac{1}{2}\operatorname{Re}\{(p_1-p_2+i\varrho_1+i\varrho_2)\log(p_1-p_2+i\varrho_1+i\varrho_2)$$
$$-(p_1-p_2+i\varrho_1-i\varrho_2)\log(p_1-p_2+i\varrho_1-i\varrho_2)\} \qquad (8.29)$$
$$= -\frac{2}{\varkappa}\operatorname{Re}\{(\lambda_1-\lambda_2)\log(\lambda_1-\lambda_2)-(\lambda_1-\bar{\lambda}_2)\log(\lambda_1-\bar{\lambda}_2)\}.$$

To return to the scattering transformation S we see that its generating function, $S_n(P_1, \ldots, P_n; \varrho_1, \ldots, \varrho_n)$, is a sum over all pairs of particles,

$$S_n(P_1, \ldots, P_n; \varrho_1, \ldots, \varrho_n) = \sum_{1 \leqslant j < k \leqslant n} S_2(P_j, P_k; \varrho_j, \varrho_k), \qquad (8.30)$$

where S_2 is related to K_2 as in (8.23). Hence the soliton scattering process reduces to a sequence of canonical transformations. Each of them involves a single pair of solitons and changes only their coordinates and phases. This is typical of factorized scattering. The generating function S_2 may be called *"the classical S-matrix" for two-particle scattering.*

This concludes our discussion of soliton scattering from the Hamiltonian standpoint.

At first glance the above picture of soliton dynamics displays a rather special type of motion in the NS model. However, we shall give evidence that *solitons provide an approximation to the general solution of the model in question*. In that case the number of solitons n must go to infinity, and the corresponding zeros, λ_j, condense towards the real axis.

To be precise, we suppose for definiteness that all the λ_j have their condensation domain in the interval $M_1 \leqslant \mu \leqslant M_2$ where they are uniformly distributed with some density $\varrho(\mu)$. This means that

$$\lambda_j = \mu_j - \frac{i\varkappa(M_2-M_1)}{2n}\varrho(\mu_j) + O\left(\frac{1}{n^2}\right), \qquad (8.31)$$

where $\varrho(\mu)$ is a smooth non-negative function vanishing outside the interval (M_1, M_2), and

$$\Delta\mu_j = \mu_{j+1} - \mu_j = \frac{M_2-M_1}{n} + O\left(\frac{1}{n^2}\right), \qquad (8.32)$$

for instance, $\mu_j = \dfrac{j-1}{n}(M_2-M_1)+M_1;\ j=1, \ldots, n$. Now insert these λ_j into the expression (8.2) for the local integrals. We have, as $n \to \infty$,

$$I_l = \sum_{j=1}^{n} \mu_j^{l-1} \varrho(\mu_j) \frac{M_2 - M_1}{n} + O\left(\frac{1}{n}\right)$$

$$= \sum_{j=1}^{n} \mu_j^{l-1} \varrho(\mu_j) \Delta\mu_j + O\left(\frac{1}{n}\right)$$

$$= \int_{M_1}^{M_2} \mu^{l-1} \varrho(\mu) d\mu + O\left(\frac{1}{n}\right). \tag{8.33}$$

Since the interval (M_1, M_2) and the function $\varrho(\mu)$ are arbitrary, (8.33) implies

$$I_l = \int_{-\infty}^{\infty} \mu^{l-1} \varrho(\mu) d\mu, \quad l = 1, 2, \ldots, \tag{8.34}$$

the familiar expression for the local integrals on the component $\mathfrak{M}_0$ (see § 7).

We have thus shown that "condensation" of solitons allows us to reconstruct the phase space of the continuous spectrum data (at least on the level of motion integrals). Of course, fixing m out of n zeros, $\lambda_1, \ldots, \lambda_n$, and condensing the others we can also obtain the expressions for the local integrals I_l on the component $\mathfrak{M}_m$.

The above argument emphasizes the general nature of soliton dynamics for our model in the rapidly decreasing case. The motion of a large number of solitons with small amplitudes models the motion associated with the continuous spectrum.

In conclusion, let us *analyze the effect of condensation of zeros for the coefficient* $a(\lambda)$ which in our case is a product of the Blaschke factors,

$$a_n(\lambda) = \prod_{j=1}^{n} \frac{\lambda - \lambda_j}{\lambda - \bar{\lambda}_j}. \tag{8.35}$$

Then

$$a_n(\lambda) = \prod_{j=1}^{n} \left(1 + \frac{\bar{\lambda}_j - \lambda_j}{\lambda - \bar{\lambda}_j}\right)$$

$$= \prod_{j=1}^{n} \left(1 - \frac{i\varkappa(M_2 - M_1)}{n} \frac{\varrho(\mu_j)}{\mu_j - \lambda} + O\left(\frac{1}{n^2}\right)\right), \tag{8.36}$$

whence for $\operatorname{Im}\lambda > 0$ we have

$$\lim_{n\to\infty} a_n(\lambda)=a_+(\lambda)=\exp\left\{i\varkappa \int_{-\infty}^{\infty} \frac{\varrho(\mu)}{\lambda-\mu}\, d\mu\right\}. \tag{8.37}$$

In a similar manner, for $\operatorname{Im}\lambda<0$, we have

$$\lim_{n\to\infty} \bar{a}_n(\bar{\lambda})=a_-(\lambda)=\exp\left\{-i\varkappa \int_{-\infty}^{\infty} \frac{\varrho(\mu)}{\lambda-\mu}\, d\mu\right\}. \tag{8.38}$$

The functions $a_\pm(\lambda)$ inherit on the real line the limiting values from their domains of analyticity so that $a_-(\lambda)=\bar{a}_+(\lambda)$. The Sochocki-Plemelj formulae yield, for real λ,

$$a_+(\lambda)a_-(\lambda)=g(\lambda), \qquad g(\lambda)=e^{2\pi\varkappa\varrho(\lambda)}, \tag{8.39}$$

so that, in particular, $|a_+(\lambda)|\leqslant 1$ and

$$\varrho(\lambda)=\frac{1}{\pi\varkappa}\log|a_+(\lambda)|. \tag{8.40}$$

By comparing (8.40) with (7.6) and using the normalization condition we deduce that the density $\varrho(\mu)$ in the motion integrals (8.34) is indeed a variable of action type.

The above discussion shows how condensation of the zeros λ_j converts the trivial scalar Riemann problem with zeros,

$$a(\lambda)\bar{a}(\lambda)=1, \tag{8.41}$$

into the regular scalar Riemann problem (8.39). Because of $\varkappa<0$, which assures $\operatorname{Im}\lambda_j>0$ in (8.31), this method only gives the Riemann problem for contracting functions $g(\lambda)$,

$$|g(\lambda)|\leqslant 1, \tag{8.42}$$

which is the case for the NS model with $\varepsilon=-1$.

This concludes our description of the NS model in the rapidly decreasing case.

§ 9. Complete Integrability in the Case of Finite Density

This section completes the Hamiltonian approach to the NS model. We shall discuss, from the Hamiltonian point of view, the characteristics of the auxiliary linear problem, $b_\varrho(\lambda)$, λ_j, γ_j, introduced in Chapter I and providing an explicit solution to the equations of motion (see §§ I.10, II.6). In particular, we shall point out some interesting distinctions from the rapidly decreasing case in the programme of constructing the action-angle variables in terms of transition coefficients and discrete spectrum data.

First, following §§ I.8–I.9, recall the definition and properties of these data. The reduced monodromy matrix

$$T_\varrho(\lambda)=\lim_{\substack{x\to+\infty\\ y\to-\infty}} E_\varrho^{-1}(x,\lambda)\,Q(\theta)\,T(x,y,\lambda)\,E_\varrho(x,\lambda)=\begin{pmatrix} a_\varrho(\lambda) & \overline{b_\varrho}(\lambda)\\ b_\varrho(\lambda) & \overline{a_\varrho}(\lambda)\end{pmatrix} \tag{9.1}$$

defines the transition coefficients $a_\varrho(\lambda)$ and $b_\varrho(\lambda)$ for the continuous spectrum. Here λ is in $\mathbb{R}'_\omega$ (i.e. λ is real and $|\lambda|>\omega$),

$$E_\varrho(x,\lambda)=\begin{pmatrix} 1 & \dfrac{i(k-\lambda)}{\omega}\\ \dfrac{i(\lambda-k)}{\omega} & 1\end{pmatrix} e^{-\frac{ikx}{2}\sigma_3}, \tag{9.2}$$

and $k(\lambda)=\sqrt{\lambda^2-\omega^2}$, $\operatorname{sign}k(\lambda)=\operatorname{sign}\lambda$ for λ in $\mathbb{R}'_\omega$. There is also the normalization condition

$$|a_\varrho(\lambda)|^2-|b_\varrho(\lambda)|^2=1. \tag{9.3}$$

The function $a_\varrho(\lambda)$ has an analytic continuation to the sheet Γ_+ of the Riemann surface Γ of the function $k(\lambda)=\sqrt{\lambda^2-\omega^2}$ specified by $\operatorname{Im}k(\lambda)\geqslant 0$, with the possible exception of the branch points $\lambda=\pm\omega$. Asymptotically as $|\lambda|\to\infty$, we have

$$a_\varrho(\lambda)=e^{\frac{i\theta}{2}}+O\left(\frac{1}{|\lambda|}\right) \tag{9.4}$$

for $\operatorname{Im}\lambda>0$ and

$$a_\varrho(\lambda)=e^{-\frac{i\theta}{2}}+O\left(\frac{1}{|\lambda|}\right) \tag{9.5}$$

for $\operatorname{Im}\lambda<0$ with the involution

$$a_\varrho(\lambda, +) = \overline{a_\varrho}(\bar{\lambda}, +). \tag{9.6}$$

The zeros λ_j of $a_\varrho(\lambda)$ on Γ_+ may only lie in the gap $-\omega<\lambda_j<\omega$ and are simple. Their number n is finite, and they constitute the discrete spectrum of the auxiliary linear problem.

The coefficient $b_\varrho(\lambda)$ satisfies the involution

$$b_\varrho(\lambda - i0) = -\overline{b_\varrho}(\lambda + i0) \tag{9.7}$$

for λ in $\mathbb{R}_\omega$. In general, $b_\varrho(\lambda)$ has no extension off the cut $\mathscr{R}_\omega = (\mathbb{R}_\omega, \pm)$ on Γ. At $\lambda = \pm\omega$, the coefficients $a_\varrho(\lambda)$ and $b_\varrho(\lambda)$ are either regular or singular simultaneously. If in the neighbourhood of $\lambda = \pm\omega$

$$b_\varrho(\lambda) = \frac{b_\pm}{k} + O(1), \tag{9.8}$$

with $b_\pm \neq 0$ (the generic case), then

$$a_\varrho(\lambda) = \pm\frac{ib_\pm}{k} + O(1) \tag{9.9}$$

with $b_\pm$ real and

$$\operatorname{sign} b_\pm = (-1)^{N_\pm}, \tag{9.10}$$

where the integers $N_\pm$ are given by the condition for the determination of signs (I.9.58).

Finally, there is the dispersion relation

$$a_\varrho(\lambda) = e^{\frac{i\theta}{2}} \prod_{j=1}^{n} \frac{\lambda + k(\lambda) - \lambda_j - k_j}{\lambda + k(\lambda) - \lambda_j + k_j}$$

$$\times \exp\left\{\frac{1}{2\pi i} \int_{\mathbb{R}_\omega} \frac{\log(1 + |b_\varrho(\mu)|^2)}{k(\mu)} \left(1 + \frac{k(\lambda)}{\mu - \lambda}\right) d\mu\right\}, \tag{9.11}$$

where

$$k_j = i\sqrt{\omega^2 - \lambda_j^2}, \quad j = 1, \ldots, n, \tag{9.12}$$

and the condition (θ)

$$\theta \equiv \frac{1}{\pi} \int\limits_{\mathbb{R}_\omega} \frac{\log(1+|b_\varrho(\lambda)|^2)}{k(\lambda)} d\lambda + 2 \sum_{j=1}^{n} \arg(\lambda_j - k_j) (\operatorname{mod} 2\pi), \tag{9.13}$$

consistent with the asymptotic behaviour (9.4)–(9.5).

The Jost solutions $T_\pm(x,\lambda)$ for λ in $\mathbb{R}_\omega$ are defined by the limits

$$\begin{aligned} T_+(x,\lambda) &= \lim_{y\to+\infty} T(x,y,\lambda) Q^{-1}(\theta) E_\varrho(y,\lambda), \\ T_-(x,\lambda) &= \lim_{y\to-\infty} T(x,y,\lambda) E_\varrho(y,\lambda). \end{aligned} \tag{9.14}$$

The first column, $T_-^{(1)}(x,\lambda)$, of $T_-(x,\lambda)$ and the second column, $T_+^{(2)}(x,\lambda)$, of $T_+(x,\lambda)$ have an analytic continuation to the sheet Γ_+. The transition coefficients for the discrete spectrum, γ_j, defined by

$$T_-^{(1)}(x,\lambda_j) = \gamma_j T_+^{(2)}(x,\lambda_j), \tag{9.15}$$

are pure imaginary and satisfy

$$\operatorname{sign} i\gamma_j = \operatorname{sign} \frac{da_\varrho}{d\lambda}(\lambda_j) = \varepsilon_j, \quad j=1,\dots,n. \tag{9.16}$$

To compute the Poisson brackets of the transition coefficients one can apply the method of § 6 almost word for word, starting from the basic formula of § 1,

$$\{T(x,y,\lambda) \underset{,}{\otimes} T(x,y,\mu)\} = [r(\lambda-\mu), T(x,y,\lambda) \otimes T(x,y,\mu)], \quad y<x. \tag{9.17}$$

The involutions (9.6)–(9.7) allow us to concentrate on the case when λ and μ are in $\mathbb{R}'_\omega$. We shall only state the final results.

For the Jost solutions $T_\pm(x,\lambda)$ we have

$$\begin{aligned} \{T_+(x,\lambda) \underset{,}{\otimes} T_+(x,\mu)\} = &-r(\lambda-\mu) T_+(x,\lambda) \otimes T_+(x,\mu) \\ &+ T_+(x,\lambda) \otimes T_+(x,\mu) r_+(\lambda,\mu), \end{aligned} \tag{9.18}$$

$$\begin{aligned} \{T_-(x,\lambda) \underset{,}{\otimes} T_-(x,\mu)\} = &\, r(\lambda-\mu) T_-(x,\lambda) \otimes T_-(x,\mu) \\ &- T_-(x,\lambda) \otimes T_-(x,\mu) r_-(\lambda,\mu), \end{aligned} \tag{9.19}$$

and

$$\{T_-(x,\lambda) \underset{,}{\otimes} T_+(x,\mu)\} = 0. \tag{9.20}$$

Here the matrices $r_\pm(\lambda,\mu)$ are given by

$$r_-(\lambda,\mu)=\lim_{y\to-\infty}(E_\varrho^{-1}(y,\lambda)\otimes E_\varrho^{-1}(y,\mu))r(\lambda-\mu)\cdot(E_\varrho(y,\lambda)\otimes E_\varrho(y,\mu))$$
$$=\lim_{y\to-\infty}(E_\varrho^{-1}(y,\lambda)E_\varrho(y,\mu)\otimes E_\varrho^{-1}(y,\mu)E_\varrho(y,\lambda))r(\lambda-\mu) \tag{9.21}$$

and

$$r_+(\lambda,\mu)=\lim_{y\to+\infty}(E_\varrho^{-1}(y,\lambda)\otimes E_\varrho^{-1}(y,\mu))(Q(\theta)\otimes Q(\theta))$$
$$\times r(\lambda-\mu)(Q^{-1}(\theta)\otimes Q^{-1}(\theta))(E_\varrho(y,\lambda)\otimes E_\varrho(y,\mu))$$
$$=\lim_{y-+\infty}(E_\varrho^{-1}(y,\lambda)E_\varrho(y,\mu)\otimes E_\varrho^{-1}(y,\mu)E_\varrho(y,\lambda))r(\lambda-\mu). \tag{9.22}$$

The last formulae involve the limits of expressions such as $\dfrac{\exp\{\pm i(k(\lambda)\pm k(\mu))y\}}{\lambda-\mu}$ for $y\to\pm\infty$ taken in the sense of generalized functions. For λ in $\mathbb{R}'_\omega$ the function $k(\lambda)$ is monotone increasing, so using (6.5) we have

$$\lim_{y\to+\infty}\text{p.v.}\frac{e^{\pm i(k(\lambda)-k(\mu))y}}{\lambda-\mu}=\lim_{y\to+\infty}\text{p.v.}\frac{e^{\pm i(k(\lambda)-k(\mu))y}}{k(\lambda)-k(\mu)}\,\frac{k(\lambda)-k(\mu)}{\lambda-\mu}$$
$$=\pm\pi i\frac{dk(\lambda)}{d\lambda}\delta(k(\lambda)-k(\mu))=\pm\pi i\delta(\lambda-\mu). \tag{9.23}$$

The other limits $\lim\limits_{y\to\infty}\text{p.v.}\dfrac{e^{\pm i(k(\lambda)+k(\mu))y}}{\lambda-\mu}$, $\lim\limits_{y\to\infty}\text{p.v.}\dfrac{e^{\pm ik(\lambda)y}}{\lambda-\mu}$, and $\lim\limits_{y\to\infty}\text{p.v.}\dfrac{e^{\pm ik(\mu)y}}{\lambda-\mu}$ vanish. We emphasize that λ and μ are in $\mathbb{R}'_\omega$, i.e. $|\lambda|,|\mu|>\omega$.

As a result, for $r_\pm(\lambda,\mu)$ we finally obtain

$$r_\pm(\lambda,\mu)=-\varkappa\begin{pmatrix}\text{p.v.}\dfrac{\alpha(\lambda,\mu)}{\lambda-\mu} & 0 & 0 & 0\\ 0 & \text{p.v.}\dfrac{\beta(\lambda,\mu)}{\lambda-\mu} & \pm\pi i\delta(\lambda-\mu) & 0\\ 0 & \mp\pi i\delta(\lambda-\mu) & \text{p.v.}\dfrac{\beta(\lambda,\mu)}{\lambda-\mu} & 0\\ 0 & 0 & 0 & \text{p.v.}\dfrac{\alpha(\lambda,\mu)}{\lambda-\mu}\end{pmatrix}, \tag{9.24}$$

where

$$\alpha(\lambda,\mu)=\frac{1}{2}+\frac{\lambda\mu-\omega^2}{2k(\lambda)k(\mu)},\qquad \beta(\lambda,\mu)=\frac{1}{2}-\frac{\lambda\mu-\omega^2}{2k(\lambda)k(\mu)}, \tag{9.25}$$

so that

$$\alpha(\lambda,\mu)+\beta(\lambda,\mu)=1. \tag{9.26}$$

For the reduced monodromy matrix $T_\varrho(\lambda)$ we find

$$\{T_\varrho(\lambda) \underset{,}{\otimes} T_\varrho(\mu)\} = r_+(\lambda,\mu)\, T_\varrho(\lambda)\otimes T_\varrho(\mu) - T_\varrho(\lambda)\otimes T_\varrho(\mu)\, r_-(\lambda,\mu). \tag{9.27}$$

We shall now state the Poisson brackets of the transition coefficients and discrete spectrum resulting from (9.18)-(9.21) and (9.24)-(9.27). Starting with the continuous spectrum, we have

$$\{a_\varrho(\lambda), a_\varrho(\mu)\} = \{a_\varrho(\lambda), \overline{a_\varrho}(\mu)\} = 0, \tag{9.28}$$

$$\{a_\varrho(\lambda), b_\varrho(\mu)\} = \frac{\varkappa(\lambda\mu-\omega^2)}{k(\lambda)k(\mu)(\lambda-\mu+i0)}\, a_\varrho(\lambda)\, b_\varrho(\mu), \tag{9.29}$$

$$\{a_\varrho(\lambda), \overline{b_\varrho}(\mu)\} = \frac{\varkappa(\lambda\mu-\omega^2)}{k(\lambda)k(\mu)(\lambda-\mu+i0)}\, a_\varrho(\lambda)\, \overline{b_\varrho}(\mu), \tag{9.30}$$

and

$$\{b_\varrho(\lambda), b_\varrho(\mu)\} = 0, \tag{9.31}$$

$$\{b_\varrho(\lambda), \overline{b_\varrho}(\mu)\} = 2\pi i \varkappa\, |a_\varrho(\lambda)|^2\, \delta(\lambda-\mu), \tag{9.32}$$

with λ and μ in $\mathbb{R}'_\omega$. The expressions (9.28)-(9.30) may be analytically extended with respect to λ to the sheet Γ_+ except the branch points.

For the discrete spectrum data we have

$$\{b_\varrho(\lambda), \lambda_j\} = \{\overline{b_\varrho}(\lambda), \lambda_j\} = 0, \tag{9.33}$$

$$\{b_\varrho(\lambda), \gamma_j\} = \{\overline{b_\varrho}(\lambda), \gamma_j\} = 0, \tag{9.34}$$

$$\{a_\varrho(\lambda), \gamma_j\} = \frac{\varkappa(\lambda\lambda_j-\omega^2)}{k(\lambda)k_j(\lambda-\lambda_j)}\, a_\varrho(\lambda)\,\gamma_j \tag{9.35}$$

and

$$\{\lambda_j, \lambda_l\} = \{\gamma_j, \gamma_l\} = 0, \tag{9.36}$$

$$\{\gamma_j, \lambda_l\} = \varkappa\, \delta_{jl}\, \gamma_j, \quad j,l = 1,\ldots,n. \tag{9.37}$$

These expressions imply that for λ in $\mathbb{R}'_\omega$ *the variables*

$$\varrho(\lambda)=\frac{1}{2\pi\varkappa}\log(1+|b_\varrho(\lambda)|^2), \qquad \varphi(\lambda)=-\arg b_\varrho(\lambda), \tag{9.38}$$

$$p_j=-\frac{1}{\varkappa}\lambda_j, \qquad q_j=\log i\varepsilon_j\gamma_j, \qquad j=1,\ldots,n, \tag{9.39}$$

form a canonical set, i.e. their non-vanishing Poisson brackets are

$$\{\varrho(\lambda),\varphi(\mu)\}=\delta(\lambda-\mu), \qquad \{p_j,q_l\}=\delta_{jl}; \quad j,l=1,\ldots,n. \tag{9.40}$$

The variables $\varrho(\lambda)$, $\varphi(\lambda)$ and q_j have the same range $0\leqslant\varrho(\lambda)<\infty$, $0\leqslant\varphi(\lambda)<2\pi$ and $-\infty<q_j<\infty$ as in the rapidly decreasing case. However, the range of the variable p_j becomes bounded: $-\frac{\omega}{\varkappa}<p_j<\frac{\omega}{\varkappa}$.

To derive (9.40) one can repeat the reasoning of § 7. An alternative method is based on (9.29) which we write as

$$\{\log a_\varrho(\lambda),\log b_\varrho(\mu)\}=\frac{\varkappa(\lambda\mu-\omega^2)}{k(\lambda)k(\mu)(\lambda-\mu+i0)}. \tag{9.41}$$

Consider the imaginary part of this relation. The normalization condition (9.3) and (9.28) yield

$$\{\arg a_\varrho(\lambda),\log|b_\varrho(\mu)|\}=0, \tag{9.42}$$

so that the imaginary part on the left of (9.41) is given by the Poisson bracket $\{\log|a_\varrho(\lambda)|,\arg b_\varrho(\mu)\}$. The imaginary part on the right is trivially evaluated by (6.19). We then obtain

$$\{\log|a_\varrho(\lambda)|,\arg b_\varrho(\mu)\}=-\pi\varkappa\delta(\lambda-\mu), \tag{9.43}$$

which is equivalent to the first formula in (9.40).

Let us point out that expressions (9.40) have been derived only for λ and μ different from the boundary points ±ω of the continuous spectrum. Therefore they need more precision. The following argument shows that this specification is non-trivial.

The condition (θ) in the new variables becomes

$$2\varkappa\int_{\mathbb{R}_\omega}\frac{\varrho(\lambda)}{k(\lambda)}d\lambda+2\sum_{j=1}^{n}\arccos\frac{\varkappa p_j}{\omega}\equiv\theta(\operatorname{mod}2\pi) \tag{9.44}$$

and the Poisson bracket of the right hand side of this identity with all other observables must vanish. Indeed, the phase θ is not a dynamical variable but rather a label for the phase space $\mathscr{M}_{\varrho,\theta}$. On the other hand, from (9.40) taken literally, we derive the Poisson brackets

$$\{\theta, \varphi(\lambda)\} = \frac{2\varkappa}{k(\lambda)} \qquad \{\theta, q_j\} = -\frac{2i\varkappa}{k_j} \tag{9.45}$$

which do not vanish. *A correct amendment to (9.40) should eliminate the "paradox".*

For this purpose note that (9.40) is interpreted in the sense of generalized functions. So, for instance, the first formula in (9.40) leads to

$$\left\{\int_{\mathbb{R}_\omega} f(\lambda)\varrho(\lambda)\,d\lambda, \varphi(\mu)\right\} = f(\mu), \tag{9.46}$$

which is, of course, valid whenever $f(\lambda)$ is a smooth function on $\mathbb{R}_\omega$ including $\lambda = \pm\omega$. However, (9.44) and the expressions for the local integrals $J_{2l,\varrho}$ given below show that we also have to deal with functions $f(\lambda)$ with singularities of the type $\frac{1}{k(\lambda)}$ at $\lambda = \pm\omega$. Thus we must scrutinize the derivation of (9.40) in the viscinity of $\lambda = \pm\omega$.

Consider once again the right hand side of (9.41) for λ in Γ_+ outside of the cut $\mathbb{R}_\omega$. For such λ we have $\operatorname{Im} k(\lambda) > 0$ so that for λ in $\mathbb{R}_\omega$

$$\frac{1}{k(\lambda + i0)} = \frac{1}{k(\lambda) + i0} \tag{9.47}$$

and the expression on the right has the imaginary part $-\pi i\delta(k(\lambda))$ which was not taken into account in (9.43). Hence (9.43) should be modified for $\lambda = \pm\omega$ by the formal expression

$$\{\log|a_\varrho(\lambda)|, \arg b_\varrho(\mu)\} = -\varkappa\pi\delta(\lambda-\mu) - \frac{\pi\varkappa(\lambda\mu - \omega^2)}{k(\mu)(\lambda-\mu)}\delta(k(\lambda)). \tag{9.48}$$

Of course, the second term on the right hand side above vanishes when paired with smooth functions $f(\lambda)$, for by the change of variable formula

$$\delta(k(\lambda)) = \frac{k(\lambda)}{\omega}(\delta(\lambda-\omega) - \delta(\lambda+\omega)) \tag{9.49}$$

we have

$$\int_{\mathbb{R}_\omega} f(\lambda)\,\delta(k(\lambda))\,d\lambda = \frac{1}{2\omega}\,(k(\omega)f(\omega) - k(-\omega)f(-\omega)) = 0, \tag{9.50}$$

since $k(\pm\omega) = 0$.

However, as we have just observed, we also need functions $f(\lambda)$ with singularities of the type $\frac{1}{k(\lambda)}$ at $\lambda = \pm\omega$. These are representable as

$$f(\lambda) = \frac{f_1(\lambda)}{k(\lambda)} + f_2(\lambda), \tag{9.51}$$

where $f_1(\lambda)$ and $f_2(\lambda)$ are smooth functions. For such functions we have

$$\int_{\mathbb{R}_\omega} f(\lambda)\,\delta(k(\lambda))\,d\lambda = \frac{1}{2\omega}\,(f_1(\omega) - f_1(-\omega)). \tag{9.52}$$

In fact, only the first term in (9.51) contributes to the integral. The change of variable $k = k(\lambda)$ gives

$$\int_{\mathbb{R}_\omega} \frac{f_1(\lambda)}{k(\lambda)}\,\delta(k(\lambda))\,d\lambda = \int_{-\infty}^{\infty} \tilde{f}_1(k)\,\delta(k)\,dk, \tag{9.53}$$

with $\tilde{f}_1(k) = \frac{f_1(\lambda(k))}{\lambda(k)}$ and $\lambda(k) = \sqrt{k^2+\omega^2}$. This function is discontinuous at $k = 0$ because $\lambda(\pm 0) = \pm\omega$ and so

$$\int_{-\infty}^{\infty} \tilde{f}_1(k)\,\delta(k)\,dk = \frac{1}{2}\,(\tilde{f}_1(+0) + \tilde{f}_1(-0)) = \frac{1}{2\omega}\,(f_1(\omega) - f_1(-\omega)). \tag{9.54}$$

Hence (9.52) provides an accurate definition for the generalized function $\delta(k(\lambda))$ on the extended space of test functions of the form (9.51).

The Poisson bracket $\{\log|a_\varrho(\lambda)|, \log|\gamma_j|\}$ vanishing for $\lambda \neq \pm\omega$ needs a similar modification. We start from (9.35) written as

$$\{\log a_\varrho(\lambda), \log\gamma_j\} = \frac{\varkappa(\lambda\lambda_j - \omega^2)}{k(\lambda)\,k_j(\lambda - \lambda_j)}, \tag{9.55}$$

and evaluate its real part to obtain

$$\{\log|a_{\varrho}(\lambda)|, \log|\gamma_j|\} = -\frac{\pi i \varkappa(\lambda\lambda_j - \omega^2)}{k_j(\lambda-\lambda_j)}\delta(k(\lambda)), \qquad (9.56)$$

$j=1,\ldots,n$.

The remaining Poisson brackets (9.38)-(9.39) need no modification.

Thus the non-vanishing Poisson brackets of $\varrho(\lambda)$, $\varphi(\lambda)$, p_j, q_j valid for all λ, $|\lambda|\geqslant\omega$, are given by

$$\{\varrho(\lambda), \varphi(\mu)\} = \delta(\lambda-\mu) - \frac{1}{k(\mu)}\delta^*(k(\lambda)), \qquad (9.57)$$

$$\{\varrho(\lambda), q_j\} = \frac{i}{k_j}\delta^*(k(\lambda)) \qquad (9.58)$$

and

$$\{p_j, q_l\} = \delta_{jl}\,; \quad j,l=1,\ldots,n. \qquad (9.59)$$

Here the generalized function $\delta^*(k(\lambda))$ is defined by

$$\delta^*(k(\lambda)) = \frac{\omega^2-\lambda\mu}{\lambda-\mu}\delta(k(\lambda)) \qquad (9.60)$$

and does not actually depend on μ. In fact, as it follows from (9.52), $\delta^*(k(\lambda))$ paired with functions $f(\lambda)$ of the form (9.51) gives

$$\int_{\mathbb{R}_\omega} f(\lambda)\delta^*(k(\lambda))\,d\lambda = \frac{f_1(\omega)+f_1(-\omega)}{2}. \qquad (9.61)$$

Observe another distinction from the rapidly decreasing case, connected with the presence of a gap in the continuous spectrum of the auxiliary linear problem. In a generic situation, $\varrho(\lambda)$ has a singularity of the type $\log|k(\lambda)|$ as $\lambda\to\pm\omega$; in this case $\varphi(\pm\omega)$ are fixed and equal to 0 or π in agreement with (9.10) and (9.38). If $\lambda=\omega$, or $\lambda=-\omega$, or both are virtual levels, then $\varrho(\lambda)$ is finite at these points, and $\varphi(\lambda)$ by virtue of the involution (9.7) takes the values $\pm\frac{\pi}{2}$. Besides, the condition (θ) is satisfied. These conditions give a complete characterization of the image of the phase space $\mathcal{M}_{\varrho,\theta}$ under the mapping $\mathcal{F}$ of Chapter II.

So the variables $\varrho(\lambda)$, $\varphi(\lambda)$, p_j, and q_j under the restrictions referred to above may be thought of as new coordinates on the phase space $\mathcal{M}_{\varrho,\theta}$.

Let us now verify that the correct Poisson brackets (9.57)-(9.59) eliminate the paradox connected with the condition (θ). Indeed from relations (9.44), (9.57)-(9.59) and the definition (9.61) we find

$$\{\theta, \varphi(\lambda)\} = \int\limits_{\mathbb{R}_\omega} \frac{\delta\theta}{\delta\varrho(\mu)} \{\varrho(\mu), \varphi(\lambda)\} d\mu = \frac{2\varkappa}{k(\lambda)} - \frac{2\varkappa}{k(\lambda)} = 0 \tag{9.62}$$

and

$$\{\theta, q_j\} = \int\limits_{\mathbb{R}_\omega} \frac{\delta\theta}{\delta\varrho(\lambda)} \{\varrho(\lambda), q_j\} d\lambda + \frac{\partial\theta}{\partial p_j} = \frac{2i\varkappa}{k_j} - \frac{2i\varkappa}{k_j} = 0. \tag{9.63}$$

One can cite quite a few other imaginary paradoxes resulting from the naive Poisson brackets (9.40). All of them are resolved by the correct modification of the Poisson brackets performed above. One example of this kind will be discussed below in connection with the higher NS equations. Let us give another example. If one computes the bracket $\{a_\varrho(\lambda), b_\varrho(\mu)\}$ by using the dispersion relation (9.11) and the naive Poisson brackets (9.40), then the result will not agree with (9.29). However, the agreement will be reached by using the correct Poisson brackets (9.57)–(9.59).

To summarize, the final Poisson brackets of the variables $\varrho(\lambda)$, $\varphi(\lambda)$, p_j, q_j are given by (9.57)–(9.59). The Poisson brackets omitted there vanish identically.

The explicit form of the final Poisson brackets does not permit to call $\varrho(\lambda)$, $\varphi(\lambda)$, p_j, q_j action-angle variables in the strict sense. So, for instance, the discrete spectrum variable q_j (generalized angle) does not Poisson commute with the continuous spectrum variable $\varrho(\lambda)$ (generalized action). Besides, the Poisson bracket (9.57) of $\varrho(\lambda)$ and $\varphi(\mu)$ has not the canonical form. These effects mark the difference from the rapidly decreasing case.

Nevertheless, the Poisson structure defined by (9.57)–(9.59) is in good agreement with the dynamics induced by the local integrals for the NS model, and in practice is no less convenient than the action-angle variables in the rapidly decreasing case. So we shall presently see that all the higher NS equations can be explicitly integrated in the new variables $\varrho(\lambda)$, $\varphi(\lambda)$, p_j, q_j.

However, the complete identification of the algebra of observables on the phase space $\mathscr{M}_{\varrho,\theta}$ is much more involved than in the rapidly decreasing case in § 7. We will not treat here this sophisticated and cumbersome problem, but only point out that conditions imposed on the admissible functionals F must ensure that Hamilton's equations of motion induced with respect to the Poisson bracket (9.57)–(9.59) in the variables $\varrho(\lambda)$, $\varphi(\lambda)$, p_j, q_j preserve the functional class described above.

Let us now turn to the Hamiltonian flows generated by the local integrals $J_{l,\varrho}$. These functionals were defined in § I.10. The trace identities cited there allow us to express $J_{l,\varrho}$ in terms of $\varrho(\lambda)$ and p_j. We have

$$J_{l,\varrho} = \int\limits_{\mathbb{R}_\omega} \lambda k^{l-2}(\lambda)\varrho(\lambda)d\lambda + \frac{(-1)^{\frac{l+1}{2}}}{\varkappa l} \sum_{j=1}^{n} (\omega^2 - \varkappa^2 p_j^2)^{\frac{l}{2}} \tag{9.64}$$

for odd $l \geq 1$ and

$$J_{l,\varrho} = \int_{\mathbb{R}_\omega} \frac{1}{k(\lambda)} \sum_{m=0}^{l/2} \binom{\frac{1}{2}}{m} \omega^{2m} k^{l-2m}(\lambda)\varrho(\lambda)\,d\lambda$$

$$-\frac{1}{l}\sum_{j=1}^{n} \frac{p_j}{\sqrt{\omega^2-\varkappa^2 p_j^2}} \sum_{m=0}^{\frac{l}{2}-1} (-1)^{\frac{l}{2}-m} \omega^{2m} \binom{-\frac{1}{2}}{m} (\omega^2-\varkappa^2 p_j^2)^{\frac{l}{2}-m} \tag{9.65}$$

for even l.

In § 4 it was shown that the integrals $J_{l,\varrho}$ for $l>1$ are admissible functionals in involution on the phase space $\mathcal{M}_{\varrho,\theta}$. They give rise to the higher NS equations for the case of finite density,

$$\frac{\partial\psi}{\partial t} = \{J_{l,\varrho}, \psi\}, \quad \frac{\partial\bar{\psi}}{\partial t} = \{J_{l,\varrho}, \bar{\psi}\}. \tag{9.66}$$

Let us verify that in the new variables $\varrho(\lambda)$, $\varphi(\lambda)$, p_j, q_j these equations can be solved explicitly.

First let l be odd. The variational derivative

$$\frac{\delta J_{l,\varrho}}{\delta\varrho(\lambda)} = \lambda k^{l-2}(\lambda) \tag{9.67}$$

is regular at $\lambda = \pm\omega$ so that the Poisson brackets (9.57)–(9.59) reduce to the naive ones (9.40). In the new variables (9.66) becomes

$$\frac{\partial\varrho(\lambda)}{\partial t} = \{J_{l,\varrho}, \varrho(\lambda)\} = 0, \quad \frac{dp_j}{dt} = \{J_{l,\varrho}, p_j\} = 0, \tag{9.68}$$

$$\frac{\partial\varphi(\lambda)}{\partial t} = \{J_{l,\varrho}, \varphi(\lambda)\} = \lambda k^{l-2}(\lambda), \tag{9.69}$$

$$\frac{dq_j}{dt} = \{J_{l,\varrho}, q_j\} = -i\lambda_j k_j^{l-2}, \quad j=1,\dots,n, \tag{9.70}$$

and is solved in closed form. The solution is conveniently written in terms of transition coefficients,

$$\begin{gathered} a_\varrho(\lambda,t) = a_\varrho(\lambda,0), \quad b_\varrho(\lambda,t) = e^{-i\lambda k^{l-2}(\lambda)t} b_\varrho(\lambda,0), \\ \gamma_j(t) = e^{-i\lambda_j k_j^{l-2}t}\gamma_j(0), \quad j=1,\dots,n. \end{gathered} \tag{9.71}$$

In particular, if $l=3$ we recover the familiar expressions (I.10.7) for the NS equation.

If $l=1$, the integrand in (9.64) has a singularity of the type $\frac{1}{k(\lambda)}$ as $\lambda \to \pm\omega$ so that at first glance we are to use the Poisson brackets (9.57)–(9.59). However, the coefficient of $\frac{1}{k(\lambda)}$ equals $f_1(\lambda)=\lambda$ and is an odd function giving no contribution into (9.61). Hence we have

$$\{J_{1,\varrho}, \varphi(\lambda)\} = \frac{\lambda}{k(\lambda)} \tag{9.72}$$

and the equation of motion

$$\frac{\partial \varphi(\lambda)}{\partial t} = \{J_{1,\varrho}, \varphi(\lambda)\} \tag{9.73}$$

has a formal solution,

$$\varphi(\lambda, t) = \varphi(\lambda, 0) + \frac{\lambda}{k(\lambda)}\, t. \tag{9.74}$$

This solution, however, is singular for $\lambda \to \pm\omega$ for any $t>0$ and thus leaves the class of admissible $\varphi(\lambda)$. In fact, in the case of virtual level $\varphi(\lambda)$ is regular at $\lambda = \pm\omega$ and in a generic situation takes the values $0, \pi$.

We have seen once again that the functional $J_{1,\varrho}=N$, the analogue of charge in the rapidly decreasing case (see § I.1), is an inadmissible functional on $\mathcal{M}_{\varrho,\theta}$.

Now let l be even. From (9.65) we have

$$\frac{\delta J_{l,\varrho}}{\delta \varrho(\lambda)} = \frac{\omega^l \binom{\frac{1}{2}}{\frac{l}{2}}}{k(\lambda)} + g_l(\lambda), \tag{9.75}$$

where

$$g_l(\lambda) = \sum_{m=0}^{\frac{l}{2}-1} \binom{\frac{1}{2}}{m} \omega^{2m} k^{l-2m-1}(\lambda). \tag{9.76}$$

Hence to write down the equations of motion we must use the Poisson brackets (9.57)–(9.59). Then

$$\{J_{l,\varrho}, \varphi(\lambda)\} = g_l(\lambda) \tag{9.77}$$

and in a similar manner,

$$\{J_{l,\varrho}, q_j\} = -ig_l(\lambda_j), \quad j=1,\ldots,n. \tag{9.78}$$

So for l even, the dynamics of the higher NS equations is given by

$$\begin{aligned} &a_\varrho(\lambda,t)=a_\varrho(\lambda,0), \quad b_\varrho(\lambda,t)=e^{-ig_l(\lambda)t}b_\varrho(\lambda,0), \\ &\gamma_j(t)=e^{-ig_l(\lambda_j)t}\gamma_j(0); \quad j=1,\ldots,n. \end{aligned} \tag{9.79}$$

In particular, for $l=2$ the functional $J_{2,\varrho}$ coincides with the momentum P (see § I.10). The function $g_2(\lambda)$ is

$$g_2(\lambda)=k(\lambda). \tag{9.80}$$

This expression is in agreement with the interpretation of momentum as a generator of translations in the spatial variable x.

Remark that if for even l the naive Poisson brackets (9.40) were used to derive the equations of motion for $\varphi(\lambda)$, then the result would be singular for $\lambda \to \pm\omega$. This would imply that the $J_{l,\varrho}$ are inadmissible. However, this is not the case, and it is the correct Poisson brackets (9.57)–(9.59) that eliminate the possible contradiction.

As in the rapidly decreasing case, the expressions (9.64)–(9.65) for the motion integrals are interpreted in terms of independent modes labelled by the continuous variable λ in $\mathbb{R}_\omega$ and by the discrete variable j. However, it should be remembered that the variables $\varrho(\lambda)$ and p_j describing the additive contribution from these modes into the motion integrals are subject to a constraint, the condition (θ). The constraint can be removed by taking the union of the phase spaces $\mathcal{M}_{\varrho,\theta}$,

$$\mathcal{M}_\varrho = \bigcup_{0 \leq \theta < 2\pi} \mathcal{M}_{\varrho,\theta}. \tag{9.81}$$

But the Poisson structure (9.57)–(9.59) on the space $\mathcal{M}_\varrho$ is then degenerate.

It is instructive to consider the energy and momentum of separate modes. For that purpose rewrite the expressions for the momentum, P, and the energy, $H_\varrho = J_{3,\varrho}$, with the replacement

$$P \to P_\varrho = P - \varrho^2\theta, \quad \varrho = \frac{\omega}{2\sqrt{\varkappa}}, \tag{9.82}$$

which does not alter the equations of motion and whose meaning will become clear a bit later. We have

$$P_\varrho = \int_{\mathbb{R}_\omega} k(\lambda)\varrho(\lambda)\,d\lambda + 2\sum_{j=1}^{n}\left(\frac{p_j}{4}\sqrt{\omega^2-\varkappa^2 p_j^2} - \varrho^2 \arccos\frac{\varkappa p_j}{\omega}\right) \tag{9.83}$$

and

$$H_\varrho = \int_{\mathbb{R}_\omega} \lambda k(\lambda)\varrho(\lambda)\,d\lambda + \frac{1}{3\varkappa}\sum_{j=1}^{n}(\omega^2-\varkappa^2 p_j^2)^{3/2}. \tag{9.84}$$

Here the principle branch of $\arccos x$ is taken: for $-1 \leqslant x \leqslant 1$ we have $0 \leqslant \arccos x \leqslant \pi$.

The energy and momentum of a single mode of the continuous spectrum with index λ, where λ is in $\mathbb{R}_\omega$, are

$$E(\lambda) = \lambda k(\lambda) \tag{9.85}$$

and

$$P(\lambda) = k(\lambda) \tag{9.86}$$

(cf. (9.80)). The momentum $P(\lambda)$ ranges on the whole axis, whereas the energy $E(\lambda)$ is positive and vanishes for $\lambda = \pm\omega$, i.e. when $P(\lambda)=0$. This is the reason why the momentum P was shifted by the constant $-\varrho^2\theta$. The dispersion law for continuous spectrum modes is

$$E = |P|\sqrt{P^2+\omega^2}. \tag{9.87}$$

The second terms in (9.83)-(9.84) constitute the contribution from the discrete spectrum modes associated with solitons. The energy and momentum of the soliton with index j are

$$E_j = \frac{1}{3\varkappa}(\omega^2-\varkappa^2 p_j^2)^{3/2}, \tag{9.88}$$

$$P_j = \frac{p_j}{2}\sqrt{\omega^2-\varkappa^2 p_j^2} - 2\varrho^2 \arccos\frac{\varkappa p_j}{\omega}. \tag{9.89}$$

When p_j varies from $-\dfrac{\omega}{\varkappa}$ to $\dfrac{\omega}{\varkappa}$, the momentum of the soliton is monotone increasing and covers the *Brillouin zone* $[-2\pi\varrho^2, 0]$. Thus we have the estimate

$$-2\pi\varrho^2 \leqslant P_j \leqslant 0. \tag{9.90}$$

The dispersion law for solitons cannot be expressed through elementary functions. Still, for $P_j \to 0$ and $P_j \to -2\pi\varrho^2$ we have, respectively,

$$E_j = \omega |P_j| + O(P_j^2) \tag{9.91}$$

and

$$E = \omega |P_j + 2\pi\varrho^2| + O((P_j + 2\pi\varrho^2)^2). \tag{9.92}$$

The first of these formulae coincides asymptotically with (9.87).

Thus in the finite density case there are two related branches of semiclassical excitation spectrum. The dispersion law for the first branch is linear for small momenta. For the second one the linearity holds for momenta in the viscinity of the boundary of the Brillouin zone. Such dispersion is typical of the so-called gapless, or Bogolyubov, excitations.

As opposed to the rapidly decreasing case, the dispersion law for the continuous spectrum modes is essentially different from that for solitons. It should not therefore be expected that the continuous spectrum modes could be obtained from solitons by condensation or in some other limit.

To conclude this section we shall consider soliton dynamics from the Hamiltonian viewpoint. The condition (θ) leads to a substantial difference from the rapidly decreasing case. Namely, although the submanifold $\Gamma_{n,\theta}$ in the phase space $\mathcal{M}_{\varrho,\theta}$ corresponding to n-soliton solutions is invariant under the dynamics, it does not inherit the Poisson structure from $\mathcal{M}_{\varrho,\theta}$.

In fact, the natural coordinates p_j, q_j, $j=1, \ldots, n$, on $\Gamma_{n,\theta}$ given by (9.39) are related by

$$2\sum_{j=1}^{n} \arccos \frac{\varkappa p_j}{\omega} \equiv \theta (\operatorname{mod} 2\pi) \tag{9.93}$$

so that the dimension of $\Gamma_{n,\theta}$ is odd. Hence the symplectic structure induced by the imbedding of $\Gamma_{n,\theta}$ into $\mathcal{M}_{\varrho,\theta}$ is degenerate and cannot be converted into a Poisson structure.

An alternative way is to ignore the condition (θ) and consider the phase space $\mathcal{M}_\varrho$. However, the Poisson structure on $\mathcal{M}_\varrho$ is degenerate and there is no associated symplectic form necessary for restricting to soliton submanifolds.

As a manifestation of these facts, the naive requirement $\varrho(\lambda)=0$ for all λ in $\mathbb{R}_\omega$ is inconsistent with the correct Poisson brackets (9.57)–(9.59). We see once again that in this case the discrete and continuous spectrum variables cannot be separated from one another in a way consistent with the Poisson structure.

This discrepancy, however, can be circumvented for all the higher NS equations. It will be instructive to consider first the one-soliton solution

$$\psi(x,t)=\varrho\,\frac{1+e^{i\theta}\,e^{\nu(x-vt-x_0)}}{1+e^{\nu(x-vt-x_0)}}\,, \tag{9.94}$$

where

$$v=\lambda_0=-\omega\cos\frac{\theta}{2}\,,\qquad x_0=\frac{1}{\nu}\log i\gamma_0\,,\qquad \nu=\omega\sin\frac{\theta}{2} \tag{9.95}$$

(see § II.8). For fixed θ, associated to this solution is a one-dimensional space $\Gamma_{1,\theta}$ with a free coordinate x_0. Nevertheless, the dynamics generated by the NS equation

$$x_0(t)=x_0+vt\,,\qquad v=\text{const}\,, \tag{9.96}$$

may be obtained in a Hamiltonian way from the Hamiltonian

$$H_{\text{sol}}^{(1)}=\frac{1}{3\varkappa}\,(\omega^2-\varkappa^2p^2)^{3/2} \tag{9.97}$$

and the Poisson brackets

$$\{p,q\}=1\,, \tag{9.98}$$

with

$$p=-\frac{v}{\varkappa}\,,\qquad q=\nu x_0\,. \tag{9.99}$$

Of course, we have to renounce the condition (θ) and θ itself is not, formally, an annihilator.

Similar arguments apply to n-soliton solutions and the associated submanifold $\Gamma_{n,\theta}$. In the coordinates p_j, q_j introduced in (9.39), the dynamics of the NS equation is defined by the Hamiltonian

$$H_{\text{sol}}^{(n)}=\frac{1}{3\varkappa}\sum_{j=1}^{n}(\omega^2-\varkappa^2p_j^2)^{3/2} \tag{9.100}$$

with the canonical Poisson brackets

$$\{p_j,q_l\}=\delta_{jl}\,;\qquad j,l=1,\ldots,n\,. \tag{9.101}$$

Moreover, the higher NS dynamics (see (9.71) and (9.79)) is given by the Hamiltonians

$$J_{l,\mathrm{sol}} = \frac{(-1)^{\frac{l+1}{2}}}{\varkappa l} \sum_{j=1}^{n} (\omega^2 - \varkappa^2 p_j^2)^{\frac{l}{2}} \tag{9.102}$$

for odd l and

$$J_{l,\varrho} = \frac{1}{l} \sum_{j=1}^{n} \sum_{m=0}^{\frac{l}{2}-1} (-1)^{\frac{l}{2}-m-1} \binom{-\frac{1}{2}}{m} \omega^{2m} \cdot p_j (\omega^2 - \varkappa^2 p_j^2)^{\frac{l-1}{2}-m}$$
$$- \frac{\omega^l}{\varkappa} \binom{\frac{1}{2}}{\frac{l}{2}} \sum_{j=1}^{n} \arccos \frac{\varkappa p_j}{\omega} \tag{9.103}$$

for even l.

For odd l these Hamiltonians result from the local integrals $J_{l,\varrho}$ by setting $\varrho(\lambda)=0$ in (9.64). For even l, one must first substract $\frac{\omega^l}{2\varkappa}\binom{1/2}{l/2}\theta$ from $J_{l,\varrho}$ in order to make the integrand in (9.65) smooth at $\lambda = \pm\omega$, and then set $\varrho(\lambda)=0$.

The Poisson brackets (9.101) result from the naive Poisson brackets (9.40) by reduction $\varrho(\lambda)=0$. As was noted before, the naive Poisson brackets are justified for functionals of the form

$$F = \int_{\mathbb{R}_\omega} f(\lambda)\varrho(\lambda)\, d\lambda + \Phi(p_1, \ldots, p_n), \tag{9.104}$$

where $f(\lambda)$ is a smooth function on $\mathbb{R}_\omega$ including $\lambda = \pm\omega$. In that sense the choice of the Poisson brackets (9.101) is consistent with the regularization of the motion integrals just described. Clearly, the quantity θ formally vanishes in the regularization and so is trivially in involution with p_j and q_j.

Let us now discuss soliton scattering from the Hamiltonian point of view. In § II.8 it was shown that soliton scattering only shifts the coordinates q_j,

$$q_j^{(+)} = q_j^{(-)} + \Delta q_j, \tag{9.105}$$

where

$$\Delta q_j = 2\sum_{l=1}^{j-1} \log \frac{\omega}{\varkappa} \frac{1-\cos\dfrac{\theta_j+\theta_l}{2}}{p_l-p_j} - 2\sum_{l=j+1}^{n} \log \frac{\omega}{\varkappa} \frac{1-\cos\dfrac{\theta_j+\theta_l}{2}}{p_j-p_l} \tag{9.106}$$

and

$$\theta_j = 2\arccos\frac{\varkappa p_j}{\omega}, \quad j=1,\dots,n. \tag{9.107}$$

It is assumed that $p_1 > p_2 > \dots > p_n$ (cf. the analogous formulae (8.10)–(8.13) for the rapidly decreasing case).

However, *in contrast to the rapidly decreasing case the soliton scattering transformation is no longer canonical with respect to the Poisson bracket (9.101).* It is best verified for the case of two solitons. We have

$$\Delta q_2 = -\Delta q_1 = \log\frac{\omega}{\varkappa} \frac{1-\cos\dfrac{\theta_1+\theta_2}{2}}{p_1-p_2} \tag{9.108}$$

and the right hand side is not a function of the difference $p_1 - p_2$ only, hence there can be no expression of the form

$$\Delta q_1 = \frac{\partial K(p_1,p_2)}{\partial p_1}, \quad \Delta q_2 = \frac{\partial K(p_1,p_2)}{\partial p_2}. \tag{9.109}$$

The fact that soliton scattering is not canonical with respect to the Poisson bracket (9.101) has a natural explanation. The point is that the asymptotic variables $p_j^{(\pm)} = p_j$ and $q_j^{(\pm)}$ need not have the Poisson brackets of the form (9.101); a correct computation of their Poisson brackets should use explicit asymptotic expressions, as $|t| \to \infty$, for solutions of the NS equation in the finite density case. The Poisson brackets thus obtained differ from (9.101), and relative to these brackets soliton scattering is canonical. We do not present the corresponding computations because the description of the asymptotic dynamics of all modes in the NS model is a hard computational problem which goes out of the scope of this book.

This completes Part I devoted to the NS model under various boundary conditions: quasi-periodic, and, chiefly, rapidly decreasing and finite density. We have seen associated with this model the following interesting mathematical entities.

1. The zero curvature condition generating the equations of motion.

2. The auxiliary linear problem, with its characteristics and their interpretation in terms of spectral theory and scattering theory.

3. The inverse problem formulated as the matrix Riemann problem.

4. The existence of the r-matrix and fundamental Poisson brackets, and their role in constructing the zero curvature representation.

5. The interpretation of the transition coefficients and discrete spectrum of the auxiliary linear problem as canonical variables of action-angle type.

The description of these entities for both types of boundary conditions was sometimes almost identical, but sometimes differed substantially, especially in this section. Our presentation was intended to make it clear for the reader that the structures described above are of fairly general nature and will give rise to other interesting nonlinear equations. Part II will demonstrate it.

§ 10. Notes and References

1. An interpretation of the mapping $\mathcal{F}$ as a canonical transformation to action-angle variables was suggested for the first time in [ZF 1971] for the case of the Korteweg-de Vries equation. Namely, the pull-back of the symplectic form Ω under the mapping $\mathcal{F}^{-1}$ given by the inverse scattering problem for the one-dimensional Schrödinger operator was computed, and canonical action-angle variables were constructed. An analogous derivation for the NS model in the rapidly decreasing case for $\varepsilon=1$ was performed in [T 1973].

An alternative programme for computing the Poisson brackets of the scattering data for the KdV and NS models was first presented in [ZM 1974] (see also the book [ZMNP 1980]). However, the peculiarities of the correct Poisson brackets for the NS equation under finite density boundary conditions discussed in § 9 escaped the authors.

To express symplectic, or Poisson, structure in new variables, the above papers made extensive use of the identities for solutions of the auxiliary linear problem permitting to explicitly evaluate the occurring integrals. These identities say that certain special quartic homogeneous forms of the solutions are total derivatives. The very existence of such expressions for various models seemed to be a kind of computational miracle. The classical r-matrix provides a rational explanation (see note 3).

2. The notion of the r-matrix first appeared within the quantum version of the inverse scattering method in [STF 1979], [S 1979 a], [TF 1979], [F 1980]. This work was greatly influenced by the results of R. Baxter on exactly solved models in statistical mechanics [B 1972] (see also the treatise [B 1982]). The notion of the r-matrix in the sense adopted in this text was introduced by E. K. Sklyanin [S 1979 b] in a paper on the Landau-Lifshitz equation (see Part II) as a natural semiclassical limiting case of the quantum problem. The fundamental role of the r-matrix in the classical inverse scattering method has become generally acknowledged since (see the reviews [KS 1980], [IK 1982], [KS 1982]).

The discussion of the NS model in the rapidly decreasing case based on the r-matrix approach was given in [S 1980].

3. A simple derivation of the global relation (1.20) from the infinitesimal one (1.18) in § 1 is a major formal achievement of the r-matrix method. The first variant of proving (1.20) reproduces the corresponding quantum derivation (see, for instance, [TF 1979]). The second variant was reported in [IK 1981]. The fact that the integrand in (1.38) is a total derivative represents in abstract form the identities mentioned in note 1.

4. The role of equations (1.40)–(1.41) in defining the Poisson structure was observed in [Be 1980], [KS 1980] and [BD 1982]. In analogy to the quantum case, (1.40) is called "*the classical unitarity equation*", and (1.41) "*the classical Yang-Baxter equation*" or "*the classical triangle equation*". In the quantum case the term "Yang-Baxter equation" was introduced in [TF 1979]. Details on the history of these names can be found in the review paper [KS 1980]. The fundamental role played by the solutions of (1.40)–(1.41) in the construction of integrable systems will be clarified in Part II.

5. The Liouville-Arnold theorem and, generally, Hamiltonian mechanics of systems with finitely many degrees of freedom can be found in the textbooks by V. I. Arnold [A 1974], B. A. Dubrovin, S. P. Novikov, A. T. Fomenko [DNF 1979] and L. D. Landau, E. M. Lifshitz [LL 1965]. The first book makes use of symplectic structure whereas the other two prefer the Poisson structure.

6. The quasi-periodic case of the NS model requires a special treatment based on studying the behaviour of solutions of the auxiliary linear problem on the Riemann surface Γ of the function

$$y^2 = \prod_{n=1}^{\infty}\left(1-\frac{\lambda}{E_n}\right). \tag{10.1}$$

Here E_n are the boundary points of the gaps in the spectrum of the corresponding operator $\mathscr{L}$ which are determined by

$$p_L(E_n) = \pm 2. \tag{10.2}$$

If there are finitely many gaps, then (10.1) defines a hyperelliptic curve; the functions $\psi(x)$, $\bar{\psi}(x)$ appearing in the associated auxiliary linear problem are called *finite-gap solutions*. They have explicit expressions in terms of the Riemann theta-functions of the curve Γ. As an alternative, finite-gap functions can be defined as stationary (i.e. t-independent) solutions of the higher NS equations

$$\sum_n c_n \frac{\delta I_n}{\delta\psi(x)} = \sum_n c_n \frac{\delta I_n}{\delta\bar{\psi}(x)} = 0. \tag{10.3}$$

These are commonly called *Novikov's equations*.

The class of finite-gap initial data is invariant with respect to the NS dynamics which becomes linear on the Jacobi torus (Jacobian) of the curve Γ. Finite-gap functions $\psi(x)$, $\bar{\psi}(x)$ are dense in the space of all quasi-periodic functions. As $L \to \infty$, finite-gap solutions of the NS equation turn into soliton solutions.

The theory of finite-gap solutions of nonlinear evolution equations (in one spatial variable) originates from the work of S. P. Novikov [N 1974]. As a theory of finite-gap integration it was formulated by B. A. Dubrovin and S. P. Novikov [DN 1974], A. R. Its and V. B. Matveev [IM 1975], P. Lax [L 1975], H. McKean and P. van Moerbeke [MKM 1975], and V. A. Marchenko [M 1974]; all these papers dealt with the KdV equation. An algebraic-geometrical approach to the integration of nonlinear evolution equations in two spatial variables based on the so-called Baker-Akhiezer function was developed by I. M. Krichever in [K 1977]. This approach proved to be quite fruitful also for equations in one spatial variable. The present state of the finite-gap integration theory is reviewed in [DMN 1976], [K 1977], [D 1981], and in the books [M 1977], [ZMNP 1980], [L 1984].

Explicit formulae for the finite-gap solutions of the NS equation were first obtained in [I 1976], [IK 1976], [Ko 1976].

The canonical action-angle variables for the KdV equation in the periodic case were constructed in [FM 1976], [VN 1982]. The variables of action type are given by the A-periods of the 1-form $p_L(\lambda)\,d\lambda$ on Γ, and the conjugate angle variables are the linear coordinates on the Jacobian. The latter fact accounts for the small efficiency of this construction in the general case of infinitely many gaps discussed in [MKT 1976] and [L 1981]. The NS model is treated in [DN 1982] (as compared to the KdV equation, the question of reality is non-trivial here, see also [N 1984]).

The manifold of finite-gap functions $\psi(x)$, $\bar{\psi}(x)$ has another natural Poisson structure provided by the calculus of variations. For the case of the KdV equation it was first introduced by S. P. Novikov [N 1974]. The construction of the action-angle variables for these Poisson brackets and the relationship with the original Poisson brackets for KdV and NS are discussed in [GD 1975], [BN 1976], [Bo 1976], [GD 1979], [A 1981], [VN 1982], [DN 1982]. The analysis of Poisson structures on the manifold of finite-gap solutions has led to a general concept of algebraic-geometrical (or analytic) Poisson brackets in the theory of finite-dimensional integrable systems [VN 1984].

We have deliberately given here a large number of references to original and review papers on integrable models with periodic boundary conditions so that the reader could have guidelines in this domain, which otherwise remains practically untouched in the main text.

7. The derivation of the zero curvature representation from the r-matrix formulation of the Poisson brackets in the rapidly decreasing case of the NS model is given in [S 1980]. Our exposition follows [TF 1982].

8. The notion of the Λ-operator for the one-dimensional Schrödinger equation

$$-\frac{d^2y}{dx^2}+u(x)y=\lambda y \tag{10.4}$$

has a long history. Thus, the third order differential operator

$$\Lambda=-\frac{1}{4}\frac{d^3}{dx^3}+u(x)\frac{d}{dx}+\frac{1}{2}\frac{du(x)}{dx} \tag{10.5}$$

satisfying

$$\Lambda(y_1y_2)=\lambda\frac{d}{dx}(y_1y_2) \tag{10.6}$$

for any two solutions of (10.4) was already known to Hermite [H 1912].

Within the inverse scattering formalism for the KdV equation, the operator Λ was first used in [GGKM 1974], [IM 1975] and [MKM 1975] where a compact formulation of the higher KdV equations

$$\frac{\partial u}{\partial t}=\frac{\partial}{\partial x}\frac{\delta I_n}{\delta u(x)}=2\frac{\partial}{\partial x}(2d^{-1}\Lambda)^{n-1}u_0, \tag{10.7}$$

with $u_0(x)\equiv 1$, was established (compare (10.7) with (5.28)).

For the NS model in the rapidly decreasing case the operator Λ was first defined in [AKNS 1974] by the requirement that the squared Jost solutions of the auxiliary linear problem be its eigenfunctions. More precisely, we have

$$\Lambda F(x,\lambda)=\lambda F(x,\lambda), \tag{10.8}$$

where

$$F(x,\lambda)=\begin{pmatrix}0 & f_1^2(x,\lambda)\\ f_2^2(x,\lambda) & 0\end{pmatrix}, \tag{10.9}$$

and $f_{1,2}(x,\lambda)$ are the components of the columns of the Jost solutions $T_\pm(x,\lambda)$. The expansion theorem with respect to the functions $F(x,\lambda)$ was proved in [K 1976], [GK 1980a] and [GK 1980b].

For the general first order linear differential operator with matrix coefficients, the Λ-operator was defined in [Ne 1978]. The operator Λ also appeared in [CD 1976], [CD 1977] as a means for compact formulation of nonlinear evolution equations. The latter papers contain a generalization of the

Λ-operator for the product of solutions of two different auxiliary problems.

9. In [M 1978] it was shown that for the KdV and NS equations there exists a second Hamiltonian structure. The Λ-operator was used there to construct in the rapidly decreasing case an infinite sequence of Hamiltonian vector fields in involution with respect to both structures.

In [KR 1978] the result was analyzed from the inverse scattering point of view. For the KdV and NS equations, the paper presented a hierarchy of symplectic structures generated by the Λ-operator by showing that in terms of the canonical action-angle variables the action of the Λ-operator reduces to multiplication by the spectral parameter λ.

In the review paper [K 1980] the operator Λ with its Hamiltonian interpretation was reported for other integrable nonlinear equations.

We should also mention the investigation of the abstract Hamiltonian formulation for a pair of compatible Poisson brackets in [GD 1980], [GD 1981]. It was shown that two Poisson brackets are compatible if and only if their Jacobi operators have zero Nijenhuis bracket, well known in differential geometry. In this case the associated Λ-operator, which is the ratio of the Jacobi operators, gives rise to a whole hierarchy of Poisson structures (see Chapter IV of Part II).

10. The method for computing the Poisson brackets of transition coefficients outlined in §§ 6, 9 is one of the methodological achievements of the r-matrix approach. It is based on the expression (1.20) for the Poisson brackets of the transition matrix $T(x,y,\lambda)$, which does not depend on the boundary conditions. The latter are only important in the limit as $x,y\to\pm\infty$, and manifest themselves in the matrix factors $E(y,\lambda)$ of $T(x,y,\lambda)$ which are to be reduced out.

This kind of computation was done for the first time in [F 1980] for the quantum NS model and applied to the classical model in [S 1980].

Observe the fundamental role of the Poisson bracket (6.22) for the transition coefficients. Namely, write it as

$$\{\log a(\lambda), b(\mu)\} = \frac{\varkappa}{\lambda-\mu}\, b(\mu) \tag{10.10}$$

and expand both sides in inverse powers of λ. Since $\frac{1}{i}\log a(\lambda)$ is the generating function for the local integrals I_n, we have

$$\{I_n, b(\mu)\} = -i\mu^{n-1} b(\mu). \tag{10.11}$$

These formulae determine the time dynamics of $b(\mu)$ with respect to the higher NS equations: for the n-th NS equation

$$\frac{\partial b}{\partial t}(\mu)=\{I_n, b(\mu)\} \tag{10.12}$$

from (10.11) we find

$$b(\mu, t)=e^{-i\mu^{n-1}t}b(\mu). \tag{10.13}$$

This observation underlies the extension of the inverse scattering method to the quantum situation in [SF 1978].

11. Relations (7.45)–(7.46) used in § 7 were proved in § II.2 only when $\psi(x)$, $\bar{\psi}(x)$ lie in $L_1(-\infty, \infty)$. Nevertheless, it is not hard to see that if $\psi(x)$, $\bar{\psi}(x)$ are Schwartz functions, the limiting values in these formulae are taken in the sense of Schwartz.

12. Canonical variables of action-angle type for the NS model in the rapidly decreasing case were defined in [T 1973] for $\varepsilon=1$ and in [ZM 1974] for $\varepsilon=\pm 1$.

13. As was noticed in § 7, the description of the image of the algebra of observables on the phase space $\mathcal{M}_0$ under the mapping $\mathcal{F}$ is a difficult problem even in the simplest case $\varepsilon=1$. Perhaps, it is too restrictive to consider only real-analytic functionals (with variational derivatives in Schwartz space). Alternative conditions on admissible functionals could be provided by a suitable extension to functionals $F(\psi, \bar{\psi})$ of the definition of a Schwartz function. A rigorous study of this question is an interesting problem of global analysis.

14. Obviously, the phase space $\mathcal{M}_0$ is connected. On the other hand, if $\varepsilon=-1$, a domain in $\mathcal{M}_0$ defined by the condition (A) – the submanifold $\tilde{\mathcal{M}}_0$ breakes up into components

$$\tilde{\mathcal{M}}_0=\bigcup_{n=0}^{\infty}\mathfrak{M}_n \tag{10.14}$$

and hence is disconnected. The point is that the condition (A) forbids the zeros λ_j to reach the real axis or to acquire multiplicity. The intersection in $\mathcal{M}_0$ of the reasonably completed $\mathfrak{M}_n$ would correspond precisely to real or multiple zeros. To construct a suitable global topology on $\mathcal{M}_0$ and, in particular, to define correctly the "sheets" $\mathfrak{M}_n$ is an unsolved and, as we believe, fairly interesting problem.

Nevertheless, we point out again that $\tilde{\mathcal{M}}_0$ is open and dense in $\mathcal{M}_0$ and quite suffices for describing the dynamics of the NS model. In particular, multiple and real zeros for soliton solutions can be obtained by taking an appropriate limit of the explicit formulae.

15. At the end of § 7 we discussed how the hierarchy of Poisson structures on the algebra of observables can be expressed in the coordinates $\varrho(\lambda), \varphi(\lambda); \varrho_j, \varphi_j, p_j, q_j$. This can also be done directly, using property (10.8)

of the operator Λ. For the KdV and NS models the corresponding computation (in terms of symplectic, rather than Poisson, structure) was performed in [KR 1978].

16. The Hamiltonian interpretation of soliton scattering in the rapidly decreasing case presented in § 8 was first proposed in [KMF 1976] and applied to semiclassical quantization. In particular, the expression $\exp iK(p_1, \ldots, p_n; \varrho_1, \ldots, \varrho_n)$ is the semiclassical approximation for the n-soliton quantum scattering matrix.

17. It is highly desirable to develope a matrix analogue of the method of condensation of zeros outlined in § 8 for the scalar Riemann problem. More precisely, we would like to obtain the solution of the regular Riemann problem (see §§ II.1–II.2) with matrix $G(\lambda)$ of the form

$$G(\lambda) = \begin{pmatrix} 1 & -\bar{b}(\lambda) \\ -b(\lambda) & 1 \end{pmatrix}, \tag{10.15}$$

as a limit, as $n \to \infty$, of the solutions of the trivial Riemann problem with n zeros. The difficult problem is to define condensation of the projection operators P_j involved in the corresponding Blaschke-Potapov factors.

A solution of this problem would enable us to determine the asymptotic behaviour of the general solution $\psi(x, t)$ of the NS equation as $t \to \pm\infty$ by condensing the explicit formulae for the n-soliton case of § II.5.

18. The naive Poisson brackets for the NS model in the case of finite density were derived in [ZM 1974] and [KMF 1976]. The construction of correct Poisson brackets is also important for the KdV equation (see [FT 1985]) and for the Toda model (see Part II). In general, such a modification of the Poisson brackets arises whenever the continuous spectrum of the auxiliary linear problem does not cover the whole axis (i.e., there is a gap).

19. The correct Poisson brackets (9.57)–(9.59) may be interpreted as the *Poisson-Dirac brackets* induced by the naive Poisson brackets (9.40) under the constraints

$$\theta = c_1, \quad \varphi(\omega) + \varphi(-\omega) = c_2, \tag{10.16}$$

with c_1, c_2 arbitrary constants.

Recall the definition of these brackets for systems with finitely many degrees of freedom described by canonical coordinates p_j, q_j,

$$\{p_j, q_l\} = \delta_{jl}, \quad j, l = 1, \ldots, n, \tag{10.17}$$

with the constraints

$$\Phi_k(p, q) = c_k; \quad k = 1, \ldots, m, \quad m < n. \tag{10.18}$$

If the matrix M of the Poisson brackets is non-degenerate,

$$M_{ij}=\{\Phi_i, \Phi_j\}, \qquad i,j=1,\dots,m, \tag{10.19}$$

the Poisson-Dirac brackets are given by

$$\{f,g\}_*=\{f,g\}+\sum_{i,j=1}^{m}\{f,\Phi_i\}M^{ij}\{g,\Phi_j\}, \tag{10.20}$$

where the M^{ij} are the matrix elements of the matrix inverse to M (see [D 1964]).

The Poisson brackets (9.57)-(9.59) result from a formal extension of (10.20) to the infinite-dimensional situation.

20. The problem of identifying the algebra of observables and the topology of the phase space $\mathscr{M}_{\varrho,\theta}$ in the coordinates $\varrho(\lambda)$, $\varphi(\lambda)$, p_j, q_j is still more complicated than in the rapidly decreasing case and has not been discussed in the literature.

21. The interpretation of the excitation spectrum branches for the case of finite density was first suggested in [KMF 1976]. The momentum shift $P\to P_\varrho=P-\varrho^2\theta$ used in the text was also introduced in that paper.

References

[A 1974] Arnold, V. I.: Mathematical Methods of Classical Mechanics. Moscow, Nauka 1974 [Russian]; English transl.: Graduate Texts in Mathematics 60, New York-Berlin-Heidelberg, Springer 1978

[A 1981] Alber, S. I.: On stationary problems for equations of Korteweg-de Vries type. Comm. Pure Appl. Math. *34*, 259-272 (1981)

[AKNS 1974] Ablowitz, M. J., Kaup, D. J., Newell, A. C., Segur, H.: The inverse scattering transform - Fourier analysis for nonlinear problems. Stud. Appl. Math. *53*, 249-315 (1974)

[B 1972] Baxter, R. J.: Partition function of the eight-vertex lattice model. Ann. of Physics (N.Y.) *70*, 193-228 (1972)

[B 1982] Baxter, R. J.: Exactly solved models in statistical mechanics. London, Academic Press 1982

[BD 1982] Belavin, A. A., Drinfeld, V. G.: Solutions of the classical Yang-Baxter equation for simple Lie algebras. Funk. Anal. Prilož. *16* (3), 1-29 (1982) [Russian]; English transl. in Funct. Anal. Appl. *16*, 159-180 (1982)

[Be 1980] Belavin, A. A.: Discrete groups and integrability of quantum systems. Funk. Anal. Prilož. *14* (4), 18-26 (1980) [Russian]; English transl. in Funct. Anal. Appl. *14*, 260-267 (1980)

[BN 1976] Bogoyavlensky, O. I., Novikov, S. P.: The connection between the Hamiltonian formalisms of stationary and non-stationary problems. Funk. Anal. Prilož. *10* (1), 9-13 (1976) [Russian]; English transl. in Funct. Anal. Appl. *10*, 8-11 (1976)

[Bo 1976] Bogoyavlensky, O. I.: The integrals of higher stationary KdV equations and the eigenvalues of the Hill operator. Funk. Anal. Prilož. *10* (2), 9-12 (1976) [Russian]; English transl. in Funct. Anal. Appl. *10*, 92-95 (1976)

[CD 1976] Calogero, F., Degasperis, A.: Nonlinear evolution equations solvable by the inverse spectral transform. I. Nuovo Cimento *32B*, 201-242 (1976)

[CD 1977] Calogero, F., Degasperis, A.: Nonlinear evolution equations solvable by the inverse spectral transform. II. Nuovo Cimento *39B*, 1-54 (1977)

[D 1964] Dirac, P. A. M.: Lectures on quantum mechanics. Belfer Grad. School of Science, Yeshiva University, N.-Y. 1964

[D 1981] Dubrovin, B. A.: Theta functions and nonlinear equations. Uspekhi Mat. Nauk *36* (2), 11-80 (1981) [Russian]; English transl. in Russian Math. Surveys *36* (2), 11-92 (1982)

[DMN 1976] Dubrovin, B. A., Matveev, V. B., Novikov, S. P.: Nonlinear equations of Korteweg-de Vries type, finite-zone linear operators and abelian varieties. Uspekhi Mat. Nauk *31* (1), 55-136 (1976) [Russian]; English transl. in Russian Math. Surveys *31* (1), 59-146 (1976)

[DN 1974] Dubrovin, B. A., Novikov, S. P.: Periodic and conditionally periodic analogues of the many-soliton solutions of the Korteweg-de Vries equation. Zh. Eksp. Teor. Fiz. *67*, 2131-2144 (1974) [Russian]; English transl. in Sov. Phys. JETP *40*, 1058-1063

[DN 1982] Dubrovin, B. A., Novikov, S. P.: Algebraic-geometrical Poisson brackets for real finite-gap solutions of the sine-Gordon and nonlinear Schrödinger equations. Dokl. Akad. Nauk SSSR *267*, 1295-1300 (1982) [Russian]; English transl. in Sov. Math. Dokl. *26*, 760-765 (1983)

[DNF 1979] Dubrovin, B. A., Novikov, S. P., Fomenko, A. T.: Modern Geometry. Methods and Applications. Moscow, Nauka 1979 [Russian]; English translation of Part I: Graduate Texts in Mathematics 93; Part II: Graduate Texts in Mathematics 104, New York-Berlin-Heidelberg-Tokyo, Springer 1984, 1985

[F 1980] Faddeev, L. D.: Quantum completely integrable models in field theory. In: Mathematical Physics Review. Sect. C.: Math. Phys. Rev. *1*, 107-155, Harwood Academic (1980)

[FM 1976] Flaschka, H., McLaughlin, D.: Canonically conjugate variables for the Korteweg-de Vries equation and the Toda lattice with periodic boundary conditions. Prog. of Theor. Phys. *55*, 438-456 (1976)

[FT 1985] Faddeev, L. D., Takhtajan, L. A.: Poisson structure for the KdV equation. Lett. Math. Phys. *10*, 183-188 (1985)

[GD 1975] Gelfand, I. M., Dikiĭ, L. A.: Asymptotic behaviour of the resolvent of Sturm-Liouville equations and the algebra of the Korteweg-de Vries equations. Uspekhi Mat. Nauk *30* (5), 67-100 (1975) [Russian]; English transl. in Russian Math. Surveys *30* (5), 77-113 (1975)

[GD 1979] Gelfand, I. M., Dikiĭ, L. A.: Integrable nonlinear equations and the Liouville theorem. Funk. Anal. Prilož. *13* (1), 8-20 (1979) [Russian]; English transl. in Funct. Anal. Appl. *13*, 6-15 (1979)

[GD 1980] Gelfand, I. M., Dorfman, I. Ya.: Schouten bracket and Hamiltonian operators. Funk. Anal. Prilož. *14* (3), 71-74 (1980) [Russian]; English transl. in Funct. Anal. Appl. *14*, 223-226 (1980)

[GD 1981] Gelfand, I. M., Dorfman, I. Ya.: Hamiltonian operators and infinite-dimensional Lie algebras. Funk. Anal. Prilož. *15* (3), 23-40 (1981) [Russian]; English transl. in Funct. Anal. Appl. *15*, 173-187 (1981)

[GGKM 1974] Gardner, C. S., Greene, J. M., Kruskal, M. D., Miura, R. M.: Korteweg-de Vries equation and generalizations. VI. Methods for exact solution. Comm. Pure Appl. Math. *27*, 97-133 (1974)

[GK 1980a] Gerdjikov, V. S., Khristov, E. Kh.: On the evolution equations solvable through the inverse scattering method. I: Spectral Theory. Bulg. J. Phys. *7*, 28–41 (1980) [Bulgarian]

[GK 1980b] Gerdjikov, V. S., Khristov, E. Kh.: On the evolution equations solvable through the inverse scattering method. II: Hamiltonian structure and Bäcklund transformations. Bulg. J. Phys. *7*, 119–133 (1980) [Bulgarian]

[H 1912] Hermite, Ch.: Sur l'équation de Lamé. – Cours d'analyse de l'École polytechn. Paris, 1872–1873, 32-e leçon; Oeuvres, T. III, 118–122, Paris 1912

[I 1976] Its, A. R.: Inversion of hyperelliptic integrals and integration of nonlinear differential equations. Vestnik Leningrad Univ., ser. mat.-mech.-astr. *7* (2), 39–46 (1976) [Russian]

[IK 1976] Its, A. R., Kotlyarov, V. P.: Explicit formulas for solutions of the nonlinear Schrödinger equation. Dokl. Akad. Nauk Ukr. SSR, ser. A., *11*, 965–968 (1976) [Russian]

[IK 1981] Izergin, A. G., Korepin, V. E.: The inverse scattering method approach to the quantum Shabat-Mikhailov model. Comm. Math. Phys. *79*, 303–316 (1981)

[IK 1982] Izergin, A. G., Korepin, V. E.: The quantum inverse scattering method. Physics of elementary particles and atomic nuclei v. *13*, no. 3, 501–541 (1982) [Russian]; English transl. in Soviet J. Particles and Nuclei *13* (3), 207–223 (1982)

[IM 1975] Its, A. R., Matveev, V. B.: Schrödinger operators with finite-band spectrum and *N*-soliton solutions of the Korteweg-de Vries equation. Teor. Mat. Fiz. *23* (1), 51–68 (1975) [Russian]; English transl. in Theor. Math. Phys. *23*, 343–355 (1976)

[K 1976] Kaup, D. J.: Closure of the squared Zakharov-Shabat eigenstates. J. Math. Anal. Appl. *54*, 849–864 (1976)

[K 1977] Krichever, I. M.: Methods of algebraic geometry in the theory of nonlinear equations. Uspekhi Mat. Nauk *32* (6), 183–208 (1977) [Russian]; English transl. in Russian Math. Surveys *32* (6), 185–213 (1977)

[K 1980] Kulish, P. P.: Generating operators for integrable nonlinear evolution equations. In: Boundary-value problems of mathematical physics and related questions in function theory. 12. Zapiski Nauchn. Semin. LOMI *96*, 105–112 (1980) [Russian]; English transl. in J. Sov. Math. *21*, 717–723 (1983)

[KMF 1976] Kulish, P. P., Manakov, S. V., Faddeev, L. D.: Comparison of the exact quantum and quasi-classical results for the nonlinear Schrödinger equation. Teor. Mat. Fiz. *28* (1), 38–45 (1976) [Russian]; English transl. in Theor. Math. Phys. *28*, 615–620 (1977)

[Ko 1976] Kotlyarov, V. P.: Periodic problem for the nonlinear Schrödinger equation. In: Problems of mathematical physics and functional analysis (seminar notes) *1*, 121–131, Kiev, Naukova Dumka, 1976 [Russian]

[KR 1978] Kulish, P. P., Reyman, A. G.: A hierarchy of symplectic forms for the Schrödinger and the Dirac equations on the line. In: Problems in quantum field theory and statistical physics. I. Zapiski Nauchn. Semin. LOMI *77*, 134–147 (1978) [Russian]. English transl. in J. Sov. Math. *22*, 1627–1637 (1983)

[KS 1980] Kulish, P. P., Sklyanin, E. K.: Solutions of the Yang-Baxter equation. In: Differential geometry, Lie groups and mechanics. III. Zapiski Nauchn. Semin. LOMI *95*, 129–160 (1980) [Russian]; English transl. in J. Sov. Math. *19* (5), 1596–1620 (1982)

[KS 1982] Kulish, P. P., Sklyanin, E. K.: Quantum spectral transform method. Recent developments. Lecture Notes in Physics, vol. 151, 61–119, Berlin–Heidelberg–New York, Springer 1982

[L 1975] Lax, P. D.: Periodic solutions of the KdV equation. Comm. Pure Appl. Math. *28*, 141–188 (1975)

[L 1981] Levitan, B. M.: Almost periodicity of infinite-zone potentials. Izv. Akad. Nauk SSSR ser. mat. *45* (2), 291–320 (1981) [Russian]; English transl. in Math. USSR - Izv. *18*, 249–273 (1982)

[L 1984] Levitan, B. M.: Inverse Sturm-Liouville Problems. Moscow, Nauka 1984 [Russian]

[LL 1965] Landau, L. D., Lifshitz, E. M.: Mechanics. Course of Theoretical Physics, vol. I. Moscow, Nauka 1965 [Russian]; English transl. Oxford-London-New York-Paris, Pergamon Press 1960

[M 1974] Marchenko, V. A.: The periodic Korteweg-de Vries problem. Mat. Sb. *95*, 331–356 (1974) [Russian]

[M 1977] Marchenko, V. A.: Sturm-Liouville Operators and their Applications. Kiev, Naukova Dumka 1977 [Russian]

[M 1978] Magri, F.: A simple model of the integrable Hamiltonian equation. J. Math. Phys. *19*, 1156–1162 (1978)

[MKM 1975] McKean, H. P., van Moerbeke, P.: The spectrum of Hill's equation. Invent. Math. *30*, 217–274 (1975)

[MKT 1976] McKean, H. P., Trubowitz, E.: Hill's operator and hyperelliptic function theory in the presence of infinitely many branch points. Comm. Pure Appl. Math. *29*, 143–226 (1976)

[N 1974] Novikov, S. P.: The periodic problem for the Korteweg-de Vries equation. Funk. Anal. Prilož. *8* (3), 54–66 (1974) [Russian]; English transl. in Func. Anal. Appl. *8*, 236–246 (1974)

[N 1984] Novikov, S. P.: Algebraic-topological approach to reality problems. Real action variables in the theory of finite-gap solutions of the sine-Gordon equation. In: Differential geometry, Lie groups and mechanics. VI. Zapiski Nauchn. Semin. LOMI *133*, 177–196 (1984) [Russian]

[Ne 1978] Newell, A. C.: Near-integrable systems, nonlinear tunneling and solutions in slowly changing media. In: Calogero, F. (ed.). Nonlinear evolution equations solvable by the spectral transform. Research Notes in Mathematics *26*, 127–179, London, Pitman 1978

[S 1979 a] Sklyanin, E. K.: The inverse scattering method and the quantum nonlinear Schrödinger equation. Dokl. Akad. Nauk SSSR *244*, 1337–1341 (1979) [Russian]; English transl. in Sov. Phys. Dokl. *24*, 107–110 (1979)

[S 1979 b] Sklyanin, E. K.: On complete integrability of the Landau-Lifshitz equation. Preprint LOMI, E-3-79, Leningrad 1979

[S 1980] Sklyanin, E. K.: Quantum version of the inverse scattering problem method. In: Differential geometry, Lie groups and mechanics. III. Zapiski Nauchn. Semin. LOMI *95*, 55–128 (1980) [Russian]; English transl. in J. Sov. Math. *19*, 1546–1595 (1982)

[SF 1978] Sklyanin, E. K., Faddeev, L. D.: The quantum-mechanical approach to completely integrable models of field theory. Dokl. Akad. Nauk SSSR *243*, 1430–1433 (1978) [Russian]; English transl. in Sov. Phys. Dokl. *23*, 902–906 (1978)

[STF 1979] Sklyanin, E. K., Takhtajan, L. A., Faddeev, L. D.: The quantum method of the inverse problem. I. Teor. Mat. Fiz. *40* (2) 194–220 (1979) [Russian]; English transl. in Theor. Math. Phys. *40* (2) 688–706 (1980)

[T 1973] Takhtajan, L. A.: Hamiltonian systems connected with the Dirac equation, in: Differential geometry, Lie groups and mechanics. I. Zapiski Nauchn. Semin. LOMI *37*, 66–76 (1973) [Russian]; English transl. in J. Sov. Math. *8*, 219–228 (1977)

[TF 1979] Takhtajan, L. A., Faddeev, L. D.: The quantum inverse problem method and the XYZ Heisenberg model. Uspekhi Mat. Nauk *34* (5), 13–63 (1979) [Russian]; English transl. in Russian Math. Surveys *34* (5), 11–68 (1979)

[TF 1982] Takhtajan, L. A., Faddeev, L. D.: A simple connection between geometrical and Hamiltonian representations for the integrable nonlinear equations. In: Boundary-value problems of mathematical physics and related questions in function theory. 14. Zapiski Nauchn. Semin. LOMI *115*, 264–273 (1982) [Russian]

[VN 1982] Veselov, A. P., Novikov, S. P.: On Poisson brackets compatible with algebraic geometry and Korteweg-de Vries dynamics on the set of finite-gap potentials. Dokl. Akad. Nauk SSSR *266*, 533–537 (1982) [Russian]; English transl. in Sov. Math. Dokl. *26*, 357–362 (1983)

[VN 1984] Veselov, A. P., Novikov, S. P.: Poisson brackets and complex tori. Trudy Mat. Inst. Steklov *165*, 49–61 (1984) [Russian]

[ZF 1971] Zakharov, V. E., Faddeev, L. D.: Korteweg-de Vries equation, a completely integrable Hamiltonian system. Funk. Anal. Prilož. *5* (4), 18–27 (1971) [Russian]; English transl. in Func. Anal. Appl. *5*, 280–287 (1972)

[ZM 1974] Zakharov, V. E., Manakov, S. V.: On the complete integrability of the nonlinear Schrödinger equation. Teor. Mat. Fiz. *19*, 332–343 (1974) [Russian]; English transl. in Theor. Math. Phys. *19*, 551–560 (1975)

[ZMNP 1980] Zakharov, V. E., Manakov, S. V., Novikov, S. P., Pitaievski, L. P.: Theory of Solitons. The Inverse Problem Method. Moscow, Nauka 1980 [Russian]; English transl.: New York, Plenum 1984

Part Two
General Theory of Integrable Evolution Equations

The inverse scattering method outlined in Part I for the NS model would not be worth much attention if it had no other applications. It is well known, however, that the NS model is by no means an exception: applications of the inverse scattering method are numerous and far from exhausted.

The present Part II treats several other typical examples, which will allow us in the end to form a general view of the applicability of the method. Naturally, these models are discussed in lesser detail since the basic notions and techniques of the inverse scattering method have already been introduced and worked out for the case of the NS model.

Chapter I
Basic Examples and Their General Properties

In this chapter we shall give a list of typical examples and establish their general properties: the zero curvature representation and the Hamiltonian formulation. Then, motivated by these examples, we shall outline a general scheme for constructing integrable equations and their solutions based on the matrix Riemann problem. A detailed study of the most important models and the Hamiltonian interpretation of the general scheme will be presented in the following chapters. The examples to be considered fall into two classes: dynamical systems generated by partial differential evolution equations (continuous models), and evolution systems of difference type (lattice models).

§ 1. Formulation of the Basic Continuous Models

We shall now proceed directly to the examples and exhibit the corresponding phase spaces (sets of dynamical variables), the equations of motion, and their interpretation as the zero curvature conditions.

1. The continuous isotropic Heisenberg ferromagnet model (HM model)

The phase space of the model consists of vector-valued functions $\vec{S}(x)=(S_1(x), S_2(x), S_3(x))$ with values in the unit sphere S^2 in $\mathbb{R}^3$,

$$\vec{S}^2(x)=\sum_{a=1}^{3} S_a^2(x)=1\,, \tag{1.1}$$

satisfying certain boundary conditions (see below). The equations of motion are

$$\frac{\partial \vec{S}}{\partial t}=\vec{S}\wedge\frac{\partial^2 \vec{S}}{\partial x^2}\,, \tag{1.2}$$

where $\wedge$ indicates the wedge (vector) product in $\mathbb{R}^3$. Clearly, the constraint (1.1) is preserved by these equations. The model is $O(3)$-invariant: if $\vec{S}(x, t)$ is a solution of the equations of motion, and R is an arbitrary orthogonal matrix in $\mathbb{R}^3$ which does not depend on x and t, then $R\vec{S}(x, t)$ is also a solution.

Typical boundary conditions are

a) the periodic boundary conditions

$$\vec{S}(x+2L)=\vec{S}(x); \tag{1.3}$$

b) the rapidly decreasing boundary conditions

$$\lim_{|x|\to\infty} \vec{S}(x)=\vec{S}_0, \tag{1.4}$$

where in view of the $O(3)$-invariance there is no loss of generality in fixing the constant vector $\vec{S}_0$ to be

$$\vec{S}_0=(0, 0, 1). \tag{1.5}$$

It is also assumed that the limiting values are approached sufficiently fast, for instance, in the sense of Schwartz.

More general boundary conditions are the quasi-periodicity conditions

$$\vec{S}(x+2L)=R\vec{S}(x), \tag{1.6}$$

where the matrix R is fixed and lies in the group $O(3)$, and their limiting case, as $L\to\infty$ (the analogue of the finite density conditions). However, these will not be discussed here.

The above model occurs in solid state physics and describes the classical spin $\vec{S}$ distributed along the line, i.e. the one-dimensional continuous magnet.

The Poisson structure on phase space is given by the Poisson brackets

$$\{S_a(x), S_b(y)\}=-\varepsilon_{abc}S_c(x)\delta(x-y), \tag{1.7}$$

where S_a, $a=1, 2, 3$, are the components of the vector $\vec{S}$, and ε_{abc} is a totally skew-symmetric rank 3 tensor, $\varepsilon_{123}=1$. From the Lie-group point of view, the Poisson bracket (1.7) is a restriction of the general Lie-Poisson bracket associated with the so-called current group, the group of matrix-valued functions $g(x)$ with values in $O(3)$, to the symplectic orbit defined by (1.1). Of course, the non-degeneracy of this Poisson structure can be verified directly by using (1.1), without recourse to the general theory.

The HM equation can be written in Hamiltonian form

$$\frac{\partial \vec{S}}{\partial t} = \{H, \vec{S}\}, \tag{1.8}$$

where

$$H = \frac{1}{2} \int \left(\frac{\partial \vec{S}}{\partial x}\right)^2 dx, \tag{1.9}$$

the integral being taken over the fundamental domain for the boundary conditions of type a) or over the whole real line for the case b).

Other physically interesting integrals of the motion are the momentum – the generator of the x-translations

$$P = \int \frac{S_1 \dfrac{\partial S_2}{\partial x} - S_2 \dfrac{\partial S_1}{\partial x}}{1+S_3}\, dx \tag{1.10}$$

and the total spin in the case of the periodic boundary conditions,

$$\vec{M} = \int_{-L}^{L} \vec{S}(x)\, dx. \tag{1.11}$$

The components, M_a, of the total spin induce a Hamiltonian action of the Lie algebra of the group $O(3)$, their Poisson brackets being

$$\{M_a, M_b\} = -\varepsilon_{abc} M_c \tag{1.12}$$

(cf. (1.7)).

In the rapidly decreasing case only the regularized third spin component survives as an observable,

$$M_3 = \int_{-\infty}^{\infty} (S_3(x) - 1)\, dx. \tag{1.13}$$

The quantities M_1 and M_2, though formally defined, are inadmissible functionals since the corresponding Hamiltonian flows do not respect the boundary conditions (1.4)–(1.5).

The expression for the momentum (1.10) has a geometrical meaning which will be discussed at the end of the section. It will then become clear that in the periodic case the momentum is $O(3)$-invariant, which is not obvious from (1.10).

Equation (1.2) is representable as the zero curvature condition for the connection $(U(x,t,\lambda), V(x,t,\lambda))$ of the form

$$U(\lambda)=\frac{\lambda}{2i}S, \qquad V(\lambda)=\frac{i\lambda^2}{2}S+\frac{\lambda}{2}\frac{\partial S}{\partial x}S. \tag{1.14}$$

Here

$$S=\vec{S}\cdot\vec{\sigma}=\sum_{a=1}^{3} S_a\sigma_a \tag{1.15}$$

is a traceless Hermitian matrix satisfying

$$S^2=I \tag{1.16}$$

(see (1.1)), and the σ_a's are the Pauli matrices. In fact, the zero curvature condition

$$\frac{\partial U}{\partial t}-\frac{\partial V}{\partial x}+[U, V]=0 \tag{1.17}$$

together with (1.16) is equivalent to

$$\frac{\partial S}{\partial t}=\frac{1}{2i}\left[S, \frac{\partial^2 S}{\partial x^2}\right], \tag{1.18}$$

which in turn is equivalent to the original HM equation.

Notice that the equivalence of (1.14), (1.17) and (1.18) relies solely on (1.16) and so remains true for matrices $S(x, t)$ of any dimension.

2. The sine-Gordon model (SG model)

The equation of motion is

$$\frac{\partial^2\varphi}{\partial t^2}-\frac{\partial^2\varphi}{\partial x^2}+\frac{m^2}{\beta}\sin\beta\varphi=0, \tag{1.19}$$

with $\varphi(x, t)$ a real-valued function and β, m some positive parameters. The functions $\varphi(x, t)$ and $\varphi(x, t)+\dfrac{2\pi}{\beta}$ are regarded as equivalent.

Typical boundary conditions for the initial data

$$\varphi(x)=\varphi(x, t)|_{t=0}, \qquad \pi(x)=\frac{\partial\varphi}{\partial t}(x, t)\Big|_{t=0} \tag{1.20}$$

are as follows:

a) the periodic boundary conditions

$$\varphi(x+2L)\equiv\varphi(x)\left(\mathrm{mod}\,\frac{2\pi}{\beta}\right),\qquad \pi(x+2L)=\pi(x);\tag{1.21}$$

b) the rapidly decreasing boundary conditions

$$\lim_{|x|\to\infty}\varphi(x)\equiv 0\left(\mathrm{mod}\,\frac{2\pi}{\beta}\right),\qquad \lim_{|x|\to\infty}\pi(x)=0.\tag{1.22}$$

Here the boundary values must be approached sufficiently fast, e.g. in the sense of Schwartz.

From the physical point of view, (1.19) provides a model of relativistic field theory in two space-time dimensions. The parameters m and β play the role of mass and coupling constant, respectively. The massive real scalar field $\varphi(x,t)$ carries an important characteristic, the topological charge

$$Q=\frac{\beta}{2\pi}\int\frac{\partial\varphi}{\partial x}\,dx,\tag{1.23}$$

where integration goes over the fundamental domain in the case a) and over the whole line in the case b). This quantity is conserved due to the boundary conditions, and is an integer. Mathematically, Q is the winding number (degree) of the function $\chi(x)=\exp\{i\beta\varphi(x)\}$.

The phase space of the model consists of initial data given by pairs of functions $(\pi(x),\varphi(x))$ with $\varphi(x)$ regarded mod $\frac{2\pi}{\beta}$, satisfying the boundary conditions a) or b). The Poisson structure is defined by the Poisson brackets

$$\{\varphi(x),\varphi(y)\}=\{\pi(x),\pi(y)\}=0,\qquad \{\pi(x),\varphi(y)\}=\delta(x-y)\tag{1.24}$$

and is obviously non-degenerate. Equation (1.19) can be written in Hamiltonian form

$$\frac{\partial\varphi}{\partial t}=\{H,\varphi\},\qquad \frac{\partial\pi}{\partial t}=\{H,\pi\}\tag{1.25}$$

with the Hamiltonian

$$H=\int\left(\frac{1}{2}\pi^2+\frac{1}{2}\left(\frac{\partial\varphi}{\partial x}\right)^2+\frac{m^2}{\beta^2}(1-\cos\beta\varphi)\right)dx,\tag{1.26}$$

where the domain of integration depends on the boundary conditions. The Hamiltonian H, the momentum P,

$$P = -\int \pi \frac{\partial \varphi}{\partial x} dx \tag{1.27}$$

and the boost generator K,

$$K = \int x \left(\frac{1}{2} \pi^2 + \frac{1}{2} \left(\frac{\partial \varphi}{\partial x} \right)^2 + \frac{m^2}{\beta^2} (1 - \cos \beta \varphi) \right) dx \tag{1.28}$$

induce a Hamiltonian action of the Lie algebra of the Poincare group of two-dimensional space-time. Their Poisson brackets are

$$\{H, P\} = 0, \quad \{H, K\} = P, \quad \{K, P\} = -H. \tag{1.29}$$

Equations (1.19) are representable as the zero curvature condition for the connection $(U(x, t, \lambda), V(x, t, \lambda))$ of the form

$$U(\lambda) = \frac{\beta}{4i} \pi \sigma_3 + \frac{k_0}{i} \sin \frac{\beta \varphi}{2} \sigma_1 + \frac{k_1}{i} \cos \frac{\beta \varphi}{2} \sigma_2, \tag{1.30}$$

$$V(\lambda) = \frac{\beta}{4i} \frac{\partial \varphi}{\partial x} \sigma_3 + \frac{k_1}{i} \sin \frac{\beta \varphi}{2} \sigma_1 + \frac{k_0}{i} \cos \frac{\beta \varphi}{2} \sigma_2, \tag{1.31}$$

with σ_a being, as usual, the Pauli matrices and

$$k_0 = \frac{m}{4} \left(\lambda + \frac{1}{\lambda} \right), \quad k_1 = \frac{m}{4} \left(\lambda - \frac{1}{\lambda} \right). \tag{1.32}$$

The covariant derivatives X_μ, $\mu = 0, 1$, where

$$X_0 = \frac{\partial}{\partial x_0} - V, \quad X_1 = \frac{\partial}{\partial x_1} - U; \quad x_0 = t, \ x_1 = x, \tag{1.33}$$

are manifestly Lorentz-invariant. To show this observe that $\pi = \frac{\partial \varphi}{\partial x_0}$ and combine k_0, k_1 into a Lorentz vector of length $\frac{m}{2}$: $k^2 = k_\mu k_\mu = k_0^2 - k_1^2 = \frac{m^2}{4}$, and also consider the dual vector $k_\mu = \varepsilon_{\mu\nu} k_\nu$ with components k_1, k_0.

3. The Landau-Lifshitz model of continuous anisotropic magnet (LL model)

The phase space is the same as in our first example, the isotropic magnet. Consider the Hamiltonian

$$H=\frac{1}{2}\int\left(\left(\frac{\partial\vec{S}}{\partial x}\right)^2-J(\vec{S})\right)dx, \tag{1.34}$$

where $J(\vec{S})$ is the quadratic form of a constant matrix J which without loss of generality can be assumed to be diagonal, so that

$$J(\vec{S})=J_1S_1^2+J_2S_2^2+J_3S_3^2, \quad J_1\leqslant J_2\leqslant J_3. \tag{1.35}$$

(In the rapidly decreasing case, the integrand in (1.34) must be modified by substracting $-J(\vec{S}_0)$.)

Then Hamilton's equations of motion are

$$\frac{\partial\vec{S}}{\partial t}=\vec{S}\wedge\frac{\partial^2\vec{S}}{\partial x^2}+\vec{S}\wedge J\vec{S} \tag{1.36}$$

and describe the anisotropic magnet. In solid state physics equation (1.36) is called the Landau-Lifshitz equation.

In the generic situation $J_1<J_2<J_3$, the zero curvature representation for the LL model is given by the matrices

$$U(x,t,\lambda)=\frac{1}{i}\sum_{a=1}^{3}u_a(\lambda)S_a\sigma_a, \tag{1.37}$$

$$V(x,t,\lambda)=2i\sum_{a=1}^{3}\frac{u_1(\lambda)u_2(\lambda)u_3(\lambda)}{u_a(\lambda)}S_a\sigma_a + \frac{1}{i}\sum_{a,b,c=1}^{3}u_a(\lambda)\varepsilon_{abc}S_b\frac{\partial S_c}{\partial x}\sigma_a, \tag{1.38}$$

where

$$u_1(\lambda)=\varrho\frac{1}{\operatorname{sn}(\lambda,k)}, \quad u_2(\lambda)=\varrho\frac{\operatorname{dn}(\lambda,k)}{\operatorname{sn}(\lambda,k)}, \quad u_3(\lambda)=\varrho\frac{\operatorname{cn}(\lambda,k)}{\operatorname{sn}(\lambda,k)} \tag{1.39}$$

and

$$k=\sqrt{\frac{J_2-J_1}{J_3-J_1}}, \quad 0<k<1 \tag{1.40}$$

with

$$\varrho=\tfrac{1}{2}\sqrt{J_3-J_1}, \quad \varrho>0. \tag{1.41}$$

Here $\operatorname{sn}(\lambda, k)$, $\operatorname{cn}(\lambda, k)$, and $\operatorname{dn}(\lambda, k)$ are the Jacobi elliptic functions of modulus k.

The functions $u_a(\lambda)$ are subject to quadratic relations

$$u_a^2(\lambda) - u_b^2(\lambda) = \tfrac{1}{4}(J_b - J_a); \qquad a, b = 1, 2, 3, \tag{1.42}$$

defining an elliptic curve; the spectral parameter λ plays the role of a uniformization variable. Observe that (1.39)–(1.40) is just one of the many possible parametrizations of (1.42). To derive (1.36) from the zero curvature condition it is enough to use only (1.41)–(1.42).

Later we shall see that the LL model is in a certain sense universal, so that all the above models can be obtained as its various limiting cases.

All the models discussed so far admit a zero curvature representation with 2×2 matrices $U(x, t, \lambda)$ and $V(x, t, \lambda)$. In a different way, the corresponding vector bundle (see § I.2 of Part I) has the space $\mathbb{C}^2$ as its fibre. The latter is commonly referred to as an *auxiliary space* since it determines the matrix form of the auxiliary linear problem.

The inverse scattering method is not in fact restricted to two-dimensional auxiliary spaces. There are several physically interesting models that exploit auxiliary spaces of higher dimensions. Let us give an illustrative example.

4. The vector nonlinear Schrödinger model (vector NS model)

The dynamical variables are complex vector-valued functions $\psi_a(x)$, $\bar{\psi}_a(x)$, $a = 1, \ldots, n$, describing a charged field with n colours. The equations of motion are an immediate extension of the ordinary NS equation,

$$i \frac{\partial \psi_a}{\partial t} = -\frac{\partial^2 \psi_a}{\partial x^2} + 2\varkappa \sum_{b=1}^{n} |\psi_b|^2 \psi_a . \tag{1.43}$$

The Poisson structure on phase space in given by the Poisson brackets

$$\begin{aligned} &\{\psi_a(x), \psi_b(y)\} = \{\bar{\psi}_a(x), \bar{\psi}_b(y)\} = 0, \\ &\{\psi_a(x), \bar{\psi}_b(y)\} = i\delta_{ab}\,\delta(x-y); \qquad a, b = 1, \ldots, n. \end{aligned} \tag{1.44}$$

The Hamiltonian of the model is

$$H = \int \left(\sum_{a=1}^{n} \left| \frac{\partial \psi_a}{\partial x} \right|^2 + \varkappa \left(\sum_{a=1}^{n} |\psi_a|^2 \right)^2 \right) dx, \tag{1.45}$$

where integration takes account of the boundary conditions which extend those for the ordinary NS model.

The system of equations (1.43) has a natural $U(n)$-invariance, so that the quasi-periodic boundary conditions have the form

$$\psi(x+2L)=\psi(x)\,U, \tag{1.46}$$

where $\psi(x)$ is a row-vector with components $\psi_a(x)$, $a=1,\dots,n$, and U is a constant unitary matrix in $\mathbb{C}^n$.

The matrices $U(x,t,\lambda)$ and $V(x,t,\lambda)$ in the zero curvature representation for the vector NS model are

$$U(\lambda)=U_0+\lambda U_1, \qquad V(\lambda)=V_0+\lambda V_1+\lambda^2 V_2, \tag{1.47}$$

(compare with the ordinary NS model in § I.2, Part I) where, in the block notation,

$$U_0=\sqrt{\varkappa}\begin{pmatrix} \mathbb{O} & \psi^+ \\ \psi & \mathbb{O} \end{pmatrix}, \qquad U_1=\frac{1}{2i}\,\mathrm{diag}\,(\underbrace{1,\dots,1}_{n},-1) \tag{1.48}$$

and

$$V_0=i\sqrt{\varkappa}\begin{pmatrix} \sqrt{\varkappa}\,\psi^+\psi & -\dfrac{\partial\psi^+}{\partial x} \\ \dfrac{\partial\psi}{\partial x} & -\sqrt{\varkappa}\,\psi\psi^+ \end{pmatrix}, \qquad V_1=-U_0,\ V_2=-U_1. \tag{1.49}$$

Here the column-vector $\psi^+(x)$ is Hermitian conjugate to the row-vector $\psi(x)$, and $\mathbb{O}$ indicates a zero block of dimension $n\times n$.

These formulae can be extended to other cases, for instance, when $\psi(x)$ is a $n_1\times n_2$ matrix. The most general situation is formulated in terms of homogeneous spaces for compact Lie groups.

The above example only serves to illustrate the significance of multidimensional auxiliary spaces. The analysis of the auxiliary linear problem for systems of general type of dimension greater than 2 is much more complicated that for two-dimensional systems, and will not be our concern here.

This concludes our list of basic continuous models.

We close this section with a general comment on observables such as the momentum P which is the generator of translations in the spatial variable x, i.e.

$$\{P,\varphi(x)\}=\frac{d\varphi}{dx}(x) \tag{1.50}$$

for any local observable $\varphi(x)$. We will show that P can be expressed in terms of the 2-form Ω that defines the symplectic structure.

Let M be a manifold of dimension n with symplectic form ω. Consider the phase space $\mathscr{M}$ consisting of functions $u(x)$ with values in M satisfying the periodic or rapidly decreasing boundary conditions. There is a natural Poisson structure on $\mathscr{M}$ which in local coordinates $u_a(x)$, $a=1,\ldots,n$, is given by the Poisson brackets

$$\{u_a(x), u_b(y)\} = \eta_{ab}(u(x))\delta(x-y); \quad a,b=1,\ldots,n, \tag{1.51}$$

where the η_{ab} are the coefficients of the Jacobi matrix, η, of the Poisson structure on M. The associated symplectic form Ω is

$$\Omega(u) = \int \omega(u(x))\,dx, \tag{1.52}$$

where

$$\omega(u(x)) = -\sum_{1\leqslant a<b\leqslant n} \eta^{ab}(u(x))\,du_a(x)\wedge du_b(x). \tag{1.53}$$

Here integration takes account of the boundary conditions, η^{ab} are the entries of the matrix inverse to η,

$$\sum_{c=1}^{n} \eta^{ac}\eta_{cb} = \delta_{ab}. \tag{1.54}$$

Since ω is closed, we have (at least locally in a fixed coordinate patch on M)

$$\omega = d\theta, \tag{1.55}$$

so that the 1-form θ is (locally) a primitive form for the 2-form ω, $\theta = d^{-1}\omega$. Hence, in the same coordinates on $\mathscr{M}$, there is a primitive form, Θ, for Ω,

$$\Theta(u) = \int \theta(u(x))\,dx, \quad \theta(u(x)) = \sum_{a=1}^{n} \theta^a(u(x))\,du_a(x) \tag{1.56}$$

and

$$\frac{\partial\theta^a}{\partial u_b} - \frac{\partial\theta^b}{\partial u_a} = \eta^{ab}; \quad a,b=1,\ldots,n. \tag{1.57}$$

The form Θ is invariant under the x-translations. According to the general recipes of Hamiltonian mechanics, the momentum P as a generator of the x-translations results from applying Θ to the tangent vector $\frac{du}{dx}$; in local coordinates

$$P(u) = -\sum_{a=1}^{n} \int \theta^a(u(x)) \frac{du_a(x)}{dx}\, dx. \tag{1.58}$$

It is easy to verify (1.50) for the functional P thus defined, when $\varphi(x) = u_a(x)$, directly by using (1.51), (1.54), and (1.57).

In view of the boundary conditions, the functions $u_a(x)$, $a = 1, \ldots, n$, give rise to a closed path on M, the one-dimensional cycle γ, so that (1.58) can be written as

$$P = -\int_{\gamma} \theta. \tag{1.59}$$

This relation may be reexpressed in terms of symplectic form ω alone by means of the Stokes formula. More precisely, let B_γ be a film with boundary γ contained in the same patch on M. Then

$$P = -\int_{B_\gamma} \omega. \tag{1.60}$$

Let us show that the above construction provides the desired expressions for the momentum in the previous examples. This is elementary for the NS and SG models since ω is exact, $\omega = d\theta$, with

$$\theta_{\mathrm{NS}} = \frac{1}{2i}(\psi\, d\bar{\psi} - \bar{\psi}\, d\psi), \qquad \theta_{\mathrm{SG}} = \pi\, d\varphi. \tag{1.61}$$

It then remains to compare the general formula (1.58) with (I.1.26) of Part I and (1.27).

For the HM and LL models the verification is more instructive. Namely, in this case ω coincides with the standard area form on $\mathbf{S}^2$ normalized by

$$\int_{\mathbf{S}^2} \omega = 4\pi. \tag{1.62}$$

As opposed to the previous cases, ω is not exact and hence the primitive form, $\theta = d^{-1}\omega$, exists only locally. It is not hard to see that (1.59) gives the same as (1.10) if the coordinate patch on $\mathbf{S}^2$ is chosen to be $\mathbf{S}^2 \backslash \{(0, 0, -1)\}$, and the variables S_1 and S_2 are taken for local coordinates. In fact, with these coordinates,

$$\omega = -\frac{dS_1 \wedge dS_2}{S_3}, \qquad \theta = \frac{S_2 dS_1 - S_1 dS_2}{1+S_3}, \tag{1.63}$$

where $S_3 = \sqrt{1-S_1^2-S_2^2}$.

In local coordinates on a different patch we would get a different expression for the momentum. However, in simply connected patches these expressions differ by an integral multiple of 4π, the total area of $\mathbf{S}^2$. Indeed, (1.60) implies that the ambiguity in the momentum functional stems from the ambiguity in the choice of the film B_γ. For different films, however, the values of P differ by an integral multiple of the period $\int_{\mathbf{S}^2} \omega$ of ω, which is equal to 4π. So in our case (1.60) finally becomes

$$P \equiv -\int_{B_\gamma} \omega \,(\mathrm{mod}\, 4\pi). \tag{1.64}$$

This expression makes it obvious that, under the periodic boundary conditions, the momentum is $O(3)$-invariant (mod 4π), as was claimed earlier.

Another important example of a multivalued functional will be encountered in § 5 in connection with other models.

§ 2. Examples of Lattice Models

In the preceding section we were dealing with partial differential evolution equations where the one-dimensional spatial variable x varies continuously on the circle (a finite interval with the ends identified), or on the whole real axis. In applications, however, an important part is also played by lattice models where the spatial variable takes on discrete (for instance, integral) values. These may arise as an artificial finite difference approximation for differential equations, or else may serve as a natural model for, say, crystal lattice oscillations in solid state physics. For a finite lattice, which is the analogue of the circle for continuous models, the evolution system has finitely many degrees of freedom and actually belongs to classical mechanics.

We shall therefore assume that the "discretized" spatial variable n takes integral values and ranges over either the set of all integers, $\mathbb{Z}$, (the analogue of the real axis) or its finite subset. For the most part we shall deal with an analogue of the circle, $n=1,\ldots,N$; $N+1\equiv 1$, i.e. with the periodic lattice $\mathbb{Z}_N = \mathbb{Z}/N\mathbb{Z}$. As before, the time variable t is continuous and ranges over the real line.

The zero curvature condition has a natural extension to the case of lattice models. The base of the fibre bundle is the discretized space-time $\mathbb{Z}\times\mathbb{R}^1$ or $\mathbb{Z}_N\times\mathbb{R}^1$, and the fibre is the auxiliary space $\mathbb{C}^M$. The covariant derivative $X_1=\frac{\partial}{\partial x}-U(x,t,\lambda)$ (or rather the infinitely small parallel transport in spatial direction, $\Omega_n=\exp\int_{\Delta_n} U(x,t,\lambda)\,dx$; see (I.2.14), Part I) is replaced by the matrix $L_n(t,\lambda)$ defining the transport from site n to site $n+1$. Parallel transport in temporal direction is given, as before, by the covariant derivative $\frac{\partial}{\partial t}-V(x,t,\lambda)$. The equations for the vector $F_n(t,\lambda)$ to be covariantly constant given by formulae (I.2.1)–(I.2.2) of Part I become

$$F_{n+1}=L_n(t,\lambda)\,F_n\,, \tag{2.1}$$

$$\frac{dF_n}{dt}=V_n(t,\lambda)\,F_n\,. \tag{2.2}$$

The compatibility condition for the system,

$$\frac{dL_n(t,\lambda)}{dt}+L_n(t,\lambda)\,V_n(t,\lambda)-V_{n+1}(t,\lambda)\,L_n(t,\lambda)=0\,, \tag{2.3}$$

is a zero curvature condition relative to the elementary closed path in the base space with vertices in (n,t), $(n+1,t)$, $(n+1,t+dt)$, and $(n,t+dt)$. Of course, (2.3) implies that parallel transport around any closed path is the identity. Thus (2.3) and (2.1)–(2.2) will be called respectively *the zero curvature condition and the zero curvature representation for lattice models. Equation (2.1) will play the role of the auxiliary linear problem.*

After this general introduction we shall consider a list of basic examples.

The most popular model with numerous applications is

1. The Toda model

The equations of motion have the form

$$\frac{d^2q_n}{dt^2}=e^{q_{n+1}-q_n}-e^{q_n-q_{n-1}}\,, \tag{2.4}$$

where the real variables q_n have the meaning of coordinates of classical particles with one degree of freedom. Typical boundary conditions are as follows.

a) Free end conditions: $1\leqslant n\leqslant N$,

$$q_0 = -q_{N+1} = +\infty . \tag{2.5}$$

b) Quasi-periodic boundary conditions

$$q_{n+N} = q_n + c , \tag{2.6}$$

where c is an arbitrary real constant which does not depend on t.

c) Rapidly decreasing boundary conditions

$$\lim_{n\to-\infty} q_n = 0 , \quad \lim_{n\to+\infty} q_n = c , \tag{2.7}$$

where the limiting values are approached sufficiently fast.

Actually, conditions c) are rather the analogues of the finite density boundary conditions for the NS model (see Part I, § I.1). Yet, we call them rapidly decreasing because the differences $q_n - q_{n-1}$ occurring in the equations of motion decrease rapidly as $|n| \to \infty$.

We shall be mostly interested in the boundary conditions b) and c).

The equations of motion of the Toda model are Newton's equations

$$\frac{d^2 q_n}{dt^2} = -\frac{\partial V}{\partial q_n} \tag{2.8}$$

for a system of N one-dimensional particles with the potential

$$V(q) = \sum_n (e^{q_n - q_{n-1}} - 1) , \tag{2.9}$$

where summation takes account of the boundary conditions, so that the case of $N = \infty$ is also included. These are therefore Hamilton's equations with the Hamiltonian

$$H = \sum_n \tfrac{1}{2} p_n^2 + V(q) \tag{2.10}$$

on the usual phase space with coordinates p_n, q_n and the Poisson structure

$$\{p_n, p_m\} = \{q_n, q_m\} = 0 , \quad \{p_n, q_m\} = \delta_{nm} . \tag{2.11}$$

The Toda equations allow a zero curvature representation (2.1)–(2.2) with matrices $L_n(t, \lambda)$ and $V_n(t, \lambda)$ given by

$$L_n(\lambda) = \begin{pmatrix} p_n + \lambda & e^{q_n} \\ -e^{-q_n} & 0 \end{pmatrix} , \quad V_n(\lambda) = \begin{pmatrix} 0 & -e^{q_n} \\ e^{-q_{n-1}} & \lambda \end{pmatrix} . \tag{2.12}$$

A detailed study of this important model will be carried out in Chapter III.

2. The Volterra model

The phase space of the model is composed of positive variables u_n. The equations of motion are

$$\frac{du_n}{dt} = (u_{n+1} - u_{n-1})\, u_n . \tag{2.13}$$

They were first derived to describe the population evolution in a hierarchical system of competing individuals, but also have other applications. Typically one considers the periodic,

$$u_{n+N} = u_n , \tag{2.14}$$

or the rapidly decreasing boundary conditions

$$\lim_{|n| \to \infty} u_n = 1 . \tag{2.15}$$

The equations of motion (2.13) can be written in Hamiltonian form

$$\frac{du_n}{dt} = \{H, u_n\} \tag{2.16}$$

with the Hamiltonian

$$H = \sum_n \log u_n , \tag{2.17}$$

where summation takes account of the boundary conditions. The Poisson structure is defined by the following Poisson brackets

$$\{u_n, u_m\} = u_n u_m \Bigg((\delta_{n,m+1} - \delta_{n,m-1}) \left(\frac{u_n + u_m}{2} - 2 \right) + \frac{1}{2}\, \delta_{n,m+2}\, u_{n-1} - \frac{1}{2}\, \delta_{n,m-2}\, u_{m-1} \Bigg) . \tag{2.18}$$

Notice that (2.18) has a far less familiar appearance than, say, (2.11). Nevertheless, it really defines a Poisson bracket; the verification of the Jacobi identity, though tiresome, is elementary.

The equations of motion (2.13) allow a zero curvature representation with matrices $L_n(t,\lambda)$ and $V_n(t,\lambda)$ given by

$$L_n(\lambda)=\begin{pmatrix}\lambda & u_n\\ -1 & 0\end{pmatrix},\quad V_n(\lambda)=\begin{pmatrix}u_n & \lambda u_n\\ -\lambda & -\lambda^2+u_{n-1}\end{pmatrix}. \tag{2.19}$$

We shall not elaborate this model any further leaving it as an instructive illustration with an interesting Poisson structure.

3. The lattice isotropic Heisenberg magnet model (LHM model)

Here the spin variables $\vec{S}_n=(S_n^1, S_n^2, S_n^3)$, $\vec{S}_n^2=s^2$, are defined on a one-dimensional lattice (chain). This kind of model is physically more natural than the continuous model of § 1. The variables $\vec{S}_n$ under the periodic,

$$\vec{S}_{n+N}=\vec{S}_n, \tag{2.20}$$

or the rapidly decreasing,

$$\lim_{|n|\to\infty}\vec{S}_n=s\vec{S}_0, \tag{2.21}$$

boundary conditions make up the phase space of the model with the Poisson structure defined by the following Poisson brackets

$$\{S_n^a, S_m^b\}=-\varepsilon^{abc}\delta_{nm}S_n^c, \tag{2.22}$$

$a, b, c=1, 2, 3$ (cf. (1.7)). An integrable model is associated with the Hamiltonian

$$H=-2\sum_n \log\left(\frac{s^2+\vec{S}_n\cdot\vec{S}_{n+1}}{2s^2}\right), \tag{2.23}$$

which leads to the equations of motion

$$\frac{d\vec{S}_n}{dt}=2\vec{S}_n\wedge\left(\frac{\vec{S}_{n+1}}{s^2+\vec{S}_n\cdot\vec{S}_{n+1}}+\frac{\vec{S}_{n-1}}{s^2+\vec{S}_{n-1}\cdot\vec{S}_n}\right). \tag{2.24}$$

The latter admit a zero curvature representation with matrices $L_n(t,\lambda)$ and $V_n(t,\lambda)$ given by

$$L_n(\lambda) = I + \frac{\lambda}{2i} S_n, \tag{2.25}$$

$$V_n(\lambda) = \frac{\lambda v_n^{(+)}}{2i+s\lambda} + \frac{\lambda v_n^{(-)}}{2i-s\lambda}, \tag{2.26}$$

where

$$v_n^{(\pm)} = \alpha_n \left(1 \pm \frac{S_n}{s}\right)\left(1 \pm \frac{S_{n-1}}{s}\right), \tag{2.27}$$

and

$$\alpha_n = \frac{is^2}{s^2 + \vec{S}_{n-1}\cdot\vec{S}_n}, \quad S_n = \vec{S}_n\cdot\vec{\sigma} = \sum_{a=1}^{3} S_n^a \sigma_a. \tag{2.28}$$

In fact, the right hand side of the zero curvature condition

$$\frac{dL_n(\lambda)}{dt} = V_{n+1}(\lambda) L_n(\lambda) - L_n(\lambda) V_n(\lambda) \tag{2.29}$$

after dividing out the common factor $\frac{\lambda}{2i}$ is a rational function with simple poles at $\lambda = \pm\frac{2i}{s}$. The residues at these poles vanish owing to the special choice of the matrices $v_n^{(\pm)}$. By using the explicit expression for α_n and the elementary relation

$$S_n S_{n+1} = \vec{S}_n\cdot\vec{S}_{n+1} + i(\vec{S}_n \wedge \vec{S}_{n+1})\cdot\vec{\sigma} \tag{2.30}$$

the constant term (the value at $\lambda = \infty$) is shown to coincide with the right hand side of (2.24).

Observe that the auxiliary linear problem

$$F_{n+1} = L_n(\lambda) F_n = F_n + \frac{\lambda}{2i} S_n F_n \tag{2.31}$$

is a naive difference approximation to the auxiliary linear problem for the continuous HM model. This is, however, no longer so for the Hamiltonian and the equation in t in the zero curvature representation. Nevertheless, such a sophistication is justified. Indeed, in the first instance the naive Hamiltonian

$$\tilde{H} = -\sum_n \vec{S}_n\cdot\vec{S}_{n+1} \tag{2.32}$$

gives the equations of motion

$$\frac{d\vec{S}_n}{dt} = \vec{S}_n \wedge (\vec{S}_{n+1} + \vec{S}_{n-1}), \tag{2.33}$$

which have no zero curvature representation. Secondly, our lattice model also provides a finite-difference analogue of the HM model reproducing the latter in the continuum limit.

This limit is taken, as usual, by condensing the lattice. Let Δ denote the lattice spacing and let $x = n\Delta$. Assuming that the length of the vector $\vec{S}_n$ equals Δ,

$$\vec{S}_n^2 = s^2 = \Delta^2 \tag{2.34}$$

we set, for $\Delta \to 0$,

$$\vec{S}_n = \Delta \vec{S}(x), \tag{2.35}$$

where $\vec{S}(x)$ is a smooth vector-valued function of length 1. Then

$$\vec{S}_{n+1} = \Delta \vec{S}(x) + \Delta^2 \frac{d\vec{S}}{dx}(x) + \frac{\Delta^3}{2} \frac{d^2\vec{S}}{dx^2}(x) + \ldots, \tag{2.36}$$

so that

$$\frac{s^2 + \vec{S}_n \cdot \vec{S}_{n+1}}{2s^2} = 1 + \frac{\Delta^2}{4} \vec{S}(x) \cdot \frac{d^2\vec{S}}{dx^2}(x) + \ldots = 1 - \frac{\Delta^2}{4} \left(\frac{d\vec{S}}{dx}(x)\right)^2 + \ldots, \tag{2.37}$$

where we used that $\vec{S}^2(x) = 1$ which implies

$$\vec{S} \cdot \frac{d\vec{S}}{dx} = 0, \quad \vec{S} \cdot \frac{d^2\vec{S}}{dx^2} = -\left(\frac{d\vec{S}}{dx}\right)^2. \tag{2.38}$$

From (2.37) it follows that

$$H = \Delta \int \frac{1}{2} \left(\frac{d\vec{S}}{dx}\right)^2 dx + O(\Delta^2) \tag{2.39}$$

and upon time rescaling $t \mapsto \Delta \cdot t$ we come down to the Hamiltonian and equations of motion for the HM model.

The limit of the zero curvature representation is analyzed in a similar way. The continuum limit of the Poisson brackets results from (2.22) according to

$$\frac{\delta_{nm}}{\Delta} \cong \delta(x-y), \quad x=n\Delta, \quad y=m\Delta. \tag{2.40}$$

The anisotropic magnet – the LL model – is also a continuum limit of the corresponding lattice model to be defined in Chapter III. The latter will turn out to be a fairly universal model.

4. The LNS$_1$ model

We will show that upon a slight modification, the lattice magnet model just described can be interpreted as a difference approximation to the NS model.

We begin with phase space. The natural variables for the NS model on the lattice, ψ_n, $\bar{\psi}_n$, have the Poisson brackets

$$\{\psi_n, \psi_m\} = \{\bar{\psi}_n, \bar{\psi}_m\} = 0, \quad \{\psi_n, \bar{\psi}_m\} = i\delta_{nm}. \tag{2.41}$$

Let

$$S_n^+ = S_n^1 + iS_n^2 = \frac{2\cdot(-1)^n}{\sqrt{g}} \psi_n \sqrt{1+\frac{g}{4}|\psi_n|^2},$$

$$S_n^- = S_n^1 - iS_n^2 = \frac{2\cdot(-1)^n}{\sqrt{g}} \bar{\psi}_n \sqrt{1+\frac{g}{4}|\psi_n|^2}, \tag{2.42}$$

$$S_n^3 = \frac{2}{|g|}\left(1+\frac{g}{2}|\psi_n|^2\right).$$

For $g<0$, the variables ψ_n, $\bar{\psi}_n$ are supposed to vary in the disk $|\psi_n|^2 \leqslant -\frac{4}{g}$. Then the variables $\vec{S}_n$ range over the sphere of radius s in $\mathbb{R}^3$,

$$s^2 = \frac{4}{g^2}. \tag{2.43}$$

For $g>0$ these variables range over the upper sheet of the two-sheeted hyperboloid

$$(S_n^3)^2 - (S_n^1)^2 - (S_n^2)^2 = \frac{4}{g^2}. \tag{2.44}$$

The Poisson brackets (2.41) imply

$$\{S_n^a, S_m^b\} = -f^{abc}\delta_{nm}S_n^c, \tag{2.45}$$

where f^{abc} are the structure constants of the Lie algebra of the group $SU(2)$ ($f^{abc} = \varepsilon^{abc}$) or $SU(1,1)$, if $g<0$ or $g>0$, respectively. In what follows we shall concentrate on the compact case, $g<0$.

Expressions (2.42) have a simple geometrical origin. Consider the mapping of $\mathbb{C}^1 \cup \{\infty\}$ onto the sphere $\mathbf{S}^2$ given by

$$\begin{aligned} S_+ &= S_1 + iS_2 = \frac{2z}{1+|z|^2}, \\ S_- &= S_1 - iS_2 = \frac{2\bar{z}}{1+|z|^2}, \\ S_3 &= \frac{1-|z|^2}{1+|z|^2} \end{aligned} \tag{2.46}$$

(the inverse stereographic projection). This mapping converts the standard symplectic form (area form) ω on $\mathbf{S}^2$ (see § 1) into

$$\omega_* = \frac{2}{i}\frac{dz \wedge d\bar{z}}{(1+|z|^2)^2}. \tag{2.47}$$

The latter can be reduced to canonical form

$$\omega_* = \frac{1}{i}\, d\psi \wedge d\bar{\psi} \tag{2.48}$$

by the scaling transformation

$$z = f(|\psi|^2)\psi, \quad \bar{z} = f(|\psi|^2)\bar{\psi}, \tag{2.49}$$

where

$$f(x) = \frac{1}{\sqrt{2-x}}. \tag{2.50}$$

Expressions (2.42) are the result of these transformations (with a sphere of radius $s = -\dfrac{2}{g}$ in place of $\mathbf{S}^2$) combined with the alternation of sign,

$$\psi_n \to (-1)^n \psi_n, \quad \bar{\psi}_n \to (-1)^n \bar{\psi}_n, \tag{2.51}$$

which obviously preserves the Poisson brackets (2.41). The significance of the last operation will become clear a bit later.

Let us now insert the expressions (2.42) for $\vec{S}_n$ into the LHM equations of motion. It will be convenient to use, instead of (2.24), the equations of motion generated by the Hamiltonian

$$H_{\text{reg}} = -sH - 4\sum_n (S_n^3 - s), \tag{2.52}$$

with H given by (2.23). The right hand side of the associated Hamilton equations differs from (2.24) by the factor $-s$ and the summand $4\vec{S}_n \wedge \vec{S}_0$. In the variables ψ_n, $\bar{\psi}_n$ these equations are equivalent to

$$i\frac{d\psi_n}{dt} = 4\psi_n + \frac{P_{n,n+1}}{Q_{n,n+1}} + \frac{P_{n,n-1}}{Q_{n,n-1}}, \tag{2.53}$$

where

$$\begin{aligned} P_{n,n+1} = -\Bigg(&\psi_n + \psi_{n+1}\sqrt{1+\frac{g}{4}|\psi_n|^2}\cdot\sqrt{1+\frac{g}{4}|\psi_{n+1}|^2} + \frac{g}{2}\psi_n|\psi_{n+1}|^2 \\ &+ \frac{g}{8}(|\psi_n|^2\psi_{n+1} + \psi_n^2\bar{\psi}_{n+1})\frac{\sqrt{1+\frac{g}{4}|\psi_{n+1}|^2}}{\sqrt{1+\frac{g}{4}|\psi_n|^2}}\Bigg) \end{aligned} \tag{2.54}$$

and

$$\begin{aligned} Q_{n,n+1} = 1 + \frac{g}{4}\Bigg(&|\psi_n|^2 + |\psi_{n+1}|^2 + (\psi_n\bar{\psi}_{n+1} + \bar{\psi}_n\psi_{n+1}) \\ &\times\sqrt{1+\frac{g}{4}|\psi_n|^2}\sqrt{1+\frac{g}{4}|\psi_{n+1}|^2} + \frac{g}{2}|\psi_n|^2|\psi_{n+1}|^2\Bigg). \end{aligned} \tag{2.55}$$

The resulting equation for ψ_n is fairly combersome. However, in the continuum limit,

$$x = n\Delta, \quad g = \varkappa\Delta, \quad \psi_n = \sqrt{\Delta}\,\psi(x), \tag{2.56}$$

where $\psi(x)$ is a smooth function, it turns into the NS equation. In fact, using the elementary relations

$$\sqrt{1+\frac{g}{4}|\psi_n|^2} = 1 + \frac{g}{8}|\psi_n|^2 + O(\Delta^3), \qquad Q_{n,n+1} = 1 + g|\psi_n|^2 + O(\Delta^3), \tag{2.57}$$

we can easily write (2.53) as

$$i\frac{d\psi_n}{dt} = -(\psi_{n+1}+\psi_{n-1}-2\psi_n)+2g|\psi_n|^2\psi_n+O\left(\Delta^{3+\frac{1}{2}}\right) \tag{2.58}$$

which upon rescaling $t\mapsto\Delta^2 t$ goes over into the NS equation in the limit as $\Delta\to 0$.

Thus the LHM model expressed in the new variables can indeed be considered as a difference approximation to the NS model; as such, it will be called the LNS_1 model.

We emphasize that the passage from the LHM model to the HM limiting case is essentially different from the passage to the NS model. Also, it is important that the sign in (2.42) is alternated, for otherwise the continuum limit would not give the nonlinear term $2\varkappa|\psi|^2\psi$.

Clearly, the LNS_1 model has a zero curvature representation with matrices $L_n(t,\lambda)$ and $\tilde{V}_n(t,\lambda)$ given by

$$\tilde{V}_n(t,\lambda)=-s\,V_n(t,\lambda)+2iS_0, \qquad S_0=\vec{S}_0\cdot\vec{\sigma}, \tag{2.59}$$

where $\vec{S}_n$ is replaced by ψ_n, $\bar{\psi}_n$ according to (2.42). It will be instructive to relate these matrices to those for the NS model.

Consider the auxiliary linear problem for the LHM model

$$F_{n+1}=L_n(\lambda)F_n. \tag{2.60}$$

Insert into $L_n(\lambda)$ the expression (2.42) for $\vec{S}_n$, replace λ by $\dfrac{2g}{\lambda}$ and set

$$G_n(\lambda)=\sigma_3^n\left(\frac{\lambda}{2i}\right)^n F_n\left(\frac{2g}{\lambda}\right). \tag{2.61}$$

For the vector $G_n(\lambda)$ in (2.60) we get the equation

$$G_{n+1}=\tilde{L}_n(\lambda)G_n, \tag{2.62}$$

with

$$\begin{aligned}\tilde{L}_n(\lambda)&=\frac{\lambda}{2i}\sigma_3^{n+1}L_n\left(\frac{2g}{\lambda}\right)\sigma_3^n\\ &=I+\begin{pmatrix}\frac{g}{2}|\psi_n|^2 & -\sqrt{g}\,\bar{\psi}_n\sqrt{1+\frac{g}{4}|\psi_n|^2}\\ -\sqrt{g}\,\psi_n\sqrt{1+\frac{g}{4}|\psi_n|^2} & \frac{g}{2}|\psi_n|^2\end{pmatrix}+\frac{\lambda\sigma_3}{2i}.\end{aligned} \tag{2.63}$$

The last formula shows that $\tilde{L}_n(\lambda)$ is obtained from $L_n(\lambda)$ by changing the spectral parameter and performing a lattice analogue of gauge transformation. This transformation, in particular, compensates for the alternation of sign in (2.51). The auxiliary linear problem (2.62) modified by the replacements (2.56) and $\lambda \mapsto \Delta \cdot \lambda$ turns into the NS auxiliary linear problem

$$\frac{dG}{dx} = \left(\frac{\lambda\sigma_3}{2i} + \sqrt{\varkappa}\begin{pmatrix} 0 & \bar{\psi}(x) \\ \psi(x) & 0 \end{pmatrix}\right) G \tag{2.64}$$

in the continuum limit as $\Delta \to 0$.

Similar arguments apply to the continuum limit of both the equation in t with the matrix $\tilde{V}_n(t,\lambda)$ and the Poisson brackets. In the Hamiltonian H_{reg} only the first two terms of the Taylor series of $\log(1+x)$ at $x=0$ must be retained. Finally we find

$$H_{\text{reg}} = \Delta^2 \int \left(\left| \frac{d\psi}{dx} \right|^2 + \varkappa |\psi|^4 \right) dx + O(\Delta^3). \tag{2.65}$$

The LNS_1 model for $g>0$ is analyzed in a similar manner. It is associated with the $SU(1,1)$ magnet model, so that the sphere $\mathbf{S}^2$ is replaced by a sheet (e.g. the upper sheet) of the two-sheeted hyperboloid, a model for the Lobachevski plane.

There is also another difference approximation to the NS model which we will now define.

5. The LNS_2 model

The equations of motion are

$$i\frac{d\psi_n}{dt} = 2\psi_n - \psi_{n-1} - \psi_{n+1} + \varkappa|\psi_n|^2(\psi_{n-1} + \psi_{n+1}). \tag{2.66}$$

In the continuum limit

$$x = n\Delta, \qquad \psi_n \cong \Delta\,\psi(x) \tag{2.67}$$

upon rescaling $t \to \Delta^2 t$, (2.66) clearly goes into the NS equations of motion. The phase space of the model consists of functions ψ_n, $\bar{\psi}_n$ subject to certain boundary conditions (for instance, periodic or rapidly decreasing). The Poisson structure is given by the following Poisson brackets

$$\{\psi_n, \psi_m\} = \{\bar{\psi}_n, \bar{\psi}_m\} = 0, \qquad \{\psi_n, \bar{\psi}_m\} = i(1 - \varkappa|\psi_n|^2)\delta_{nm} \tag{2.68}$$

and the Hamiltonian for the model is

$$H=\sum_n\left(-\psi_n(\bar{\psi}_{n+1}+\bar{\psi}_{n-1})-\frac{2}{\varkappa}\log(1-\varkappa|\psi_n|^2)\right), \tag{2.69}$$

where summation takes account of the boundary conditions. If $\varkappa>0$, it is supposed that ψ_n, $\bar{\psi}_n$ vary in the disk $|\psi_n|^2\leqslant\frac{1}{\varkappa}$. The Poisson brackets $\frac{1}{\Delta}\{,\}$ and the Hamiltonian $\frac{1}{\Delta^3}H$ turn into their counterparts for the NS model, as $\Delta\to 0$.

The symplectic form on $\mathbb{C}^1$ associated with the Poisson brackets (2.68) is given by

$$\omega=\frac{1}{i}\frac{dz\wedge d\bar{z}}{1-\varkappa|z|^2} \tag{2.70}$$

and is different from both the canonical form (2.48) and the form (2.47) induced by the area form on $\mathbf{S}^2$. In the representation theory for the groups $SU(2)$ and $SU(1,1)$ one encounters the forms

$$\omega_l=\frac{dz\wedge d\bar{z}}{i(1-\varkappa|z|^2)^{2+2l}}, \tag{2.71}$$

$l=0,\frac{1}{2},1,\ldots$, defined on the sphere $\mathbf{S}^2$ for $\varkappa<0$, or on the Lobachevski plane for $\varkappa>0$. The form ω fits into the family ω_l by setting formally $l=-\frac{1}{2}$.

The LNS_2 model admits a zero curvature representation with matrices $L_n(t,\lambda)$ and $V_n(t,\lambda)$ of the form

$$L_n(\lambda)=\begin{pmatrix}\lambda & \sqrt{\varkappa}\,\bar{\psi}_n\\ \sqrt{\varkappa}\,\psi_n & \lambda^{-1}\end{pmatrix}, \tag{2.72}$$

$$V_n(\lambda)=i\begin{pmatrix}1+\varkappa\bar{\psi}_n\psi_{n-1}-\lambda^2 & \sqrt{\varkappa}\left(\frac{1}{\lambda}\bar{\psi}_{n-1}-\lambda\bar{\psi}_n\right)\\ \sqrt{\varkappa}\left(\frac{1}{\lambda}\psi_n-\lambda\psi_{n-1}\right) & -1-\varkappa\psi_n\bar{\psi}_{n-1}+\lambda^{-2}\end{pmatrix}. \tag{2.73}$$

Replacing t by $\Delta^2 t$ and λ by $e^{-\frac{i\lambda\Delta}{2}}$ and taking the continuum limit, we recover the zero curvature representation for the NS model.

Compared with (2.53)-(2.55), the equations of motion (2.66) have the advantage of apparent simplicity. Yet, the disguised symmetry of the LNS_1 model relative to the action of the group $O(3)$ on its phase space induced by the natural action of $O(3)$ on the phase space of the LHM model shows that the LNS_1 model is also a natural and interesting one. We shall see in Chapter III that from the Hamiltonian standpoint it is even closer to the NS model.

§ 3. Zero Curvature Representation as a Method for Constructing Integrable Equations

The preceding sections contained an extensive list of integrable equations. Now the natural question is: given a nonlinear evolution equation, how can one determine whether there exists a corresponding zero curvature representation? Unfortunately, no answer has been given so far, and there is little hope for such an answer in general form. A more realistic approach is to develope principles for classifying integrable equations. The zero curvature representation inherent in all of the above models may be taken as a foundation of the classification scheme. In this section we outline the scheme for continuous models. In Chapter IV we shall present a more elegant Hamiltonian formulation based on Lie-algebraic methods.

A distinctive property of the continuous models discussed in § 1 (with the exception of the LL model) is the rational dependence of the matrices $U(x,t,\lambda)$ and $V(x,t,\lambda)$ on the spectral parameter λ. Matrices of this kind can be resolved into partial fractions,

$$U(x,t,\lambda)=\sum_k \sum_{s=1}^{n_k} \frac{U_{k,s}(x,t)}{(\lambda-\lambda_k)^s}+\sum_{s=0}^{n_\infty} U_s(x,t)\lambda^s \tag{3.1}$$

and

$$V(x,t,\lambda)=\sum_l \sum_{s=1}^{m_l} \frac{V_{l,s}(x,t)}{(\lambda-\mu_l)^s}+\sum_{s=0}^{m_\infty} V_s(x,t)\lambda^s, \tag{3.2}$$

where the coefficients $U_{k,s}(x,t)$, $U_s(x,t)$ and $V_{l,s}(x,t)$, $V_s(x,t)$ are matrices in the auxiliary space $\mathbb{C}^n$.

Consider now the zero curvature condition

$$\frac{\partial U}{\partial t}-\frac{\partial V}{\partial x}+[U,V]=0, \tag{3.3}$$

resolve it into partial fractions and require that the coefficients of all the poles vanish. Then for the matrices $U_{k,s}$, U_s and $V_{l,s}$, V_s we get a system of nonlinear differential equations (in general supplemented by algebraic equations). The very construction implies the zero curvature representation

$$\frac{\partial F}{\partial x} = U(x, t, \lambda) F, \tag{3.4}$$

$$\frac{\partial F}{\partial t} = V(x, t, \lambda) F, \tag{3.5}$$

where (3.4) is the auxiliary linear problem. This completes the abstract description of the general nonlinear equations associated with the zero curvature condition.

Let us count the number of unknown functions in (3.3). Let the number of poles of the matrix functions $U(x, t, \lambda)$ and $V(x, t, \lambda)$ counting their multiplicities be N_1 and N_2 respectively. Then we have N_1+N_2+2 matrix parameters $U_{k,s}$, U_s and $V_{l,s}$, V_s. There are altogether N_1+N_2+1 equations in (3.3) since the constant terms U_0 and V_0 of the expansions (3.1)–(3.2) are subject to the single equation

$$\frac{\partial U_0}{\partial t} - \frac{\partial V_0}{\partial x} + [U_0, V_0] = 0. \tag{3.6}$$

Thus the number of unknown matrices exceeds by 1 the number of equations. The fact that (3.3) is under-determined is tied to the gauge freedom in choosing the matrices $U(x, t, \lambda)$ and $V(x, t, \lambda)$,

$$U \to U^{\Omega} = \frac{\partial \Omega}{\partial x} \Omega^{-1} + \Omega U \Omega^{-1}, \tag{3.7}$$

$$V \to V^{\Omega} = \frac{\partial \Omega}{\partial t} \Omega^{-1} + \Omega V \Omega^{-1}, \tag{3.8}$$

where $\Omega(x, t)$ does not depend on λ. This transformation leaves (3.3) invariant and preserves the pole structure (divisor) of $U(x, t, \lambda)$ and $V(x, t, \lambda)$ relative to λ. By using a gauge transformation, one of the matrix parameters, say, $U_0(x, t)$ can be fixed; the number of unknowns in (3.3) will then be equal to the number of equations.

A particular choice of the gauge transformation to fix the form of the matrices $U(x, t, \lambda)$ and $V(x, t, \lambda)$, and the subsequent parametrization of their entries may lead to equations that differ in form but are essentially equivalent. Such equations are called *gauge equivalent*. In the next section we will show, as a noteworthy illustration, that the NS equation (for $\varkappa = -1$) and the HM equation are gauge equivalent.

The equations induced by (3.3) are often called integrable; following the tradition, we have included the term in the title of this section. It should be emphasized, however, that the proof of complete integrability of a particular equation in the sense of Hamiltonian mechanics is in each case a nontrivial dynamical problem. Part I devoted to the NS equation gives an example of this kind of investigation. Nevertheless, the zero curvature representation proves to be indispensable as a first step in treating each specific equation. Equations admitting this type of representation will therefore be called integrable in the kinematical sense.

The above scheme gives rise to kinematically integrable equations of fairly general structure. Realistic applications involve, as a rule, a smaller number of unknown functions. Thus the equations stated in § 1 are obtained from the general system (3.3) by reduction, i.e. by constraining the matrix elements of $U(x, t, \lambda)$ and $V(x, t, \lambda)$ in a way compatible with the system. In other words, the reduction problem amounts to the determination of invariant submanifolds for (3.3).

Since (3.3) is expressed in terms of commutators, the obvious reduction is to require that the matrices $U(x, t, \lambda)$ and $V(x, t, \lambda)$ belong to some representation of a given Lie algebra. Less trivial reductions also exist, and their full description is an important part of the classification of integrable systems. The Hamiltonian interpretation of the zero curvature representation will lead in Chapter IV to a series of interesting reductions.

To illustrate the possible reduced types of the matrices $U(x, t, \lambda)$ and $V(x, t, \lambda)$ we shall exhibit several further evolution equations admitting a zero curvature representation and having interesting application. In addition, we shall also give their Hamiltonian formulation.

Historically, the first model treated by the inverse scattering method was

1. The Korteweg-de Vries equation (KdV model)

$$\frac{\partial u}{\partial t} - 6u\frac{\partial u}{\partial x} + \frac{\partial^3 u}{\partial x^3} = 0, \tag{3.9}$$

where $u(x, t)$ is a real-valued function. The matrices $U(x, t, \lambda)$ and $V(x, t, \lambda)$ of the corresponding zero curvature representation are

$$U(\lambda) = \frac{\lambda}{2i}\sigma_3 + \begin{pmatrix} 0 & 1 \\ u & 0 \end{pmatrix} \tag{3.10}$$

and

$$V(\lambda)=\frac{\lambda^3}{2i}\sigma_3+\lambda^2\begin{pmatrix}0 & 1\\ u & 0\end{pmatrix}+\frac{\lambda}{i}\begin{pmatrix}u & 0\\ \dfrac{\partial u}{\partial x} & -u\end{pmatrix}+\begin{pmatrix}-\dfrac{\partial u}{\partial x} & 2u\\ 2u^2-\dfrac{\partial^2 u}{\partial x^2} & \dfrac{\partial u}{\partial x}\end{pmatrix}. \tag{3.11}$$

It is more usual to write the auxiliary linear problem for the KdV equation,

$$\frac{dF}{dx}=U(x,\lambda)F, \tag{3.12}$$

as the one-dimensional Schrödinger equation

$$-\frac{d^2y}{dx^2}+u(x)y=Ey, \qquad E=\frac{\lambda^2}{4}. \tag{3.13}$$

The relationship between (3.12) and (3.13) is given by

$$F=\begin{pmatrix}y\\ \dfrac{dy}{dx}+\dfrac{i\lambda}{2}y\end{pmatrix}. \tag{3.14}$$

It is now appropriate to compare the auxiliary linear problems for the KdV model (3.12) and the NS model (I.2.22), Part I.

It becomes clear that the NS model is generic in its class, with auxiliary space $\mathbb{C}^2$ and the Lie algebra of the group $SU(2)$ or $SU(1,1)$, whereas the KdV model is obtained by a further reduction. That is why we have chosen the NS equation to be the basic model of our book.

For various types of boundary condition, the KdV model is a Hamiltonian system. So, in the rapidly decreasing case the phase space consists of real-valued Schwartz functions $u(x)$; the Poisson structure is defined by the Poisson brackets

$$\{u(x),u(y)\}=\frac{1}{2}\left(\frac{\partial}{\partial y}-\frac{\partial}{\partial x}\right)\delta(x-y). \tag{3.15}$$

The KdV equation can be written in Hamiltonian form,

$$\frac{\partial u}{\partial t}=\{H,u\}=\frac{\partial}{\partial x}\frac{\delta H}{\delta u}, \tag{3.16}$$

with

$$H(u)=\int_{-\infty}^{\infty}\left(\frac{1}{2}\left(\frac{\partial u}{\partial x}\right)^2+u^3\right)dx. \tag{3.17}$$

Observe that the Poisson structure (3.15) is degenerate and has a one-dimensional annihilator spanned by the observable

$$Q=\int_{-\infty}^{\infty} u(x)\,dx. \tag{3.18}$$

Hence, symplectic structure is only defined on the level surfaces $Q=\text{const}$. The corresponding 2-form Ω *is*

$$\Omega=\frac{1}{2}\int_{-\infty}^{\infty} du(x)\wedge(\partial^{-1}du)(x)\,dx, \tag{3.19}$$

where ∂^{-1} denotes the integration operator. The momentum functional

$$P=-\frac{1}{2}\int_{-\infty}^{\infty} u^2(x)\,dx \tag{3.20}$$

is deduced from Ω by applying the construction of § 1.

2. The N-wave model

The model arises as the zero curvature condition for the connection $(U(x,t,\lambda), V(x,t,\lambda))$ whose coefficients have one pole in common, say, at $\lambda=\infty$:

$$U(x,t,\lambda)=U_0+\lambda U_1, \qquad V(x,t,\lambda)=V_0+\lambda V_1. \tag{3.21}$$

In a generic situation, the gauge ambiguity can be eliminated by requiring that U_1 and V_1 be diagonal matrices with distinct eigenvalues, while U_0 and V_0 have zero diagonal parts. The diagonal and off-diagonal parts in (3.3) can be decoupled, so that we are in a position to impose the following reduction: U_1 and V_1 do not depend on x, t. Then (3.3) becomes

$$[U_1, V_0]=[V_1, U_0], \tag{3.22}$$

$$\frac{\partial U_0}{\partial t}-\frac{\partial V_0}{\partial x}+[U_0, V_0]=0. \tag{3.23}$$

The algebraic equation (3.22) is trivially solved,

$$U_0=[U_1, W], \qquad V_0=[V_1, W], \tag{3.24}$$

where W is a matrix with zero diagonal, so (3.23) defines an evolution system with a quadratic nonlinearity,

$$\left[U_1, \frac{\partial W}{\partial t}\right]-\left[V_1, \frac{\partial W}{\partial x}\right]+[[U_1, W],[V_1, W]]=0. \tag{3.25}$$

The resulting system has interesting applications if a further reduction is made, namely

$$U_1^*=-U_1, \qquad V_1^*=-V_1, \qquad W^*=-JWJ, \tag{3.26}$$

where J is a diagonal matrix, $J^2=I$. This reduces by half the number of unknown functions in (3.25). The first nontrivial case corresponds to the auxiliary space $\mathbb{C}^3$ and describes the simplest nonlinear interaction of three wave packets. In the general case the number of elementary waves is $N=\frac{n(n-1)}{2}$, n being the dimension of the auxiliary space $\mathbb{C}^n$.

Let us state (3.25) more explicitly for the case of skew-Hermitian reduction, $J=I$. Let

$$U_1=i\,\mathrm{diag}(a_1, \ldots, a_n), \qquad V_1=i\,\mathrm{diag}(b_1, \ldots, b_n), \tag{3.27}$$

$$w_{jk}=\frac{\psi_{jk}}{\sqrt{a_j-a_k}}, \tag{3.28}$$

where $j<k$ and it is assumed that $a_1>a_2>\ldots>a_n$. For the functions $\psi_{jk}(x, t)$ we find a system of hyperbolic type

$$\frac{\partial \psi_{jk}}{\partial t}=v_{jk}\frac{\partial \psi_{jk}}{\partial x}+\frac{1}{i}\sum_{l=k+1}^{n} v_{jkl}\psi_{jl}\bar{\psi}_{kl} + \frac{1}{i}\sum_{l=j+1}^{k-1} v_{jlk}\psi_{jl}\psi_{lk}+\frac{1}{i}\sum_{l=1}^{j-1} v_{ljk}\bar{\psi}_{lj}\psi_{lk}, \tag{3.29}$$

where $j<k$, $j, k=1, \ldots, n$, and

$$v_{jk}=\frac{b_j-b_k}{a_j-a_k}, \qquad v_{jkl}=\frac{a_j b_l-a_l b_j+a_l b_k-a_k b_l+a_k b_j-a_j b_k}{\sqrt{(a_j-a_k)(a_j-a_l)(a_k-a_l)}}. \tag{3.30}$$

This is a Hamiltonian system. For the rapidly decreasing boundary conditions, the phase space is parametrized by a set of $n(n-1)$ Schwartz functions $(\psi_{jk}(x), \bar{\psi}_{jk}(x);\ 1 \leqslant j < k \leqslant n)$. The Poisson structure is given by the following Poisson brackets:

$$\begin{aligned} &\{\psi_{jk}(x), \psi_{lm}(y)\} = \{\bar{\psi}_{jk}(x), \bar{\psi}_{lm}(y)\} = 0, \\ &\{\psi_{jk}(x), \bar{\psi}_{lm}(y)\} = i\delta_{jl}\delta_{km}\delta(x-y); \quad \begin{matrix} 1 \leqslant j < k \leqslant n, \\ 1 \leqslant l < m \leqslant n, \end{matrix} \end{aligned} \tag{3.31}$$

and (3.29) can be written in Hamiltonian form

$$\frac{\partial \psi_{jk}}{\partial t} = \{H, \psi_{jk}\}, \qquad \frac{\partial \bar{\psi}_{jk}}{\partial t} = \{H, \bar{\psi}_{jk}\} \tag{3.32}$$

with the Hamiltonian

$$\begin{aligned} H = \int_{-\infty}^{\infty} \Bigg\{ &\frac{i}{2} \sum_{1 \leqslant j < k \leqslant n} v_{jk} \left(\frac{\partial \psi_{jk}}{\partial x} \bar{\psi}_{jk} - \psi_{jk} \frac{\partial \bar{\psi}_{jk}}{\partial x} \right) \\ &+ \sum_{1 \leqslant j < k < l \leqslant n} v_{jkl} (\psi_{jk}\psi_{kl}\bar{\psi}_{jl} + \bar{\psi}_{jk}\bar{\psi}_{kl}\psi_{jl}) \Bigg\} dx. \end{aligned} \tag{3.33}$$

3. The chiral field equations

The term chiral field in modern literature is used to denote a function on space-time with values in a nonlinear manifold M. Actually, such fields are studied when M is a homogeneous space of a Lie group G which will be assumed compact. If $M = G$, one commonly speaks of a principal chiral field.

The equations of motion for the principal chiral field $g(x, t)$ are

$$\frac{\partial^2 g}{\partial t^2} - \frac{\partial^2 g}{\partial x^2} = \frac{\partial g}{\partial t} g^{-1} \frac{\partial g}{\partial t} - \frac{\partial g}{\partial x} g^{-1} \frac{\partial g}{\partial x}. \tag{3.34}$$

They are conveniently written in terms of the new independent matrices

$$l_0(x, t) = \frac{\partial g}{\partial t} g^{-1}, \qquad l_1(x, t) = \frac{\partial g}{\partial x} g^{-1}. \tag{3.35}$$

These matrices lie in the Lie algebra $\mathfrak{G}$ of the group G and are called the *left currents* for $g(x, t)$. The equations of motion then become

$$\frac{\partial l_1}{\partial t}-\frac{\partial l_0}{\partial x}+[l_1,l_0]=0\,,\qquad \frac{\partial l_0}{\partial t}-\frac{\partial l_1}{\partial x}=0\,. \tag{3.36}$$

The first one is a zero curvature condition implied by (3.35), the second one follows from (3.34).

The zero curvature representation is given by the matrices $U(x,t,\lambda)$ and $V(x,t,\lambda)$ with two simple poles. There is no loss of generality in fixing the poles at $\lambda=\pm 1$. The gauge ambiguity is eliminated by the condition that the constant terms of these matrices vanish. Then $U(x,t,\lambda)$ and $V(x,t,\lambda)$ have the form

$$U(\lambda)=\frac{U_+}{1-\lambda}-\frac{U_-}{1+\lambda}\,, \tag{3.37}$$

$$V(\lambda)=\frac{V_+}{1-\lambda}-\frac{V_-}{1+\lambda}\,, \tag{3.38}$$

and a relativistically-invariant reduction gives

$$U_+=V_+\,,\qquad U_-=-V_-\,. \tag{3.39}$$

After that the zero curvature condition reduces to (3.36) with the identification

$$U_+=\frac{l_0+l_1}{2}\,,\qquad U_-=\frac{l_0-l_1}{2}\,. \tag{3.40}$$

Equations (3.36) are the Euler-Lagrange equations for the action functional

$$S(g)=\int\int \operatorname{tr}(l_1^2-l_0^2)\,dx\,dt\,, \tag{3.41}$$

where the integral with respect to x is specified by the boundary conditions and the integral with respect to t is taken over the interval $t_1\leqslant t\leqslant t_2$. The action $S(g)$ has a simple geometric origin. The pull-back of the Maurer-Cartan form $\theta=dg\cdot g^{-1}$ under the mapping $g(x,t)$ is a matrix-valued 1-form $\Theta=l_0\,dt+l_1\,dx$. The local inner product of such forms is given by the Killing form tr, and the integral represents the inner product of 1-forms relative to the Minkowski metric on $\mathbb{R}^2$.

The equations of motion for chiral fields with values in a homogeneous space M of the group G are more involved. However, as a rule, these can be obtained by reducing the principal chiral field equations.

For instance, the constraint

$$g = I - 2P, \tag{3.42}$$

where P is a projection operator, $P^2 = P$, or equivalently

$$g^2 = I, \tag{3.43}$$

is compatible with (3.34). These projection operators are parametrized by various homogeneous spaces. The simplest case occurs when $G = SO(N)$ and P is a one-dimensional projection operator whose range is spanned by a unit vector $\vec{n}(x,t)$ in $\mathbb{R}^N$. For this vector we have the equation

$$\left(\frac{\partial^2}{\partial t^2} - \frac{\partial^2}{\partial x^2}\right)\vec{n} + \left(\left(\frac{\partial \vec{n}}{\partial t}\right)^2 - \left(\frac{\partial \vec{n}}{\partial x}\right)^2\right)\vec{n} = 0. \tag{3.44}$$

This is the so-called $\vec{n}$-field equation (or the nonlinear σ-model) on the unit sphere $\mathbf{S}^{N-1}$ in $\mathbb{R}^N$, the simplest homogeneous space for the group G, $\mathbf{S}^{N-1} = SO(N)/SO(N-1)$. (More precisely, the above reduction gives us the $\vec{n}$-field on the real projective space $\mathbb{R}\mathbb{P}^{N-1} = \mathbf{S}^{N-1}/\mathbb{Z}_2$).

The Hamiltonian formulation of the chiral field equations and its geometric interpretation will be discussed in § 5.

4. The two-dimensional Toda model

The equations of motion are

$$\begin{gathered}\left(\frac{\partial^2}{\partial t^2} - \frac{\partial^2}{\partial x^2}\right)\varphi_a = e^{\varphi_{a+1}-\varphi_a} - e^{\varphi_a - \varphi_{a-1}}, \\ a = 1, \ldots, n, \quad \varphi_{n+1} = \varphi_1,\end{gathered} \tag{3.45}$$

where $\varphi_a(x,t)$ are real-valued functions. If the φ_a do not depend on x, (3.45) turns into the equations of motion for the periodic Toda lattice (see § 2); this accounts for the name of the model.

In the rapidly decreasing case the phase space of the model consists of real-valued Schwartz functions $\{\varphi_a(x), \pi_a(x);\ a = 1, \ldots, n\}$ with the usual Poisson structure given by the Poisson brackets

$$\begin{gathered}\{\varphi_a(x), \varphi_b(y)\} = \{\pi_a(x), \pi_b(y)\} = 0, \\ \{\pi_a(x), \varphi_b(y)\} = \delta_{ab}\,\delta(x-y); \quad a, b = 1, \ldots, n.\end{gathered} \tag{3.46}$$

Equations (3.45) can be written in Hamiltonian form with the Hamiltonian

$$H = \int_{-\infty}^{\infty} \sum_{a=1}^{n} \left(\frac{1}{2}\pi_a^2 + \frac{1}{2}\left(\frac{\partial \varphi_a}{\partial x}\right)^2 + e^{\varphi_a - \varphi_{a-1}} - 1\right) dx. \tag{3.47}$$

The model admits a zero curvature representation with matrices $U(x, t, \lambda)$ and $V(x, t, \lambda)$ given by

$$U(\lambda) = \frac{1}{2} \sum_{a=1}^{n} \left(\pi_a h_a - e^{\frac{1}{2}(\varphi_{a+1}-\varphi_a)} \left(\lambda e_a + \frac{1}{\lambda} e_{-a}\right)\right), \tag{3.48}$$

$$V(\lambda) = \frac{1}{2} \sum_{a=1}^{n} \left(\frac{\partial \varphi_a}{\partial x} h_a - e^{\frac{1}{2}(\varphi_{a+1}-\varphi_a)} \left(\lambda e_a - \frac{1}{\lambda} e_{-a}\right)\right), \tag{3.49}$$

where the $e_{\pm a}$ are the root vectors that correspond to admissible roots of the Lie algebra A_{n-1} (i.e., to the simple roots and the minimal root), and the h_a are the basis diagonal matrices in the vector representation,

$$\begin{aligned} &(e_a)_{ij} = \delta_{ai}\delta_{a+1,j}, \quad (e_{-a})_{ij} = \delta_{a+1,i}\delta_{aj}, \quad (h_a)_{ij} = \delta_{ai}\delta_{aj}; \\ &\delta_{a+n,j} = \delta_{a,j}, \quad \delta_{i,a+n} = \delta_{i,a}, \quad a = 1, \ldots, n. \end{aligned} \tag{3.50}$$

The matrices $U(x, t, \lambda)$ and $V(x, t, \lambda)$ given by (3.48)–(3.49) can be derived from general matrices with simple poles at $\lambda = 0$ and $\lambda = \infty$ by using the reduction

$$U(\zeta\lambda) = Z^{-1} U(\lambda) Z, \quad V(\zeta\lambda) = Z^{-1} V(\lambda) Z, \tag{3.51}$$

where $\zeta = e^{\frac{2\pi i}{n}}$ is a primitive n-th root of unity, and Z is a diagonal matrix,

$$Z_{ij} = \zeta^i \delta_{ij}, \quad i, j = 1, \ldots, n. \tag{3.52}$$

This is the so-called $\mathbb{Z}_n$-reduction accompanied by fixing a relativistic gauge.

The two-dimensional Toda model in turn admits some interesting Hamiltonian reductions. So in the case of $n=2$, reducing $\pi_1 = -\pi_2 = \pi$, $\varphi_1 = -\varphi_2 = \varphi$ gives the equation

$$\frac{\partial^2 \varphi}{\partial t^2} - \frac{\partial^2 \varphi}{\partial x^2} + 2 \operatorname{sh} 2\varphi = 0, \tag{3.53}$$

which results from the sine-Gordon equation (see § 1) by setting $m=2$, $\beta = 2i$. If $n=3$, letting $\varphi_1 = -\varphi_3 = \varphi$, $\varphi_2 = 0$, $\pi_1 = -\pi_3 = \pi$, $\pi_2 = 0$ gives an equation for one real-valued field $\varphi(x, t)$,

$$\frac{\partial^2 \varphi}{\partial t^2} - \frac{\partial^2 \varphi}{\partial x^2} = e^{-\varphi} - e^{2\varphi}. \tag{3.54}$$

This completes our list of examples of integrable equations.

§ 4. Gauge Equivalence of the NS Model ($\varkappa = -1$) and the HM Model

Here we shall illustrate the notion of gauge equivalence by relating the NS model for $\varkappa = -1$ to the HM model. The matrices $U(x, t, \lambda)$ and $V(x, t, \lambda)$ of the corresponding zero curvature representations are

$$U_{\mathrm{NS}}(\lambda) = \frac{\lambda \sigma_3}{2i} + U_0, \qquad U_0 = i(\bar{\psi}\sigma_+ + \psi\sigma_-), \tag{4.1}$$

$$U_{\mathrm{HM}}(\lambda) = \frac{\lambda}{2i} S, \tag{4.2}$$

where

$$S = \sum_{a=1}^{3} S_a \sigma_a, \qquad S^* = S, \qquad S^2 = I \tag{4.3}$$

and

$$V_{\mathrm{NS}}(\lambda) = \frac{i\lambda^2}{2}\sigma_3 - \lambda U_0 + V_0, \tag{4.4}$$

$$V_0 = i|\psi|^2\sigma_3 + \frac{\partial\bar{\psi}}{\partial x}\sigma_+ - \frac{\partial\psi}{\partial x}\sigma_-, \tag{4.5}$$

$$V_{\mathrm{HM}} = \frac{i\lambda^2}{2} S + \frac{\lambda}{2}\frac{\partial S}{\partial x} S \tag{4.6}$$

(see § I.2 of Part I, and § 1). The labels NS and HM in this notation serve to distinguish between the models.

These expressions show that the matrices $U_{\mathrm{NS}}(\lambda)$, $V_{\mathrm{NS}}(\lambda)$ have the same poles as $U_{\mathrm{HM}}(\lambda)$, $V_{\mathrm{HM}}(\lambda)$; yet, in contrast to the NS model, the HM matrices contain no constant terms. We shall find a gauge transformation with matrix $\Omega(x, t)$ from the NS model to the HM model which kills these constant terms.

Let $\psi(x, t)$ be a solution to the NS equation. Then the skew-Hermitian matrices $U_0(x, t)$ and $V_0(x, t)$ given by (4.1) and (4.5) satisfy the zero curvature condition. Let $\Omega(x, t)$ be a unitary matrix satisfying the following consistent system:

$$\frac{\partial\Omega}{\partial x} = U_0(x, t)\Omega, \tag{4.7}$$

$$\frac{\partial \Omega}{\partial t} = V_0(x, t)\Omega. \tag{4.8}$$

Consider the gauge transformation induced by $\Omega^{-1}(x, t)$,

$$U_{\mathrm{NS}}^{\Omega^{-1}}(\lambda) = -\Omega^{-1}\frac{\partial \Omega}{\partial x} + \Omega^{-1} U_{\mathrm{NS}}(\lambda)\Omega, \tag{4.9}$$

$$V_{\mathrm{NS}}^{\Omega^{-1}}(\lambda) = -\Omega^{-1}\frac{\partial \Omega}{\partial t} + \Omega^{-1} V_{\mathrm{NS}}(\lambda)\Omega. \tag{4.10}$$

From (4.1) and (4.7) we find

$$U_{\mathrm{NS}}^{\Omega^{-1}}(\lambda) = \frac{\lambda}{2i} S = U_{\mathrm{HM}}(\lambda), \tag{4.11}$$

where $S(x, t)$ is given by

$$S(x, t) = \Omega^{-1}(x, t)\sigma_3\Omega(x, t) \tag{4.12}$$

and obviously satisfies (4.3).

In a similar manner we obtain

$$V_{\mathrm{NS}}^{\Omega^{-1}}(\lambda) = \frac{i\lambda^2}{2} S - \lambda\Omega^{-1}\frac{\partial \Omega}{\partial x}. \tag{4.13}$$

This expression is to be modified. From (4.12) we have

$$\frac{\partial S}{\partial x} = \left[S, \Omega^{-1}\frac{\partial \Omega}{\partial x}\right], \tag{4.14}$$

and the differential equation (4.7) implies

$$\Omega^{-1}\frac{\partial \Omega}{\partial x} = \Omega^{-1} U_0 \Omega. \tag{4.15}$$

Since σ_3 anticommutes with $U_0(x, t)$, it follows that S anticommutes with $\Omega^{-1}\frac{\partial \Omega}{\partial x}$. Hence we finally deduce

$$\Omega^{-1}\frac{\partial \Omega}{\partial x} = \frac{1}{2} S \frac{\partial S}{\partial x} = -\frac{1}{2}\frac{\partial S}{\partial x} S \tag{4.16}$$

and

$$V_{\mathrm{NS}}^{\Omega^{-1}}(\lambda) = \frac{i\lambda^2}{2} S + \frac{\lambda}{2} \frac{\partial S}{\partial x} S = V_{\mathrm{HM}}(\lambda). \tag{4.17}$$

Thus the matrix $S(x,t)$ constructed from the NS solution $\psi(x,t)$ satisfies the HM equation.

The above formulae allow us to express the densities of the local integrals for the NS model in terms of $S(x,t)$ thus obtaining the densities of integrals of the motion for the HM model. So, for instance, writing (4.14) as

$$\frac{\partial S}{\partial x} = 2\Omega^{-1}\sigma_3 U_0 \Omega \tag{4.18}$$

we have

$$\frac{1}{2}\left(\frac{\partial \vec{S}}{\partial x}\right)^2 = \frac{1}{4}\operatorname{tr}\left(\frac{\partial S}{\partial x}\right)^2 = -\operatorname{tr} U_0^2 = 2|\psi|^2, \tag{4.19}$$

so that the density of the Hamiltonian for the HM model equals twice the charge (number of particles) density for the NS model.

In a similar way one easily gets the momentum density for the NS model:

$$\frac{1}{2i}\left(\bar{\psi}\frac{\partial \psi}{\partial x} - \psi\frac{\partial \bar{\psi}}{\partial x}\right) = \frac{i}{2}\operatorname{tr}\sigma_3 U_0 \frac{\partial U_0}{\partial x} = \frac{i}{8}\operatorname{tr}\frac{\partial S}{\partial x} S \frac{\partial^2 S}{\partial x^2}$$

$$= -\frac{1}{4}\frac{\partial \vec{S}}{\partial x}\cdot \vec{S}\wedge \frac{\partial^2 \vec{S}}{\partial x^2}. \tag{4.20}$$

By comparing (4.19) with (4.20) we find

$$-\frac{\partial}{\partial x}\arg\psi = \frac{1}{\left(\frac{\partial \vec{S}}{\partial x}\right)^2}\left(\frac{\partial \vec{S}}{\partial x}\cdot \vec{S}\wedge\frac{\partial^2 \vec{S}}{\partial x^2}\right). \tag{4.21}$$

Later we shall give a simple differential-geometric argument showing that the right hand side of (4.21) is, up to a total derivative, the momentum density for the HM model.

The above transformation from the NS model to the HM model is in fact reversible. To construct a gauge transformation from the HM model to the NS model, i.e. to recover $\Omega(x,t)$ from a given matrix $S(x,t)$, we proceed as follows.

Consider a matrix $S(x)$ satisfying (4.3) and reduce it to diagonal form

$$S(x)=\Omega^{-1}(x)\sigma_3\Omega(x) \tag{4.22}$$

by a unitary matrix $\Omega(x)$. This determines $\Omega(x)$ up to a left diagonal unitary factor. The latter can be chosen in such a way that

$$\sigma_3\frac{\partial\Omega}{\partial x}\Omega^{-1}+\frac{\partial\Omega}{\partial x}\Omega^{-1}\sigma_3=0, \tag{4.23}$$

which implies that the skew-Hermitian matrix $\frac{\partial\Omega}{\partial x}\Omega^{-1}$ has zero diagonal part. So we can set

$$U_0(x)=\frac{\partial\Omega}{\partial x}\Omega^{-1}=i\begin{pmatrix}0 & \bar{\psi}(x)\\ \psi(x) & 0\end{pmatrix}, \tag{4.24}$$

thus defining the functions $\psi(x)$, $\bar{\psi}(x)$ for the given $S(x)$. We then have

$$U^{\Omega}_{\mathrm{HM}}(\lambda)=\frac{\partial\Omega}{\partial x}\Omega^{-1}+\Omega U_{\mathrm{HM}}(\lambda)\Omega^{-1}=\frac{\lambda\sigma_3}{2i}+U_0(x)=U_{\mathrm{NS}}(\lambda). \tag{4.25}$$

Up to this point we did not assume that S is a solution of the equations of motion, so that the mapping $F\colon \vec{S}(x)\to(\psi(x),\bar{\psi}(x))$ was defined for fixed t.

Suppose now that $S(x,t)$ satisfies the HM equation. Then the connection $(U^{\Omega}_{\mathrm{HM}}(x,t,\lambda),\ V^{\Omega}_{\mathrm{HM}}(x,t,\lambda))$ has zero curvature,

$$\frac{\partial U^{\Omega}_{\mathrm{HM}}}{\partial t}-\frac{\partial V^{\Omega}_{\mathrm{HM}}}{\partial x}+[U^{\Omega}_{\mathrm{HM}},V^{\Omega}_{\mathrm{HM}}]=0. \tag{4.26}$$

In view of the properties of $\Omega(x,t)$ established earlier, the matrix

$$V^{\Omega}_{\mathrm{HM}}(\lambda)=\frac{\partial\Omega}{\partial t}\Omega^{-1}+\Omega V_{\mathrm{HM}}(\lambda)\Omega^{-1} \tag{4.27}$$

can be written in the form

$$V^{\Omega}_{\mathrm{HM}}(\lambda)=\frac{i\lambda^2}{2}\sigma_3-\lambda U_0+\frac{\partial\Omega}{\partial t}\Omega^{-1}. \tag{4.28}$$

Let us express $\frac{\partial\Omega}{\partial t}\Omega^{-1}$ in terms of $\psi(x,t)$, $\bar{\psi}(x,t)$ by using the zero curvature condition (4.26). The latter is a polynomial in λ of degree three. The coefficients of λ^3 and λ^2 vanish identically; the vanishing of the coefficient of λ leads to

$$\frac{\partial \Omega}{\partial t}\Omega^{-1}=\frac{1}{i}\sigma_3\frac{\partial U_0}{\partial x}+ic(x,t)\sigma_3, \tag{4.29}$$

with $c(x, t)$ a real-valued function. The diagonal part of the constant term in (4.26) gives

$$\frac{\partial}{\partial x}(c-|\psi|^2)=0. \tag{4.30}$$

As a result, $\frac{\partial \Omega}{\partial t}\Omega^{-1}$ can be written as

$$\frac{\partial \Omega}{\partial t}\Omega^{-1}=V_0(x,t)+i\alpha(t)\sigma_3, \tag{4.31}$$

with $V_0(x, t)$ given by (4.5) and $\alpha(t)$ a real-valued function.

Notice now that (4.23) still allows an arbitrariness in $\Omega(x, t)$ of the form $\Omega \rightarrow \exp\{i\beta(t)\sigma_3\}\Omega$ where $\beta(t)$ is a real-valued function. If we require $\beta(t)$ to satisfy

$$\frac{d\beta}{dt}(t)=\alpha(t), \tag{4.32}$$

then Ω can be modified so that for the new Ω the second term on the right hand side of (4.31) vanishes. It then results that

$$V_{\mathrm{HM}}^{\Omega}(\lambda)=V_{\mathrm{NS}}(\lambda), \tag{4.33}$$

and hence $\psi(x, t)$ *satisfies the NS equation.*

This completes the formulation of the gauge equivalence between the NS and HM models. In Chapter II this gauge transformation will be studied from the Hamiltonian point of view.

To conclude the section, let us discuss the mapping F: $\vec{S}(x) \rightarrow (\psi(x), \bar{\psi}(x))$ defined above from the geometric standpoint. We shall concentrate, for definiteness, on the case of the periodic boundary conditions

$$\vec{S}(x+2L)=\vec{S}(x), \tag{4.34}$$

so that the vector function $\vec{S}(x)$ determines a closed path on the sphere $\mathbf{S}^2$, i.e. a 1-cycle γ. The mapping F takes γ into a path on the complex plane $\mathbb{C}^1$ which in general need not be closed. More precisely, we will show that

$$U_0(L)=e^{\frac{ip}{2}\sigma_3}\, U_0(-L)\, e^{-\frac{ip}{2}\sigma_3}, \tag{4.35}$$

where p is the momentum value for the HM model evaluated at $\vec{S}(x)$ (see § 1), or

$$\psi(L)=e^{-ip}\,\psi(-L). \tag{4.36}$$

This allows us to write

$$p=-\int_{-L}^{L}\frac{d}{dx}\arg\psi(x)\,dx, \tag{4.37}$$

which yields, by (4.21), a new expression for the momentum density in the HM model.

We shall give a geometric proof of (4.35). Consider the Hopf fibre bundle $\mathbf{S}^3\cong SU(2)\to\mathbf{S}^2$ defined by the mapping

$$\Omega\to\vec{S}, \qquad S=\vec{S}\cdot\vec{\sigma}=\Omega^{-1}\sigma_3\Omega, \tag{4.38}$$

where Ω is a matrix in $SU(2)$ and $\vec{S}$ is a vector in $\mathbf{S}^2$. A right-invariant 1-form A on $SU(2)$,

$$A=\frac{1}{4\pi}\operatorname{tr}(d\Omega\cdot\Omega^{-1}\sigma_3), \tag{4.39}$$

gives rise to a $U(1)$ connection in this fibre bundle. Its curvature is a horizontal 2-form whose projection to the base space, the sphere $\mathbf{S}^2$, coincides with the area form $\frac{1}{4\pi}\omega$. The equation (4.23) for the determination of $\Omega(x)$ which also defines the mapping F can be interpreted as the requirement that the path γ is lifted horizontally to the total space of the bundle. The end-points of the lifted path are related by the holonomy transformation

$$\Omega(L)=e^{2\pi i\alpha\sigma_3}\,\Omega(-L). \tag{4.40}$$

A theorem on holonomy of $U(1)$ connections then implies

$$\alpha = \frac{1}{4\pi} \int_{B_\gamma} \omega, \tag{4.41}$$

where B_γ is a film in S^2 spanned by the 1-cycle γ. Hence (see § 1) $4\pi\alpha$ is equal to the momentum of the field $\vec{S}(x)$. Now, taking x as the initial point on γ we can write (4.40) as

$$\Omega(x+2L) = e^{\frac{ip}{2}\sigma_3}\Omega(x). \tag{4.42}$$

Then (4.35) is an immediate consequence of this relation.

§ 5. Hamiltonian Formulation of the Chiral Field Equations and Related Models

The chiral field equations and the corresponding zero curvature representation were defined in § 3. Here the associated models will be discussed from the Hamiltonian viewpoint. We begin with the principal chiral field model.

The equations of motion are

$$\frac{\partial^2 g}{\partial t^2} - \frac{\partial^2 g}{\partial x^2} = \frac{\partial g}{\partial t} g^{-1} \frac{\partial g}{\partial t} - \frac{\partial g}{\partial x} g^{-1} \frac{\partial g}{\partial x}, \tag{5.1}$$

where the function $g(x, t)$ takes values in a compact Lie group G. It will be convenient to use the left currents of the field g,

$$l_0 = \frac{\partial g}{\partial t} g^{-1}, \quad l_1 = \frac{\partial g}{\partial x} g^{-1}, \tag{5.2}$$

as dynamical variables, so that the equations of motion become

$$\frac{\partial l_1}{\partial t} - \frac{\partial l_0}{\partial x} + [l_1, l_0] = 0, \tag{5.3}$$

$$\frac{\partial l_0}{\partial t} - \frac{\partial l_1}{\partial x} = 0. \tag{5.4}$$

In the Lie algebra $\mathfrak{g}$ of G we fix a basis t^a, $a = 1, \ldots, n$; $n = \dim \mathfrak{g}$, normalized with respect to the Killing form – the matrix trace in the adjoint representation,

$$\operatorname{tr} t^a t^b = -\tfrac{1}{2}\delta^{ab}. \tag{5.5}$$

Then the structure constants f^{abc} which enter into the basic commutation relations

$$[t^a, t^b] = f^{abc} t^c \tag{5.6}$$

will make up a totally antisymmetric tensor. Here and in what follows we adopt the usual convention on summation over repeated indices.

Let l_μ^a denote the coefficients of l_μ with respect to the basis t^a,

$$l_\mu = l_\mu^a t^a, \quad \mu = 0, 1. \tag{5.7}$$

With this notation the action functional $S(g)$ is

$$S(g) = \frac{1}{2} \int \int_{t_1}^{t_2} \sum_{a=1}^{n} ((l_0^a)^2 - (l_1^a)^2)\, dx\, dt. \tag{5.8}$$

We regard $q^a(x) = l_1^a(x)$ as a set of generalized coordinates for the chiral field. Using (5.3) we can write down their time derivatives

$$\dot{q}^a = \frac{\partial q^a}{\partial t} = (\nabla_1 l_0)^a = \frac{\partial l_0^a}{\partial x} - f^{abc} q^b l_0^c, \tag{5.9}$$

where ∇_1 is the covariant derivative determined by the connection $l_1 = q^a t^a$. The canonically conjugate momentum is

$$\pi^a(x) = \frac{\delta S}{\delta \dot{q}^a(x)} = -(\nabla_1^{-1} l_0)^a(x), \tag{5.10}$$

where the inverse operator ∇_1^{-1} is chosen to be skew-symmetric. The variables $\pi^a(x)$ and $q^a(x)$ have canonical Poisson brackets,

$$\begin{gathered} \{q^a(x), q^b(y)\} = \{\pi^a(x), \pi^b(y)\} = 0, \\ \{\pi^a(x), q^b(y)\} = \delta^{ab} \cdot \delta(x-y); \quad a, b = 1, \ldots, n, \end{gathered} \tag{5.11}$$

and the Hamiltonian H results form the Lagrangian $\mathscr{L}$ for the action

$$S = \int_{t_1}^{t_2} \mathscr{L}\, dt \tag{5.12}$$

by means of the usual Legendre transformation

$$H = \int \pi^a \dot{q}^a \, dx - \mathscr{L} = \frac{1}{2} \int \sum_{a=1}^{n} ((l_0^a)^2 + (l_1^a)^2) \, dx . \tag{5.13}$$

The Poisson brackets (5.11) can easily be written in terms of currents. In fact, by using

$$l_0 = -\nabla_1 \pi \tag{5.14}$$

and the Jacobi identity for the structure constants f^{abc}, (5.11) yields

$$\{l_0^a(x), l_0^b(y)\} = -f^{abc} l_0^c(x) \delta(x-y), \tag{5.15}$$

$$\{l_0^a(x), l_1^b(y)\} = -f^{abc} l_1^c(x) \delta(x-y) - \delta^{ab} \delta'(x-y), \tag{5.16}$$

$$\{l_1^a(x), l_1^b(y)\} = 0, \tag{5.17}$$

where $\delta'(x-y)$ is the derivative of $\delta(x-y)$.

The Poisson brackets (5.15)-(5.17) are in fact the Lie-Poisson brackets associated with an infinite-dimensional Lie algebra which is a semidirect product of an abelian algebra $\mathscr{A}(\mathfrak{g})$ with generators $l_1^a(x)$ and a current algebra $C(\mathfrak{g})$ over $\mathfrak{g}$ with generators $l_0^a(x)$. The action of $C(\mathfrak{g})$ on $\mathscr{A}(\mathfrak{g})$ is an extension of the natural action (by local rotations) by means of the Maurer-Cartan 2-cocycle $\delta^{ab} \delta'(x-y)$.

The Poisson brackets (5.16)-(5.17) can be derived from the Poisson brackets for $g(x)$ and $l_0^a(x)$,

$$\{g(x), l_0^a(y)\} = -t^a g(x) \delta(x-y), \tag{5.18}$$

$$\{g(x), g(y)\} = 0 \tag{5.19}$$

by differentiating with respect to x and expressing through $l_1(x)$ according to (5.2). The left hand side of (5.18) is a matrix composed of the Poisson brackets of the matrix elements of $g(x)$ and $l_0^a(y)$. Formula (5.19) means that all the Poisson brackets of the entries of $g(x)$ vanish.

On an equal footing with the left currents l_μ we could consider the *right currents* of the field g,

$$r_0 = -g^{-1} \frac{\partial g}{\partial t}, \quad r_1 = -g^{-1} \frac{\partial g}{\partial x}. \tag{5.20}$$

Obviously,

$$r_\mu = -g^{-1} l_\mu g, \quad \mu = 0, 1, \tag{5.21}$$

and from (5.18) we have

$$\{g(x), r_0^a(y)\} = g(x) t^a \delta(x-y), \tag{5.22}$$

where we used the decomposition

$$r_\mu = r_\mu^a t^a, \quad \mu = 0, 1. \tag{5.23}$$

Hence the Poisson brackets of the right currents are

$$\{r_0^a(x), r_0^b(y)\} = -f^{abc} r_0^c(x) \delta(x-y), \tag{5.24}$$

$$\{r_0^a(x), r_1^b(y)\} = -f^{abc} r_1^c(x) \delta(x-y) - \delta^{ab} \delta'(x-y), \tag{5.25}$$

$$\{r_1^a(x), r_1^b(y)\} = 0. \tag{5.26}$$

Finally, from (5.15)-(5.19) and (5.21) we find the Poisson brackets of the left and right currents

$$\{r_0^a(x), l_0^b(y)\} = \{r_1^a(x), l_1^b(y)\} = 0, \tag{5.27}$$

$$\{l_0^a(x), r_1(y)\} = \tilde{t}^a(y) \delta'(x-y), \tag{5.28}$$

$$\begin{aligned} \{l_1^a(x), r_0(y)\} &= [\tilde{t}^a(x), r_1(x)] \delta(x-y) + \tilde{t}^a(y) \delta'(x-y) \\ &= \tilde{t}^a(x) \delta'(x-y), \end{aligned} \tag{5.29}$$

where we have set $\tilde{t}^a(x) = g^{-1}(x) t^a g(x)$.

So far, there was no question of boundary conditions so that the above discussion was of formal nature. There are several ways to impose boundary conditions: they may be stated in terms of either the variables $g(x)$, $l_0(x)$ or their currents $l_0(x)$, $l_1(x)$ or else $r_0(x)$, $r_1(x)$. For definiteness, we shall parametrize our phase space by the left currents with the periodic boundary conditions

$$l_\mu(x+2L) = l_\mu(x), \quad \mu = 0, 1. \tag{5.30}$$

In the expressions for the action (5.8) and the Hamiltonian (5.13), the integral is taken over the fundamental domain $-L \leqslant x \leqslant L$.

The Poisson structure (5.15)-(5.17) is degenerate. In fact, the monodromy matrix U of the connection $l_1(x)$,

$$U = \tilde{g}(L, -L) = \widehat{\exp} \int_{-L}^{L} l_1(x)\, dx, \tag{5.31}$$

is in involution with all the generators $l_\mu^a(x)$. To show this, let us formally compute the Poisson bracket of the functional $l_0^a(y)$ and the matrix $\tilde{g}(x,-L)$, where $\tilde{g}(x,y)$ is defined by

$$\tilde{g}(x,y)=\overset{\curvearrowleft}{\exp}\int_y^x l_1(x')\,dx' \tag{5.32}$$

(cf. the definition of the transition matrix in Part I). This matrix satisfies the differential equation

$$\frac{\partial\tilde{g}}{\partial x}=l_1(x)\tilde{g}\,,\qquad \frac{\partial\tilde{g}}{\partial y}=-\tilde{g}\,l_1(y)\,. \tag{5.33}$$

Then, assuming $-L<y<L$ and using (5.16) and (5.33) we have

$$\begin{aligned}\{\tilde{g}(x,-L),l_0^a(y)\} &= \int_{-L}^{x}\tilde{g}(x,z)\{l_1(z),l_0^a(y)\}\,\tilde{g}(z,-L)\,dz\\ &= \int_{-L}^{x}\tilde{g}(x,z)([l_1(z),t^a]\,\delta(y-z)+t^a\,\delta'(y-z))\,\tilde{g}(z,-L)\,dz\\ &= -t^a\,\tilde{g}(x,-L)\,\delta(x-y)\,,\end{aligned} \tag{5.34}$$

where the last identity results from integration by parts. We point out that this expression coincides with the Poisson bracket (5.18) if $g(x)$ is replaced by $\tilde{g}(x,-L)$.

Setting $x=L$ in (5.34) we find that $l_0^a(y)$ is in involution with the monodromy matrix U. For $l_1^a(y)$, the same is obvious from (5.17).

We shall now use $\tilde{g}(x,-L)$ to define the analogues of the right currents,

$$\tilde{r}_\mu=-\tilde{g}^{-1}l_\mu\tilde{g}\,,\qquad \mu=0,1\,. \tag{5.35}$$

Their Poisson brackets with the left currents, l_μ, have the same form as in (5.27)-(5.29), with r_μ replaced by $\tilde{r}_\mu$. Hence the quantities

$$\tilde{R}^a=\int_{-L}^{L}\tilde{r}_0^a(x)\,dx\,, \tag{5.36}$$

as well as the matrix elements of U, are in involution with all the $l_\mu^a(x)$.

The above computations remained, however, on a formal level. In order to determine the functionals that span the annihilator of the Poisson structure, it is necessary to indicate admissible functionals compatible with the boundary conditions (5.30). If such a functional depends only on the monodromy matrix U, then it must be invariant under the adjoint action of G, i.e. under the transformations $U \to a U a^{-1}$ for all a in G (cf. the determination of admissible functionals in § III.2 of Part I). The admissibility conditions for functionals depending on $r_0(x)$ are more complicated, namely, their densities should be periodic functions of x and, moreover, these functionals must be invariant under the substitution $\tilde{g}(x, -L) \to \tilde{g}(x, -L)C$ for any matrix C. In a generic situation, when U has distinct eigenvalues, it can be shown that the number of generating functionals for the annihilator of our Poisson structure is twice the dimension of the Cartan subalgebra of $\mathfrak{g}$. This completes our discussion of phase space in terms of the left currents $l_\mu(x)$.

Consider now the parametrization by $g(x)$ and $l_0(x)$. Under the periodic boundary conditions

$$g(x+2L) = g(x) \tag{5.37}$$

the Poisson structure defined by (5.15) and (5.18)–(5.19) is nondegenerate. In terms of $l_1(x)$, equation (5.37) becomes

$$U = I, \tag{5.38}$$

so that, in particular, all the quantities $\tilde{R}^a$ are admissible functionals on the phase space described by the variables $g(x)$ and $l_0(x)$. This is no surprise since in this phase space the principal chiral field model is $G \times G$-invariant. The group $G \times G$ acts on the phase space in a Hamiltonian way, with the generators

$$L^a = \int_{-L}^{L} l_0^a(x)\,dx, \qquad R^a = \int_{-L}^{L} r_0^a(x)\,dx \tag{5.39}$$

whose Poisson brackets are

$$\{L^a, L^b\} = -f^{abc} L^c, \tag{5.40}$$

$$\{R^a, R^b\} = -f^{abc} R^c, \tag{5.41}$$

$$\{L^a, R^b\} = 0. \tag{5.42}$$

There is an involution, $g \to g^{-1}$, which takes left currents into right ones.

Assuming (5.38) we can go from the first parametrization of the phase space to the second one by integrating (5.33),

$$g(x) = \tilde{g}(x, -L)\, g(-L). \tag{5.43}$$

This operation gives rise to a new dynamical variable ("constant of integration") $g(-L)$ which is in involution with all the $l_\mu(x)$ but not with the former annihilator $\tilde{R}^a$. A similarity transformation

$$g^{-1}(-L)\tilde{R}g(-L)=R=R^a t^a \tag{5.44}$$

takes $\tilde{R}^a$ into the right currents.

Our aim in presenting all these details was to draw the reader's attention to some non-obvious properties of the standard Poisson structure of the principal chiral field model.

The above model admits an interesting reduction which leads to a Hamiltonian system whose Poisson structure is different from the one given by (5.15)-(5.17). Restricting, for simplicity, our attention to the case of $G=SU(2)$ and the periodic boundary conditions for the currents $l_\mu(x)$, we set

$$S=\frac{l_0+l_1}{2}, \quad T=\frac{l_0-l_1}{2}. \tag{5.45}$$

In these variables equations (5.3)-(5.4) become

$$\frac{\partial S}{\partial t}=\frac{\partial S}{\partial x}-[S,T], \tag{5.46}$$

$$\frac{\partial T}{\partial t}=-\frac{\partial T}{\partial x}+[S,T]. \tag{5.47}$$

(With the notation of § 3 we have $S=U_+$, $T=U_-$.)

The reduction

$$S^2=T^2=I \tag{5.48}$$

defines an invariant submanifold for the system (5.46)-(5.47). We consider it as a new phase space with the Poisson structure defined by the following Poisson brackets:

$$\{S^a(x),S^b(y)\}=-\varepsilon^{abc}S^c(x)\delta(x-y), \tag{5.49}$$

$$\{T^a(x),T^b(y)\}=-\varepsilon^{abc}T^c(x)\delta(x-y), \tag{5.50}$$

$$\{S^a(x),T^b(y)\}=0. \tag{5.51}$$

The phase space thus defined is a direct product, $\mathcal{O}\times\mathcal{O}$, where $\mathcal{O}$ is a symplectic orbit for the current algebra of the group $SU(2)$, specified by $S^2=I$, so that the Poisson structure (5.49)–(5.51) is nondegenerate. Equations (5.46)–(5.47) can be written in Hamiltonian form

$$\frac{\partial S}{\partial t}=\{H,S\},\qquad \frac{\partial T}{\partial t}=\{H,T\} \tag{5.52}$$

with the Hamiltonian

$$H(S,T)=P(T)-P(S)-2\int_{-L}^{L}\operatorname{tr} S\,T\,dx, \tag{5.53}$$

where P, the momentum functional on the phase space $\mathcal{O}$, was defined in § 1. Notice that H provides an example of a multi-valued functional defined up to an integral multiple of 8π (see § 1). However, its variational derivatives are obviously single-valued periodic functions.

We now turn to the Hamiltonian formulation of the $\vec{n}$-field model; for simplicity we will concentrate on the case of the sphere $\mathbf{S}^2$. The phase space of the model consists of vector-valued functions $\vec{\pi}(x)$, $\vec{n}(x)$ with values in $\mathbb{R}^3$ satisfying the periodic boundary conditions

$$\vec{\pi}(x+2L)=\vec{\pi}(x),\qquad \vec{n}(x+2L)=\vec{n}(x) \tag{5.54}$$

with the constraints

$$\vec{n}^2=1,\qquad \vec{\pi}\cdot\vec{n}=0. \tag{5.55}$$

The equations of motion

$$\frac{\partial\vec{n}}{\partial t}=\vec{\pi}, \tag{5.56}$$

$$\frac{\partial\vec{\pi}}{\partial t}=\frac{\partial^2\vec{n}}{\partial x^2}+\left(\left(\frac{\partial\vec{n}}{\partial x}\right)^2-\vec{\pi}^2\right)\vec{n} \tag{5.57}$$

are generated by the Hamiltonian

$$H=\int_{-L}^{L}\left(\frac{1}{2}\vec{\pi}^2+\frac{1}{2}\left(\frac{\partial\vec{n}}{\partial x}\right)^2\right)dx \tag{5.58}$$

and the Poisson brackets

$$\{n^a(x), n^b(y)\} = 0, \tag{5.59}$$

$$\{\pi^a(x), \pi^b(y)\} = (n^a(x)\pi^b(x) - n^b(x)\pi^a(x))\,\delta(x-y), \tag{5.60}$$

$$\{\pi^a(x), n^b(y)\} = (\delta^{ab} - n^a(x)n^b(x))\,\delta(x-y), \quad a, b = 1, 2, 3. \tag{5.61}$$

The Poisson structure induced by these Poisson brackets is consistent with the constraints (5.55).

The Poisson brackets (5.59)-(5.61) can be simplified by using, instead of $\vec{\pi}(x)$, the variable $\vec{l}(x)$,

$$\vec{l} = \vec{n} \wedge \vec{\pi}, \quad \vec{l}^2 = \vec{\pi}^2. \tag{5.62}$$

As a result we find the following Poisson brackets:

$$\{l^a(x), l^b(y)\} = -\varepsilon^{abc}\, l^c(x)\,\delta(x-y), \tag{5.63}$$

$$\{l^a(x), n^b(y)\} = -\varepsilon^{abc}\, n^c(x)\,\delta(x-y), \tag{5.64}$$

$$\{n^a(x), n^b(y)\} = 0, \tag{5.65}$$

which are characteristic of the current algebra of the group $E(3)$. The phase space of the model is a symplectic orbit for the algebra $C(e(3))$ defined by

$$\vec{n}^2 = 1, \quad \vec{l} \cdot \vec{n} = 0 \tag{5.66}$$

so that the Poisson structure in question is nondegenerate.

The $\vec{n}$-field model is in this case $O(3)$-invariant. The generators of the action of $O(3)$ on phase space are given by the quantities

$$L^a = \int_{-L}^{L} l^a(x)\,dx \tag{5.67}$$

with the Poisson brackets

$$\{L^a, L^b\} = -\varepsilon^{abc} L^c. \tag{5.68}$$

To conclude this section, we shall discuss yet another chiral field model having an interesting topological origin. Its construction is made possible by the fact that besides the ordinary action $S(g)$, on compact Lie groups there is another bi-invariant functional $W(g)$ which does not depend on the metric in $\mathbb{R}^2$.

In order to define it consider a right-invariant 3-form Ω on a compact Lie group G given at $g = I$ by

$$\Omega(x, y, z) = \operatorname{tr}([x, y]z), \tag{5.69}$$

so that

$$\Omega = \operatorname{tr}\theta \wedge \theta \wedge \theta, \tag{5.70}$$

where θ is the Maurer-Cartan form on G,

$$\theta = dg \cdot g^{-1}. \tag{5.71}$$

The form Ω is bi-invariant and closed,

$$d\Omega = 0, \tag{5.72}$$

but not exact (the existence of such a form means that for G compact the cohomology group $H^3(G, \mathbb{R})$ is nontrivial).

Let $g(x, t)$ be a principal chiral field, i.e. a mapping $g: \mathbb{R}^2 \to G$. We shall assume that it can be compactified by extending it to a mapping from $\mathbf{S}^2$ to G; this gives a 2-cycle γ in G. We can cover this cycle by a simply-connected coordinate patch on G and find a local primitive form ω for Ω within this patch,

$$\omega = d^{-1}\Omega. \tag{5.73}$$

Let

$$W(g) = \int_\gamma \omega. \tag{5.74}$$

By definition, $W(g)$ is a multi-valued functional, as ω does not extend to a 2-form on the whole of G. Let us analyze the nature of its multi-valuedness.

Let B_γ be a film spanned by γ (as is well known, the homotopy group $\pi_2(G)$ is trivial). By the Stokes formula,

$$W(g) = \int_{B_\gamma} \Omega. \tag{5.75}$$

A different film B'_γ would give an extra summand $\int_B \Omega$, where $B = B'_\gamma - B_\gamma$ is a 3-dimensional cycle in G. Suppose that the homology group $H_3(G)$ has a single generator B_0, i.e. $H_3(G) = \mathbb{Z}$. Then the ambiguity in the definition of $W(g)$ amounts to adding an integral multiple of $\int_{B_0} \Omega$, the period of Ω (cf. the definition of the momentum functional for the HM model in § 1). In this case $W(g)$ exists on an equal footing with the elementary multi-valued function $\log z$.

The above assumption $H_3(G)=\mathbb{Z}$ holds for any simple Lie group. In the simplest case of $G=SU(2)$, Ω coincides with the volume element on the group.

The key property of $W(g)$ is that its variation is well defined and single-valued. For the proof we shall make use of a formula for the variation of an integral under a variation of the closed surface of integration $\gamma(s)$,

$$\frac{d}{ds}\int_{\gamma(s)}\omega\bigg|_{s=0}=\int_{\gamma} i_{\xi}\,d\omega, \tag{5.76}$$

where $\gamma=\gamma(0)$, ξ is the vector field of a variation on γ and $i_{\xi}\,d\omega$ the contraction of ξ with $d\omega$. In our case $\xi=\delta g$, so that (5.74) gives

$$\delta W(g)=\int_{g(S^2)} i_{\xi}\,d\omega=\int_{S^2}\operatorname{tr}([l_1,l_0]\,\delta g\cdot g^{-1})\,dx\,dt, \tag{5.77}$$

with l_{μ} denoting as usual the left currents of the field.

Next we define a modified action functional for the chiral field g by

$$S_{\alpha}(g)=S(g)+\alpha W(g), \tag{5.78}$$

where $S(g)$ is given by (5.8) and α is a real constant. The Euler-Lagrange equations for $S_{\alpha}(g)$ written in terms of the left currents are

$$\frac{\partial l_1}{\partial t}-\frac{\partial l_0}{\partial x}+[l_1,l_0]=0, \tag{5.79}$$

$$\frac{\partial l_0}{\partial t}-\frac{\partial l_1}{\partial x}+\alpha[l_1,l_0]=0. \tag{5.80}$$

They are called the *modified principal chiral field equations.*

Between local solutions of the ordinary and modified principal chiral field equations there is a simple one-to-one correspondence. Namely, let l_0 and l_1 be some solutions of (5.79)–(5.80). Then the matrices

$$\tilde{l}_0=l_0-\alpha l_1, \tag{5.81}$$

$$\tilde{l}_1=l_1-\alpha l_0 \tag{5.82}$$

satisfy (5.3)–(5.4). The inverse transformation is given by

$$l_0=\frac{1}{1-\alpha^2}(\tilde{l}_0+\alpha\tilde{l}_1), \tag{5.83}$$

$$l_1 = \frac{1}{1-\alpha^2}(\tilde{l}_1 + \alpha \tilde{l}_0). \tag{5.84}$$

Now, let $U(x, t, \lambda)$ and $V(x, t, \lambda)$ be the coefficients of the zero curvature representation for the principal chiral field model (see § 3). Consider a compatible system of equations,

$$\frac{\partial F}{\partial x} = U(x, t, \lambda) F, \tag{5.85}$$

$$\frac{\partial F}{\partial t} = V(x, t, \lambda) F, \tag{5.86}$$

where $F(x, t, \lambda)$ is a matrix-valued function with values in the group G, and set

$$\tilde{g}(x, t) = F(x, t, \lambda)|_{\lambda=0}, \tag{5.87}$$

$$g(x, t) = F(x, t, \lambda)|_{\lambda=\alpha}. \tag{5.88}$$

Then the matrices $\tilde{g}$ and g satisfy the ordinary and the modified principal chiral field equations, respectively. Thus the zero curvature representation (3.37)–(3.40) serves for the modified equations as well.

There are, however, certain points in which the ordinary model differs from the modified one. Firstly, since

$$W(g) = -W(g^{-1}) \tag{5.89}$$

the invariance of the ordinary model under the change $g \to g^{-1}$ breaks down in the modified one. Secondly, these models have different Poisson structures.

In fact, (5.79)–(5.80) are Hamilton's equations with the Hamiltonian H which coincides with the principal chiral field Hamiltonian (5.13), relative to the following Poisson brackets:

$$\{l_0^a(x), l_0^b(y)\} = -f^{abc}(l_0^c(x) + \alpha l_1^c(x))\,\delta(x-y), \tag{5.90}$$

$$\{l_0^a(x), l_1^b(y)\} = -f^{abc} l_1^c(x)\,\delta(x-y) - \delta^{ab}\delta'(x-y), \tag{5.91}$$

$$\{l_1^a(x), l_1^b(y)\} = 0. \tag{5.92}$$

These only differ from the Poisson brackets (5.15)–(5.17) by an extra term $-\alpha f^{abc} l_1^c(x)\,\delta(x-y)$ on the right hand side of (5.90). A similar modification of the Poisson brackets for the right currents has the form

$$\{r_0^a(x), r_0^b(y)\} = -f^{abc}(r_0^c(x) - \alpha r_1^c(x))\delta(x-y), \tag{5.93}$$

$$\{r_0^a(x), r_1^b(y)\} = -f^{abc} r_1^c(x)\delta(x-y) - \delta^{ab}\delta'(x-y), \tag{5.94}$$

$$\{r_1^a(x), r_1^b(y)\} = 0. \tag{5.95}$$

Notice that the Poisson brackets of $g(x)$ and $l_0(x)$, as well as those of $g(x)$ and $r_0(x)$, are given by (5.18) and (5.22) as before.

By repeating the reasoning which has led to (5.27)–(5.29) we see that the quantities

$$L^a(x) = l_0^a(x) - \alpha l_1^a(x), \tag{5.96}$$

$$R^a(x) = r_0^a(x) + \alpha r_1^a(x) \tag{5.97}$$

have the following Poisson brackets:

$$\{L^a(x), L^b(y)\} = -f^{abc} L^c(x)\delta(x-y) + 2\alpha\delta^{ab}\delta'(x-y), \tag{5.98}$$

$$\{R^a(x), R^b(y)\} = -f^{abc} R^c(x)\delta(x-y) - 2\alpha\delta^{ab}\delta'(x-y), \tag{5.99}$$

$$\{L^a(x), R^b(y)\} = 0. \tag{5.100}$$

The arguments from perturbation theory show that if $\alpha \neq 0$, the variables $L^a(x)$ and $R^a(x)$ can be used to parametrize the phase space. In this way the phase space of the model is associated with the direct sum of two centrally extended current algebras $C_\pm(\mathfrak{g})$ built up from the algebra $\mathfrak{g}$ by means of the 2-cocycles $\pm 2\alpha\delta^{ab}\delta'(x-y)$. It is for the sake of this elegant and unexpected interpretation that we have discussed the modified principal chiral field model. It has also brought yet another illustration of the utility of multi-valued functionals.

§ 6. The Riemann Problem as a Method for Constructing Solutions of Integrable Equations

Here we resume the discussion started in § 3 of the general properties of equations representable as a zero curvature condition,

$$\frac{\partial U(\lambda)}{\partial t} - \frac{\partial V(\lambda)}{\partial x} + [U(\lambda), V(\lambda)] = 0, \tag{6.1}$$

$U(x, t, \lambda)$ and $V(x, t, \lambda)$ being rational matrix-valued functions of the spectral parameter λ,

$$U(x, t, \lambda) = \sum_{i=1}^{m} \sum_{r=1}^{n_i} \frac{U_{i,r}(x, t)}{(\lambda - \lambda_i)^r} + \sum_{k=0}^{n_\infty} \lambda^k U_k(x, t), \tag{6.2}$$

$$V(x, t, \lambda) = \sum_{j=1}^{\tilde{m}} \sum_{s=1}^{\tilde{n}_j} \frac{V_{j,s}(x, t)}{(\lambda - \mu_j)^s} + \sum_{l=0}^{\tilde{n}_\infty} \lambda^l V_l(x, t). \tag{6.3}$$

In § 3 we saw that (6.1) is a system of nonlinear equations for the matrix coefficients $U_{i,r}(x, t)$, $U_k(x, t)$ and $V_{j,s}(x, t)$, $V_l(x, t)$. In a generic situation the structure of this system depends only on the *pole divisors*, $\mathfrak{U} = \{(\lambda_i, n_i), i = 1, \ldots, m; (\infty, n_\infty)\}$ and $\mathfrak{B} = \{(\mu_j, \tilde{n}_j), j = 1, \ldots, \tilde{m}; (\infty, \tilde{n}_\infty)\}$, of the matrices $U(x, t, \lambda)$ and $V(x, t, \lambda)$. More specialized systems result from reductions which amount to a priori hypotheses on the structure of the matrix coefficients of $U(x, t, \lambda)$ and $V(x, t, \lambda)$ that are consistent with (6.1). In this section we shall approach the problem of constructing as large a class as possible of solutions of the general equation (6.1) with the given divisors $\mathfrak{U}$ and $\mathfrak{B}$. We shall see that the kinematic integrability of a nonlinear equation allows one to produce a rich variety of its solutions.

Suppose we are given a solution of (6.1), i.e. matrices $U_0(x, t, \lambda)$ and $V_0(x, t, \lambda)$ with pole divisors $\mathfrak{U}$ and $\mathfrak{B}$ respectively. For example, we could take a "trivial" solution given by

$$U_0(x, t, \lambda) = U_0(x, \lambda), \quad V_0(x, t, \lambda) = V_0(t, \lambda), \tag{6.4}$$

where

$$[U_0(x, \lambda), V_0(t, \lambda)] = 0. \tag{6.5}$$

Let $F_0(x, t, \lambda)$ denote a nondegenerate matrix solution of the compatible system of equations

$$\frac{\partial F_0}{\partial x} = U_0(x, t, \lambda) F_0, \tag{6.6}$$

$$\frac{\partial F_0}{\partial t} = V_0(x, t, \lambda) F_0. \tag{6.7}$$

We will show that, given $U_0(x, t, \lambda)$ and $V_0(x, t, \lambda)$, one can produce a whole family of local solutions to (6.1) *defined in some domain $\mathscr{D}$ of the variables x and t on the plane $\mathbb{R}^2$. This family is parametrized by a closed oriented contour Γ on the extended complex plane $\bar{\mathbb{C}} = \mathbb{C} \cup \{\infty\}$ and a smooth bounded* *nondegenerate matrix-valued function $G(\lambda)$ defined on Γ.* In the case when $\mathfrak{U}$ or $\mathfrak{B}$ intersects Γ, it is required that

$$G(\lambda)=I+O(|\lambda-\lambda_0|^{n_0}) \tag{6.8}$$

for λ in the vicinity of the intersection point (λ_0, n_0) in $\mathfrak{U}$ or $\mathfrak{B}$. Given these data, in order to construct a solution of (6.1) we consider the regular Riemann problem on Γ,

$$G(x,t,\lambda)=G_+(x,t,\lambda)\,G_-(x,t,\lambda), \tag{6.9}$$

with

$$G(x,t,\lambda)=F_0(x,t,\lambda)\,G(\lambda)\,F_0^{-1}(x,t,\lambda), \tag{6.10}$$

where the matrix-functions $G_+(\lambda)$ and $G_-(\lambda)$ have an analytic continuation into the interior or the exterior of Γ and are nondegenerate in these domains. The variables x and t play the role of parameters in this problem. We assume this Riemann problem has a solution when the parameters range over some domain $\mathscr{D}$.

Differentiating (6.9) with respect to x and using (6.6) and (6.10) we find that, for λ in Γ,

$$\frac{\partial G_+}{\partial x}G_- + G_+\frac{\partial G_-}{\partial x} = U_0 G_+ G_- - G_+ G_- U_0, \tag{6.11}$$

or

$$U(x,t,\lambda) = -G_+^{-1}\left(\frac{\partial G_+}{\partial x} - U_0 G_+\right) = \left(\frac{\partial G_-}{\partial x} + G_- U_0\right)G_-^{-1}. \tag{6.12}$$

Here and below, for notational simplicity, we shall occasionally leave out the dependence on x and t. From (6.12) it follows that the matrix $U(x,t,\lambda)$ thus defined has an analytic continuation into $\bar{\mathbb{C}}\backslash\mathfrak{U}$. Let us show that for x, t in $\mathscr{D}$, $U(x,t,\lambda)$ is actually a rational function of λ with pole divisor $\mathfrak{U}$.

In fact, if a point (λ_0, n_0) in $\mathfrak{U}$ does not lie on Γ, then clearly $U(x,t,\lambda)$ has a pole at $\lambda=\lambda_0$ of the same order n_0 as that of $U_0(x,t,\lambda)$. If λ_0 belongs to Γ, the function $F_0(x,t,\lambda)$ has an essential singularity on Γ. Condition (6.8), however, ensures that the functions $G(x,t,\lambda)$, $G_\pm(x,t,\lambda)$, as well as $\frac{\partial G_\pm}{\partial x}(x,t,\lambda)$, are regular at $\lambda=\lambda_0$. Thus in this case $U(x,t,\lambda)$ also has a pole of order n_0 at $\lambda=\lambda_0$. To complete the proof it suffices to appeal to the Liouville theorem.

In a similar manner, by differentiating (6.9) with respect to t we derive that the matrix

$$V(x,t,\lambda) = -G_+^{-1}\left(\frac{\partial G_+}{\partial t} - V_0 G_+\right) = \left(\frac{\partial G_-}{\partial t} + G_- V_0\right) G_-^{-1} \tag{6.13}$$

for x, t in $\mathscr{D}$ is a rational function of λ with pole divisor $\mathfrak{B}$.

From (6.12)–(6.13) if follows that the matrix functions

$$F_+ = G_+^{-1} F_0, \quad F_- = G_- F_0 \tag{6.14}$$

for x and t as above satisfy a system of equations

$$\frac{\partial F_\pm}{\partial x} = U(x,t,\lambda) F_\pm, \tag{6.15}$$

$$\frac{\partial F_\pm}{\partial t} = V(x,t,\lambda) F_\pm \tag{6.16}$$

which is therefore a compatible system. We conclude that $U(x,t,\lambda)$ *and* $V(x,t,\lambda)$ *given by (6.12)–(6.13) provide a solution of (6.1) with the given pole divisors* $\mathfrak{U}$ *and* $\mathfrak{B}$ *for* x, t *in the domain* $\mathscr{D}$.

The above construction of solutions of zero curvature equations is colloquially called "*the dressing of the trial solution* $U_0(x,t,\lambda)$, $V_0(x,t,\lambda)$". It is based on the relationship between the zero curvature representation and the matrix Riemann problem established for the NS model in Part I. There remains a highly nontrivial problem of determining whether the solution of the zero curvature equation obtained by the dressing procedure belongs to a given functional class: the choice of the trial matrices $U_0(x,t,\lambda)$, $V_0(x,t,\lambda)$, the contour Γ and the matrix $G(\lambda)$ needs special analysis in each particular case. The results of Chapter II, Part I, can be summarized as solving this problem for the NS model in the case of rapidly decreasing or finite density boundary conditions.

The Riemann problem (6.9) has more than one solution: along with $G_+(x,t,\lambda)$, $G_-(x,t,\lambda)$ the matrices $G_+(x,t,\lambda)\Omega^{-1}(x,t)$, $\Omega(x,t)G_-(x,t,\lambda)$ also satisfy (6.9), where $\Omega(x,t)$ is any nondegenerate matrix that does not depend on λ. The solution of the zero curvature equation is then modified by a gauge transformation with matrix $\Omega(x,t)$ so that the solutions $U(x,t,\lambda)$ and $V(x,t,\lambda)$ are replaced by $U^\Omega(x,t,\lambda)$ and $V^\Omega(x,t,\lambda)$. This ambiguity is eliminated by normalizing the Riemann problem, i.e. by fixing the value of one of the matrices $G_\pm(x,t,\lambda)$ at some point, say, at $\lambda = \infty$. The Riemann problem then has a unique solution. In particular, setting $G_\pm(\infty) = G(\infty) = I$ leads to the so-called *normalization to unity*, which already occurred in the study of the NS model. In the general case it is convenient to take one of the points of the divisor $\mathfrak{U}$ or $\mathfrak{B}$ as the normalization point. Some specific examples of normalization will be met in the next two chapters.

Alongside the regular Riemann problem, the dressing procedure can also make use of the Riemann problem with zeros so that the matrices $G_{\pm}(x, t, \lambda)$ may have in their domains of analyticity a finite set of zeros (degeneracy points) $\lambda = \lambda_j^{(\pm)}, j = 1, \ldots, N_{\pm}$, which do not depend on x and t and do not belong to $\mathfrak{U}$ or $\mathfrak{B}$. In the case of simple zeros on which we shall concentrate here, i.e. when $G_{\pm}^{-1}(x, t, \lambda)$ have simple poles at $\lambda = \lambda_j^{(\pm)}$, the Riemann problem data should be supplemented by a list of subspaces,

$$N_j^{(+)}(x, t) = \operatorname{Im} G_{+}(x, t, \lambda_j^{(+)}), \quad N_j^{(-)}(x, t) = \operatorname{Ker} G_{-}(x, t, \lambda_j^{(-)}), \quad (6.17)$$

$j = 1, \ldots, N_{\pm}$, which together with the normalization ensure the uniqueness of the solution to the Riemann problem with zeros. On the condition that the dependence of the subspaces $N_j^{(\pm)}(x, t)$ on x and t is consistent with equations (6.6)-(6.7),

$$N_j^{(\pm)}(x, t) = F_0(x, t, \lambda_j^{(\pm)}) N_j^{(\pm)}, \quad j = 1, \ldots, N_{\pm}, \quad (6.18)$$

where $N_j^{(\pm)}$ do not depend on x and t, the matrices $U(x, t, \lambda)$ and $V(x, t, \lambda)$ in (6.12)-(6.13) will not acquire unwanted poles at $\lambda = \lambda_j^{(\pm)}$, so that $\mathfrak{U}$ and $\mathfrak{B}$ will remain their pole divisors. In the case when $G(\lambda) = I$, the Riemann problem with zeros reduces to a system of linear algebraic equations and can be solved in closed form. This leads to a rich set of solutions of (6.1) including soliton solitons.

A particular case of this construction was already present in the investigation of the NS model. The proofs given there make essentially no use of the specific form of the model and can be extended to the general zero curvature equation discussed here.

The dressing procedure just described is suited for the general zero curvature equation (6.1). If we are dealing with a reduced system (for instance, we may assume that $U(\lambda)$ and $V(\lambda)$ belong to some complex Lie algebra), then the contour Γ, the matrix $G(\lambda)$, and other data should be taken consistent with the reduction. For the NS model these reduction constraints were imposed in the form of involution conditions. Other examples will be given in the chapters that follow.

To conclude the discussion of the dressing procedure we point out that the whole scheme extends word for word to the case of the lattice zero curvature equation

$$\frac{dL_n(\lambda)}{dt} = V_{n+1}(\lambda) L_n(\lambda) - L_n(\lambda) V_n(\lambda), \quad (6.19)$$

as derived in § 2. Namely, the scheme is based, as before, on the Riemann problem

$$G(n,t,\lambda)=G_+(n,t,\lambda)\,G_-(n,t,\lambda) \tag{6.20}$$

with parameters n and t where

$$G(n,t,\lambda)=F_0(n,t,\lambda)\,G(\lambda)\,F_0^{-1}(n,t,\lambda), \tag{6.21}$$

and the matrix $F_0(n,t,\lambda)$ is determined by the trial solution $L_n^0(t,\lambda)$, $V_n^0(t,\lambda)$ of (6.19),

$$F_0(n+1,t,\lambda)=L_n^0(t,\lambda)\,F_0(n,t,\lambda), \tag{6.22}$$

$$\frac{dF_0}{dt}(n,t,\lambda)=V_n^0(t,\lambda)\,F_0(n,t,\lambda). \tag{6.23}$$

With these definitions the matrices $F_\pm(n,t,\lambda)$,

$$F_+=G_+^{-1}F_0, \qquad F_-=G_-F_0, \tag{6.24}$$

satisfy the equations

$$F_\pm(n+1,t,\lambda)=L_n(t,\lambda)\,F_\pm(n,t,\lambda) \tag{6.25}$$

and

$$\frac{dF_\pm}{dt}(n,t,\lambda)=V_n(t,\lambda)\,F_\pm(n,t,\lambda), \tag{6.26}$$

where

$$L_n(t,\lambda)=G_+^{-1}(n+1)\,L_n^0\,G_+(n)=G_-(n+1)\,L_n^0\,G_-^{-1}(n) \tag{6.27}$$

and

$$V_n(t,\lambda)=-G_+(n)\left(\frac{dG_+(n)}{dt}-V_n^0G_+(n)\right)=\left(\frac{dG_-(n)}{dt}+G_-(n)\,V_n^0\right)G_-^{-1}(n) \tag{6.28}$$

(cf. (6.12)–(6.13)).

The derivation of the differential equation with respect to t is identical to the derivation of the corresponding equation (6.16) in the continuous case. To derive the difference equation one should compare the Riemann problems (6.20) for the values n and $n+1$ and reduce $G(\lambda)$ out of them (the analogue of differentiation with respect to x in the continuous case).

Thus the matrices $L_n(t,\lambda)$ and $V_n(t,\lambda)$ constructed in this way satisfy (6.19) and have the same pole divisors as $L_n^0(t,\lambda)$ and $V_n^0(t,\lambda)$.

§ 7. A Scheme for Constructing the General Solution of the Zero Curvature Equation. Concluding Remarks on Integrable Equations

We describe here a general local solution of the zero curvature equation

$$\frac{\partial U(\lambda)}{\partial t} - \frac{\partial V(\lambda)}{\partial x} + [U(\lambda), V(\lambda)] = 0 \tag{7.1}$$

with the given pole divisors $\mathfrak{U}$ and $\mathfrak{V}$. It is convenient to assume that neither $\mathfrak{U}$ nor $\mathfrak{V}$ contains the point $\lambda = \infty$ and that the matrices $U(\lambda)$, $V(\lambda)$ vanish at $\lambda = \infty$ (this can be achieved by a linear-fractional transformation of the variable λ and a gauge transformation).

Let $U(x, t, \lambda)$ and $V(x, t, \lambda)$ satisfy (7.1) and let their pole divisors coincide with $\mathfrak{U}$ and $\mathfrak{V}$, respectively. We define a matrix $F(x, t, \lambda)$ to be a solution of the compatible system of equations

$$\frac{\partial F}{\partial x} = U(x, t, \lambda) F, \tag{7.2}$$

$$\frac{\partial F}{\partial t} = V(x, t, \lambda) F \tag{7.3}$$

with the initial condition

$$F(x, t, \lambda)|_{x=t=0} = I. \tag{7.4}$$

The matrix $F(x, t, \lambda)$ is an analytic function in the region $\bar{\mathbb{C}} \setminus (\mathfrak{U} \cup \mathfrak{V})$ and has essential singularities at the points of $\mathfrak{U}$ and $\mathfrak{V}$. We shall describe a simple method for identifying the principal parts of this matrix at singular points.

Consider a family of regular Riemann problems indexed by points ν in $\bar{\mathbb{C}}$, the corresponding contour Γ_ν being a small circle of radius ε_ν around ν. The Riemann problem of index ν amounts to decomposing the matrix $F(x, t, \lambda)$ into a product,

$$F(x, t, \lambda) = P_\nu(x, t, \lambda) Q_\nu(x, t, \lambda), \tag{7.5}$$

where the factors P_ν and Q_ν have an analytic continuation respectively inside and outside the contour Γ_ν given by $|\lambda - \nu| = \varepsilon_\nu$, with the normalization condition

$$Q_\nu(x, t, \lambda)|_{\lambda=\infty} = I. \tag{7.6}$$

For x and t sufficiently small, this Riemann problem has a unique solution because, in view of (7.4), $F(x, t, \lambda)$ is close to I. The function $Q_\nu(x, t, \lambda)$ is the principal (singular) part of $F(x, t, \lambda)$ at $\lambda = \nu$.

If ν does not belong to the divisors $\mathfrak{U}$ and $\mathfrak{B}$, then $F(x, t, \lambda)$ is regular at $\lambda = \nu$ and so for such ν we have the identity

$$Q_\nu(x, t, \lambda) = I. \tag{7.7}$$

Thus, out of the infinite set of Riemann problems (7.5)–(7.6) there are only a finite number of nontrivial ones indexed by the points ν that belong to $\mathfrak{U} + \mathfrak{B}$.

We shall now define a mapping

$$(U(x, t, \lambda), V(x, t, \lambda)) \to (U_\nu(x, t, \lambda), V_\nu(x, t, \lambda)) \tag{7.8}$$

which sends $U(\lambda)$ *and* $V(\lambda)$ *into a set of matrices,* $U_\nu(\lambda)$ *and* $V_\nu(\lambda)$, *indexed by points* ν *in* $\mathfrak{U} + \mathfrak{B}$. The matrices $U_\nu(\lambda)$ and $V_\nu(\lambda)$ are rational functions of λ with poles only at $\lambda = \nu$ of the same multiplicity as for $U(\lambda)$ and $V(\lambda)$, respectively, and without constant terms. In addition, if ν belongs to $\mathfrak{U}$ but not to $\mathfrak{B}$ (or vice versa), then $V_\nu(\lambda)$ (or respectively $U_\nu(\lambda)$) vanishes by the Liouville theorem. This set of matrices reflects the decomposition of $\mathfrak{U}$ and $\mathfrak{B}$ into a sum of elementary divisors,

$$\mathfrak{U} = \sum_i \mathfrak{U}_i, \qquad \mathfrak{B} = \sum_j \mathfrak{B}_j, \tag{7.9}$$

where $\mathfrak{U}_i = \{\lambda_i, n_i\}$, $\mathfrak{B}_j = \{\mu_j, \tilde{n}_j\}$. More explicitly,

$$U_\nu = \frac{\partial Q_\nu}{\partial x} Q_\nu^{-1}, \tag{7.10}$$

$$V_\nu = \frac{\partial Q_\nu}{\partial t} Q_\nu^{-1}. \tag{7.11}$$

Equations (7.5) imply the following relations on the contour Γ_ν:

$$U_\nu = U^{P_\nu^{-1}} = P_\nu^{-1} U P_\nu - P_\nu^{-1} \frac{\partial P_\nu}{\partial x}, \tag{7.12}$$

$$V_\nu = V^{P_\nu^{-1}} = P_\nu^{-1} V P_\nu - P_\nu^{-1} \frac{\partial P_\nu}{\partial t}, \tag{7.13}$$

which ensure the previously stated properties of $U_\nu(x, t, \lambda)$ and $V_\nu(x, t, \lambda)$.

It follows from (7.10)–(7.11) that $U_\nu(\lambda)$ and $V_\nu(\lambda)$ satisfy the zero curvature equation for every ν. Besides, if $\mathfrak{U}$ and $\mathfrak{B}$ do not intersect, there is a separation of variables, namely, $U_\nu(\lambda)$ depends only on x and $V_\nu(\lambda)$ only on t. In fact, let ν belong, say, to $\mathfrak{U}$ but not to $\mathfrak{B}$; then $V_\nu(\lambda)$ vanishes and the zero curvature equation reduces to

$$\frac{\partial U_\nu}{\partial t} = 0. \tag{7.14}$$

The opposite case of $V_\nu(\lambda)$ is proved in a similar manner. The case of intersecting $\mathfrak{U}$ and $\mathfrak{B}$ will be analyzed later.

Thus the mapping (7.8) leads to separation of variables in the zero curvature equation. The method used for that purpose may be called the "*undressing procedure*" in contrast to the dressing procedure of the previous section.

We will now show that the mapping (7.8) has an inverse. Suppose we are given matrices $U_\nu(x,\lambda)$ and $V_\nu(x,\lambda)$ without constant terms and elementary pole divisors $\mathfrak{U}_\nu$ and $\mathfrak{B}_\nu$ such that if $\nu=\lambda_i$, then $\mathfrak{U}_\nu=\{\lambda_i, n_i\}$, $\mathfrak{B}_\nu=0$, whereas if $\nu=\mu_j$, then $\mathfrak{U}_\nu=0$, $\mathfrak{B}_\nu=\{\mu_j, \tilde{n}_j\}$. Let the matrices $F_\nu(x,t,\lambda)$ be defined as solutions of the compatible system

$$\frac{\partial F_\nu}{\partial x} = U_\nu(x,\lambda)\, F_\nu, \tag{7.15}$$

$$\frac{\partial F_\nu}{\partial t} = V_\nu(t,\lambda)\, F_\nu, \tag{7.16}$$

normalized by the condition

$$F_\nu(x,t,\lambda)|_{x=t=0} = I. \tag{7.17}$$

Associated with these data is the following Riemann problem. The contour Γ of the problem consists of a set of circles Γ_ν of small radii ε_ν centered at the points ν of $\mathfrak{U}+\mathfrak{B}$ where the divisors $\mathfrak{U}$ and $\mathfrak{B}$ are given by (7.9); the matrix to be factorized equals $F_\nu(x,t,\lambda)$ on the circle Γ_ν. In other words, we look for a matrix $F(x,t,\lambda)$ analytic outside of Γ and satisfying

$$F_\nu(x,t,\lambda) = G_\nu(x,t,\lambda)\, F(x,t,\lambda) \tag{7.18}$$

on Γ_ν, where $G_\nu(x,t,\lambda)$ can be analytically continued inside Γ_ν. In addition,

$$F(x,t,\lambda)|_{\lambda=\infty} = I. \tag{7.19}$$

This problem has a unique solution for small x and t.

We set

$$U(x,t,\lambda)=\frac{\partial F}{\partial x}F^{-1}, \qquad V(x,t,\lambda)=\frac{\partial F}{\partial t}F^{-1}. \tag{7.20}$$

These matrices obviously satisfy (7.1). We will show that their pole divisors coincide with $\mathfrak{U}$ and $\mathfrak{V}$, respectively. This is clear from the following relations on Γ_ν:

$$U(\lambda)=G_\nu^{-1}U_\nu G_\nu-G_\nu^{-1}\frac{\partial G_\nu}{\partial x}, \tag{7.21}$$

$$V(\lambda)=G_\nu^{-1}V_\nu G_\nu-G_\nu^{-1}\frac{\partial G_\nu}{\partial t}, \tag{7.22}$$

which result from (7.15)–(7.18) and (7.18) by using the familiar trick of differentiating the equation of the Riemann problem with respect to the parameters x and t.

Clearly, this construction provides the general solution of the zero curvature equation locally in x and t.

If the divisors $\mathfrak{U}$ and $\mathfrak{V}$ intersect, complete simplification cannot be achieved at their common points. In this case a similar construction gives a partial separation of variables and reduces the order of equation (7.1).

As an illustration we shall consider the zero curvature representation for the principal chiral field model (see § 3). At first sight (see (3.37)–(3.40)), the associated divisors $\mathfrak{U}$ and $\mathfrak{V}$ are not separated but rather coincide with each other. Separation, however, is achieved in the light-cone coordinates

$$\xi=\frac{t+x}{2}, \qquad \eta=\frac{t-x}{2}, \tag{7.23}$$

in which the equations of motion take the form

$$\frac{\partial^2 g}{\partial\xi\partial\eta}=\frac{1}{2}\left(\frac{\partial g}{\partial\xi}g^{-1}\frac{\partial g}{\partial\eta}+\frac{\partial g}{\partial\eta}g^{-1}\frac{\partial g}{\partial\xi}\right). \tag{7.24}$$

In fact, the covariant differentiation operators in these coordinates are $\dfrac{\partial}{\partial\xi}-\dfrac{A(\xi,\eta)}{1-\lambda}$ and $\dfrac{\partial}{\partial\eta}-\dfrac{B(\xi,\eta)}{1+\lambda}$, where

$$A = l_0 + l_1 = \frac{\partial g}{\partial \xi} g^{-1}, \quad B = l_0 - l_1 = \frac{\partial g}{\partial \eta} g^{-1}. \tag{7.25}$$

The matrices $U(\xi, \eta, \lambda)$ and $V(\xi, \eta, \lambda)$ with

$$U(\lambda) = \frac{A}{1-\lambda}, \quad V(\lambda) = \frac{B}{1+\lambda}, \tag{7.26}$$

satisfy the zero curvature equation

$$\frac{\partial U(\lambda)}{\partial \eta} - \frac{\partial V(\lambda)}{\partial \xi} + [U(\lambda), V(\lambda)] = 0, \tag{7.27}$$

which is already fit for the undressing procedure that consists in constructing the matrices $U_1(\xi, \lambda)$ and $V_{-1}(\eta, \lambda)$.

It turns out that

$$U_1(\xi, \lambda) = \frac{A(\xi, 0)}{1-\lambda}, \quad V_{-1}(\eta, \lambda) = \frac{B(0, \eta)}{1+\lambda}, \tag{7.28}$$

that is U_1 and V_{-1} are determined by the initial data on the characteristics $\eta = 0$ and $\xi = 0$, respectively.

For the proof let us consider, for instance, equation (7.5) for $\nu = 1$,

$$F(\xi, \eta, \lambda) = P_1(\xi, \eta, \lambda) Q_1(\xi, \eta, \lambda) \tag{7.29}$$

and show that

$$F(\xi, 0, \lambda) = Q_1(\xi, 0, \lambda). \tag{7.30}$$

The first equation in (7.28) is an immediate consequence of (7.30) and the equation

$$\frac{\partial F}{\partial \xi} = U(\xi, \eta, \lambda) F. \tag{7.31}$$

To prove (7.30) it is enough to show that $F(\xi, 0, \lambda)$ has no singularity at $\lambda = -1$. This, however, clearly follows from the differential equation (7.31) at $\eta = 0$ with the initial condition

$$F(\xi, 0, \lambda)|_{\xi=0} = I. \tag{7.32}$$

The second formula in (7.28) is proved in a similar way.

So, in order to produce the general local solution of the principal chiral field model we must perform the following operations.

1. Construct the matrices

$$F_1(\xi,\lambda)=\overset{\curvearrowleft}{\exp}\int_0^{\xi} U_1(\xi',\lambda)\,d\xi' \tag{7.33}$$

and

$$F_{-1}(\eta,\lambda)=\overset{\curvearrowleft}{\exp}\int_0^{\eta} V_{-1}(\eta',\lambda)\,d\eta', \tag{7.34}$$

i.e., solve two systems of first order ordinary linear differential equations.

2. Determine the matrix $F(\xi,\eta,\lambda)$ from its principal parts $F_1(\xi,\lambda)$ at $\lambda=1$ and $F_{-1}(\eta,\lambda)$ at $\lambda=-1$. This is done by means of the Riemann problem

$$F_1(\xi,\lambda)=G_1(\xi,\eta,\lambda)F(\xi,\eta,\lambda) \tag{7.35}$$

for $|\lambda-1|=\varepsilon$ and

$$F_{-1}(\eta,\lambda)=G_{-1}(\xi,\eta,\lambda)F(\xi,\eta,\lambda) \tag{7.36}$$

for $|\lambda+1|=\varepsilon$, where $F(\xi,\eta,\lambda)$ is analytic in the exterior of the circles indicated above and normalized to I for $\lambda=\infty$. Then a solution of (7.24) with the initial data $A(\xi)$ and $B(\eta)$ on the characteristics is given by

$$g(\xi,\eta)=F(\xi,\eta,\lambda)|_{\lambda=0}. \tag{7.37}$$

We have thus obtained an analogue of the D'Alembert representation for the general solution of the principal chiral field equation, as a nonlinear superposition of waves propagating along the characteristics. The Riemann problem (7.35)–(7.36) plays the role of a nonlinear analogue of the superposition principle.

We have seen that the requirement of kinematic integrability of a nonlinear system, i.e. the possibility to write it down in the form of a zero curvature equation, allows us to present a vast class of its solutions: explicit solutions of soliton type provided by the dressing procedure, local solutions of the general form provided by the undressing procedure, etc. The zero curvature equation may be exploited for a further study of the associated nonlinear system. For example, it can be used to define a sequence of integrals of the motion in the usual (not necessarily Hamiltonian) sense as functionals preserved by the equations of motion.

In fact, in view of the periodic boundary conditions

$$U(x+2L,\lambda)=U(x,\lambda), \tag{7.38}$$

$$V(x+2L,\lambda)=V(x,\lambda) \tag{7.39}$$

the monodromy matrix

$$T_L(t,\lambda)=\overset{\curvearrowleft}{\exp}\int_{-L}^{L} U(x,t,\lambda)\,dx \tag{7.40}$$

satisfies

$$\frac{\partial T_L(t,\lambda)}{\partial t}=[V(L,t,\lambda),T(t,\lambda)] \tag{7.41}$$

(cf. § I.2 of Part I). Hence its invariants, $\operatorname{tr} T_L^k(\lambda)$ for $k=1,\ldots,n$, are the generating functions for the integrals (n is the dimension of the auxiliary space). The local integrals result from the asymptotic expansions of these functionals in λ about the poles of $U(x,t,\lambda)$. One example of such an expansion was already discussed in Part I; other examples will be treated in the next chapter. A detailed analysis of the construction of local integrals of the motion in the general case is fairly tedious and will not be given here.

In the case of models on the lattice, a counterpart of the matrix $T_L(t,\lambda)$ is the matrix

$$T_N(t,\lambda)=\prod_{n=1}^{\overset{\curvearrowleft}{N}} L_n(t,\lambda), \tag{7.42}$$

which will receive closer attention in Chapter III.

In Part I we saw for the case of the NS model that from the Hamiltonian point of view the zero curvature condition is a secondary matter: it is a consequence of the fundamental Poisson brackets which incorporate the matrix $U(x,\lambda)$ of the auxiliary linear problem and the r-matrix. The models to be considered are also Hamiltonian, so we shall again use the r-matrix formalism. This approach culminates in the Lie-algebraic classification of the fundamental Poisson brackets in Chapter IV. We decided, however, to include a general discussion of the zero curvature equation in order to demonstrate the immediate benefit of this representation for nonlinear integrable equations.

§ 8. Notes and References

There is an extensive literature devoted to the models defined in this chapter. Here we shall only refer to papers in which these models are treated via the inverse scattering method, i. e. via the zero curvature representation; as a rule, the boundary conditions imposed are those of rapid decrease. For further references the reader should consult the corresponding sections of Chapters II, III; papers that discuss these models from the Hamiltonian standpoint are also cited there.

1. The HM model was incorporated into the inverse scattering method in [T 1977].

2. The SG model was actively studied during the years 1973–1975. The zero curvature representation for the SG equation in light-cone coordinates

$$\frac{\partial^2 \varphi}{\partial \xi \partial \eta} + \frac{m^2}{\beta} \sin \beta \varphi = 0 \tag{8.1}$$

was obtained in [AKNS 1973], [T 1974], and in the form (1.19) in [ZTF 1974], [TF 1974]. Of course, these two zero curvature representations are formally equivalent: the difference between the above papers is that they are dealing with different auxiliary linear problems. For more references see Chapter II.

3. The inverse scattering method was applied to the LL model in [Sk 1979], [BR 1981].

4. The vector NS model with two colours was immersed into the inverse scattering method in [M 1973]; the results of this paper immediately extend to an arbitrary number of colours. The model was generalized to the case of homogeneous spaces of Lie groups in [FK 1983].

5. The general observation at the end of § 1 concerning the connection between the momentum functional and the symplectic form on phase space resulted from a discussion with A. G. Reyman.

6. The first integrable model on the lattice was the Toda model that describes anharmonic oscillations of a one-dimensional crystal lattice (see [T 1970]). The inverse scattering method for obtaining its exact solution was applied by S. V. Manakov [M 1974] and H. Flaschka [F 1974a], [F 1974b]. For further references see Chapter III.

7. A reformulation of the zero curvature condition for lattice models was given by M. Ablowitz and G. Ladic [AL 1976].

8. The second model of § 2 was introduced by V. Volterra in [V 1931], whence its name. Besides applications in ecology, it is encountered in plasma physics in the studies of the fine spectral structure of the Langmuir oscillations. The inverse scattering method for the model was developed in [M 1974]. For the Poisson structure of the Volterra model see Chapter III.

9. The LHM model and the related auxiliary linear problem were introduced in [Sk 1982].

10. The LNS_1 model and the related auxiliary linear problem (2.62)–(2.63) first appeared in [IK 1981].

11. The LNS_2 model was defined in [AL 1976], in which the inverse scattering method was applied. The Hamiltonian formulation of the model was given in [KM 1979], [GIK 1984].

12. The relationship between the LNS_1 and LHM models, in particular the operation of alternation of sign in (2.51), was discussed in [TTF 1983].

13. For information on representation theory of the groups $SU(2)$ and $SU(1, 1)$ see, for instance, [GGP 1966].

14. The idea to exploit the general zero curvature representation for studying nonlinear equations is due to V. E. Zakharov and A. B. Shabat [ZS 1979] and is currently referred to as the Zakharov-Shabat scheme. For the case of two-dimensional auxiliary space, the general zero curvature representation was previously studied in [AKNS 1974]. The role of gauge transformation in this approach was discussed in [ZT 1979] for the case of the HM and NS models.

15. A systematic study of various reductions in the zero curvature representation was carried out by A. V. Mikhailov [Mi 1981].

16. As was noted in the Introduction, the celebrated KdV equation appears in the main text only in Chapter I of Part II. Traditionally, this equation is treated by a Lax representation [L 1968]. The zero curvature representation for KdV was given in [N 1974]. Other references concerning this equation were already given in Chapter III of Part I.

17. The passage from equation (3.13) to (3.12) is a special case of the Frobenius transformation that carries an ordinary linear differential equation of order n into a system of n first order linear equations. Clearly, such systems play a unifying role in the zero curvature representation. Higher order differential equations considered in [GD 1975], [GD 1977] may thus be regarded as reductions of the Zakharov-Shabat scheme [DS 1981], [DS 1984].

18. The Poisson structure (3.15) was introduced in [ZF 1971] and [G 1971]. In [FT 1985], [BFT 1986] some important peculiarities were pointed out connected with the Hamiltonian interpretation of the KdV equation.

19. The N-wave model studied by V. E. Zakharov and S. V. Manakov [ZM 1975] was the first one to display all the advantages of the zero curvature representation. It was also basic for developing the matrix Riemann problem method [S 1975], [S 1979] which gives an alternative to the Gelfand-Levitan-Marchenko integral equations formalism.

20. In the stationary case (i.e. when there is no dependence on t), the N-wave equations reduce to Euler's equations that generalize the equations of motion of a three-dimensional rigid body. This remarkable fact was observed by S. V. Manakov [M 1976]. Many other examples of this kind have been discovered since, see [MF 1978], [P 1981], [V 1983], [B 1984]. These systems are naturally incorporated into a general scheme based on affine Lie algebras [RSF 1979], [RS 1979], [R 1980], [AM 1980].

21. The chiral field is a popular field-theoretic model of geometric origin (see, for instance, [DNF 1979]). The zero curvature representation for the n-field was obtained in [P 1976], and for the principal chiral field in [ZM 1978]. We refer to the matrices l_μ as the left currents because the Hamiltonians $\int l_0^a(x)A^a(x)\,dx$ with respect to the Poisson structure (5.15), (5.18)–(5.19) give rise to the action of the group G on itself by left translations. For similar reasons the matrices $r_0=-g^{-1}\dfrac{\partial g}{\partial t}$, $r_1=-g^{-1}\dfrac{\partial g}{\partial x}$ are called the right currents.

22. The two-dimensional extension of the Toda model was defined in [Mi 1979] where the associated zero curvature representation was obtained. In a different form it was considered in [LS 1979]; a generalization to other root systems was given in [MOP 1981]. Independently of the two-dimensional Toda model, equation (3.54) occurred in [ZS 1971], [DB 1977]. A zero curvature representation for the model was found in [Mi 1979].

23. The exposition in § 4 follows [ZT 1979]. Formulae such as (4.19)–(4.21) were also derived in [L 1977] where a relationship between the HM and NS models was established without recourse to the inverse scattering method. Notice that although the LHM and LNS_1 models essentially coincide in the lattice case, their relationship in the continuous case is given by a gauge transformation. This is so because the HM and NS models are different continuum limits of the same lattice model.

24. For the Hopf fibre bundle and the theorem on holonomy see, for instance, the textbook [DNF 1979]. The idea to use this theorem for the proof of (4.21) was suggested by A. G. Reyman.

25. The Lie-Poisson brackets that define a Poisson structure on the dual space of a Lie algebra will be discussed in detail in Chapter IV.

26. The extended current algebra that appeared in § 5 is very popular at present in mathematical physics (see [RS 1980], [FK 1980], [DKM 1981], [DJKM 1981], [Se 1981]). Some of the applications to integrable equations will be discussed in Chapter IV. The term Maurer-Cartan cocycle was introduced in [VGG 1973] in connection with the representation theory of current groups.

27. The Poisson brackets (5.15) and (5.18)–(5.19) define the usual Poisson structure on the cotangent bundle $T^*C(G)$ of the current group $C(G)$. The relationship between these Poisson brackets and the brackets (5.15)–(5.17) is an illustration of Hamiltonian reduction of $T^*C(G)$ under the action of G. For the notion of Hamiltonian reduction see, for example, [A 1974].

28. The second Hamiltonian formulation of the chiral field model in § 5 was proposed and elaborated in [FR 1986].

29. The Poisson brackets (5.59)–(5.61) for the n-field model may be obtained, as the Poisson-Dirac brackets, from the canonical brackets

$$\{\pi^a(x), \pi^b(y)\} = \{n^a(x), n^b(y)\} = 0,$$
$$\{\pi^a(x), n^b(y)\} = \delta_{ab}\delta(x-y) \tag{8.2}$$

with the constraints

$$\vec{n}^2 = 1, \quad \vec{\pi}\cdot\vec{n} = 0. \tag{8.3}$$

30. The role of multi-valued functionals such as (5.74), regarded as the action functionals in mechanics and field theory, was revealed in the general form by S. P. Novikov [N 1982 a, b]. The functional $W(g)$ is a typical illustration of the general construction of [N 1982 a, b] which raised the problem of studying a model with the action functional $S(g)+\alpha W(g)$. The momentum functional for the HM model in § 1 provides another interesting example, but is an observable rather than action.

31. The simple topological properties of Lie groups used in § 5 can be found, for instance, in [A 1969]. Formula (5.76) is derived from the Stokes formula by differentiating with respect to s.

32. Multi-valued action functionals were implicit in the physics literature in the work of G. Wess and B. Zumino [WZ 1971] in the specific case of the principal chiral field model in four-dimensional space-time; the geometric origin of the ambiguity of action was discovered by E. Witten in [W 1983]. We also point out that in the model studied in [BT 1977], [Bu 1978], the multi-valued action (5.78) was actually used as a kinetic term. The model in question is a matrix generalization of the SG model and is integrable via the inverse scattering method. It describes a field $g(x, t)$ with values in the group $SO(n)$; the equations of motion are

$$\frac{\partial}{\partial t}\left(\frac{\partial g}{\partial t} g^{-1}\right) - \frac{\partial}{\partial x}\left(\frac{\partial g}{\partial x} g^{-1}\right) + \left[\frac{\partial g}{\partial x} g^{-1}, \frac{\partial g}{\partial t} g^{-1}\right] + [A, gBg^{-1}] = 0, \tag{8.4}$$

where A and B are real diagonal matrices that do not depend on x and t. The matrix SG model is gauge equivalent to the principal chiral field model on the group $SO(n)$ (see [Bu 1978]).

33. The relationship between the modified and ordinary principal chiral field equations was established in [VT 1984].

34. A geometric interpretation of the symplectic structure connected with the modified Poisson brackets (5.90)–(5.92) was given in [R 1984].

35. The appearance of two independent extended current algebras in the modified chiral field model with $\alpha = \pm 1$ was first observed in [W 1984].

36. The Riemann problem method for constructing solutions of the general zero curvature equation in § 6 and the term "dressing procedure" are due to V. E. Zakharov and A. B. Shabat [ZS 1979]. A discussion of reductions in the Riemann problem can be found in [Mi 1981]. For the use of the

dressing procedure for constructing soliton solutions of various equations see [ZM 1978], [ZMNP 1980].

37. The scheme for constructing the general local solution of the zero curvature equation presented in § 7 is due to I. M. Krichever (see [K 1983]) and was originally applied in [K 1980] for solving the principal chiral field model and the SG model in light-cone coordinates.

38. The undressing scheme actually made use of the group $GL(n)$ over the adele ring of the field of rational functions over $\mathbb{C}$. The utility of this language in presenting the generalities of the inverse scattering method is illustrated in [C 1983].

39. The Riemann problem (7.5) is a matrix extension of the Weierstrass method of separating the principal parts of the function $F(x, t, \lambda)$, whereas the Riemann problem (7.18) is a matrix analogue of the multiplicative Cousin problem of determining a function from its principal parts.

40. There is an extensive literature dealing with the construction of integrals of the motion for integrable equations: [GD 1975], [ZM 1975], [GD 1977], [C 1979 a, b], [Ne 1979], [W 1979], [DS 1981], [C 1981], [W 1981], [TF 1982], [KR 1983], [R 1983], [DS 1984], [VV 1985].

41. In this book we only consider integrable equations in one spatial variable. The inverse scattering method also applies to equations in two spatial dimensions, the best known being the Kadomtsev-Petviashvili equation

$$\frac{\partial}{\partial x}\left(\frac{\partial u}{\partial t} - 6u\frac{\partial u}{\partial x} + \frac{\partial^3 u}{\partial x^3}\right) = 3\varepsilon\frac{\partial^2 u}{\partial y^2}, \qquad \varepsilon = \pm 1, \tag{8.5}$$

which has numerous physical applications. The auxiliary linear problem for (8.5) has the form

$$\sqrt{\varepsilon}\,\frac{\partial F}{\partial y} - \frac{\partial^2 F}{\partial x^2} - u(x, y)F = \lambda F \tag{8.6}$$

(see [D 1974], [ZS 1974]). There exists a vast literature for the Kadomtsev-Petviashvili equation; we only mention the textbook [ZMNP 1980] and the papers [KN 1978], [DJKM 1981], [M 1981], [FA 1983], [AYF 1983].

References

[A 1969] Adams, J. F.: Lectures on Lie groups. New York–Amsterdam, Benjamin 1969

[A 1974] Arnold, V. I.: Mathematical Methods of Classical Mechanics. Moscow: Nauka 1974 [Russian]; English transl.: Graduate Texts in Mathematics 60, New York–Berlin–Heidelberg, Springer 1978

[AKNS 1973] Ablowitz, M. J., Kaup, D. J., Newell, A. C., Segur, H.: Method for solving the Sine-Gordon equation. Phys. Rev. Lett. *30*, 1262–1264 (1973)

[AKNS 1974] Ablowitz, M. J., Kaup, D. J., Newell, A. C., Segur, H.: The inverse scattering transform – Fourier analysis for nonlinear problems. Stud. Appl. Math. *53*, 249–315 (1974)

[AL 1976] Ablowitz, M. J., Ladik, J. F.: Nonlinear differential-difference equations and Fourier analysis. J. Math. Phys. *17*, 1011–1018 (1976)

[AM 1980] Adler, M., van Moerbeke, P.: Completely integrable systems, Euclidean Lie algebras and curves. Adv. Math. *38*, 267–317 (1980)

[AYF 1983] Ablowitz, M. J., van Yaakov, D., Fokas, A. S.: On the inverse scattering transform for Kadomtsev-Petviashvili equation. Stud. Appl. Math. *69*, 135–143 (1983)

[B 1984] Bogoyavlensky, O. I.: Integrable equations on Lie algebras arising in problems of mathematical physics. Izv. Akad. Nauk SSSR Ser. Mat. *48*, 883–938 (1984) [Russian]

[BFT 1986] Buslaev, V. S., Faddeev, L. D., Takhtajan, L. A.: Scattering theory for the Korteweg-de Vries equation and its Hamiltonian interpretation. Physica *D 18*, 255–266 (1986)

[BR 1981] Borovik, A. E., Robuk, V. N.: Linear pseudopotentials and conservation laws for the Landau-Lifshitz equation describing the nonlinear dynamics of a ferromagnet with uniaxial anisotropy. Teor. Mat. Fiz. *46*, 371–381 (1981) [Russian]; English transl. in Theor. Math. Phys. *46*, 242–248 (1981)

[BT 1977] Budagov, A. S., Takhtajan, L. A.: A nonlinear one-dimensional model of classical field theory with internal degrees of freedom. Dokl. Akad. Nauk SSSR *235*, 805–808 (1977) [Russian]

[Bu 1978] Budagov, A. S.: A completely integrable model of classical field theory with nontrivial particle interaction in two space-time dimensions. In: Problems in quantum field theory and statistical physics. I. Zap. Nauchn. Semin. LOMI *77*, 24–56 (1978) [Russian]

[C 1979 a] Cherednik, I. V.: Local conservation laws for principal chiral fields ($d=1$). Teor. Mat. Fiz. *38*, 179–185 (1979) [Russian]; English transl. in Theor. Math. Phys. *38*, 120–124 (1979)

[C 1979 b] Cherednik, I. V.: Conservation laws and elements of scattering theory for principal chiral fields ($d=1$). Teor. Mat. Fiz. *41*, 236–244 (1979) [Russian]; English transl. in Theor. Math. Phys. *41*, 997–1002 (1980)

[C 1981] Cherednik, I. V.: Algebraic aspects of two-dimensional chiral fields. Itogi Nauki Tekh., Ser. Sovrem. Probl. Mat. *17*, 175–218 (1981) [Russian]

[C 1983] Cherednik, I. V.: On the definition of τ-function for generalized affine Lie algebras. Funk. Anal. Prilož. *17* (3), 93–95 (1983) [Russian]

[D 1974] Druma, V. S.: Solution of the two-dimensional Korteweg-de Vries (KdV) equation. Pisma Zh. Exp. Teor. Fiz. *19*, 753–755 (1974) [Russian]; English transl. in Sov. Phys. JETP Letters *19*, 387–388 (1974)

[DB 1977] Dodd, R. K., Bullough, R. K.: Polynomial conserved densities for the Sine-Gordon equations. Proc. Roy. Soc. (London) *A352*, 481–503 (1977)

[DJKM 1981] Date, E., Jimbo, M., Kashiwara, M., Miwa, T.: Operator approach to the Kadomtsev-Petviashvili equation. Transformation groups for soliton equations. III. J. Phys. Soc. Japan *50*, 3806–3812 (1981)

[DKM 1981] Date, E., Kashiwara, M., Miwa, T.: Vertex operators and τ-functions. Transformation groups for soliton equations. II. Proc. Japan Acad. *57A*, 387–392 (1981)

[DNF 1979] Dubrovin, B. A., Novikov, S. P., Fomenko, A. T.: Modern Geometry. Methods and Applications. Moscow: Nauka 1979 [Russian]; English translation of Part I: Graduate Texts in Mathematics 93; Part II: Graduate Texts in Mathematics 104, New York–Berlin–Heidelberg–Tokyo, Springer 1984, 1985

[DS 1981] Drinfeld, V. G., Sokolov, V. V.: Equations of Korteweg-de Vries type and simple Lie algebras. Dokl. Akad. Nauk SSSR *258*, 11–16 (1981) [Russian]; English transl. in Sov. Math. Dokl. *23*, 457–461 (1981)

[DS 1984] Drinfeld, V. G., Sokolov, V. V.: Lie algebras and equations of Korteweg-de Vries type. Itogi Nauki Tekh., Ser. Sovrem. Probl. Mat. *24*, 81–180, Moscow, VINITI 1984 [Russian]

[F 1974a] Flaschka, H.: The Toda lattice. I. Existence of integrals. Phys. Rev. *B9*, 1924–1925 (1974)

[F 1974b] Flaschka, H.: On the Toda lattice. II. Inverse scattering solution. Prog. Theor. Phys. *51*, 703–716 (1974)

[FA 1983] Fokas, A. C., Ablowitz, M. J.: On the inverse scattering of the time dependent Schrödinger equation and the associated Kadomtsev-Petviashvili (I) equation. Stud. Appl. Math. *69*, 211–228 (1983)

[FK 1980] Frenkel, I. B., Kac, V. G.: Basic representation of affine Lie algebras and dual resonance models. Invent. Math. *62*, 23–66 (1980)

[FK 1983] Fordy, A. P., Kulish, P. P.: Nonlinear Schrödinger equations and simple Lie algebras. Comm. Math. Phys. *89*, 427–443 (1983)

[FR 1986] Faddeev, L. D., Rechetikhin, N. Yu.: Integrability of the principal chiral field model in 1+1 dimension. Ann. Phys. (N.Y.) *167*, 227–256 (1986)

[FT 1985] Faddeev, L. D., Takhtajan, L. A.: Poisson structure for the KdV equation. Lett. Math. Phys. *10*, 183–188 (1985)

[G 1971] Gardner, C. S.: Korteweg-de Vries equation and generalizations. IV. The Korteweg-de Vries equation as a Hamiltonian system. J. Math. Phys. *12*, 1548–1551 (1971)

[GD 1975] Gelfand, I. M., Dikiĭ, L. A.: Asymptotic behaviour of the resolvent of Sturm-Liouville equations and the algebra of the Korteweg-de Vries equations. Usp. Mat. Nauk *30* (5), 67–100 (1975) [Russian]; English transl. in Russian Math. Surveys *30* (5), 77–113 (1975)

[GD 1977] Gelfand, I. M., Dikiĭ, L. A.: The resolvent and Hamiltonian systems. Funk. Anal. Prilož. *11* (2), 11–17 (1977) [Russian]; English transl. in Funct. Anal. Appl. *11*, 93–104 (1977)

[GGP 1966] Gelfand, I. M., Graev, M. I., Piatetski-Shapiro, I. I.: Representation theory and automorphic functions. Moscow, Nauka 1966 [Russian]; English transl.: Philadelphia, Saunders 1969

[GIK 1984] Gerdjikov, V. S., Ivanov, M. I., Kulish, P. P.: Expansions over the "squared" solutions and difference evolution equations. J. Math-Phys. *25*, 25–34 (1984)

[IK 1981] Izergin, A. G., Korepin, V. E.: A lattice model associated with the nonlinear Schrödinger equation. Dokl. Akad. Nauk SSSR *259*, 76–79 (1981) [Russian]

[K 1980] Krichever, I. M.: Analogue of the D'Alembert formula for the equations of principal chiral field and the sine-Gordon equation. Dokl. Akad. Nauk SSR *253*, 288–292 (1980) [Russian]; English transl. in Sov. Math. Dokl. *22*, 79–84 (1981)

[K 1983] Krichever, I. M.: Nonlinear equations and elliptic curves. Itogi Nauki Tekh., Ser. Sovrem. Probl. Math. *23*, 79–136, Moscow, VINITI 1983 [Russian]

[KM 1979] Kako, F., Mugibayashi, N.: Complete integrability of general nonlinear differential-difference equations solvable by the inverse method. II. Prog. Theor. Phys. *61*, 776–790 (1979)

[KN 1978] Krichever, I. M., Novikov, S. P.: Holomorphic bundles over Riemann surfaces and the Kadomtsev-Petviashvili equation. Funk. Anal. Prilož. *12* (4), 41–52 (1978) [Russian]; English transl. in Funct. Anal. Appl. *12*, 276–286 (1978)

[KR 1983] Kulish, P. P., Reyman, A. G.: Hamiltonian structure of polynomial bundles. In: Differential geometry, Lie groups and mechanics. V. Zap. Nauchn. Semin. LOMI *123*, 67–76 (1983) [Russian]

[L 1968] Lax, P. D.: Integrals of nonlinear equations of evolution and solitary waves. Comm. Pure and Appl. Math. *21*, 467–490 (1968)

[L 1977] Lakshmanan, M.: Continuum spin system as an exactly solvable dynamical system. Phys. Lett. *61A*, 53–54 (1977)

[LS 1979] Leznov, A. N., Saveliev, M. A.: Representation of zero curvature for the system of nonlinear partial differential equations $\varkappa_{\alpha z\bar{z}} = (\exp k\chi)_\alpha$ and its integrability. Lett. Math. Phys. *3*, 489–494 (1979)

[M 1973] Manakov, S. V.: On the theory of two-dimensional stationary self-focusing of electro-magnetic waves. Zh. Exp. Teor. Fiz. *65*, 505–516 (1973) [Russian]; English transl. in Sov. Phys. JETP *38*, 248–253 (1974)

[M 1974] Manakov, S. V.: Complete integrability and stochastization of discrete dynamical systems. Zh. Exp. Teor. Fiz. *67*, 543–555 (1974) [Russian]; English transl. in Sov. Phys. JETP *40*, 269–274 (1975)

[M 1976] Manakov, S. V.: Note on the integration of Euler's equations of the dynamics of an n-dimensional rigid body. Funk. Anal. Prilož. *10* (4), 93–94 (1976) [Russian]; English transl. in Funct. Anal. Appl. *10*, 328–329 (1976)

[M 1981] Manakov, S. V.: The inverse scattering transform for the time-dependent Schrödinger equation and Kadomtsev-Petviashvili equation. Physica *D3*, 420–427 (1981)

[MF 1978] Miscenko, A. S., Fomenko, A. T.: Euler equations on finite-dimensional Lie groups. Izv. Akad. Nauk SSSR Ser. Mat. *42*, 396–415 (1978) [Russian]; English transl. in Math. USSR, Izv. *12*, 371–389 (1978)

[Mi 1979] Mikhailov, A. V.: Integrability of the two-dimensional generalization of the Toda chain. Pisma Zh. Exp. Teor. Fiz. *30*, 443–448 (1979) [Russian]; English transl. in Sov. Phys. JETP Letters *30*, 414–418 (1979)

[Mi 1981] Mikhailov, A. V.: The reduction problem and the inverse scattering method. Physica *D3*, 73–107 (1981)

[MOP 1981] Mikhailov, A. V., Olshanetsky, M. A., Perelomov, A. M.: Two-dimensional generalized Toda lattice. Comm. Math. Phys. *79*, 473–488 (1981)

[N 1974] Novikov, S. P.: The periodic problem for the Korteweg-de Vries equation, Funk. Anal. Prilož. *8* (3) 54–66 (1974) [Russian]; English transl. in Funct. Anal. Appl. *8*, 236–246 (1974)

[N 1982a] Novikov, S. P.: Hamiltonian formalism and a multi-valued analogue of Morse theory. Usp. Mat. Nauk *37* (5), 3–49 (1982) [Russian]; English transl. in Russian Math. Surveys *37* (5), 1–56 (1982)

[N 1982b] Novikov, S. P.: Hamiltonian formalism and variational-topological methods for finding periodic trajectories of conservative dynamical systems. Soviet Scientific Reviews, Sect. C, Math. Physics Reviews *3*, 3–51 (1982)

[Ne 1979] Newell, A. C.: The general structure of integrable evolution equations. Proc. Royal Soc. (London) *A365*, 283–311 (1979)

[P 1976] Pohlmeyer, K.: Integrable Hamiltonian systems and interaction through quadratic constraints. Comm. Math. Phys. *46*, 207–223 (1976)

[P 1981] Perelomov, A. M.: Several remarks on the integrability of the equations of motion of a rigid body in ideal fluid. Funk. Anal. Prilož. *15* (2), 83–85 (1981) [Russian]; English transl. in Funct. Anal. Appl. *15*, 144–146 (1981)

[R 1980] Reyman, A. G.: Integrable Hamiltonian systems connected with graded Lie algebras. In: Differential geometry, Lie groups and mechanics. III. Zap. Nauchn. Semin. LOMI *95*, 3–54 (1980) [Russian]; English transl. in J. Sov. Math. *19*, 1507–1545 (1982)

[R 1983] Reyman, A. G.: A unified Hamiltonian system on polynomial bundles and the structure of stationary problems. In: Problems in quantum field theory and statistical physics. 4. Zap. Nauchn. Semin. LOMI *131*, 118–127 (1983) [Russian]

[R 1984] Ramadas, T. R.: The Wess-Zumino term and fermionic solutions. Comm. Math. Phys. *93*, 355–365 (1984)

[RS 1979] Reyman, A. G., Semenov-Tian-Shansky, M. A.: Reduction of Hamiltonian systems, affine Lie algebras and Lax equations. I. Inventiones math. *54*, 81–100 (1979)

[RS 1980] Reyman, A. G., Semenov-Tian-Shansky, M. A.: Current algebras and non-linear partial differential equations. Dokl. Akad. Nauk SSSR *251*, 1310–1314 (1980) [Russian]; English transl. in Sov. Math. Dokl. *21*, 630–634 (1980)

[RSF 1979] Reyman, A. G., Semenov-Tian-Shansky, M. A., Frenkel, I. B.: Graded Lie algebras and completely integrable dynamical systems. Dokl. Akad. Nauk SSSR *247*, 802–805 (1979) [Russian]; English transl. in Sov. Math. Dokl. *20*, 811–814 (1979)

[S 1975] Shabat, A. B.: The inverse scattering problem for a system of differential equations. Funk. Anal. Prilož. *9* (3), 75–78 (1975) [Russian]

[S 1979] Shabat, A. B.: An inverse scattering problem. Differencialnye Uravneniya *15*, 1824–1834 (1979) [Russian]; English transl. in Diff. Equations *15*, 1299–1307 (1980)

[Se 1981] Segal, G.: Unitary representations of some infinite dimensional groups. Comm. Math. Phys. *80*, 301–342 (1981)

[Sk 1979] Sklyanin, E. K.: On complete integrability of the Landau-Lifshitz equation. Preprint LOMI E-3-79, Leningrad 1979

[Sk 1982] Sklyanin, E. K.: Algebraic structures connected with the Yang-Baxter equation. Funk. Anal. Prilož. *16* (4), 27–34 (1982) [Russian]; English transl. in Funct. Anal. Appl. *16*, 263–270 (1982)

[T 1970] Toda, M.: Waves in nonlinear lattice. Prog. Theor. Phys. Suppl. *45*, 174–200 (1970)

[T 1974] Takhtajan, L. A.: Exact theory of propagation of ultra-short optical pulses in two-level media. Zh. Exp. Teor. Fiz. *66*, 476–489 (1974) [Russian]

[T 1977] Takhtajan, L. A.: Integration of the continuous Heisenberg spin chain through the inverse scattering method. Phys. Lett. *64A*, 235–237 (1977)

[TF 1974] Takhtajan, L. A., Faddeev, L. D.: Essentially nonlinear one-dimensional model of classical field theory. Teor. Mat. Fiz. *21* (2), 160–174 (1974) [Russian]; English transl. in Theor. Math. Phys. *21*, 1046–1057 (1974)

[TF 1982] Takhtajan, L. A., Faddeev, L. D.: A simple connection between geometrical and Hamiltonian representations for the integrable nonlinear equations. In: Boundary-value problems of mathematical physics and related questions in function theory. 14. Zapiski Nauchn. Semin. LOMI *115*, 264–273 (1982) [Russian]

[TTF 1983] Tarasov, V. O., Takhtajan, L. A., Faddeev, L. D.: Local Hamiltonians for integrable quantum models on the lattice. Teor. Mat. Fiz. *57*, 163–181 (1983) [Russian]; English transl. in Theor. Math. Phys. *57*, 1059–1073 (1983)

[V 1931] Volterra, V.: Leçons sur la théorie Mathématique de la Lutte pour la Vie. Paris, Gauthier-Villars 1931

[V 1983] Veselov, A. P.: The Landau-Lifshitz equation and integrable systems of classical mechanics. Dokl. Akad. Nauk SSSR *270*, 1094–1097 (1983) [Russian]; English transl. in Sov. Phys. Dokl. *28*, 458–459 (1983)

[VGG 1973] Vershik, A. M., Gelfand, I. M., Graev, M. I.: Representations of the group $SL(2, R)$, where R is a ring of functions. Usp. Mat. Nauk *28* (5), 83–128 (1973) [Russian]; English transl. in Russian Math. Surveys *28* (5), 87–132 (1973)

[VT 1984] Veselov, A. P., Takhtajan, L. A.: The integrability of Novikov's equations for the principal chiral fields with a multi-valued Lagrangian. Dokl. Akad. Nauk SSSR *279*, 1097–1110 (1984) [Russian]

[VV 1985] Vladimirov, V. S., Volovich, I. V.: Local and nonlocal currents for nonlinear equations. Teor. Mat. Fiz. *62*, 3–29 (1985) [Russian]

[W 1979] Wilson, G.: Commuting flows and conservation laws for Lax equations. Math. Proc. Cambridge Phil. Soc. *86*, 131–143 (1979)

[W 1981] Wilson, G.: On two constructions of conservation laws for Lax equations. Quart. J. Math. Oxford *32*, 491–512 (1981)

[W 1983] Witten, E.: Global aspects of current algebra. Nucl. Phys. *B223*, 422–432 (1983)

[W 1984] Witten, E.: Nonabelian bosonization in two dimensions. Comm. Math. Phys. *92*, 455–472 (1984)

[WZ 1971] Wess, J., Zumino, B.: Consequences of anomalous Ward identities. Phys. Lett. *37B*, 95–97 (1971)

[ZF 1971] Zakharov, V. E., Faddeev, L. D.: Korteweg-de Vries equation, a completely integrable Hamiltonian system. Funk. Anal. Prilož. *5* (4) 18–27 (1971) [Russian]; English transl. in Funct. Anal. Appl. *5*, 280–287 (1971)

[ZM 1975] Zakharov, V. E., Manakov, S. V.: The theory of resonant interaction of wave packets in nonlinear media. Zh. Eksp. Teor. Fiz. *69*, 1654–1673 (1975) [Russian]; English transl. in Sov. Phys. JETP *42*, 842–850 (1976)

[ZM 1978] Zakharov, V. E., Mikhailov, A. V.: Relativistically invariant two-dimensional models of field theory which are integrable by means of the inverse scattering problem method. Zh. Exp. Teor. Fiz. *74*, 1953–1973 (1978) [Russian]; English transl. in Sov. Phys. JETP *47*, 1017–1027 (1978)

[ZMNP 1980] Zakharov, V. E., Manakov, S. V., Novikov, S. P., Pitaievski, L. P.: Theory of Solitons. The Inverse Problem Method. Moscow, Nauka 1980 [Russian]; English transl.: New York, Plenum 1984

[ZS 1971] Zhiber, A. V., Shabat, A. B.: Klein-Gordon equations with a nontrivial group. Dokl. Akad. Nauk SSSR *247*, 1103–1106 (1971) [Russian]

[ZS 1974] Zakharov, V. E., Shabat, A. B.: Integration of the nonlinear equations of mathematical physics by the method of the inverse scattering problem. I. Funk. Anal. Prilož. *8* (3), 43–53 (1974) [Russian]; English transl. in Funct. Anal. Appl. *8*, 226–235 (1974)

[ZS 1979] Zakharov, V. E., Shabat, A. B.: Integration of the nonlinear equations of mathematical physics by the method of the inverse scattering problem. II. Funk. Anal. Prilož. *13* (3), 13–22 (1979) [Russian]; English transl. in Funct. Anal. Appl. *13*, 166–174 (1979)

[ZT 1979] Zakharov, V. E., Takhtajan, L. A.: Equivalence of the nonlinear Schrödinger equation and the Heisenberg ferromagnet equation. Teor. Mat. Fiz. *38* (1), 26–35 (1979) [Russian]; English transl. in Theor. Math. Phys. *38*, 17–23 (1979)

[ZTF 1974] Zakharov, V. E., Takhtajan, L. A., Faddeev, L. D.: A complete description of solution of the "sine-Gordon" equation. Dokl. Akad. Nauk SSSR *219*, 1334–1337 (1974) [Russian]; English transl. in Sov. Phys. Dokl. *19*, 824–826 (1975)

Chapter II
Fundamental Continuous Models

We shall give a complete list of results pertaining to two fundamental continuous models, the HM and SG models. For the rapidly decreasing boundary conditions we shall analyze the mapping $\mathscr{F}$ from the initial data of the auxiliary linear problem to the transition coefficients and the discrete spectrum, and show how to solve the inverse problem, i.e. how to construct the mapping $\mathscr{F}^{-1}$. We shall see that these models allow an r-matrix approach, which will enable us to show that $\mathscr{F}$ is a canonical transformation to variables of action-angle type. It will thus be proved that the HM and SG models are completely integrable Hamiltonian systems. We shall also present a Hamiltonian interpretation of the change to light-cone coordinates in the SG model. To conclude this chapter, we shall explain that in some sense the LL model is the most universal integrable system with two-dimensional auxiliary space.

§ 1. The Auxiliary Linear Problem for the HM Model

The problem mentioned in the title reads (see § I.1)

$$\frac{dF}{dx}=\frac{\lambda}{2i}S(x)F, \tag{1.1}$$

where $S(x)$ is a Hermitian traceless 2×2 matrix satisfying

$$S^2(x)=I. \tag{1.2}$$

We shall only consider the rapidly decreasing case

$$\lim_{|x|\to\infty} S(x)=\sigma_3, \tag{1.3}$$

where the boundary values are attained in the sense of Schwartz.

It was shown in § I.4 that HM and SG are gauge equivalent models. Specifically, by means of the gauge transformation

$$\tilde{F}(x,\lambda)=\Omega(x)F(x,\lambda), \tag{1.4}$$

the auxiliary linear problem (1.1) is reduced to a form typical for the NS model,

$$\frac{d\tilde{F}}{dx}=\left(\frac{\lambda}{2i}\sigma_3+U_0(x)\right)\tilde{F}, \tag{1.5}$$

where

$$U_0(x)=\frac{d\Omega}{dx}(x)\Omega^{-1}(x). \tag{1.6}$$

The unitary matrix $\Omega(x)$ is determined from

$$S(x)=\Omega^{-1}(x)\sigma_3\Omega(x) \tag{1.7}$$

with the condition that the diagonal of $U_0(x)$ in (1.6) be zero,

$$U_0(x)=i\begin{pmatrix} 0 & \bar{\psi}(x) \\ \psi(x) & 0 \end{pmatrix}, \tag{1.8}$$

which corresponds to the NS model in the rapidly decreasing case for $\varkappa=-1$ (see § I.4).

Thus, information concerning the auxiliary linear problem (1.1) can be derived from the analysis of problem (1.5) carried out in Chapter I of Part I. Nevertheless, as the SG model is of importance in its own right, we shall give an independent treatment of (1.1). In doing so we shall, of course, compare the corresponding results. In that case the quantities in question will be labelled by the symbols NS or SG respectively.

1. The transition matrix and Jost solutions

The transition matrix $T(x,y,\lambda)$ is defined to be a solution of the differential equation (1.1) with the initial condition

$$T(x,y,\lambda)|_{x=y}=I. \tag{1.9}$$

It can be expressed as

$$T(x,y,\lambda)=\widehat{\exp}\,\frac{\lambda}{2i}\int_y^x S(z)\,dz, \tag{1.10}$$

so that

$$T(x, y, \lambda)|_{\lambda=0} = I. \tag{1.11}$$

The matrix $T(x, y, \lambda)$ is unimodular and is an entire matrix-valued function of λ. The relation

$$\bar{S}(x) = -\sigma_2 S(x) \sigma_2 \tag{1.12}$$

implies the involution property

$$T(x, y, \lambda) = \sigma_2 \bar{T}(x, y, \bar{\lambda}) \sigma_2. \tag{1.13}$$

We have the relationship

$$T^{\mathrm{HM}}(x, y, \lambda) = \Omega^{-1}(x) T^{\mathrm{NS}}(x, y, \lambda) \Omega(y). \tag{1.14}$$

As $|x| \to \infty$, the auxiliary linear problem (1.1) turns into the differential equation

$$\frac{dE}{dx} = \frac{\lambda}{2i} \sigma_3 E, \tag{1.15}$$

which can be solved explicitly,

$$E(x, \lambda) = e^{\frac{\lambda x}{2i} \sigma_3}. \tag{1.16}$$

The Jost solutions $T_{\pm}(x, \lambda)$ for real λ are defined to be the limits

$$T_{\pm}(x, \lambda) = \lim_{y \to \pm\infty} T(x, y, \lambda) E(y, \lambda). \tag{1.17}$$

The matrices $T_{\pm}(x, \lambda)$ are unimodular and satisfy the differential equation (1.1), the involution relation

$$\overline{T_{\pm}}(x, \lambda) = \sigma_2 T_{\pm}(x, \lambda) \sigma_2, \tag{1.18}$$

and

$$T_{\pm}(x, \lambda)|_{\lambda=0} = I. \tag{1.19}$$

They have the asymptotic behaviour

$$T_{\pm}(x, \lambda) = E(x, \lambda) + o(1), \tag{1.20}$$

as $x \to \pm\infty$, respectively.

An alternative definition of the Jost solutions can be given through the integral equations

$$T_-(x,\lambda) = E(x,\lambda) + \frac{\lambda}{2i}\int_{-\infty}^{x} E(x-y,\lambda)(S(y)-\sigma_3)\,T_-(y,\lambda)\,dy \tag{1.21}$$

and

$$T_+(x,\lambda) = E(x,\lambda) - \frac{\lambda}{2i}\int_{x}^{\infty} E(x-y,\lambda)(S(y)-\sigma_3)\,T_+(y,\lambda)\,dy. \tag{1.22}$$

For λ real, these are Volterra integral equations, and so their iterations are absolutely convergent. The analysis of the iterations shows that $T_\pm(x,\lambda)$ can be represented as

$$T_-(x,\lambda) = E(x,\lambda) + \frac{\lambda}{2i}\int_{-\infty}^{x} \Gamma_-(x,y)\,E(y,\lambda)\,dy \tag{1.23}$$

and

$$T_+(x,\lambda) = E(x,\lambda) + \frac{\lambda}{2i}\int_{x}^{\infty} \Gamma_+(x,y)\,E(y,\lambda)\,dy. \tag{1.24}$$

Inserting these integral representations into (1.1) we find that the matrices $\Gamma_\pm(x,y)$ satisfy the Goursat problem – the partial differential equation

$$\frac{\partial \Gamma_\pm}{\partial x}(x,y) + S(x)\,\frac{\partial \Gamma_\pm}{\partial y}\,\sigma_3 = 0 \tag{1.25}$$

for $\pm(y-x)>0$ with the boundary conditions

$$\lim_{y\to\pm\infty} \Gamma_\pm(x,y) = 0, \tag{1.26}$$

$$\mp(\Gamma_\pm(x,x) - S(x)\,\Gamma_\pm(x,x)\,\sigma_3) = S(x) - \sigma_3. \tag{1.27}$$

We have the relations

$$T_-^{\mathrm{HM}}(x,\lambda) = \Omega^{-1}(x)\,T_-^{\mathrm{NS}}(x,\lambda), \tag{1.28}$$

$$T_+^{\mathrm{HM}}(x,\lambda) = \Omega^{-1}(x)\,T_+^{\mathrm{NS}}(x,\lambda)\,\Omega_0 \tag{1.29}$$

and

$$\Gamma_-^{NS}(x,y) = -\Omega(x)\frac{\partial \Gamma_-^{HM}}{\partial y}(x,y)\sigma_3, \tag{1.30}$$

$$\Gamma_+^{NS}(x,y) = -\Omega(x)\frac{\partial \Gamma_+^{HM}}{\partial y}(x,y)\Omega_0^{-1}\sigma_3. \tag{1.31}$$

It is assumed that $\Omega(x)$ is normalized as follows:

$$\lim_{x\to-\infty} \Omega(x) = I. \tag{1.32}$$

The latter requirement combined with (1.7) and the condition that $U_0(x)$ has zero diagonal determine $\Omega(x)$ uniquely. We also have

$$\lim_{x\to+\infty} \Omega(x) = \Omega_0, \tag{1.33}$$

where Ω_0 is a unitary diagonal matrix. Relations (1.30)–(1.31) result from comparing the integral representations (1.23)–(1.24) with (I.5.10), (I.5.16) of Part I and using the formulae

$$\Gamma_-(x,x) = (\Omega^{-1}(x) - I)\sigma_3, \tag{1.34}$$

$$\Gamma_+(x,x) = (I - \Omega^{-1}(x)\Omega_0)\sigma_3, \tag{1.35}$$

which are derived by comparing the limits of both sides of (1.28)–(1.29) as $|\lambda|\to\infty$.

We point out that relations (1.34)–(1.35) agree with the boundary conditions (1.26)–(1.27); also, relations (1.30)–(1.31) and the differential equation (1.25) agree with the differential equation (I.8.15) of Part I for the kernels $\Gamma_\pm^{NS}(x,y)$, where one must set $U_\pm = 0$.

By comparing (1.19) and (1.28) we find an expression for $\Omega(x)$

$$\Omega(x) = T_-^{NS}(x,\lambda)|_{\lambda=0}. \tag{1.36}$$

The integral representations (1.23)–(1.24) and formulae (1.34)–(1.35) imply the analyticity properties of the columns, $T_\pm^{(l)}(x,\lambda)$, $l=1,2$, of the Jost solutions. The columns $T_-^{(1)}(x,\lambda)$ and $T_+^{(2)}(x,\lambda)$ can be analytically continued into the upper half-plane of the variable λ whereas $T_+^{(1)}(x,\lambda)$ and $T_-^{(2)}(x,\lambda)$ can be analytically continued into the lower half-plane. They have the following asymptotic behaviour, as $|\lambda|\to\infty$,

$$e^{\frac{i\lambda x}{2}}\, T_-^{(1)}(x,\lambda) = \Omega^{-1}(x)\begin{pmatrix}1\\0\end{pmatrix} + O\left(\frac{1}{|\lambda|}\right), \tag{1.37}$$

$$e^{-\frac{i\lambda x}{2}} T_+^{(2)}(x,\lambda) = \Omega^{-1}(x)\Omega_0 \begin{pmatrix} 0 \\ 1 \end{pmatrix} + O\left(\frac{1}{|\lambda|}\right) \tag{1.38}$$

for $\operatorname{Im}\lambda \geqslant 0$ and

$$e^{\frac{i\lambda x}{2}} T_+^{(1)}(x,\lambda) = \Omega^{-1}(x)\Omega_0 \begin{pmatrix} 1 \\ 0 \end{pmatrix} + O\left(\frac{1}{|\lambda|}\right), \tag{1.39}$$

$$e^{-\frac{i\lambda x}{2}} T_-^{(2)}(x,\lambda) = \Omega^{-1}(x) \begin{pmatrix} 0 \\ 1 \end{pmatrix} + O\left(\frac{1}{|\lambda|}\right) \tag{1.40}$$

for $\operatorname{Im}\lambda \leqslant 0$.

2. The reduced monodromy matrix and transition coefficients

The reduced monodromy matrix $T(\lambda)$ is defined for real λ as the ratio of two Jost solutions,

$$T(\lambda) = T_+^{-1}(x,\lambda)\, T_-(x,\lambda) \tag{1.41}$$

and can be expressed as a limit,

$$T(\lambda) = \lim_{\substack{x \to +\infty \\ y \to -\infty}} E(-x,\lambda)\, T(x,y,\lambda)\, E(y,\lambda). \tag{1.42}$$

The matrix $T(\lambda)$ is unimodular, obeys the involution

$$\bar{T}(\lambda) = \sigma_2 T(\lambda) \sigma_2 \tag{1.43}$$

and satisfies

$$T(\lambda)|_{\lambda=0} = I. \tag{1.44}$$

It can be written in the form

$$T(\lambda) = \begin{pmatrix} a(\lambda) & -\bar{b}(\lambda) \\ b(\lambda) & \bar{a}(\lambda) \end{pmatrix}, \tag{1.45}$$

where the coefficients $a(\lambda)$ and $b(\lambda)$ (transition coefficients for the continuous spectrum) satisfy the normalization relation

$$|a(\lambda)|^2 + |b(\lambda)|^2 = 1 \tag{1.46}$$

and the constraints

$$a(0)=1, \quad b(0)=0. \tag{1.47}$$

From (1.28)-(1.29) we deduce the relation

$$T^{HM}(\lambda)=\Omega_0^{-1}T^{NS}(\lambda), \tag{1.48}$$

which gives

$$\Omega_0^{-1}=\lim_{|\lambda|\to\infty} T^{HM}(\lambda). \tag{1.49}$$

From this and (1.44), by virtue of (1.36), we find

$$T^{NS}(\lambda)|_{\lambda=0}=\Omega_0, \tag{1.50}$$

so that

$$b^{NS}(0)=0, \tag{1.51}$$

and the matrix Ω_0 has the form

$$\Omega_0=\begin{pmatrix} a^{NS}(0) & 0 \\ 0 & \bar{a}^{NS}(0)\end{pmatrix} \tag{1.52}$$

and is unimodular. Writing (1.48) explicitly, we have

$$a^{HM}(\lambda)=\frac{a^{NS}(\lambda)}{a^{NS}(0)}, \quad b^{HM}(\lambda)=\frac{b^{NS}(\lambda)}{\bar{a}^{NS}(0)}. \tag{1.53}$$

Thus, under the rapidly decreasing boundary conditions the gauge transformation converts the HM model into the NS model with transition coefficients satisfying an additional constraint (1.51).

The analytic properties of the transition coefficients $a(\lambda)$ and $b(\lambda)$ are similar to those for the NS model. The function $a(\lambda)$ has an analytic continuation into the upper half-plane of the variable λ with the asymptotic behaviour

$$a(\lambda)=\omega_0+O\left(\frac{1}{|\lambda|}\right), \tag{1.54}$$

as $|\lambda|\to\infty$, where ω_0 is the upper diagonal element of Ω_0^{-1}, $|\omega_0|=1$. As in the NS case, we shall adopt the condition (A) for the zeros of $a(\lambda)$, which reads that all the zeros λ_j are simple and $\operatorname{Im}\lambda_j>0$. It follows that their total number n is finite, and for real λ we have a strict inequality

$$|b(\lambda)|<1. \tag{1.55}$$

The numbers $\lambda_j, \bar{\lambda}_j, j=1, \ldots, n$, identify the discrete spectrum of the auxiliary linear problem (1.1). The associated transition coefficients $\gamma_j, \bar{\gamma}_j$ are defined by

$$T_-^{(1)}(x, \lambda_j) = \gamma_j T_+^{(2)}(x, \lambda_j), \quad j=1, \ldots, n. \tag{1.56}$$

We have the relations

$$\lambda_j^{\mathrm{HM}} = \lambda_j^{\mathrm{NS}}, \quad \gamma_j^{\mathrm{HM}} = \frac{1}{\bar{a}^{\mathrm{NS}}(0)} \gamma_j^{\mathrm{NS}}, \quad j=1, \ldots, n. \tag{1.57}$$

The coefficient $b(\lambda)$ is a Schwartz function; in general, it need not have an analytic continuation off the real line. In the case when, for some $q>0$, the matrix $S(x)$ equals its asymptotic value σ_3 for $|x| \geqslant q$, $b(\lambda)$ (as well as $a(\lambda)$) extends analytically to the whole complex plane. In that case

$$\gamma_j = b(\lambda_j), \quad j=1, \ldots, n. \tag{1.58}$$

The function $a(\lambda)$ is uniquely determined by the coefficient $b(\lambda)$ and the zeros $\lambda_1, \ldots, \lambda_n$. The corresponding dispersion relation is

$$a(\lambda) = \omega_0 \prod_{j=1}^{n} \frac{\lambda - \lambda_j}{\lambda - \bar{\lambda}_j} \exp\left\{\frac{1}{2\pi i} \int_{-\infty}^{\infty} \frac{\log(1 - |b(\mu)|^2)}{\mu - \lambda - i0} d\mu\right\}, \tag{1.59}$$

where

$$\omega_0 = \prod_{j=1}^{n} \frac{\bar{\lambda}_j}{\lambda_j} \exp\left\{-\frac{1}{2\pi i} \int_{-\infty}^{\infty} \frac{\log(1 - |b(\mu)|^2)}{\mu} d\mu\right\}. \tag{1.60}$$

The integral on the right hand side of the last formula is absolutely convergent in view of (1.47).

The above discussion can be interpreted as a description of the mapping

$$\mathscr{F}: (S(x)) \to (b(\lambda), \bar{b}(\lambda); \lambda_j, \bar{\lambda}_j, \gamma_j, \bar{\gamma}_j, j=1, \ldots, n). \tag{1.61}$$

In the next section we shall see that $\mathscr{F}$ has an inverse, and in § 3 this mapping will be shown to define a canonical transformation to variables of action-angle type.

3. Time evolution of the transition coefficients

We consider the evolution of the transition coefficients assuming that $S(x, t)$ satisfies the HM equation. Using the zero curvature condition from

§ I.1 and reproducing the arguments of § I.7, Part I, we can derive the evolution equations

$$\frac{\partial T_\pm}{\partial t}(x,\lambda)=V(x,\lambda)\,T_\pm(x,\lambda)-\frac{i\lambda^2}{2}\,T_\pm(x,\lambda)\sigma_3 \tag{1.62}$$

and

$$\frac{\partial T}{\partial t}(\lambda,t)=\frac{i\lambda^2}{2}[\sigma_3,\,T(\lambda,t)], \tag{1.63}$$

which lead to the following dependence of the transition coefficients on time t:

$$a(\lambda,t)=a(\lambda,0),\qquad b(\lambda,t)=e^{-i\lambda^2 t}\,b(\lambda,0) \tag{1.64}$$

and

$$\lambda_j(t)=\lambda_j(0),\qquad \gamma_j(t)=e^{-i\lambda_j^2 t}\gamma_j(0),\qquad j=1,\ldots,n. \tag{1.65}$$

These formulae coincide with the expressions for the dynamics of the NS model. Obviously, they are consistent with (1.53) and (1.57).

As in the NS model, the coefficient $a(\lambda)$ is a generating function for integrals of the motion. We conclude this section with a method for selecting a family of local integrals. As before, the latter are understood to be functionals of the form

$$F=\int_{-\infty}^{\infty} f(x)\,dx, \tag{1.66}$$

where the density $f(x)$ is a polynomial in the matrix elements of $S(x)$ and their derivatives with respect to x.

4. Local integrals of the motion

We start with the asymptotic expansion of the transition matrix $T(x,y,\lambda)$ as $|\lambda|\to\infty$. We write it as

$$T(x,y,\lambda)=(I+W(x,\lambda))\exp Z(x,y,\lambda)(I+W(y,\lambda))^{-1}, \tag{1.67}$$

where $W(x,\lambda)$ is an off-diagonal matrix and $Z(x,y,\lambda)$ is a diagonal matrix satisfying

$$Z(x,y,\lambda)|_{x=y}=0. \tag{1.68}$$

From here on in asymptotic expansions we shall often leave out terms of the order $O(|\lambda|^{-\infty})$ as defined in § I.3, Part I. Inserting (1.67) into (1.1) and splitting the diagonal and off-diagonal parts we come down to a differential equation of Riccati type for $W(x,\lambda)$

$$\frac{dW}{dx}+i\lambda S_3\sigma_3 W+\frac{i\lambda}{2}(S_1\sigma_1+S_2\sigma_2)-\frac{i\lambda}{2}W(S_1\sigma_1+S_2\sigma_2)W=0. \quad (1.69)$$

The matrix $Z(x,y,\lambda)$ is given by

$$Z(x,y,\lambda)=-\frac{i\lambda}{2}\int_y^x (S_3(x')\sigma_3+(S_1(x')\sigma_1+S_2(x')\sigma_2)W(x',\lambda))\,dx'. \quad (1.70)$$

The difference between our case and the NS model is that the asymptotic expansion in powers of λ^{-1} for $W(x,\lambda)$ contains a constant term. In fact, (1.14) shows that this term comes from the off-diagonal part of $\Omega(x)$. The expansion for $Z(x,y,\lambda)$ begins with the term $\frac{\lambda}{2i}(x-y)\sigma_3$ and also contains a constant term associated with the diagonal parts of $\Omega(x)$ and $\Omega(y)$.

Thus the asymptotic expansion of $W(x,\lambda)$ has the form

$$W(x,\lambda)=\sum_{n=0}^{\infty}\frac{W_n(x)}{\lambda^n}. \quad (1.71)$$

Substituting (1.71) into (1.69) we get a recursion relation for the coefficients $W_n(x)$, $n\geqslant 1$:

$$\begin{aligned} &iS_3\sigma_3 W_{n+1}-\frac{i}{2}W_0(S_1\sigma_1+S_2\sigma_2)W_{n+1}+\frac{i}{2}W_{n+1}(S_1\sigma_1+S_2\sigma_2)W_0\\ &=-\frac{dW_n}{dx}+\frac{i}{2}\sum_{k=1}^{n}W_k(S_1\sigma_1+S_2\sigma_2)W_{n+1-k}, \end{aligned} \quad (1.72)$$

with the initial term $W_0(x)$ determined from the equation

$$2S_3\sigma_3 W_0-W_0(S_1\sigma_1+S_2\sigma_2)W_0+S_1\sigma_1+S_2\sigma_2=0. \quad (1.73)$$

We introduce the diagonal matrix

$$Q=(S_1\sigma_1+S_2\sigma_2)W_0, \quad (1.74)$$

and write the above equation as

$$Q^2+2S_3\sigma_3 Q+(S_3^2-1)I=0, \qquad (1.75)$$

or

$$(Q+S_3\sigma_3)^2=I. \qquad (1.76)$$

Equation (1.76) has four solutions. The solution we need is uniquely determined from the relation

$$Q+S_3\sigma_3=\sigma_3, \qquad (1.77)$$

which accounts for the appearance of the term $\frac{\lambda}{2i}(x-y)\sigma_3$ in the asymptotic expansion for $Z(x,y,\lambda)$.

The matrix $W_0(x)$ has the form

$$W_0(x)=i\frac{S_2(x)\sigma_1-S_1(x)\sigma_2}{1+S_3(x)}, \qquad (1.78)$$

so that the recursion relation (1.72) can be written as

$$W_{n+1}=i\sigma_3\frac{dW_n}{dx}+\frac{i}{2}\sum_{k=1}^{n}W_k(S_1\sigma_2-S_2\sigma_1)W_{n+1-k}, \quad n=0,1,\ldots. \qquad (1.79)$$

The coefficients $W_n(x)$ are skew-Hermitian matrices,

$$W_n(x)=\begin{pmatrix} 0 & -\overline{w_n}(x) \\ w_n(x) & 0 \end{pmatrix}. \qquad (1.80)$$

In terms of the $w_n(x)$, the recursion relation (1.79) and the initial condition (1.78) take the form

$$w_{n+1}(x)=-i\frac{dw_n}{dx}(x)-\frac{S_1(x)-iS_2(x)}{2}\sum_{k=1}^{n}w_k(x)w_{n+1-k}(x), \quad n=0,1,\ldots \qquad (1.81)$$

and

$$w_0(x)=\frac{S_1(x)+iS_2(x)}{1+S_3(x)}. \qquad (1.82)$$

From (1.70)-(1.72) and (1.80) we obtain the asymptotic expansion for $Z(x,y,\lambda)$,

$$Z(x,y,\lambda)=\frac{\lambda}{2i}(x-y)\sigma_3+\sum_{n=0}^{\infty}\frac{Z_n(x,y)}{\lambda^n}, \tag{1.83}$$

where the matrices $Z_n(x,y)$ have the form

$$Z_n(x,y)=\begin{pmatrix} z_n(x,y) & 0 \\ 0 & -\bar{z}_n(x,y)\end{pmatrix}, \tag{1.84}$$

with

$$z_n(x,y)=-\frac{i}{2}\int_y^x (S_1(x')-iS_2(x'))\,w_{n+1}(x')\,dx'. \tag{1.85}$$

The asymptotic expansion, as $|\lambda|\to\infty$, of the reduced monodromy matrix follows by taking the limit according to (1.42). Recalling that $W(x,\lambda)$ vanishes as $|x|\to\infty$, we get

$$T(\lambda)=e^{P(\lambda)}+O(|\lambda|^{-\infty}), \tag{1.86}$$

where the diagonal matrix $P(\lambda)$ has the form

$$P(\lambda)=i\begin{pmatrix} p(\lambda) & 0 \\ 0 & -\bar{p}(\lambda)\end{pmatrix}, \tag{1.87}$$

and

$$p(\lambda)=\sum_{n=0}^{\infty}\frac{I_n}{\lambda^n} \tag{1.88}$$

with

$$I_n=\lim_{\substack{x\to+\infty\\ y\to-\infty}}\frac{1}{i}z_n(x,y)=-\frac{1}{2}\int_{-\infty}^{\infty}(S_1(x)-iS_2(x))\,w_{n+1}(x)\,dx. \tag{1.89}$$

We emphasize that the property of the matrix $P(\lambda)$ in the asymptotic expansion (1.86) to be diagonal is in accord with the fact that the coefficient $b(\lambda)$ is a Schwartz function whose contribution is of the order $O(|\lambda|^{-\infty})$. Since $T(\lambda)$ is a unimodular matrix, we have

$$\operatorname{tr}P(\lambda)=0, \tag{1.90}$$

so that the coefficients I_n are real.

We have thus established the expansion

$$\frac{1}{i}\log a(\lambda)=\sum_{n=0}^{\infty}\frac{I_n}{\lambda^n}, \tag{1.91}$$

as $|\lambda|\to\infty$, where the real-valued functionals I_n are given by (1.89), (1.81)-(1.82) and are integrals of the motion for the HM model. The momentum P and the Hamiltonian H defined in § I.1 coincide with the first two of them,

$$P=-2I_0, \quad H=-2I_1. \tag{1.92}$$

The functionals I_n have the form (1.66); their densities are rational functions of the $S_a(x)$, $a=1,2,3$, and their derivatives at x. It can be shown that the density of I_n for $n\geqslant 1$ is actually a polynomial, up to a total derivative of a Schwartz function. Hence $\frac{1}{i}\log a(\lambda)$ is indeed a generating function for the local integrals.

The functionals I_n can be expressed through the transition coefficients and discrete spectrum of the auxiliary linear problem (1.1). For that purpose expansion (1.91) should be compared with the dispersion relation (1.59)-(1.60). Expanding the denominator $\frac{1}{\mu-\lambda}$ in (1.59) into a geometric progression we find

$$I_0=\arg\omega_0=\frac{1}{2\pi}\int_{-\infty}^{\infty}\frac{\log(1-|b(\lambda)|^2)}{\lambda}\,d\lambda-2\sum_{j=1}^{n}\arg\lambda_j \tag{1.93}$$

and

$$I_l=\frac{1}{2\pi}\int_{-\infty}^{\infty}\lambda^{l-1}\log(1-|b(\lambda)|^2)\,d\lambda+\sum_{j=1}^{n}\frac{\bar{\lambda}_j^l-\lambda_j^l}{il}, \quad l=1,2,\ldots, \tag{1.94}$$

which are the trace identities for the HM model.

These identities are in agreement with the equivalence between the HM and NS models,

$$I_l^{\mathrm{HM}}=-I_l^{\mathrm{NS}}, \quad l=1,2,\ldots. \tag{1.95}$$

We observed in § I.1 that for the periodic boundary conditions the HM model has additional integrals of the motion, M_a, $a=1,2,3$, which play the role of components of the total spin that generate the action of the rotation group. These integrals are not contained in the family $\{I_n\}_{n=0}^{\infty}$. Moreover, the functionals

$$M_a = \int_{-\infty}^{\infty} S_a(x)\,dx, \qquad a = 1, 2, \tag{1.96}$$

are not even admissible since the associated Hamiltonian flows violate the boundary conditions of rapid decrease (1.3) (cf. the regularized charge functional in § I.1, Part I). The admissible (regularized) functional is given by

$$M_3 = \int_{-\infty}^{\infty} (S_3(x) - 1)\,dx. \tag{1.97}$$

We shall express it in terms of the transition coefficients and the discrete spectrum.

For this differentiate (1.1) with respect to λ to obtain

$$\frac{\partial \dot{T}}{\partial x} = \frac{\lambda}{2i} S\dot{T} + \frac{1}{2i} ST, \tag{1.98}$$

the dot indicating the derivative. For the matrix

$$\dot{T}(x, y) = \dot{T}(x, y, \lambda)|_{\lambda = 0} \tag{1.99}$$

we have

$$\frac{\partial \dot{T}}{\partial x}(x, y) = \frac{1}{2i} S(x), \qquad \dot{T}(x, y)|_{x = y} = 0, \tag{1.100}$$

so that

$$\dot{T}(x, y) = \frac{1}{2i} \int_{y}^{x} S(x')\,dx', \tag{1.101}$$

and taking the limit according to (1.42) we find

$$\dot{T}(\lambda)|_{\lambda = 0} = \frac{1}{2i} \int_{\infty}^{\infty} (S(x) - \sigma_3)\,dx. \tag{1.102}$$

We then derive the required expressions,

$$M_3 = 2i\dot{a}(0) = \frac{1}{\pi} \int_{-\infty}^{\infty} \frac{\log(1 - |b(\lambda)|^2)}{\lambda^2}\,d\lambda + 2i \sum_{j=1}^{n} \frac{\lambda_j - \bar{\lambda}_j}{|\lambda_j|^2} \tag{1.103}$$

and

$$M_+ = \frac{M_1 + iM_2}{2} = 2i\dot{b}(0), \quad M_- = \frac{M_1 - iM_2}{2} = -2i\dot{\bar{b}}(0). \tag{1.104}$$

The convergence of the integral in (1.103) in the neighbourhood of $\lambda = 0$ is ensured by (1.47).

In § 3, starting from (1.104), we shall see once again that the functionals $M_\pm$ are inadmissible.

This completes our analysis of the auxiliary linear problem and the mapping $\mathscr{F}$ for the HM model.

§ 2. The Inverse Problem for the HM Model

Here we shall describe the mapping $\mathscr{F}^{-1}$, i.e. solve the inverse problem of reconstructing $S(x)$ from the transition coefficients and the discrete spectrum. Two methods will be developed, one based on the matrix Riemann problem, the other on the Gelfand-Levitan-Marchenko formulation. At the end of the section we shall discuss soliton dynamics for the HM model.

1. The Riemann problem

It is based on the formula relating the Jost solutions,

$$T_-(x,\lambda) = T_+(x,\lambda)\,T(\lambda), \tag{2.1}$$

which can be written as

$$F_-(x,\lambda) = F_+(x,\lambda)\,G(\lambda), \tag{2.2}$$

where the matrices $F_\pm(x,\lambda)$ are composed of the columns of the solutions $T_\pm(x,\lambda)$,

$$F_+(x,\lambda) = \frac{1}{a(\lambda)}\left(T_-^{(1)}(x,\lambda),\, T_+^{(2)}(x,\lambda)\right), \tag{2.3}$$

$$F_-(x,\lambda) = \left(T_+^{(1)}(x,\lambda),\, T_-^{(2)}(x,\lambda)\right), \tag{2.4}$$

and $G(\lambda)$ has the form

$$G(\lambda) = \begin{pmatrix} 1 & -\bar{b}(\lambda) \\ -b(\lambda) & 1 \end{pmatrix}. \tag{2.5}$$

We define the matrices

$$G_-(x,\lambda)=F_-(x,\lambda)E^{-1}(x,\lambda) \tag{2.6}$$

and

$$G_+(x,\lambda)=E(x,\lambda)F_+^{-1}(x,\lambda), \tag{2.7}$$

which can be analytically continued into the lower and upper halfs of the λ-plane, respectively. Then (2.2) implies the relation underlying the Riemann problem,

$$G_+(x,\lambda)G_-(x,\lambda)=G(x,\lambda), \tag{2.8}$$

with

$$G(x,\lambda)=E(x,\lambda)G(\lambda)E^{-1}(x,\lambda). \tag{2.9}$$

We list the properties of the matrices $G(x,\lambda)$ and $G_\pm(x,\lambda)$ which follow from the discussion in § 1.

I. *The matrix $G(x,\lambda)$ is Hermitian,*

$$G^*(x,\lambda)=G(x,\lambda), \tag{2.10}$$

and satisfies

$$G(x,\lambda)|_{\lambda=0}=I, \tag{2.11}$$

$$\lim_{|\lambda|\to\infty} G(x,\lambda)=I, \tag{2.12}$$

where the limiting values are attained in the sense of Schwartz.

II. *For every x, the matrices $G_\pm(x,\lambda)$ belong to the rings $\mathfrak{R}_\pm^{(2\times 2)}$, are Hermitian conjugates of each other,*

$$G_+(x,\lambda)=G_-^*(x,\bar{\lambda}), \tag{2.13}$$

and have limits as $|\lambda|\to\infty$,

$$\lim_{|\lambda|\to\infty} G_\pm(x,\lambda)=\Omega_\pm(x), \tag{2.14}$$

where $\Omega_\pm(x)$ are unitary matrices related by

$$\Omega_+(x)=\Omega_-^*(x)=\begin{pmatrix}\omega_0 & 0\\ 0 & 1\end{pmatrix}\Omega(x), \tag{2.15}$$

so that

$$S(x)=\Omega_{+}^{-1}(x)\sigma_3\Omega_{+}(x)=\Omega_{-}(x)\sigma_3\Omega_{-}^{-1}(x). \tag{2.16}$$

III. *The matrices* $G_{\pm}(x,\lambda)$ *satisfy the relation*

$$G_{\pm}(x,\lambda)|_{\lambda=0}=I. \tag{2.17}$$

IV. *The matrices* $G_{+}(x,\lambda)$ *and* $G_{-}(x,\lambda)$ *are nondegenerate in their domains of analyticity with the exception of the points* $\lambda=\lambda_j$ and $\lambda=\bar{\lambda}_j$, *respectively, where*

$$\operatorname{Im} G_{+}(x,\lambda_j)=N_j^{(+)}(x) \tag{2.18}$$

and

$$\operatorname{Ker} G_{-}(x,\bar{\lambda}_j)=N_j^{(-)}(x), \quad j=1,\dots,n. \tag{2.19}$$

Here $N_j^{(+)}(x)$ and $N_j^{(-)}(x)$ are one-dimensional subspaces of $\mathbb{C}^2$ spanned respectively by the vectors

$$\begin{pmatrix} 1 \\ -\gamma_j e^{i\lambda_j x} \end{pmatrix} \quad \text{and} \quad \begin{pmatrix} \bar{\gamma}_j e^{-i\bar{\lambda}_j x} \\ 1 \end{pmatrix}.$$

Observe that the properties of $G_{\pm}(x,\lambda)$ for the HM and NS models differ only by the normalization conditions: these matrices are normalized to I for $\lambda=0$ or $\lambda=\infty$, respectively.

We now proceed to solve the inverse problem. *Suppose we are given functions* $b(\lambda)$, $\bar{b}(\lambda)$ *and a set of numbers* $\lambda_j, \bar{\lambda}_j, \gamma_j, \bar{\gamma}_j, j=1,\dots,n$, *with the following properties.*

I′. *The function* $b(\lambda)$ *lies in Schwartz space and satisfies the conditions*

$$b(0)=0, \quad |b(\lambda)|<1. \tag{2.20}$$

II′. *The numbers* λ_j, $\operatorname{Im}\lambda_j>0$, *are pairwise distinct and* $\gamma_j\neq 0$, $j=1,\dots,n$.

Starting from these data we construct the matrix $G(x,\lambda)$ and the set of subspaces $N_j^{(\pm)}(x)$ and consider the Riemann problem

$$G(x,\lambda)=G_{+}(x,\lambda)G_{-}(x,\lambda), \tag{2.21}$$

where the matrices $G_{\pm}(x,\lambda)$ belong to the rings $\mathfrak{R}_{\pm}^{(2\times 2)}$, are normalized to I for $\lambda=0$, and satisfy conditions (2.18)–(2.19).

Then we can make the following assertions.

I″. *The Riemann problem in question has a unique solution.*

II″. *The matrices* $F_\pm(x,\lambda)$ *constructed from* $G_\pm(x,\lambda)$ *according to (2.6)–(2.7) satisfy the auxiliary linear problem*

$$\frac{dF_\pm(x,\lambda)}{dx}=\frac{\lambda}{2i}S(x)F_\pm(x,\lambda), \tag{2.22}$$

where $S(x)$ *is given by (2.14), (2.16).*

III″. $S(x)$ *is a Hermitian traceless matrix satisfying* $S^2(x)=I$ *and*

$$\lim_{|x|\to\infty} S(x)=\sigma_3, \tag{2.23}$$

where the limiting values are approached in the sense of Schwartz.

IV″. *The functions* $a(\lambda)$ *and* $b(\lambda)$ *with* $a(\lambda)$ *given by (1.59)–(1.60) play the role of transition coefficients for the auxiliary linear problem (2.22); its discrete spectrum consists of the eigenvalues* $\lambda_1,\ldots,\lambda_n;\ \bar{\lambda}_1,\ldots,\bar{\lambda}_n$ *with the transition coefficients* $\gamma_1,\ldots,\gamma_n;\ \bar{\gamma}_1,\ldots,\bar{\gamma}_n$. *The solutions* $G_\pm(x,\lambda)$ *are composed of the Jost solutions* $T_\pm(x,\lambda)$ *of the auxiliary linear problem according to (2.3)–(2.4) and (2.6)–(2.7).*

Let us comment on the proof of these assertions.

The uniqueness theorem for the Riemann problem follows in the usual way from the Liouville theorem and the normalization condition (2.17) (see the corresponding argument in § II.2, Part I). Since $G(x,\lambda)$ is a Hermitian matrix, this implies (2.13).

To prove the existence theorem, it suffices to use the transformation

$$\tilde{G}_+(x,\lambda)=G_+(x,\lambda)\Omega_+^{-1}(x), \tag{2.24}$$

$$\tilde{G}_-(x,\lambda)=\Omega_-^{-1}(x)G_-(x,\lambda), \tag{2.25}$$

where $\Omega_\pm(x)$ are the limiting values of the matrices $G_\pm(x,\lambda)$, as $|\lambda|\to\infty$. By (2.13), $\Omega_\pm(x)$ satisfy

$$\Omega_+(x)=\Omega_-^*(x) \tag{2.26}$$

and are unitary matrices. Hence, $\tilde{G}_\pm(x,\lambda)$ satisfy equation (2.21) and conditions (2.18)–(2.19), as before. This transformation, therefore, reduces the Riemann problem with the normalization to unity at $\lambda=0$ to that with the normalization to unity at $\lambda=\infty$. The existence theorem for the latter problem was proved in § II.2, Part I. The inverse transformation is given by

$$G_+(x,\lambda)=\tilde{G}_+(x,\lambda)\tilde{G}_+^{-1}(x,0), \tag{2.27}$$

$$G_-(x,\lambda)=\tilde{G}_-^{-1}(x,0)\tilde{G}_-(x,\lambda). \tag{2.28}$$

To derive the differential equation of item II″, we write the Riemann problem (2.21) in the form

$$F_-(x,\lambda)=F_+(x,\lambda)\,G(\lambda) \tag{2.29}$$

and differentiate this with respect to x writing it down as

$$U(x,\lambda)=\frac{\partial F_+(x,\lambda)}{\partial x}F_+^{-1}(x,\lambda)=\frac{\partial F_-(x,\lambda)}{\partial x}F_-^{-1}(x,\lambda). \tag{2.30}$$

As in § II.2, Part I, we see that $U(x,\lambda)$ is an entire function of λ. Using that $E(x,\lambda)F_+^{-1}(x,\lambda)$ and $F_-(x,\lambda)E^{-1}(x,\lambda)$ belong to the rings $\mathfrak{R}_\pm^{(2\times 2)}$, we exploit the asymptotic formulae (2.14) and the Liouville theorem to conclude that

$$U(x,\lambda)=\frac{\lambda}{2i}S(x)+C(x), \tag{2.31}$$

where the matrix $S(x)$ is given by (2.16). From (2.17) it follows that

$$C(x)=0 \tag{2.32}$$

so that we recover (2.22). The Hermiticity of $S(x)$ follows from the unitarity of $\Omega_\pm(x)$.

For the proof of the remaining statements of items III″–IV″, it is enough to make use of the relationship between the Riemann problems for the HM and NS models,

$$G^{\mathrm{NS}}(\lambda)=\begin{pmatrix}\bar\omega_0 & 0\\ 0 & 1\end{pmatrix}G^{\mathrm{HM}}(\lambda)\begin{pmatrix}\omega_0 & 0\\ 0 & 1\end{pmatrix}, \tag{2.33}$$

and

$$G_+^{\mathrm{NS}}(x,\lambda)=\begin{pmatrix}\bar\omega_0 & 0\\ 0 & 1\end{pmatrix}G_+^{\mathrm{HM}}(x,\lambda)\,\Omega_+^{-1}(x)\begin{pmatrix}\omega_0 & 0\\ 0 & 1\end{pmatrix}, \tag{2.34}$$

$$G_-^{\mathrm{NS}}(x,\lambda)=\begin{pmatrix}\bar\omega_0 & 0\\ 0 & 1\end{pmatrix}\Omega_-^{-1}(x)\,G_-^{\mathrm{HM}}(x,\lambda)\begin{pmatrix}\omega_0 & 0\\ 0 & 1\end{pmatrix}, \tag{2.35}$$

with

$$\omega_0=\bar a^{\mathrm{NS}}(0),\qquad |\omega_0|=1. \tag{2.36}$$

To conclude the discussion of the general properties of the Riemann problem, we note that, as in the case of the NS model, time evolution of the transition coefficients leads to the zero curvature representation for the HM model. This fact proves that if the transition coefficients depend on time according to (1.64)–(1.65), the matrix $S(x, t)$ constructed from them satisfies the HM equation.

2. The Gelfand-Levitan-Marchenko formulation

This formulation is also based on (2.1) which is now restated as

$$\frac{1}{\lambda a(\lambda)} T_{-}^{(1)}(x, \lambda) = \frac{1}{\lambda} T_{+}^{(1)}(x, \lambda) + \frac{r(\lambda)}{\lambda} T_{+}^{(2)}(x, \lambda) \tag{2.37}$$

and

$$\frac{1}{\lambda a(\lambda)} T_{+}^{(2)}(x, \lambda) = \frac{\tilde{r}(\lambda)}{\lambda} T_{-}^{(1)}(x, \lambda) + \frac{1}{\lambda} T_{-}^{(2)}(x, \lambda), \tag{2.38}$$

where

$$r(\lambda) = \frac{b(\lambda)}{a(\lambda)}, \qquad \tilde{r}(\lambda) = \frac{\bar{b}(\lambda)}{a(\lambda)}. \tag{2.39}$$

Performing the Fourier transform in these relations and using the integral representations (1.23)–(1.24), the analyticity properties of the Jost solutions, and the involutions for the kernels $\Gamma_{\pm}(x, y)$,

$$\overline{\Gamma_{\pm}}(x, y) = -\sigma_2 \Gamma_{\pm}(x, y) \sigma_2, \tag{2.40}$$

we derive the Gelfand-Levitan-Marchenko integral equations from the right

$$\Gamma_{+}(x, y) + \sigma_3 K(x+y) + \int_x^{\infty} \Gamma_{+}(x, z) K'(z+y)\, dz = 0 \tag{2.41}$$

and from the left

$$\Gamma_{-}(x, y) + \sigma_3 \tilde{K}(x+y) + \int_{-\infty}^{x} \Gamma_{-}(x, z) \tilde{K}'(z+y)\, dz = 0. \tag{2.42}$$

The prime here indicates the derivative with respect to the argument, $K(x)$ and $\tilde{K}(x)$ are off-diagonal matrices of the form

$$K(x) = \begin{pmatrix} 0 & -\bar{k}(x) \\ k(x) & 0 \end{pmatrix}, \qquad \tilde{K}(x) = \begin{pmatrix} 0 & \tilde{k}(x) \\ -\bar{\tilde{k}}(x) & 0 \end{pmatrix}, \tag{2.43}$$

where

$$k(x)=\frac{1}{2\pi i}\int_{-\infty}^{\infty}\frac{r(\lambda)}{\lambda}e^{\frac{i\lambda x}{2}}d\lambda-\sum_{j=1}^{n}\frac{m_j}{\lambda_j}e^{\frac{i\lambda_j x}{2}}, \tag{2.44}$$

$$\tilde{k}(x)=-\frac{1}{2\pi i}\int_{-\infty}^{\infty}\frac{\tilde{r}(\lambda)}{\lambda}e^{-\frac{i\lambda x}{2}}d\lambda+\sum_{j=1}^{n}\frac{\tilde{m}_j}{\lambda_j}e^{-\frac{i\lambda_j x}{2}}, \tag{2.45}$$

with

$$m_j=\frac{\gamma_j}{\dot{a}(\lambda_j)}, \quad \tilde{m}_j=\frac{1}{\gamma_j\dot{a}(\lambda_j)}, \quad j=1,\dots,n, \tag{2.46}$$

the dot indicating the derivative with respect to λ. The convergence of the integrals in the viscinity of $\lambda=0$ in these formulae is ensured by (2.20).

Notice that as the functions $\Omega_{\pm}(x)$ were already utilized in Subsection I, we use the notation $K(x)$, $\tilde{K}(x)$ for the kernels of the Gelfand-Levitan-Marchenko equations.

We have the relations

$$\frac{dK^{\mathrm{HM}}(x)}{dx}=\Omega_0^{-1}K^{\mathrm{NS}}(x)\Omega_0 \tag{2.47}$$

and

$$\frac{d\tilde{K}^{\mathrm{HM}}(x)}{dx}=\tilde{K}^{\mathrm{NS}}(x), \tag{2.48}$$

where, of course, the matrices $K^{\mathrm{NS}}(x)$ and $\tilde{K}^{\mathrm{NS}}(x)$ coincide with $\Omega(x)$ and $\tilde{\Omega}(x)$ given by (II.4.18) and (II.4.22), Part I.

The matrix $S(x)$ can be expressed in terms of the kernels $\Gamma_{\pm}(x,y)$ as

$$S(x)=B_{\pm}(x)\sigma_3 B_{\pm}^{-1}(x), \tag{2.49}$$

with

$$B_{\pm}(x)=I\mp\Gamma_{\pm}(x,x)\sigma_3. \tag{2.50}$$

We shall now outline a method for solving the inverse problem.

The data prescribed are the functions $r(\lambda)$, $\tilde{r}(\lambda)$ and the set of quantities $\{\lambda_j;\, m_j, \tilde{m}_j, j=1,\dots,n\}$ subject to the following conditions.

I. *$r(\lambda)$, $\tilde{r}(\lambda)$ are Schwartz functions satisfying*

$$r(0)=\tilde{r}(0)=0 \tag{2.51}$$

with the relation

$$|r(\lambda)| = |\tilde{r}(\lambda)|. \tag{2.52}$$

II. *The relation*

$$\frac{\tilde{r}(\lambda)}{\bar{r}(\lambda)} = \frac{\bar{a}(\lambda)}{a(\lambda)} \tag{2.53}$$

holds, where $a(\lambda)$ *is given by*

$$a(\lambda) = \omega_0 \prod_{j=1}^{n} \frac{\lambda - \lambda_j}{\lambda - \bar{\lambda}_j} \exp\left\{\frac{1}{2\pi i} \int_{-\infty}^{\infty} \frac{\log(1+|r(\mu)|^2)}{\lambda - \mu + i0} \, d\mu\right\} \tag{2.54}$$

and

$$\omega_0 = \prod_{j=1}^{n} \frac{\bar{\lambda}_j}{\lambda_j} \exp\left\{\frac{1}{2\pi i} \int_{-\infty}^{\infty} \frac{\log(1+|r(\lambda)|^2)}{\lambda} \, d\lambda\right\}. \tag{2.55}$$

III. *The numbers* λ_j *are pairwise distinct and satisfy* $\operatorname{Im}\lambda_j > 0$; *the quantities* m_j *and* $\tilde{m}_j$ *satisfy the relations*

$$m_j \tilde{m}_j = \frac{1}{\dot{a}^2(\lambda_j)}, \qquad j = 1, \ldots, n. \tag{2.56}$$

Starting from these data we construct the kernels $K(x)$ and $\tilde{K}(x)$ and consider the integral equations (2.41)–(2.42). We claim that

I′. *Equations (2.41) and (2.42) are uniquely solvable in the spaces* $L_1^{(2\times 2)}(x, \infty)$ *and* $L_1^{(2\times 2)}(-\infty, x)$, *respectively. Their matrix-valued solutions* $\Gamma_\pm(x, y)$ *satisfy the involution (2.40) and are functions of Schwartz type as* $x, y \to \pm\infty$.

II′. *The matrices* $T_\pm(x, \lambda)$ *constructed from* $\Gamma_\pm(x, y)$ *as prescribed by (1.23)–(1.24) satisfy the differential equations*

$$\frac{d}{dx} T_\pm(x, \lambda) = \frac{\lambda}{2i} S_\pm(x) T_\pm(x, \lambda), \tag{2.57}$$

where the matrices $S_\pm(x)$ *are given by (2.49)–(2.50).*

III′. $S_\pm(x)$ *are Hermitian matrices satisfying*

$$\operatorname{tr} S_\pm(x) = 0, \qquad S_\pm^2(x) = I \tag{2.58}$$

and

$$\lim_{x \to \pm\infty} S_\pm(x) = \sigma_3, \tag{2.59}$$

where the limiting values are approached in the sense of Schwartz.

IV′. *There is the consistency relation*

$$S_+(x)=S_-(x)=S(x). \tag{2.60}$$

V′. *The functions $a(\lambda)$ and $b(\lambda)=a(\lambda)r(\lambda)$ are the transition coefficients for the auxiliary linear problem (2.57). The discrete spectrum of the problem consists of the eigenvalues $\lambda_j, \bar{\lambda}_j$ with transition coefficients $\gamma_j, \bar{\gamma}_j$ where $\gamma_j=m_j\dot{a}(\lambda_j)$, $j=1,\ldots,n$.*

The proof of these assertions follows the pattern of § II.7, Part I. We shall therefore content ourselves with proving the statement of item II′ and the unitarity property of $B_\pm(x)$, which are specific features of the HM model.

Consider, for definiteness, equation (2.42) and differentiate it with respect to x. Using (2.47) and (2.50) we find

$$\frac{\partial \Gamma_-^{\mathrm{HM}}(x,y)}{\partial x}+B_-(x)\sigma_3\tilde{K}^{\mathrm{NS}}(x+y)+\int_{-\infty}^{x}\frac{\partial \Gamma_-^{\mathrm{HM}}(x,z)}{\partial x}\tilde{K}^{\mathrm{NS}}(z+y)\,dz=0. \tag{2.61}$$

By comparing (2.61) with the equation for the kernel $\Gamma_-^{\mathrm{NS}}(x,y)$,

$$\Gamma_-^{\mathrm{NS}}(x,y)+\tilde{K}^{\mathrm{NS}}(x+y)+\int_{-\infty}^{x}\Gamma_-^{\mathrm{NS}}(x,z)\tilde{K}^{\mathrm{NS}}(z+y)\,dz=0 \tag{2.62}$$

(see § II.4, Part I) we conclude that

$$\frac{\partial \Gamma_-^{\mathrm{HM}}(x,y)}{\partial x}=B_-(x)\sigma_3\Gamma_-^{\mathrm{NS}}(x,y). \tag{2.63}$$

Differentiating (2.42) with respect to y and integrating by parts in the resulting equation we find

$$\frac{\partial \Gamma_-^{\mathrm{HM}}(x,y)}{\partial y}\sigma_3-B_-(x)\tilde{K}^{\mathrm{NS}}(x+y)+\int_{-\infty}^{x}\frac{\partial \Gamma_-^{\mathrm{HM}}(x,z)}{\partial z}\sigma_3\tilde{K}^{\mathrm{NS}}(z+y)\,dz=0. \tag{2.64}$$

Comparing this with (2.62) we get

$$\frac{\partial \Gamma_-^{\mathrm{HM}}(x,y)}{\partial y}\sigma_3=-B_-(x)\Gamma_-^{\mathrm{NS}}(x,y). \tag{2.65}$$

Thus the matrix $\Gamma_-^{\mathrm{HM}}(x,y)$ satisfies an overdetermined system of differential equation, (2.63) and (2.65). The compatibility condition for this system is

$$B_-(x)\sigma_3 \frac{\partial \Gamma_-^{\mathrm{NS}}(x,y)}{\partial y} = -\frac{dB_-(x)}{dx}\Gamma_-^{\mathrm{NS}}(x,y)\sigma_3 - B_-(x)\frac{\partial \Gamma_-^{\mathrm{NS}}(x,y)}{\partial x}\sigma_3 . \tag{2.66}$$

Let us compare it with the equation for $\Gamma_-^{\mathrm{NS}}(x,y)$

$$\frac{\partial \Gamma_-^{\mathrm{NS}}(x,y)}{\partial x} + \sigma_3 \frac{\partial \Gamma_-^{\mathrm{NS}}(x,y)}{\partial y}\sigma_3 - U_0(x)\Gamma_-^{\mathrm{NS}}(x,y) = 0, \tag{2.67}$$

where

$$U_0(x) = \Gamma_-^{\mathrm{NS}}(x,x) - \sigma_3 \Gamma_-^{\mathrm{NS}}(x,x)\sigma_3 \tag{2.68}$$

has the form (1.8) (see § I.8 of Part I, where one must set $U_- = 0$). Since the matrix $\Gamma_-^{\mathrm{NS}}(x,y)$ cannot be degenerate for all $y \leqslant x$ (otherwise it would vanish identically by virtue of the involution $\bar{\Gamma} = \sigma_2 \Gamma \sigma_2$), a comparison of (2.66) and (2.67) yields a differential equation for $B_-(x)$,

$$\frac{dB_-(x)}{dx} = -B_-(x)\, U_0(x). \tag{2.69}$$

The unitarity of $B_-(x)$ now follows from the fact that $U_0(x)$ is skew-Hermitian, together with the normalization

$$\lim_{x \to -\infty} B_-(x) = I, \tag{2.70}$$

which is a consequence of (2.50).

For the derivation of (2.57) it is enough to observe that (2.63) and (2.65) imply the differential equation (1.25) for $\Gamma_-^{\mathrm{HM}}(x,y)$ which ensures the validity of (2.57).

The case of $\Gamma_+^{\mathrm{HM}}(x,y)$ is treated in a similar manner.

This completes the discussion of the two methods for constructing the mapping $\mathscr{F}^{-1}$.

3. Soliton solutions

Soliton solutions for the HM model occur when

$$b(\lambda) = 0. \tag{2.71}$$

For such scattering data both the Riemann problem and the Gelfand-Levitan-Marchenko equation reduce to linear algebraic equations and can be solved in closed form. We shall report the relevant results using the Riemann problem (2.21) where we now have

$$G(x,\lambda)=I. \tag{2.72}$$

Let us first consider the case of $n=1$, when there is a single pair of zeros $\lambda_0, \bar{\lambda}_0$, $\operatorname{Im}\lambda_0>0$, and numbers $\gamma_0, \bar{\gamma}_0$, $\gamma_0 \neq 0$. A solution to the Riemann problem is given by

$$G_+(x,\lambda)=B(x,\lambda)B^{-1}(x,0), \tag{2.73}$$

$$G_-(x,\lambda)=B(x,0)B^{-1}(x,\lambda), \tag{2.74}$$

where $B(x,\lambda)$ is the Blaschke-Potapov matrix factor introduced in § II.2, Part I,

$$B(x,\lambda)=I+\frac{\bar{\lambda}_0-\lambda_0}{\lambda-\bar{\lambda}_0}P(x), \tag{2.75}$$

and $P(x)$ is the projection operator

$$P(x)=\frac{1}{1+|\gamma_0(x)|^2}\begin{pmatrix}|\gamma_0(x)|^2 & \overline{\gamma_0}(x)\\ \gamma_0(x) & 1\end{pmatrix} \tag{2.76}$$

with $\gamma_0(x)=e^{i\lambda_0 x}\gamma_0$. The corresponding matrix $S(x)$ is given by

$$S(x)=B(x,0)\sigma_3 B^{-1}(x,0). \tag{2.77}$$

Thus the components of the vector $\vec{S}(x)=(S_1(x), S_2(x), S_3(x))$ can be expressed as follows:

$$S_3(x) = 1-\frac{8(\operatorname{Im}\lambda_0)^2|\gamma_0(x)|^2}{|\lambda_0|^2(1+|\gamma_0(x)|^2)^2}, \tag{2.78}$$

$$\begin{aligned} S_+(x) &= \frac{S_1(x)+iS_2(x)}{2} \\ &= \frac{\gamma_0(x)}{|\lambda_0|^2}\,\frac{(|\lambda_0|^2|\gamma_0(x)|^2-|\lambda_0|^2+\lambda_0^2-\bar{\lambda}_0^2|\gamma_0(x)|^2)}{(1+|\gamma_0(x)|^2)^2}, \end{aligned}$$

$$S_-(x) = \frac{S_1(x)-iS_2(x)}{2}=\overline{S_+}(x). \tag{2.79}$$

We consider the evolution of the matrix $S(x)$ according to the HM equation. For that purpose $\gamma_0(x)$ should be replaced by $\gamma_0(x, t)$, as prescribed by (1.65),

$$\gamma_0(x, t) = e^{-i\lambda_0^2 t}\gamma_0(x). \tag{2.80}$$

With the definition

$$u = 2\,\mathrm{Im}\,\lambda_0, \quad v = 2\,\mathrm{Re}\,\lambda_0, \quad x_0 = \frac{1}{\mathrm{Im}\,\lambda_0}\log|\gamma_0|, \quad \varphi_0 = \arg\gamma_0, \tag{2.81}$$

we can write (2.78)–(2.79) as

$$S_3(x, t) = 1 - \frac{2u^2}{(u^2+v^2)\,\mathrm{ch}^2\left\{\frac{u}{2}(x - vt - x_0)\right\}}, \tag{2.82}$$

$$S_+(x, t) = \frac{u e^{i\left(\varphi_0 + \frac{vx}{2} + \frac{(u^2-v^2)}{4}t\right)}}{(u^2+v^2)\,\mathrm{ch}^2\left\{\frac{u}{2}(x - vt - x_0)\right\}} \times \left(-u\,\mathrm{sh}\left\{\frac{u}{2}(x - vt - x_0)\right\} + iv\,\mathrm{ch}\left\{\frac{u}{2}(x - vt - x_0)\right\}\right),$$

$$S_-(x, t) = \overline{S_+}(x, t). \tag{2.83}$$

These formulae show that the solution $\vec{S}(x, t)$ represents a solitary wave localized along the direction

$$x(t) = x_0 + vt \tag{2.84}$$

whose center moves with constant velocity v. By the definition given in § II.5, Part I, such a solution should be called a *soliton for the HM model.* The soliton $\vec{S}(x, t)$ is specified by 4 real parameters: the velocity v, the initial center of inertia coordinate x_0, the initial phase φ_0, and the amplitude A of the coefficient $S_3(x, t)$,

$$A = \frac{v^2 - u^2}{u^2 + v^2}. \tag{2.85}$$

In addition to the translational motion, $\vec{S}(x, t)$ oscillates in x and t with frequencies $\frac{v}{2}$ and $\frac{u^2-v^2}{4}$, respectively.

We now proceed to the general case of n pairs of zeros, $\lambda_j, \bar{\lambda}_j$, $\operatorname{Im}\lambda_j > 0$, and coefficients $\gamma_j, \bar{\gamma}_j$, $\gamma_j \neq 0, j=1, \ldots, n$. The solution of the Riemann problem has the form

$$G_+(x, \lambda) = \Pi(x, \lambda)\Pi^{-1}(x, 0), \tag{2.86}$$

$$G_-(x, \lambda) = \Pi(x, 0)\Pi^{-1}(x, \lambda), \tag{2.87}$$

where $\Pi(x, \lambda)$ is an ordered product of the Blaschke-Potapov factors,

$$\Pi(x, \lambda) = \overset{\curvearrowright}{\prod_{j=1}^{n}} B_j(x, \lambda), \tag{2.88}$$

$$B_j(x, \lambda) = I + \frac{\bar{\lambda}_j - \lambda_j}{\lambda - \bar{\lambda}_j} P_j(x), \tag{2.89}$$

and the projection operators $P_j(x)$ are uniquely determined by the set of numbers $\gamma_l(x) = e^{i\lambda_l x}\gamma_l$, $l=1, \ldots, n$. A method for computing them was presented in §§ II.2, II.5 of Part I.

The matrix $S(x)$ is given by

$$S(x) = \Pi(x, 0)\sigma_3\Pi^{-1}(x, 0), \tag{2.90}$$

where $\Pi(x, 0)$ is a unitary matrix satisfying

$$\det \Pi(x, 0) = \bar{\omega}_0 = \prod_{j=1}^{n} \frac{\lambda_j}{\bar{\lambda}_j} \tag{2.91}$$

(see § II.5, Part I). Using the notation

$$\Pi(x, 0) = \begin{pmatrix} A(x) & B(x) \\ C(x) & D(x) \end{pmatrix}, \tag{2.92}$$

for the components of the vector $\vec{S}(x)$ we have

$$\begin{aligned} S_3 &= \omega_0(AD + BC), \\ S_+ &= \omega_0 CD, \quad S_- = \bar{S}_+ = -\omega_0 AB. \end{aligned} \tag{2.93}$$

In contrast to our treatment of the NS model, no explicit formulae for the matrix $S(x)$ will be given here. Instead, we will proceed directly to its time evolution. To do so we replace $\gamma_j(x)$ by $\gamma_j(x, t)$ as prescribed by (1.65),

$$\gamma_j(x, t) = e^{-i\lambda_j^2 t}\gamma_j(x), \quad j = 1, \ldots, n. \tag{2.94}$$

The resulting $S(x, t)$ is called the *n-soliton solution. In a generic situation it decays, as $t \to \pm\infty$, into a sum of spacelike-separated solitons*

$$\vec{S}(x, t) = \sum_{j=1}^{n} \vec{S}_j^{(\pm)}(x, t) - (n-1)\vec{S}_0 + O(e^{-c|t|}). \tag{2.95}$$

Here $\vec{S}_j^{(\pm)}(x, t)$ are solitons with parameters u_j, v_j, $x_{0j}^{(\pm)}$, $\varphi_{0j}^{(\pm)}$ determined by (2.81) from the data λ_j, $\gamma_j^{(\pm)}$ with

$$\gamma_j^{(+)} = \gamma_j \prod_{v_k < v_j} \frac{\lambda_k}{\bar{\lambda}_k} \cdot \frac{\lambda_j - \bar{\lambda}_k}{\lambda_j - \lambda_k} \prod_{v_k > v_j} \frac{\bar{\lambda}_k}{\lambda_k} \cdot \frac{\lambda_j - \lambda_k}{\lambda_j - \bar{\lambda}_k} \tag{2.96}$$

and

$$\gamma_j^{(-)} = \gamma_j \prod_{v_k < v_j} \frac{\bar{\lambda}_k}{\lambda_k} \cdot \frac{\lambda_j - \lambda_k}{\lambda_j - \bar{\lambda}_k} \prod_{v_k > v_j} \frac{\lambda_k}{\bar{\lambda}_k} \cdot \frac{\lambda_j - \bar{\lambda}_k}{\lambda_j - \lambda_k}, \tag{2.97}$$

with $c = \frac{1}{2}\min u_j \cdot \min_{j \neq k} |v_j - v_k|$ and $\vec{S}_0 = (0, 0, 1)$. By a generic situation we mean that all velocities v_j are distinct. To verify these formulae if suffices to use the results of § II.5, Part I. In fact, it was shown that along the trajectory C_j given by

$$x - v_j t = \text{const} \tag{2.98}$$

the matrix $\Pi(x, t, \lambda)$ has the asymptotic behaviour

$$\Pi(x, t, \lambda) = \begin{pmatrix} \prod\limits_{\pm(v_k - v_j) > 0} \dfrac{\lambda - \lambda_k}{\lambda - \bar{\lambda}_k} & 0 \\ 0 & \prod\limits_{\pm(v_k - v_j) < 0} \dfrac{\lambda - \lambda_k}{\lambda - \bar{\lambda}_k} \end{pmatrix} \tilde{B}_j^{(\pm)}(x, t, \lambda) + O(e^{-c|t|}), \tag{2.99}$$

as $t \to \pm\infty$, where the Blaschke-Potapov factor $\tilde{B}_j^{(\pm)}(x, t, \lambda)$ is determined by the parameters λ_j, $\tilde{\gamma}_j^{(\pm)}$,

$$\tilde{\gamma}_j^{(\pm)} = \gamma_j \prod_{\pm(v_j - v_k) > 0} \frac{\lambda_j - \bar{\lambda}_k}{\lambda_j - \lambda_k} \prod_{\pm(v_j - v_k) < 0} \frac{\lambda_j - \lambda_k}{\lambda_j - \bar{\lambda}_k}, \tag{2.100}$$

while along the straight lines other than C_j, $j=1, \ldots, n$, the matrix $\Pi(x, t, \lambda)$ is diagonal with the order $O(e^{-c|t|})$. Setting $\lambda=0$ in (2.99) we get (2.95)-(2.97), as was to be shown.

The formulae obtained describe the theory of soliton scattering in the HM model.

In the course of scattering only the center coordinates and phases of the solitons are altered,

$$x_{0j}^{(+)}=x_{0j}^{(-)}+\Delta x_{0j}, \quad \varphi_{0j}^{(+)}=\varphi_{0j}^{(-)}+\Delta\varphi_{0j}, \tag{2.101}$$

where

$$\Delta x_{0j}=\frac{2}{\operatorname{Im}\lambda_j}\left(\sum_{v_k<v_j}\log\left|\frac{\lambda_j-\bar{\lambda}_k}{\lambda_j-\lambda_k}\right|-\sum_{v_k>v_j}\log\left|\frac{\lambda_j-\bar{\lambda}_k}{\lambda_j-\lambda_k}\right|\right) \tag{2.102}$$

and

$$\Delta\varphi_{0j}=2\left(\sum_{v_k<v_j}\left(\arg\frac{\lambda_j-\bar{\lambda}_k}{\lambda_j-\lambda_k}+2\arg\lambda_k\right)\right.$$
$$\left.-\sum_{v_k>v_j}\left(\arg\frac{\lambda_j-\bar{\lambda}_k}{\lambda_j-\lambda_k}+2\arg\lambda_k\right)\right)(\operatorname{mod}2\pi). \tag{2.103}$$

These formulae differ from the corresponding expressions for the NS model in § II.2, Part I, by the additional terms $\pm 2\arg\lambda_k$. Their interpretation is quite similar to that for the NS model in the rapidly decreasing case.

§ 3. Hamiltonian Formulation of the HM Model

We will show that the model in question is a completely integrable Hamiltonian system. The proof will consist of an explicit construction of canonical variables of action-angle type. For that purpose it will be shown that the Poisson brackets for the HM model admit an r-matrix formulation. Using this we will compute all the Poisson brackets of the transition coefficients and the discrete spectrum. An explicit expression for the local integrals of the motion in terms of action-angle variables and an interpretation of the independent modes that occur will be given. We will also clarify the Hamiltonian meaning of the gauge transformation from the HM model to the NS model. To conclude this section, we will discuss the process of soliton scattering from the Hamiltonian point of view.

1. The fundamental Poisson brackets and the r-matrix

Consider the basic Poisson brackets for the HM model,

$$\{S_a(x), S_b(y)\} = -\varepsilon_{abc} S_c(x)\delta(x-y) \tag{3.1}$$

(see § I.1) and write them in terms of the matrix $S(x) = \sum_{a=1}^{3} S_a(x)\sigma_a$:

$$\{S(x) \underset{,}{\otimes} S(y)\} = -\sum_{a,b,c=1}^{3} \varepsilon_{abc} S_c(x)(\sigma_a \otimes \sigma_b)\delta(x-y). \tag{3.2}$$

Using the multiplication formula for the Pauli matrices

$$\alpha_a \sigma_b = \delta_{ab} I + i\varepsilon_{abc}\sigma_c \tag{3.3}$$

and the definition of the permutation matrix P from § III.1, Part I,

$$P = \frac{1}{2}\left(I \otimes I + \sum_{a=1}^{3} \sigma_a \otimes \sigma_a\right), \tag{3.4}$$

we can express the matrices $\sigma_a \otimes \sigma_b - \sigma_b \otimes \sigma_a$ occurring on the right-hand side of (3.2) as

$$\sigma_a \otimes \sigma_b - \sigma_b \otimes \sigma_a = i\varepsilon_{abc} P(I \otimes \sigma_c - \sigma_c \otimes I). \tag{3.5}$$

Hence the Poisson bracket $\{S(x) \underset{,}{\otimes} S(y)\}$ can be written in the form

$$\begin{aligned}\{S(x) \underset{,}{\otimes} S(y)\} &= iP(S(x) \otimes I - I \otimes S(x))\,\delta(x-y) \\ &= i[P, S(x) \otimes I]\,\delta(x-y).\end{aligned} \tag{3.6}$$

Let us now express the Poisson brackets of the coefficients of the auxiliary linear problem $U(x,\lambda) = \frac{\lambda}{2i} S(x)$. To do so we multiply both sides of (3.6) by $-\frac{\lambda\mu}{4}$ and transform the right hand side as follows:

$$\begin{aligned}\frac{\lambda\mu}{4i}[P, S(x) \otimes I] &= \frac{\lambda\mu}{2(\lambda-\mu)}[P, U(x,\lambda) \otimes I - U(x,\mu) \otimes I] \\ &= \frac{\lambda\mu}{2(\lambda-\mu)}[P, U(x,\lambda) \otimes I + I \otimes U(x,\mu)].\end{aligned} \tag{3.7}$$

This eventually gives the *fundamental Poisson brackets for the HM model,*

$$\{U(x,\lambda) \underset{,}{\otimes} U(x,\mu)\} = [r(\lambda,\mu),\, U(x,\lambda)\otimes I + I\otimes U(x,\mu)]\,\delta(x-y), \quad (3.8)$$

where

$$r(\lambda,\mu) = \frac{\lambda\mu}{2(\lambda-\mu)}\, P. \quad (3.9)$$

We have the relation

$$r^{\mathrm{HM}}(\lambda,\mu) = \frac{\lambda\mu}{2}\, r^{\mathrm{NS}}(\lambda-\mu) \quad (3.10)$$

or

$$r^{\mathrm{HM}}(\lambda,\mu) = -\frac{1}{2}\, r^{\mathrm{NS}}\left(\frac{1}{\lambda}-\frac{1}{\mu}\right), \quad (3.11)$$

so that, with $\frac{1}{\lambda}$ and $\frac{1}{\mu}$ as independent variables, the r-matrix for the HM model depends on the difference of its arguments.

Following the general reasoning of § III.1, Part I, we can use the fundamental Poisson brackets (3.8) to derive the Poisson brackets of the transition matrix,

$$\{T(x,y,\lambda) \underset{,}{\otimes} T(x,y,\mu)\} = [r(\lambda,\mu),\, T(x,y,\lambda)\otimes T(x,y,\mu)], \quad (3.12)$$

for $y<x$.

2. The Poisson brackets of the transition coefficients and the discrete spectrum

By an argument such as in § III.6, Part I, we find from (3.12) the Poisson brackets of the Jost solutions $T_{\pm}(x,\lambda)$ and of the reduced monodromy matrix $T(\lambda)$,

$$\begin{aligned}\{T_{\pm}(x,\lambda) \underset{,}{\otimes} T_{\pm}(x,\mu)\} = &\mp r(\lambda,\mu)\, T_{\pm}(x,\lambda)\otimes T_{\pm}(x,\mu) \\ &\pm T_{\pm}(x,\lambda)\otimes T_{\pm}(x,\mu)\, r_{\pm}(\lambda,\mu),\end{aligned} \quad (3.13)$$

$$\{T_{+}(x,\lambda) \underset{,}{\otimes} T_{-}(x,\mu)\} = 0, \quad (3.14)$$

$$\{T(\lambda) \underset{,}{\otimes} T(\mu)\} = r_{+}(\lambda,\mu)\, T(\lambda)\otimes T(\mu) - T(\lambda)\otimes T(\mu)\, r_{-}(\lambda,\mu). \quad (3.15)$$

Here

$$r_{\pm}(\lambda,\mu)=\frac{\lambda\mu}{2}\begin{pmatrix} \text{p.v.}\,\dfrac{1}{\lambda-\mu} & 0 & 0 & 0 \\ 0 & 0 & \pm\pi i\delta(\lambda-\mu) & 0 \\ 0 & \mp\pi i\delta(\lambda-\mu) & 0 & 0 \\ 0 & 0 & 0 & \text{p.v.}\,\dfrac{1}{\lambda-\mu} \end{pmatrix}, \tag{3.16}$$

so that we have the relations

$$r_{\pm}^{\mathrm{HM}}(\lambda,\mu)=\frac{\lambda\mu}{2}\,r_{\pm}^{\mathrm{NS}}(\lambda-\mu). \tag{3.17}$$

This gives the following expressions for the Poisson brackets of the transition coefficients and the discrete spectrum

$$\{a(\lambda),a(\mu)\}=\{a(\lambda),\bar{a}(\mu)\}=0, \tag{3.18}$$

$$\{b(\lambda),b(\mu)\}=0, \tag{3.19}$$

$$\{b(\lambda),\bar{b}(\mu)\}=\pi i\lambda^2\,|a(\lambda)|^2\,\delta(\lambda-\mu), \tag{3.20}$$

$$\{a(\lambda),b(\mu)\}=-\frac{\lambda\mu}{2(\lambda-\mu+i0)}\,a(\lambda)\,b(\mu), \tag{3.21}$$

$$\{a(\lambda),\bar{b}(\mu)\}=\frac{\lambda\mu}{2(\lambda-\mu+i0)}\,a(\lambda)\,\bar{b}(\mu) \tag{3.22}$$

and

$$\{b(\mu),\gamma_j\}=\{b(\lambda),\bar{\gamma}_j\}=0, \tag{3.23}$$

$$\{b(\lambda),\lambda_j\}=\{b(\lambda),\bar{\lambda}_j\}=0, \tag{3.24}$$

$$\{a(\lambda),\gamma_j\}=-\frac{\lambda\lambda_j}{2(\lambda-\lambda_j)}\,a(\lambda)\,\gamma_j, \tag{3.25}$$

$$\{a(\lambda),\bar{\gamma}_j\}=\frac{\lambda\bar{\lambda}_j}{2(\lambda-\bar{\lambda}_j)}\,a(\lambda)\,\bar{\gamma}_j, \tag{3.26}$$

and also

$$\{\lambda_i, \lambda_k\} = \{\lambda_j, \bar{\lambda}_k\} = 0, \tag{3.27}$$

$$\{\gamma_j, \gamma_k\} = \{\gamma_j, \bar{\gamma}_k\} = 0, \tag{3.28}$$

$$\{\lambda_j, \gamma_k\} = \frac{\lambda_j^2 \gamma_k}{2} \delta_{jk}, \tag{3.29}$$

$$\{\lambda_j, \bar{\gamma}_k\} = 0, \quad j, k = 1, \ldots, n. \tag{3.30}$$

Thus the data of the continuous and discrete spectra are in involution, and the non-vanishing Poisson brackets of the inverse problem data ($b(\lambda), \bar{b}(\lambda)$; $\lambda_j, \bar{\lambda}_j, \gamma_j, \bar{\gamma}_j, j = 1, \ldots, n$) *are given by (3.20) and (3.29).*

The Poisson brackets (3.18) show that $\log a(\lambda)$ is a generating function for *integrals of the motion in involution.* In particular,

$$\{I_k, I_l\} = 0, \tag{3.31}$$

where I_l, $l = 0, 1, \ldots$, are the local integrals of the motion for the HM model derived in § 1. In addition, from (1.103) if follows that

$$\{M_3, I_l\} = 0, \tag{3.32}$$

where M_3 is the third component of the total spin.

3. Canonical action-angle variables

Just as in § III.7, Part I, it follows from the above formulae that *the variables*

$$\varrho(\lambda) = -\frac{1}{\pi\lambda^2} \log(1 - |b(\lambda)|^2), \quad \varphi(\lambda) = -\arg b(\lambda) \tag{3.33}$$

and

$$p_j = -\frac{4 \operatorname{Re} \lambda_j}{|\lambda_j|^2}, \quad q_j = \log|\gamma_j|, \tag{3.34}$$

$$\varrho_j = \frac{4 \operatorname{Im} \lambda_j}{|\lambda_j|^2}, \quad \varphi_j = -\arg \gamma_j \tag{3.35}$$

are canonical variables, i.e. their nonvanishing Poisson brackets have the form

$$\{\varrho(\lambda), \varphi(\mu)\} = \delta(\lambda - \mu) \tag{3.36}$$

and

$$\{p_j, q_k\} = \delta_{jk}, \qquad \{\varrho_j, \varphi_k\} = \delta_{jk}, \quad j,k=1,\ldots,n. \tag{3.37}$$

The variable $\varrho(\lambda)$ is nonnegative and nonsingular in view of the condition (A) and the relation $b(0)=0$.

Thus the mapping

$$\mathscr{F}: (\vec{S}(x)) \to (b(\lambda), \bar{b}(\lambda), \lambda_j, \bar{\lambda}_j, \gamma_j, \bar{\gamma}_j) \tag{3.38}$$

defines a canonical transformation that trivializes the dynamics of the HM model. The local integrals I_l depend only on the action variables,

$$I_l = -\frac{1}{2} \int_{-\infty}^{\infty} \lambda^{l+1} \varrho(\lambda)\, d\lambda + \frac{(-1)^l}{il} \sum_{j=1}^{n} \left(\frac{p_j^2+\varrho_j^2}{4}\right)^{-l} ((p_j+i\varrho_j)^l - (p_j - i\varrho_j)^l), \tag{3.39}$$

where $l=0, 1, \ldots$ and, if $l=0$, the sum over the discrete spectrum is taken according to L'Hopital's rule.

From (3.36)–(3.37) and (3.49) it is clear that *all the higher HM equations,*

$$\frac{\partial \vec{S}}{\partial t} = \{-2I_l, \vec{S}\}, \tag{3.40}$$

are completely integrable Hamiltonian systems, and their time evolution is given by the following simple expressions

$$\begin{gathered} b(\lambda, t) = e^{-i\lambda^{l+1}t} b(\lambda, 0), \quad \lambda_j(t) = \lambda_j(0), \\ \gamma_j(t) = e^{-i\lambda_j^{l+1}t} \gamma_j(0), \quad j=1,\ldots,n. \end{gathered} \tag{3.41}$$

In particular, when $l=1$ we recover the formulae (1.64)–(1.65) for the HM equation.

The formal Hamiltonians M_1 and M_2 (see (1.96)) do not depend solely on the action variables. As is easily seen from (3.36) and (1.104), the induced equations of motion have the form

$$\frac{\partial \varrho(\lambda)}{\partial t} = \{M_\pm, \varrho(\lambda)\} = \pm i\delta(\lambda) M_\pm, \tag{3.42}$$

where $M_{\pm}=M_1\pm iM_2$. Clearly, this equation does not preserve the class of functions $\varrho(\lambda)$ smooth up to $\lambda=0$, so that the functionals $M_{\pm}$ are not admissible.

We shall now have a closer look at the principal integrals of the motion, namely, the momentum P, the Hamiltonian H, and the projection M_3 of the total spin. These can be written down using (1.103) and (3.33)–(3.35), (3.39). We have

$$-M_3=\int_{-\infty}^{\infty}\varrho(\lambda)\,d\lambda+\sum_{j=1}^{n}\varrho_j, \tag{3.43}$$

$$P=\int_{-\infty}^{\infty}\lambda\varrho(\lambda)\,d\lambda-4\sum_{j=1}^{n}\operatorname{arctg}\frac{\varrho_j}{p_j} \tag{3.44}$$

and

$$H=\int_{-\infty}^{\infty}\lambda^2\varrho(\lambda)\,d\lambda+16\sum_{j=1}^{n}\frac{\varrho_j}{p_j^2+\varrho_j^2}\,. \tag{3.45}$$

These expressions are sums over independent modes. Their first terms correspond to a wave packet of continuous spectrum modes with density $\varrho(\lambda)$. The mode labelled by λ describes a particle with momentum and energy given by

$$p(\lambda)=\lambda,\quad h(\lambda)=\lambda^2 \tag{3.46}$$

and related by the nonrelativistic dispersion law

$$h(p)=p^2\,. \tag{3.47}$$

It has mass $\frac{1}{2}$ and the projection of spin on the third axis is equal to 1.

The discrete spectrum modes in the HM model are associated with solitons. The momentum of a single mode has the form

$$P=-4\operatorname{arctg}\frac{\varrho}{p} \tag{3.48}$$

and varies in the Brillouin zone

$$|P|\leqslant 2\pi\,. \tag{3.49}$$

Its energy is given by

$$h=\frac{16\varrho}{p^2+\varrho^2} \tag{3.50}$$

and is related to the momentum P and the spin component $-M_3=\varrho$ by the dispersion law

$$h(P)=-\frac{16}{M_3}\sin^2\frac{P}{4}\,. \tag{3.51}$$

Observe that the momentum of a discrete mode is defined mod 4π in agreement with the discussion of § I.1.

As in the case of the NS model, the discrete modes for the HM model turn into the continuous modes when the λ_j accumulate towards the real line. Also, the dispersion law (3.51) turns into (3.47) upon natural linearization.

To close this subsection, we consider the gauge transformation from the HM model to the NS model from the Hamiltonian viewpoint. Using the relationship given by (1.53), (1.57), we have

$$\varrho^{\mathrm{HM}}(\lambda)=\frac{2}{\lambda^2}\varrho^{\mathrm{NS}}(\lambda),\qquad \varphi^{\mathrm{HM}}(\lambda)=\varphi^{\mathrm{NS}}(\lambda)+\arg\omega_0, \tag{3.52}$$

$$p_j^{\mathrm{HM}}=\frac{2}{|\lambda_j|^2}p_j^{\mathrm{NS}},\qquad q_j^{\mathrm{HM}}=q_j^{\mathrm{NS}}, \tag{3.53}$$

$$\varrho_j^{\mathrm{HM}}=\frac{2}{|\lambda_j|^2}\varrho_j^{\mathrm{NS}},\qquad \varphi_j^{\mathrm{HM}}=\varphi_j^{\mathrm{NS}}+\arg\omega_0,\qquad j=1,\dots,n, \tag{3.54}$$

where

$$\arg\omega_0=-\arg a^{\mathrm{NS}}(0)=-\tfrac{1}{2}P^{\mathrm{HM}}\,. \tag{3.55}$$

The comparison shows that *the gauge transformation* $\vec{S}(x)\to(\psi(x),\bar{\psi}(x))$ *sends the standard Poisson structure for the HM model into the second Poisson structure for the NS model, which belongs to the hierarchy defined in § III.5, Part I,*

$$\{,\}^{\mathrm{HM}}=\tfrac{1}{2}\{,\}_2^{\mathrm{NS}}\,. \tag{3.56}$$

Of course, the Hamiltonian is also shifted along the hierarchy, so that

$$H^{\mathrm{HM}}=2N^{\mathrm{NS}}, \tag{3.57}$$

where $N^{\rm NS}$ is the charge (number of particles) for the NS model. The last formula agrees with the local computation which gives

$$\left(\frac{d\vec{S}(x)}{dx}\right)^2 = 4|\psi(x)|^2, \tag{3.58}$$

as was shown in § I.4.

Thus the HM model can be regarded as the NS model equipped with a different Poisson structure and a different Hamiltonian.

4. Soliton scattering from the Hamiltonian viewpoint

The general n-soliton solution of the HM model is labelled by a set of parameters $\{p_j, q_j, \varrho_j, \varphi_j, j=1, \ldots, n\}$. In a generic situation (see Subsection 3 of § 2), as $t \to \pm\infty$, it decays into a sum of spacelike-separated solitons with parameters $p_j^{(\pm)}, q_j^{(\pm)}, \varrho_j^{(\pm)}, \varphi_j^{(\pm)}$ where

$$p_j^{(+)} = p_j^{(-)} = p_j, \qquad \varrho_j^{(+)} = \varrho_j^{(-)} = \varrho_j, \tag{3.59}$$

$$q_j^{(\pm)} = q_j \pm \Delta q_j, \qquad \varphi_j^{(\pm)} = \varphi_j \pm \Delta\varphi_j \tag{3.60}$$

and

$$\Delta q_j = \sum_{k=j+1}^{n} \log\left|\frac{\lambda_j - \bar{\lambda}_k}{\lambda_j - \lambda_k}\right| - \sum_{k=1}^{j-1} \log\left|\frac{\lambda_j - \bar{\lambda}_k}{\lambda_j - \lambda_k}\right|, \tag{3.61}$$

$$\Delta\varphi_j = \sum_{k=1}^{j-1}\left(\arg\frac{\lambda_j - \bar{\lambda}_k}{\lambda_j - \lambda_k} + 2\arg\lambda_k\right) - \sum_{k=j+1}^{n}\left(\arg\frac{\lambda_j - \bar{\lambda}_k}{\lambda_j - \lambda_k} + 2\arg\lambda_k\right). \tag{3.62}$$

It is assumed here that $\operatorname{Re}\lambda_1 > \operatorname{Re}\lambda_2 > \ldots > \operatorname{Re}\lambda_n$.

The transformations

$$W_{\pm}: \{p_j, q_j, \varrho_j, \varphi_j\,;\, j=1, \ldots, n\} \to \{p_j^{(\pm)}, q_j^{(\pm)}, \varrho_j^{(\pm)}, \varphi_j^{(\pm)}, j=1, \ldots, n\} \tag{3.63}$$

given by (3.59)–(3.62) are canonical with generating functions $\pm K_n(p_1, \ldots, p_n;$ $\varrho_1, \ldots, \varrho_n)$, *so that*

$$q_j^{(\pm)} = q_j \pm \frac{\partial K_n}{\partial p_j}, \qquad \varphi_j^{(\pm)} = \varphi_j \pm \frac{\partial K_n}{\partial \varrho_j}, \qquad j=1, \ldots, n,$$
$$K_n(p_1, \ldots, p_n; \varrho_1, \ldots, \varrho_n) = \sum_{1 \leq j < k \leq n} K_2(p_j, p_k; \varrho_j, \varrho_k), \tag{3.64}$$

where

$$\begin{aligned} K_2(p_1,p_2,\varrho_1,\varrho_2) &= \operatorname{Re}\{(p_1-p_2+i\varrho_1+i\varrho_2) \\ &\quad \times \log(p_1-p_2+i\varrho_1+i\varrho_2)-(p_1-p_2+i\varrho_1-i\varrho_2) \\ &\quad \times \log(p_1-p_2+i\varrho_1-i\varrho_2)\} \\ &= 4\operatorname{Re}\left\{\left(\frac{1}{\lambda_1}-\frac{1}{\lambda_2}\right)\log\left(\frac{1}{\lambda_2}-\frac{1}{\lambda_1}\right)\right. \\ &\quad \left. -\left(\frac{1}{\lambda_1}-\frac{1}{\bar{\lambda}_2}\right)\log\left(\frac{1}{\bar{\lambda}_2}-\frac{1}{\lambda_1}\right)\right\}. \end{aligned} \tag{3.65}$$

The expression for the generating function K_2 agrees with the formula for the NS model (III.8.29), Part I, and the relationship of the Poisson structures mentioned in Subsection 3.

As in the NS model, soliton scattering in the HM model is described by a canonical transformation. In the coordinates $p_j, q_j^{(\pm)}, \varrho_j, \varphi_j^{(\pm)}, j=1,\ldots,n$, that determine the asymptotic motion as $t\to\pm\infty$, *the scattering transformation*

$$S:\{p_j,\varrho_j,q_j^{(-)},\varphi_j^{(-)};j=1,\ldots,n\}\to\{p_j,\varrho_j,q_j^{(+)},\varphi_j^{(+)};j=1,\ldots,n\} \tag{3.66}$$

can be written as

$$S=W_+W_-^{-1} \tag{3.67}$$

and is obviously canonical. Its generating function S_n, the classical S-matrix for n-particle scattering, has the form

$$S_n(p_1,\ldots,p_n;\varrho_1,\ldots,\varrho_n)=2K_n(p_1,\ldots,p_n;\varrho_1,\ldots,\varrho_n). \tag{3.68}$$

So, soliton scattering in the HM model is yet another example of a factorized scattering theory. This completes our exposition of the HM model in the rapidly decreasing case.

§ 4. The Auxiliary Linear Problem for the SG Model

We shall define the basic ingredients of the auxiliary linear problem for the SG model (see § I.1)

$$\frac{dF}{dx} = U(x,\lambda)F = \frac{1}{4i}\left(\beta\pi(x)\sigma_3 + m\left(\lambda+\frac{1}{\lambda}\right)\sin\frac{\beta\varphi(x)}{2}\sigma_1\right.$$
$$\left. + m\left(\lambda-\frac{1}{\lambda}\right)\cos\frac{\beta\varphi(x)}{2}\sigma_2\right)F, \tag{4.1}$$

where $\varphi(x)$ and $\pi(x)$ are real-valued functions, m and β are positive constants. We shall only consider the rapidly decreasing case, when

$$\lim_{x\to-\infty}\varphi(x)=0, \qquad \lim_{x\to+\infty}\varphi(x)=\frac{2\pi}{\beta}Q, \qquad \lim_{|x|\to\infty}\pi(x)=0, \tag{4.2}$$

where Q is an integer (the topological charge, see § I.1), and the boundary values are attained in the sense of Schwartz.

The distinctive feature of this auxiliary linear problem is that the pole divisor $\mathfrak{U}$ of the matrix $U(x,\lambda)$ contains two points, $\lambda=0,\infty$, instead of one, $\lambda=\infty$, as was the case for the NS and HM models. This fact will receive special attention in what follows.

1. The transition matrix and the Jost solutions

The transition matrix $T(x,y,\lambda)$ is defined to be a solution of the differential equation (4.1) with the initial condition

$$T(x,y,\lambda)|_{x=y}=I. \tag{4.3}$$

It can be expressed as

$$T(x,y,\lambda)=\overset{\frown}{\exp}\int_y^x U(z,\lambda)\,dz. \tag{4.4}$$

The matrix $T(x,y,\lambda)$ is unimodular and analytic in the region $\mathbb{C}\backslash\{0\}$, and has essential singularities at $\lambda=\infty$ and $\lambda=0$. The relations

$$\bar{U}(x,\bar{\lambda})=\sigma_2 U(x,\lambda)\sigma_2 \tag{4.5}$$

and

$$U(x,-\lambda)=\sigma_3 U(x,\lambda)\sigma_3 \tag{4.6}$$

infer the involution properties of the transition matrix,

$$\bar{T}(x,y,\bar{\lambda})=\sigma_2 T(x,y,\lambda)\sigma_2 \tag{4.7}$$

and

$$T(x,y,-\lambda)=\sigma_3 T(x,y,\lambda)\sigma_3. \tag{4.8}$$

In addition, $U(x,\lambda)$ is invariant under the replacement $\pi(x)\to\pi(x)$, $\varphi(x)\to-\varphi(x)$, $\lambda\to-\frac{1}{\lambda}$. Hence we have

$$\hat{T}\left(x,y,-\frac{1}{\lambda}\right)=T(x,y,\lambda), \tag{4.9}$$

where $\hat{T}(x,y,\lambda)$ denotes the transition matrix for the data $\hat{\pi}(x)=\pi(x)$ and $\hat{\varphi}(x)=-\varphi(x)$.

For $x\to\pm\infty$, the auxiliary linear problem (4.1) goes into the differential equations

$$\frac{dE}{dx}=U_{\pm}(\lambda)E, \tag{4.10}$$

where

$$U_{-}(\lambda)=\frac{1}{i}k_1(\lambda)\sigma_2, \quad U_{+}(\lambda)=(-1)^Q U_{-}(\lambda), \tag{4.11}$$

and $k_1(\lambda)=\frac{m}{4}\left(\lambda-\frac{1}{\lambda}\right)$. These equations can be solved in closed form,

$$E_{-}(x,\lambda)=E(x,\lambda)=\frac{1}{\sqrt{2}}\begin{pmatrix}1 & i\\ i & 1\end{pmatrix}e^{\frac{1}{i}k_1(\lambda)x\sigma_3}, \tag{4.12}$$

$$E_{+}(x,\lambda)=(-1)^{\frac{Q}{2}}i\sigma_3 E(x,\lambda) \tag{4.13}$$

for Q even and

$$E_{+}(x,\lambda)=(-1)^{\frac{Q-1}{2}}E(x,\lambda) \tag{4.14}$$

for Q odd; the significance of this choice of $E_{+}(x,\lambda)$ depending on the value of $Q(\operatorname{mod}4)$ will become clear a bit later (cf. (4.2), (4.21) and (4.23)–(4.24), (4.28)).

The matrices $E_{\pm}(x,\lambda)$ are unimodular and obey the involutions

$$\bar{E}_{\pm}(x,\bar{\lambda})=\sigma_2 E_{\pm}(x,\lambda)\sigma_2, \tag{4.15}$$

$$E_{\pm}(x,-\lambda)=-\sigma_3 E_{\pm}(x,\lambda)\sigma_2, \tag{4.16}$$

$$\bar{E}_{\pm}(x,-\bar{\lambda})=-i\sigma_1 E_{\pm}(x,\lambda). \tag{4.17}$$

For real $\lambda \neq 0$, the Jost solutions $T_\pm(x,\lambda)$ are defined to be the limits

$$T_\pm(x,\lambda) = \lim_{y\to\pm\infty} T(x,y,\lambda)\,E_\pm(y,\lambda). \tag{4.18}$$

The matrices $T_\pm(x,\lambda)$ are unimodular and satisfy the differential equation (4.1) and the involution relations

$$\bar{T}_\pm(x,\lambda) = \sigma_2 T_\pm(x,\lambda)\sigma_2, \tag{4.19}$$

$$\bar{T}_\pm(x,-\lambda) = -i\sigma_1 T_\pm(x,\lambda). \tag{4.20}$$

As $x\to\pm\infty$, they have respectively the asymptotic behaviour

$$T_\pm(x,\lambda) = E_\pm(x,\lambda) + o(1). \tag{4.21}$$

To describe the analytic properties of the Jost solutions in the vicinity of $\lambda=\infty$, it will be convenient to make a gauge transformation so as to make the coefficient of λ in the auxiliary linear problem independent of x. For that purpose we write (4.1) as

$$\frac{dT_\pm}{dx} = \left(\frac{\beta\pi}{4i}\sigma_3 + \frac{m\lambda}{4i}\Omega\sigma_2\Omega^{-1} - \frac{m}{4i\lambda}\Omega^{-1}\sigma_2\Omega\right)T_\pm, \tag{4.22}$$

where

$$\Omega(x) = e^{\frac{i\beta\varphi(x)}{4}\sigma_3} \tag{4.23}$$

and let

$$T_\pm(x,\lambda) = \Omega(x)\,\tilde{T}_\pm(x,\lambda). \tag{4.24}$$

For $\tilde{T}_\pm(x,\lambda)$ we have the differential equations

$$\frac{d\tilde{T}_\pm}{dx} = \tilde{U}(x,\lambda)\,\tilde{T}_\pm, \tag{4.25}$$

with

$$\tilde{U}(x,\lambda) = U^{\Omega^{-1}}(x,\lambda) = \frac{\beta}{4i}\theta(x)\sigma_3 + \frac{m\lambda}{4i}\sigma_2 - \frac{m}{4i\lambda}\Omega^{-2}(x)\sigma_2\Omega^2(x) \tag{4.26}$$

and

$$\theta(x) = \pi(x) + \frac{d\varphi}{dx}(x). \tag{4.27}$$

The solutions $\tilde{T}_{\pm}(x, \lambda)$ have integral representations analogous to those for the NS and HM models,

$$\tilde{T}_{+}(x,\lambda)=E(x,\lambda)+\int_{x}^{\infty}\Gamma_{+}^{(1)}(x,y)\,E(y,\lambda)\,dy + \frac{1}{\lambda}\int_{x}^{\infty}\Gamma_{+}^{(2)}(x,y)\,E(y,\lambda)\,dy \tag{4.28}$$

and

$$\tilde{T}_{-}(x,\lambda)=E(x,\lambda)+\int_{-\infty}^{x}\Gamma_{-}^{(1)}(x,y)\,E(y,\lambda)\,dy + \frac{1}{\lambda}\int_{-\infty}^{x}\Gamma_{-}^{(2)}(x,y)\,E(y,\lambda)\,dy. \tag{4.29}$$

The kernels $\Gamma_{\pm}^{(l)}(x,y)$, $l=1,2$, obey the involutions

$$\bar{\Gamma}_{\pm}^{(1,2)}=\sigma_2\Gamma_{\pm}^{(1,2)}\sigma_2 \tag{4.30}$$

and

$$\bar{\Gamma}_{\pm}^{(1)}=\sigma_1\Gamma_{\pm}^{(1)}\sigma_1,\qquad \bar{\Gamma}_{\pm}^{(2)}=-\sigma_1\Gamma_{\pm}^{(2)}\sigma_1, \tag{4.31}$$

so that the matrices $\Gamma_{\pm}^{(1)}$ are diagonal whereas $\Gamma_{\pm}^{(2)}$ are off-diagonal. They satisfy the following systems of partial differential equations:

$$\frac{\partial\Gamma_{\pm}^{(1)}}{\partial x}(x,y)+\sigma_2\frac{\partial\Gamma_{\pm}^{(1)}}{\partial y}(x,y)\sigma_2-\frac{\beta\theta(x)}{4i}\sigma_3\Gamma_{\pm}^{(1)}(x,y) - \frac{m}{4i}(\sigma_2-\Omega^{-2}(x)\sigma_2\Omega^2(x))\Gamma_{\pm}^{(2)}(x,y)=0, \tag{4.32}$$

$$\frac{\partial\Gamma_{\pm}^{(2)}(x,y)}{\partial x}+\Omega^{-2}(x)\sigma_2\Omega^2(x)\frac{\partial\Gamma_{\pm}^{(2)}(x,y)}{\partial y}\sigma_2-\frac{\beta\theta(x)}{4i}\sigma_3\Gamma_{\pm}^{(2)}(x,y) - \frac{m}{4i}(\sigma_2-\Omega^{-2}(x)\sigma_2\Omega^2(x))\Gamma_{\pm}^{(1)}(x,y)=0 \tag{4.33}$$

for $\pm(y-x)>0$ with the boundary conditions

$$\lim_{y\to\pm\infty}\Gamma_{\pm}^{(l)}(x,y)=0,\qquad l=1,2, \tag{4.34}$$

$$\Gamma_{\pm}^{(1)}(x,x)-\sigma_2\Gamma_{\pm}^{(1)}(x,x)\sigma_2=\mp\frac{\beta}{4i}\theta(x)\sigma_3, \tag{4.35}$$

$$\Gamma_{\pm}^{(2)}(x,x)-\Omega^{-2}(x)\sigma_2\Omega^2(x)\Gamma_{\pm}^{(2)}(x,x)\sigma_2$$
$$=\pm\frac{m}{4i}(\Omega^{-2}(x)\sigma_2\Omega^2(x)-\sigma_2). \tag{4.36}$$

Formulae (4.32)–(4.36) result from inserting the integral representations (4.28)–(4.29) into the differential equation (4.25).

The correspondence between the integral representations (4.28)–(4.29) and the differential equations (4.32)–(4.36) is reversible. These differential equations can easily be related to Volterra integral equations thus proving the existence of the Jost solutions and their integral representations.

The properties of the columns $T_{\pm}^{(l)}(x,\lambda)$, $l=1,2$, of $T_{\pm}(x,\lambda)$ are as follows. The columns $T_{-}^{(1)}(x,\lambda)$ and $T_{+}^{(2)}(x,\lambda)$ can be analytically continued into the upper half of the λ-plane and $T_{+}^{(1)}(x,\lambda)$ and $T_{-}^{(2)}(x,\lambda)$ into the lower half-plane, with the asymptotic behaviour, as $|\lambda|\to\infty$,

$$e^{-\frac{\lambda m x}{4i}}T_{-}^{(1)}(x,\lambda)=\frac{1}{\sqrt{2}}\Omega(x)\begin{pmatrix}1\\ i\end{pmatrix}+O\left(\frac{1}{|\lambda|}\right), \tag{4.37}$$

$$e^{\frac{\lambda m x}{4i}}T_{+}^{(2)}(x,\lambda)=\frac{1}{\sqrt{2}}\Omega(x)\begin{pmatrix}i\\ 1\end{pmatrix}+O\left(\frac{1}{|\lambda|}\right) \tag{4.38}$$

for $\operatorname{Im}\lambda\geqslant 0$ and

$$e^{-\frac{\lambda m x}{4i}}T_{+}^{(1)}(x,\lambda)=\frac{1}{\sqrt{2}}\Omega(x)\begin{pmatrix}1\\ i\end{pmatrix}+O\left(\frac{1}{|\lambda|}\right), \tag{4.39}$$

$$e^{\frac{\lambda m x}{4i}}T_{-}^{(2)}(x,\lambda)=\frac{1}{\sqrt{2}}\Omega(x)\begin{pmatrix}i\\ 1\end{pmatrix}+O\left(\frac{1}{|\lambda|}\right) \tag{4.40}$$

for $\operatorname{Im}\lambda\leqslant 0$.

The behaviour near $\lambda=0$ is analyzed in a similar way. As is clear from (4.22), for that purpose one should make a gauge transformation with matrix $\Omega(x)$. It is possible, however, to use relation (4.9) instead. We then find the following asymptotic expressions:

$$e^{\frac{mx}{4i\lambda}}T_{-}^{(1)}(x,\lambda)=\frac{1}{\sqrt{2}}\Omega^{-1}(x)\begin{pmatrix}1\\ i\end{pmatrix}+O(|\lambda|), \tag{4.41}$$

$$e^{-\frac{mx}{4i\lambda}}T_{+}^{(2)}(x,\lambda)=\frac{(-1)^Q}{\sqrt{2}}\Omega^{-1}(x)\begin{pmatrix}i\\ 1\end{pmatrix}+O(|\lambda|) \tag{4.42}$$

for $\mathrm{Im}\lambda \geqslant 0$ and

$$e^{\frac{mx}{4i\lambda}} T_+^{(1)}(x,\lambda) = \frac{(-1)^Q}{\sqrt{2}} \Omega^{-1}(x) \begin{pmatrix} 1 \\ i \end{pmatrix} + O(|\lambda|), \tag{4.43}$$

$$e^{-\frac{mx}{4i\lambda}} T_-^{(2)}(x,\lambda) = \frac{1}{\sqrt{2}} \Omega^{-1}(x) \begin{pmatrix} i \\ 1 \end{pmatrix} + O(|\lambda|) \tag{4.44}$$

for $\mathrm{Im}\lambda \leqslant 0$.

Formulae (4.37)-(4.44) are consistent with the asymptotic expressions (4.21) and the boundary conditions (4.2). For λ complex, the involutions (4.19)-(4.20) become

$$\bar{T}_\pm^{(1)}(x,\lambda) = i\sigma_2 T_\pm^{(2)}(x,\bar{\lambda}), \tag{4.45}$$

$$\bar{T}_\pm^{(2)}(x,\lambda) = -i\sigma_2 T_\pm^{(1)}(x,\bar{\lambda}), \tag{4.46}$$

$$\bar{T}_\pm^{(1,2)}(x,-\bar{\lambda}) = -i\sigma_1 T_\pm^{(1,2)}(x,\lambda), \tag{4.47}$$

where λ is in the respective domain of analyticity.

2. The reduced monodromy matrix and transition coefficients

The reduced monodromy matrix is defined for real $\lambda \neq 0$ to be the ratio of the Jost solutions,

$$T(\lambda) = T_+^{-1}(x,\lambda)\, T_-(x,\lambda) \tag{4.48}$$

and can be expressed as a limit

$$T(\lambda) = \lim_{\substack{x\to+\infty \\ y\to-\infty}} E_+^{-1}(x,\lambda)\, T(x,y,\lambda)\, E_-(y,\lambda). \tag{4.49}$$

The matrix $T(\lambda)$ is unimodular and obeys the involutions

$$\bar{T}(\lambda) = \sigma_2 T(\lambda) \sigma_2 \tag{4.50}$$

and

$$\bar{T}(-\lambda) = T(\lambda). \tag{4.51}$$

It has the familiar form

$$T(\lambda) = \begin{pmatrix} a(\lambda) & -\bar{b}(\lambda) \\ b(\lambda) & \bar{a}(\lambda) \end{pmatrix}, \tag{4.52}$$

where the transition coefficients for the continuous spectrum, $a(\lambda)$ and $b(\lambda)$, satisfy the normalization relation

$$|a(\lambda)|^2+|b(\lambda)|^2=1 \tag{4.53}$$

and the symmetry relations

$$a(-\lambda)=\bar{a}(\lambda), \quad b(-\lambda)=\bar{b}(\lambda). \tag{4.54}$$

We point out the additional property (4.54) of the transition coefficients which is due to the involution (4.8).

For $a(\lambda)$ we have

$$a(\lambda)=\det(T_-^{(1)}(x,\lambda),\, T_+^{(2)}(x,\lambda)), \tag{4.55}$$

which implies that it has an analytic continuation into the upper half-plane with the asymptotic behaviour

$$a(\lambda)=1+O\left(\frac{1}{|\lambda|}\right) \tag{4.56}$$

as $|\lambda|\to\infty$ and

$$a(\lambda)=(-1)^Q+O(|\lambda|) \tag{4.57}$$

as $\lambda\to 0$. The involution (4.54) for λ complex takes the form

$$a(-\bar{\lambda})=\bar{a}(\lambda). \tag{4.58}$$

An analogous expression for $b(\lambda)$,

$$b(\lambda)=\det(T_-^{(1)}(x,\lambda),\, T_+^{(1)}(x,\lambda)), \tag{4.59}$$

shows that $b(\lambda)$ is of Schwartz type and vanishes with all derivatives at $\lambda=0$. In general, $b(\lambda)$ has no analytic extension off the real line. Such an extension exists if $\varphi(x)$ and $\pi(x)$ coincide with their asymptotic expressions for $|x|>q$ for some $q>0$. In that case the transition coefficients $a(\lambda)$ and $b(\lambda)$ are regular in the domain $\mathbb{C}\setminus\{0\}$ and have essential singularities at $\lambda=0$ and $\lambda=\infty$.

As in the case of the HM model and the NS model for $\varkappa<0$, we shall assume the condition (A) to the effect that

$$|b(\lambda)|<1 \tag{4.60}$$

and the zeros $\lambda_j, j=1,\dots,n$, of $a(\lambda)$ (their number is finite) are simple. In view of (4.58), these are symmetric with respect to the imaginary axis, so that there are pure imaginary zeros $\lambda_j=i\varkappa_j$, $\varkappa_j>0$, $j=1,\dots,n_1$, and symmetric pairs λ_k, $\lambda_{k+n_2}=-\bar{\lambda}_k$, $\operatorname{Im}\lambda_k$, $\operatorname{Re}\lambda_k>0$, $k=n_1+1,\dots,n_1+n_2$, where $n=n_1+2n_2$.

The quantities $\lambda_j, \bar{\lambda}_j$ constitute the discrete spectrum of the auxiliary linear problem (4.1). The associated transition coefficients are defined by the relations

$$T_-^{(1)}(x,\lambda_j)=\gamma_j T_+^{(2)}(x,\lambda_j), \quad j=1,\dots,n. \tag{4.61}$$

We have

$$\bar{\gamma}_j=\gamma_j, \quad j=1,\dots,n_1; \quad \bar{\gamma}_k=\gamma_{k+n_2}, \quad k=n_1+1,\dots,n_1+n_2. \tag{4.62}$$

In the case when $b(\lambda)$ admits an analytic continuation to $\mathbb{C}\backslash\{0\}$ we have

$$\gamma_j=b(\lambda_j), \quad j=1,\dots,n. \tag{4.63}$$

The function $a(\lambda)$ is uniquely determined by the coefficient $b(\lambda)$ and the zeros $\lambda_1,\dots,\lambda_n$. The corresponding dispersion relation has the familiar form

$$\begin{aligned} a(\lambda)=&\prod_{j=1}^{n_1}\frac{\lambda-i\varkappa_j}{\lambda+i\varkappa_j}\prod_{k=n_1+1}^{n_1+n_2}\frac{\lambda-\lambda_k}{\lambda-\bar{\lambda}_k}\cdot\frac{\lambda+\bar{\lambda}_k}{\lambda+\lambda_k} \\ &\times\exp\left\{\frac{1}{2\pi i}\int_{-\infty}^{\infty}\frac{\log(1-|b(\mu)|^2)}{\mu-\lambda-i0}\,d\mu\right\}. \end{aligned} \tag{4.64}$$

In particular, this implies

$$a(0)=(-1)^{n_1}, \tag{4.65}$$

or

$$Q\equiv n_1 (\operatorname{mod} 2). \tag{4.66}$$

Thus we have described the mapping

$$\begin{aligned} \mathscr{F}\colon (\pi(x),\varphi(x))\to(&b(\lambda),\bar{b}(\lambda);\varkappa_j,\lambda_k,\bar{\lambda}_k,\gamma_j,\gamma_k,\bar{\gamma}_k, \\ &j=1,\dots,n_1;\ k=n_1+1,\dots,n_1+n_2) \end{aligned} \tag{4.67}$$

from the functions $\pi(x)$ and $\varphi(x)$ to the transition coefficients and the discrete spectrum of the auxiliary linear problem (4.1). In the next section we shall see that $\mathscr{F}$ has an inverse, and in § 6 this mapping will be used to define a canonical transformation to action-angle variables for the SG model.

3. Time evolution of the transition coefficients

We shall determine the evolution of the transition coefficients when the functions $\varphi(x, t)$ and $\pi(x, t) = \frac{\partial \varphi}{\partial t}(x, t)$ that enter into the auxiliary linear problem satisfy the SG equation. To do so, we make use of the evolution equation for the transition matrix, which follows from the zero curvature representation:

$$\frac{\partial T}{\partial t}(x, y, \lambda) = V(x, \lambda)\, T(x, y, \lambda) - T(x, y, \lambda)\, V(y, \lambda), \tag{4.68}$$

where

$$V(x, t, \lambda) = \frac{1}{4i}\beta \frac{\partial \varphi}{\partial x}\sigma_3 + \frac{m}{4i}\left(\lambda - \frac{1}{\lambda}\right)\sin\frac{\beta\varphi}{2}\sigma_1 + \frac{m}{4i}\left(\lambda + \frac{1}{\lambda}\right)\cos\frac{\beta\varphi}{2}\sigma_2 \tag{4.69}$$

(see § I.1). We have

$$\begin{aligned}
&\lim_{x\to\pm\infty} E_{\pm}^{-1}(x, \lambda)\, V(x, t, \lambda)\, E_{\pm}(x, \lambda) \\
&= \frac{(-1)^{Q} m}{4i}\left(\lambda + \frac{1}{\lambda}\right)\lim_{x\to\pm\infty} E_{\pm}^{-1}(x, \lambda)\sigma_2 E_{\pm}(x, \lambda) \\
&= \frac{m}{4i}\left(\lambda + \frac{1}{\lambda}\right)\sigma_3,
\end{aligned} \tag{4.70}$$

so that letting $y \to \pm\infty$, $x \to +\infty$ in (4.68) we find the evolution equations for the Jost solutions,

$$\frac{\partial T_{\pm}(x, \lambda)}{\partial t} = V(x, \lambda)\, T_{\pm}(x, \lambda) - \frac{m}{4i}\left(\lambda + \frac{1}{\lambda}\right) T_{\pm}(x, \lambda)\sigma_3, \tag{4.71}$$

and for the reduced monodromy matrix

$$\frac{\partial T(\lambda)}{\partial t} = \frac{m}{4i}\left(\lambda + \frac{1}{\lambda}\right)[\sigma_3, T(\lambda)]. \tag{4.72}$$

Thus, the time dependence of the transition coefficients is given by

$$\begin{aligned} &a(\lambda,t)=a(\lambda,0), \quad b(\lambda,t)=e^{\frac{mi}{2}\left(\lambda+\frac{1}{\lambda}\right)t}\,b(\lambda,0), \\ &\lambda_j(t)=\lambda_j(0), \quad \gamma_j(t)=e^{\frac{mi}{2}\left(\lambda_j+\frac{1}{\lambda_j}\right)t}\,\gamma_j(0), \quad j=1,\ldots,n. \end{aligned} \tag{4.73}$$

Observe that the vanishing of $b(\lambda)$ with all its derivatives at $\lambda=0$ is consistent with the SG dynamics.

As has already become familiar, the role of the generating function for integrals of the motion in the rapidly decreasing case is played by the coefficient $a(\lambda)$. To conclude this section, we outline a method for constructing local integrals of the motion.

4. Local integrals of the motion

A distinctive feature of the SG model is that it has two families of local integrals of the motion resulting from the asymptotic expansion of the reduced monodromy matrix $T(\lambda)$ at the poles $\lambda=\infty$ and $\lambda=0$.

We start with the asymptotic expansion of the transition matrix $T(x,y,\lambda)$ as $|\lambda|\to\infty$. Make a gauge transformation with matrix $\Omega^{-1}(x)$,

$$T(x,y,\lambda)=\Omega(x)\,\tilde{T}(x,y,\lambda), \tag{4.74}$$

and express $\tilde{T}(x,y,\lambda)$ as

$$\tilde{T}(x,y,\lambda)=(I+W(x,\lambda))\exp Z(x,y,\lambda)\,(I+W(y,\lambda))^{-1} \tag{4.75}$$

$(\operatorname{mod} O(|\lambda|^{-\infty}))$, where $W(x,\lambda)$ is an off-diagonal matrix and $Z(x,y,\lambda)$ is a diagonal matrix satisfying

$$Z(x,y,\lambda)|_{x=y}=0. \tag{4.76}$$

From (4.25) it follows that

$$Z(x,y,\lambda)=\frac{1}{4i}\int_y^x\left(\beta\theta(x')\sigma_3+m\left(\lambda\sigma_2-\frac{1}{\lambda}\sigma_2 e^{i\beta\varphi(x')\sigma_3}\right)W(x',\lambda)\right)dx', \tag{4.77}$$

and $W(x,\lambda)$ satisfies a differential equation of Riccati type,

$$\begin{aligned} \frac{dW}{dx}=\frac{\beta}{2i}\theta\sigma_3 W+\frac{m}{4i}\lambda(\sigma_2-W\sigma_2 W) \\ -\frac{m}{4i\lambda}(\sigma_2 e^{i\beta\varphi\sigma_3}-W\sigma_2 e^{i\beta\varphi\sigma_3}W). \end{aligned} \tag{4.78}$$

The matrix $W(x,\lambda)$ has an asymptotic expansion of the form

$$W(x,\lambda)=\sum_{n=0}^{\infty}\frac{W_n(x)}{\lambda^n}, \tag{4.79}$$

where

$$W_0(x)=i\sigma_1 \tag{4.80}$$

and

$$\begin{aligned}W_{n+1}(x)=&\frac{2i\sigma_3}{m}\frac{dW_n(x)}{dx}-\frac{\beta\theta(x)}{m}W_n(x)\\&+\frac{i}{2}\sum_{k=1}^{n}W_k(x)\sigma_1 W_{n+1-k}(x)\\&-\frac{i}{2}\sum_{k=0}^{n-1}W_k(x)\sigma_1 e^{i\beta\varphi(x)\sigma_3}W_{n-1-k}(x)\\&-\frac{i}{2}\sigma_1 e^{i\beta\varphi(x)\sigma_3}\delta_{n,1},\qquad n=0,1,\ldots.\end{aligned} \tag{4.81}$$

The corresponding expansion for $Z(x,y,\lambda)$ has the form

$$Z(x,y,\lambda)=\frac{m\lambda(x-y)}{4i}\sigma_3+i\sum_{n=1}^{\infty}\frac{Z_n(x,y)}{\lambda^n}, \tag{4.82}$$

where

$$Z_n(x,y)=\frac{m}{4}\int_y^x\sigma_2(e^{i\beta\varphi(x')\sigma_3}W_{n-1}(x')-W_{n+1}(x'))\,dx'. \tag{4.83}$$

We observe that the choice of $W_0(x)$ (4.80) is consistent with the term $\frac{m\lambda(x-y)}{4i}\sigma_3$ in (4.82).

In view of the involution (4.7), $W_n(x)$ and $Z_n(x,y)$ have the form

$$W_n(x)=\begin{pmatrix}0 & -\bar{w}_n(x)\\ w_n(x) & 0\end{pmatrix} \tag{4.84}$$

and

$$Z_n(x,y)=\begin{pmatrix}z_n(x,y) & 0\\ 0 & -\bar{z}_n(x,y)\end{pmatrix}, \tag{4.85}$$

and relations (4.80)–(4.83) can be written as follows:

$$w_0(x) = i, \tag{4.86}$$

$$w_{n+1}(x) = \frac{2}{im}\frac{dw_n(x)}{dx} - \frac{\beta\theta(x)}{m} w_n(x) + \frac{i}{2}\sum_{k=1}^{n} w_k(x) w_{n+1-k}(x)$$
$$- \frac{i}{2} e^{-i\beta\varphi(x)} \sum_{k=0}^{n-1} w_k(x) w_{n-k-1}(x) - \frac{i}{2} e^{i\beta\varphi(x)} \delta_{n,1}, \tag{4.87}$$

$$z_n(x,y) = \frac{im}{4} \int_y^x (w_{n+1}(x') - e^{-i\beta\varphi(x')} w_{n-1}(x'))\, dx'. \tag{4.88}$$

To derive the asymptotic expansion of the reduced monodromy matrix $T(\lambda)$ we take the limits as $y \to -\infty$, $x \to +\infty$, as prescribed by the definition (4.49). Noting that the matrices $W_n(x)$, $n \geq 1$, vanish as $|x| \to \infty$, we find

$$T(\lambda) = e^{P(\lambda)} + O(|\lambda|^{-\infty}), \tag{4.89}$$

where

$$P(\lambda) = i \begin{pmatrix} p(\lambda) & 0 \\ 0 & -\bar{p}(\lambda) \end{pmatrix} \tag{4.90}$$

and

$$p(\lambda) = \lim_{\substack{x \to +\infty \\ y \to -\infty}} \left(\sum_{n=1}^{\infty} \frac{z_n(x,y)}{\lambda^n} - \frac{m}{4\lambda}(x-y) \right). \tag{4.91}$$

We emphasize that the property of $P(\lambda)$ to be a diagonal matrix is in agreement with the fact that $b(\lambda)$ is a function of Schwartz type.

From (4.86)–(4.88) we deduce that $p(\lambda)$ has an asymptotic expansion

$$p(\lambda) = \sum_{n=1}^{\infty} \frac{I_n}{\lambda^n}, \tag{4.92}$$

with

$$I_1 = -\frac{\beta^2}{4m} \int_{-\infty}^{\infty} \left(\frac{1}{2}\left(\pi(x) + \frac{d\varphi}{dx}(x)\right)^2 + \frac{m^2}{\beta^2}(1 - \cos\beta\varphi(x)) \right) dx \tag{4.93}$$

and for arbitrary $n > 1$

$$I_n = \frac{im}{4} \int_{-\infty}^{\infty} (w_{n+1}(x) - e^{-i\beta\varphi(x)} w_{n-1}(x))\, dx. \tag{4.94}$$

By virtue of $\operatorname{tr} P(\lambda) = 0$, which is a consequence of the fact that $T(\lambda)$ is unimodular, the quantities I_n are real.

By comparing (4.52) and (4.89)–(4.90), (4.92) we obtain the required expansion

$$\log a(\lambda) = i \sum_{n=1}^{\infty} \frac{I_n}{\lambda^n}, \tag{4.95}$$

as $|\lambda| \to \infty$, which gives the first sequence of local integrals of the motion for the SG model.

To derive the asymptotic expansion for $\lambda \to 0$ it suffices to use (4.9) (cf. Subsection 1). As a result we have

$$\log a(\lambda) = i \sum_{n=0}^{\infty} I_{-n} \lambda^n, \tag{4.96}$$

where

$$I_0 \equiv \pi Q \pmod{2\pi} \tag{4.97}$$

and

$$I_{-n}(\pi, \varphi) = (-1)^n I_n(\pi, -\varphi), \tag{4.98}$$

$n = 1, 2, \ldots$. In particular, for the momentum P and the Hamiltonian H of the SG model we have

$$P = \frac{2m}{\beta^2} (I_{-1} + I_1) \tag{4.99}$$

and

$$H = \frac{2m}{\beta^2} (I_{-1} - I_1). \tag{4.100}$$

Finally, comparing the asymptotic expansions (4.95)–(4.96) with the dispersion relation (4.64) we obtain the trace identities for the SG model

$$\operatorname{sign} l \cdot I_l = \frac{1}{2\pi} \int_{-\infty}^{\infty} \log(1 - |b(\lambda)|^2) \lambda^{l-1} d\lambda + \sum_{j=1}^{n} \frac{\bar{\lambda}_j^l - \lambda_j^l}{il}, \qquad l = -\infty, \ldots, \infty. \tag{4.101}$$

From (4.54) it follows that I_l vanishes for l even, so that the corresponding density in (4.94) is a total derivative of a Schwartz function. Hence the final form of the trace identities is

$$\operatorname{sign}(2m+1)\, I_{2m+1} = \frac{1}{\pi}\int_0^\infty \log(1-|b(\lambda)|^2)\lambda^{2m}\, d\lambda$$
$$-\frac{(-1)^m 2}{2m+1}\sum_{j=1}^{n_1}\varkappa_j^{2m+1}$$
$$+\frac{2}{2m+1}\sum_{k=n_1+1}^{n_1+n_2}\frac{\bar{\lambda}_k^{2m+1}-\lambda_k^{2m+1}}{i},$$
$$m=-\infty,\ldots,\infty. \tag{4.102}$$

This completes our analysis of the auxiliary linear problem and the mapping $\mathscr{F}$ for the SG model.

§ 5. The Inverse Problem for the SG Model

In this section we shall describe the mapping $\mathscr{F}^{-1}$, i.e. we present the solution of the inverse problem which amounts to reconstructing the functions $\pi(x)$ and $\varphi(x)$ from the transition coefficients and the discrete spectrum. We shall outline two methods, one based on the Riemann problem, the other on the Gelfand-Levitan-Marchenko formulation. We shall close the section with a discussion of soliton dynamics.

1. The Riemann problem

It is based on the relationship between the Jost solutions for real $\lambda \neq 0$,

$$T_-(x,\lambda) = T_+(x,\lambda)\, T(\lambda), \tag{5.1}$$

which can be written as

$$F_-(x,\lambda) = F_+(x,\lambda)\, G(\lambda), \tag{5.2}$$

where the matrices $F_\pm(x,\lambda)$ are composed of the columns of the solutions $T_\pm(x,\lambda)$ according to

$$F_+(x,\lambda) = \frac{1}{a(\lambda)}\left(T_-^{(1)}(x,\lambda),\, T_+^{(2)}(x,\lambda)\right), \tag{5.3}$$

$$F_-(x,\lambda)=(T_+^{(1)}(x,\lambda),\,T_-^{(2)}(x,\lambda)), \tag{5.4}$$

and

$$G(\lambda)=\begin{pmatrix} 1 & -\bar{b}(\lambda) \\ -b(\lambda) & 1 \end{pmatrix}. \tag{5.5}$$

The matrices $F_+^{-1}(x,\lambda)$ and $F_-(x,\lambda)$ have an analytic continuation into the half-planes $\operatorname{Im}\lambda>0$ and $\operatorname{Im}\lambda<0$, respectively, but $\lambda=0$ and $\lambda=\infty$ are their essentially singular points (see Subsection 1 of § 4). With the definition

$$G_+(x,\lambda)=e^{\frac{k_1(\lambda)x}{i}\sigma_3}F_+^{-1}(x,\lambda) \tag{5.6}$$

and

$$G_-(x,\lambda)=F_-(x,\lambda)e^{-\frac{k_1(\lambda)x}{i}\sigma_3}, \tag{5.7}$$

so that $G_\pm(x,\lambda)$ have finite limits as $\lambda\to 0$ and $|\lambda|\to\infty$ in the respective half-planes, we can write (5.2) as

$$G_+(x,\lambda)G_-(x,\lambda)=G(x,\lambda), \tag{5.8}$$

with

$$G(x,\lambda)=e^{\frac{k_1(\lambda)x}{i}\sigma_3}G(\lambda)e^{-\frac{k_1(\lambda)x}{i}\sigma_3} =\begin{pmatrix} 1 & -e^{-\frac{im}{2}\left(\lambda-\frac{1}{\lambda}\right)x}\bar{b}(\lambda) \\ -e^{\frac{im}{2}\left(\lambda-\frac{1}{\lambda}\right)x}b(\lambda) & 1 \end{pmatrix}. \tag{5.9}$$

Relation (5.8) is fundamental to the Riemann problem for the SG model. Before stating it explicitly, we shall give a list of the properties of $G(x,\lambda)$ and $G_\pm(x,\lambda)$ which follow from the results of § 4.

I. *The matrix* $G(x,\lambda)$ *is Hermitian,*

$$G^*(x,\lambda)=G(x,\lambda), \tag{5.10}$$

satisfies the involution

$$\bar{G}(x,-\lambda)=G(x,\lambda) \tag{5.11}$$

and the constraints

$$G(x,\lambda)|_{\lambda=0}=I, \qquad \left.\frac{\partial^k G(x,\lambda)}{\partial\lambda^k}\right|_{\lambda=0}=0, \quad k=1,2,\ldots, \tag{5.12}$$

$$\lim_{|\lambda|\to\infty} G(x,\lambda)=I, \tag{5.13}$$

where the limiting values are approached in the sense of Schwartz.

II. *The matrices $G_+(x,\lambda)$ and $G_-(x,\lambda)$ admit an analytic continuation into the upper and lower half-planes, respectively, and satisfy the involutions*

$$G_+^*(x,\lambda)=G_-(x,\bar{\lambda}), \tag{5.14}$$

$$\bar{G}_+(x,-\bar{\lambda})=iG_+(x,\lambda)\sigma_1, \tag{5.15}$$

$$\bar{G}_-(x,-\bar{\lambda})=-i\sigma_1 G_-(x,\lambda). \tag{5.16}$$

III. *In their domains of analyticity, $G_\pm(x,\lambda)$ have the asymptotic behaviour*

$$G_+(x,\lambda)=\mathscr{E}^{-1}\Omega^{-1}(x)\left(I+O\left(\frac{1}{|\lambda|}\right)\right), \tag{5.17}$$

$$G_-(x,\lambda)=\Omega(x)\mathscr{E}\left(I+O\left(\frac{1}{|\lambda|}\right)\right) \tag{5.18}$$

as $|\lambda|\to\infty$ and

$$G_+(x,\lambda)=(-\sigma_3)^Q\mathscr{E}^{-1}\Omega(x)(I+O(|\lambda|)), \tag{5.19}$$

$$G_-(x,\lambda)=\Omega^{-1}(x)\mathscr{E}(-\sigma_3)^Q(I+O(|\lambda|)) \tag{5.20}$$

as $\lambda\to 0$ with

$$\mathscr{E}=\frac{1}{\sqrt{2}}\begin{pmatrix}1 & i\\ i & 1\end{pmatrix}. \tag{5.21}$$

IV. *The matrices $G_+(x,\lambda)$ and $G_-(x,\lambda)$ are nondegenerate in their domains of analyticity except for the points $\lambda=\lambda_j$ and $\lambda=\bar{\lambda}_j$, respectively, where*

$$\operatorname{Im} G_+(x,\lambda_j)=N_j^{(+)}(x), \tag{5.22}$$

$$\operatorname{Ker} G_-(x,\bar{\lambda}_j)=N_j^{(-)}(x), \quad j=1,\ldots,n. \tag{5.23}$$

Here $N_j^{(+)}(x)$ and $N_j^{(-)}(x)$ are one-dimensional subspaces in $\mathbb{C}^2$ spanned respectively by the vectors

$$\begin{pmatrix} 1 \\ -e^{\frac{im}{2}\left(\lambda_j-\frac{1}{\lambda_j}\right)x}\gamma_j \end{pmatrix} \quad \text{and} \quad \begin{pmatrix} e^{-\frac{im}{2}\left(\bar{\lambda}_j-\frac{1}{\bar{\lambda}_j}\right)x}\bar{\gamma}_j \\ 1 \end{pmatrix}.$$

The subspaces $N_j^{(\pm)}(x)$ *obey the involutions*

$$\bar{N}_j^{(\pm)}(x)=N_j^{(\pm)}(x) \tag{5.24}$$

for $j=1,\ldots,n_1$ *and*

$$\bar{N}_{k+n_2}^{(\pm)}(x)=N_k^{(\pm)}(x) \tag{5.25}$$

for $k=n_1+1,\ldots,n_1+n_2$. The bar indicates complex conjugation in $\mathbb{C}^2$.

The distinction of the properties I–IV listed above from those for the HM and NS models is that there exist two normalization points, $\lambda=0$ and $\lambda=\infty$, (see (5.17)–(5.20)) and an additional involution $\lambda\to-\lambda$ (see (5.11), (5.15)–(5.16), (5.24)–(5.26)).

We now proceed to solve the inverse problem. *Suppose we are given positive parameters* m,β, *functions* $b(\lambda)$, $\bar{b}(\lambda)$ *and a set of quantities* λ_j, $\bar{\lambda}_j$, γ_j, $\bar{\gamma}_j$, $j=1,\ldots,n$, *with the following properties.*

I′. *The function* $b(\lambda)$ *lies in Schwartz space, vanishes with all derivatives at* $\lambda=0$ *and satisfies the inequality*

$$|b(\lambda)|<1 \tag{5.26}$$

and the involution

$$b(-\lambda)=\bar{b}(\lambda). \tag{5.27}$$

II′. *The quantities* λ_j *are distinct and* $\lambda_j=i\varkappa_j$, $\varkappa_j>0$ *for* $j=1,\ldots,n_1$ *and* $\lambda_{k+n_2}=-\bar{\lambda}_k$, $\operatorname{Im}\lambda_k$, $\operatorname{Re}\lambda_k>0$, $k=n_1+1,\ldots,n_1+n_2$ *where* $n=n_1+2n_2$. Also, $\gamma_j=\bar{\gamma}_j\neq0, j=1,\ldots,n_1$ *and* $\gamma_{k+n_2}=\bar{\gamma}_k\neq0$, $k=n_1+1,\ldots,n_1+n_2$.

Given these data, construct the matrix $G(x,\lambda)$ satisfying conditions I, and a set of subspaces $N_j^{(\pm)}(x)$ satisfying (5.24)–(5.25). The Riemann problem is parametrized by the variable x and has the form

$$G(x,\lambda)=G_{\pm}(x,\lambda)\,G_{-}(x,\lambda). \tag{5.28}$$

Here the matrix-valued functions $G_{\pm}(x,\lambda)$ have an analytic continuation into the domains $\pm\operatorname{Im}\lambda\geqslant0$ and satisfy (5.22)–(5.23) there. Also, at $\lambda=0$ and $\lambda=\infty$ they are normalized as prescribed by (5.17)–(5.20), where Q is replaced by n_1 and $\Omega(x)$ is a continuous diagonal matrix function to be determined, so that

$$\lim_{x\to-\infty} \Omega(x)=I. \tag{5.29}$$

Thus, the Riemann problem for the SG model involves nontrivial normalization conditions: the values of $G_\pm(x,\lambda)$ at singular points are not fixed, but rather their matrix structure is stipulated.

We claim the following.

I″. *The Riemann problem stated above has a unique solution.*

II″. *The matrices $F_\pm(x,\lambda)$ constructed from the solutions $G_\pm(x,\lambda)$ according to (5.6)–(5.7) satisfy the auxiliary linear equation*

$$\frac{dF_\pm(x,\lambda)}{dx}=\frac{1}{4i}\left(\beta\pi(x)\sigma_3+m\left(\lambda+\frac{1}{\lambda}\right)\sin\frac{\beta\varphi(x)}{2}\sigma_1\right.$$
$$\left.+m\left(\lambda-\frac{1}{\lambda}\right)\cos\frac{\beta\varphi(x)}{2}\sigma_2\right)F_\pm(x,\lambda), \tag{5.30}$$

with $\pi(x)$ and $\varphi(x)$ real-valued functions and

$$\Omega(x)=e^{\frac{i\beta\varphi(x)}{4}\sigma_3}. \tag{5.31}$$

III″. *The functions $\pi(x)$, $\varphi(x)$ satisfy the rapidly decreasing boundary conditions*

$$\lim_{|x|\to\infty}\pi(x)=0,\qquad \lim_{x\to-\infty}\varphi(x)=0,\qquad \lim_{x\to+\infty}\varphi(x)=\frac{2\pi}{\beta}Q, \tag{5.32}$$

where Q is an integer, $Q\equiv n_1 \pmod 2$.

IV″. *The functions $a(\lambda)$ and $b(\lambda)$, where $a(\lambda)$ is given by (4.64), are the transition coefficients for the auxiliary linear problem (5.30); the discrete spectrum consists of the quantities $\lambda_1,\ldots,\lambda_n,\bar\lambda_1,\ldots,\bar\lambda_n$ with transition coefficients $\gamma_1,\ldots,\gamma_n,\bar\gamma_1,\ldots,\bar\gamma_n$. The matrices $F_\pm(x,\lambda)$ are composed of the Jost solutions $T_\pm(x,\lambda)$ of the auxiliary linear problem as in (5.3)–(5.4).*

We shall comment on the proof of these assertions.

The uniqueness theorem for the Riemann problem is proved by recourse to the Liouville theorem. More precisely, let $G_\pm(x,\lambda)$ and $\tilde G_\pm(x,\lambda)$ be two solutions of (5.28). For real λ we have

$$\Phi(x,\lambda)=G_+^{-1}(x,\lambda)\tilde G_+(x,\lambda)=G_-(x,\lambda)\tilde G_-^{-1}(x,\lambda). \tag{5.33}$$

A standard argument shows that $\Phi(x,\lambda)$ has no singularities at $\lambda=\lambda_j,\bar\lambda_j$ and hence is an entire function. From (5.17)–(5.20) it follows that

$$\Phi(x,\lambda)|_{\lambda=0}=\Omega^{-1}(x)\tilde\Omega(x) \tag{5.34}$$

and

$$\Phi(x,\lambda)|_{\lambda=\infty}=\Omega(x)\tilde{\Omega}^{-1}(x), \tag{5.35}$$

hence (the Liouville theorem)

$$\Phi(x,\lambda)=\Omega^{-1}(x)\tilde{\Omega}(x)=\Omega(x)\tilde{\Omega}^{-1}(x), \tag{5.36}$$

that is

$$\Omega^2(x)=\tilde{\Omega}^2(x). \tag{5.37}$$

Using (5.29) and the fact that $\Omega(x)$, $\tilde{\Omega}(x)$ are diagonal continuous nonsingular matrices we deduce that

$$\Omega(x)=\tilde{\Omega}(x), \tag{5.38}$$

whence

$$G_{\pm}(x,\lambda)=\tilde{G}_{\pm}(x,\lambda). \tag{5.39}$$

Notice that the proof made use of (5.29), i.e. we were dealing with the whole family of the Riemann problems (5.28) parametrized by x. For x fixed, the Riemann problem (5.28) clearly has, besides $G_{\pm}(x,\lambda)$, the solution $-G_{\pm}(x,\lambda)$.

To prove the existence theorem, we will show how our Riemann problem can be reduced to the Riemann problem with normalization to unity at $\lambda=\infty$ studied in § II.2 of Part I.

Let $\hat{G}_{\pm}(x,\lambda)$ be a solution of the following Riemann problem

$$G(x,\lambda)=\hat{G}_{+}(x,\lambda)\hat{G}_{-}(x,\lambda), \tag{5.40}$$

where $G(x,\lambda)$ is the matrix of the Riemann problem (5.28) and a) the matrices $\hat{G}_{\pm}(x,\lambda)$ extend analytically to the half-planes $\pm\operatorname{Im}\lambda\geqslant 0$ and are normalized to I at $\lambda=\infty$,

$$\hat{G}_{\pm}(x,\lambda)=I+O\left(\frac{1}{|\lambda|}\right); \tag{5.41}$$

b) the matrices $\hat{G}_{\pm}(x,\lambda)$ are nondegenerate everywhere except $\lambda=\lambda_j$ and $\lambda=\bar{\lambda}_j$, respectively, and

$$\operatorname{Im}\hat{G}_{+}(x,\lambda_j)=N_j^{(+)}(x), \tag{5.42}$$

$$\operatorname{Ker}\hat{G}_{-}(x,\bar{\lambda}_j)=N_j^{(-)}(x), \quad j=1,\ldots,n, \tag{5.43}$$

where $\lambda_j, \bar{\lambda}_j$ and the subspaces $N_j^{(\pm)}(x)$ belong to the Riemann problem (5.28). In § II.2, Part I, it was shown that the Riemann problem stated there is uniquely solvable.

Define the matrix $\Omega^2(x)$ by

$$\Omega^2(x) = \mathscr{E}(-\sigma_3)^{n_1} \hat{G}_+(x,0)\mathscr{E}^{-1} = \mathscr{E}(-\sigma_3)^{n_1} \hat{G}_-^{-1}(x,0)\mathscr{E}^{-1} \tag{5.44}$$

and let us verify that it is a diagonal matrix. Indeed, the involutions

$$\hat{G}_+^*(x,\bar{\lambda}) = \hat{G}_-(x,\lambda) \tag{5.45}$$

and

$$\overline{\hat{G}_\pm(x,-\bar{\lambda})} = \hat{G}_\pm(x,\lambda) \tag{5.46}$$

with the requirement

$$\hat{G}_+(x,0)\,\hat{G}_-(x,0) = I \tag{5.47}$$

imply that the matrix $\hat{G}_+(x,0)$ is unitary and real, hence orthogonal. Its determinant is given by

$$\det \hat{G}_+(x,0) = a(0) = (-1)^{n_1}, \tag{5.48}$$

so that the matrix $(-\sigma_3)^{n_1}\hat{G}_+(x,0)$ is unimodular, i.e.

$$(-\sigma_3)^{n_1}\hat{G}_+(x,0) = \begin{pmatrix} \cos\alpha(x) & -\sin\alpha(x) \\ \sin\alpha(x) & \cos\alpha(x) \end{pmatrix} \tag{5.49}$$

and the specific form of $\mathscr{E}$ leads to

$$\Omega^2(x) = e^{i\alpha(x)\sigma_3}. \tag{5.50}$$

Let us now consider the asymptotic behaviour of $\Omega^2(x)$ as $x \to -\infty$. Arguing as in § II.2, Part I, we can show that

$$\lim_{x\to-\infty} \hat{G}(x,\lambda) = \begin{pmatrix} a(\lambda) & -\dfrac{\bar{b}(\lambda)}{a(\lambda)} e^{-\frac{im}{2}\left(\lambda - \frac{1}{\lambda}\right)x} \\ 0 & 1 \end{pmatrix}, \tag{5.51}$$

whence, in particular, we get

$$\lim_{x\to-\infty} \hat{G}(x,0)=(-\sigma_3)^{n_1}, \tag{5.52}$$

so that

$$\lim_{x\to-\infty} \Omega^2(x)=I. \tag{5.53}$$

Hence $\Omega(x)$ is uniquely determined as a continuous diagonal square root of $\Omega^2(x)$ that satisfies (5.29).

It is now obvious that the matrices

$$G_+(x,\lambda)=\hat{G}_+(x,\lambda)\mathscr{E}^{-1}\Omega^{-1}(x), \tag{5.54}$$

$$G_-(x,\lambda)=\Omega(x)\mathscr{E}\hat{G}_-(x,\lambda) \tag{5.55}$$

provide a solution to the Riemann problem for the SG model in terms of the Riemann problem with the standard normalization.

To derive the differential equation of the auxiliary linear problem of item II″, we write the Riemann problem (5.28) in the form (5.2) and differentiate it with respect to x. We get

$$U(x,\lambda)=\frac{\partial F_+(x,\lambda)}{\partial x}F_+^{-1}(x,\lambda)=\frac{\partial F_-(x,\lambda)}{\partial x}F_-^{-1}(x,\lambda). \tag{5.56}$$

In the usual way we conclude that $U(x,\lambda)$ is an entire matrix-function. It follows from (5.17)–(5.20) that

$$U(x,\lambda)=\frac{m\lambda}{4i}\Omega(x)\sigma_2\Omega^{-1}(x)+C_0(x)+O\left(\frac{1}{|\lambda|}\right), \tag{5.57}$$

as $|\lambda|\to\infty$, and

$$U(x,\lambda)=-\frac{m}{4i\lambda}\Omega^{-1}(x)\sigma_2\Omega(x)+\tilde{C}_0(x)+O(|\lambda|), \tag{5.58}$$

as $\lambda\to 0$. By the Liouville theorem we then find

$$U(x,\lambda)=C(x)+\frac{m\lambda}{4i}\Omega(x)\sigma_2\Omega^{-1}(x)-\frac{m}{4i\lambda}\Omega^{-1}(x)\sigma_2\Omega(x), \tag{5.59}$$

with

$$C(x)=C_0(x)=\tilde{C}_0(x). \tag{5.60}$$

Let us now determine the matrix structure of $C(x)$. Relations (5.10)–(5.11) and the uniqueness theorem imply that the solutions $G_{\pm}(x,\lambda)$ obey the involutions (5.14)–(5.15), hence

$$U^*(x,\lambda) = -U(x,\bar{\lambda}), \tag{5.61}$$

$$\bar{U}(x,-\bar{\lambda}) = \sigma_1 U(x,\lambda)\sigma_1, \tag{5.62}$$

so that

$$C^*(x) = -C(x) \tag{5.63}$$

and

$$\bar{C}(x) = \sigma_1 C(x)\sigma_1. \tag{5.64}$$

This allows us to introduce a real-valued function $\pi(x)$ by

$$C(x) = \frac{\beta}{4i}\pi(x)\sigma_3. \tag{5.65}$$

Setting $\varphi(x) = \frac{2}{\beta}\alpha(x)$ we see that $U(x,\lambda)$ takes the form (5.30).

The proof of assertions III″–IV″ follows the same lines as in § II.2, Part I.

As in the case of the NS and HM models, the time evolution (4.73) of the transition coefficients leads to the zero curvature representation for the SG model. This ensures that the functions $\varphi(x,t)$ and $\pi(x,t) = \frac{\partial\varphi}{\partial t}(x,t)$ recovered from such transition coefficients satisfy the SG equation.

2. The Gelfand-Levitan-Marchenko formulation

The second approach to solving the inverse problem is also based on (5.1) which must now be written as

$$\frac{1}{a(\lambda)} T_{-}^{(1)}(x,\lambda) = T_{+}^{(1)}(x,\lambda) + r(\lambda) T_{+}^{(2)}(x,\lambda) \tag{5.66}$$

and

$$\frac{1}{a(\lambda)} T_{+}^{(2)}(x,\lambda) = \tilde{r}(\lambda) T_{-}^{(1)}(x,\lambda) + T_{-}^{(2)}(x,\lambda), \tag{5.67}$$

where

$$r(\lambda) = \frac{b(\lambda)}{a(\lambda)}, \quad \tilde{r}(\lambda) = \frac{\bar{b}(\lambda)}{a(\lambda)}. \tag{5.68}$$

Instead of the Fourier transform used for the NS and HM models in the rapidly decreasing case, we shall utilize the integral transformations defined by the kernel $\exp\left\{\frac{mi}{4}\left(\lambda-\frac{1}{\lambda}\right)x\right\}$, already encountered when studying the finite density case of the NS model (see § II.7, Part I). The associated completeness relations are

$$\int_{-\infty}^{\infty} e^{\frac{mi}{4}\left(\lambda-\frac{1}{\lambda}\right)x}\, d\lambda=\frac{8\pi}{m}\,\delta(x), \tag{5.69}$$

$$\int_{-\infty}^{\infty} e^{\frac{mi}{4}\left(\lambda-\frac{1}{\lambda}\right)x}\,\frac{d\lambda}{\lambda}=0, \qquad \int_{-\infty}^{\infty} e^{\frac{mi}{4}\left(\lambda-\frac{1}{\lambda}\right)x}\,\frac{d\lambda}{\lambda^2}=\frac{8\pi}{m}\,\delta(x). \tag{5.70}$$

To derive a system of the Gelfand-Levitan-Marchenko integral equations from the right, consider (5.66) expressed in terms of $\tilde{T}_{\pm}(x,\lambda)$ and perform the following operations. Insert (4.28)–(4.29) into (5.66), multiply successively both sides of the resulting equation by $\exp\left\{\frac{mi}{4}\left(\lambda-\frac{1}{\lambda}\right)y\right\}$, $\frac{1}{\lambda}\exp\left\{\frac{mi}{4}\left(\lambda-\frac{1}{\lambda}\right)y\right\}$ and integrate over λ from $-\infty$ to ∞. Using (5.69)–(5.70), the analyticity properties of the Jost solutions, and the involutions (4.30)–(4.31), we finally obtain

$$\begin{aligned}\Gamma_{+}^{(1)}(x,y)+K_{+}^{(0)}(x+y)&+\int_x^{\infty}\Gamma_{+}^{(1)}(x,z)K_{+}^{(0)}(z+y)\,dz\\&+\int_x^{\infty}\Gamma_{+}^{(2)}(x,z)K_{+}^{(1)}(z+y)\,dz=0,\end{aligned} \tag{5.71}$$

$$\begin{aligned}\Gamma_{+}^{(2)}(x,y)+K_{+}^{(1)}(x+y)&+\int_x^{\infty}\Gamma_{+}^{(1)}(x,z)K_{+}^{(1)}(z+y)\,dz\\&+\int_x^{\infty}\Gamma_{+}^{(2)}(x,z)K_{+}^{(2)}(z+y)\,dz=0, \qquad y\geqslant x,\end{aligned} \tag{5.72}$$

where

$$K_{+}^{(0,2)}(x)=ik_{+}^{(0,2)}(x)\sigma_3, \qquad K_{+}^{(1)}(x)=k_{+}^{(1)}(x)\sigma_1, \tag{5.73}$$

and

$$k_{+}^{(l)}(x)=\frac{m}{8\pi}\int_{-\infty}^{\infty} r(\lambda)\,e^{\frac{mi}{4}\left(\lambda-\frac{1}{\lambda}\right)x}\,\frac{d\lambda}{\lambda^l}+\frac{m}{4i}\sum_{j=1}^{n}\frac{m_j}{\lambda_j^l}\,e^{\frac{mi}{4}\left(\lambda_j-\frac{1}{\lambda_j}\right)x}, \qquad l=0,1,2, \tag{5.74}$$

with

$$m_j = \frac{\gamma_j}{\dot{a}(\lambda_j)}, \quad j=1,\dots,n. \tag{5.75}$$

The dot in (5.75) indicates the derivative with respect to λ.

In a similar manner, from (5.67) we derive a system of the Gelfand-Levitan-Marchenko integral equations from the left

$$\Gamma_-^{(1)}(x,y)+K_-^{(0)}(x+y)+\int_{-\infty}^{x}\Gamma_-^{(1)}(x,z)K_-^{(0)}(z+y)\,dz$$
$$+\int_{-\infty}^{x}\Gamma_-^{(2)}(x,z)K_-^{(1)}(z+y)\,dz=0, \tag{5.76}$$

$$\Gamma_-^{(2)}(x,y)+K_-^{(1)}(x+y)+\int_{-\infty}^{x}\Gamma_-^{(1)}(x,z)K_-^{(1)}(z+y)\,dz$$
$$+\int_{-\infty}^{x}\Gamma_-^{(2)}(x,z)K_-^{(2)}(z+y)\,dz=0, \quad y\leqslant x, \tag{5.77}$$

where

$$K_-^{(0,2)}(x)=-ik_-^{(0,2)}(x)\sigma_3, \quad K_-^{(1)}(x)=k_-^{(1)}(x)\sigma_1, \tag{5.78}$$

and

$$k_-^{(l)}(x)=\frac{m}{8\pi}\int_{-\infty}^{\infty}\tilde{r}(\lambda)e^{-\frac{mi}{4}\left(\lambda-\frac{1}{\lambda}\right)x}\frac{d\lambda}{\lambda^l}+\frac{m}{4i}\sum_{j=1}^{n}\frac{\tilde{m}_j}{\lambda_j^l}e^{-\frac{mi}{4}\left(\lambda_j-\frac{1}{\lambda_j}\right)x}, \quad l=0,1,2, \tag{5.79}$$

with

$$\tilde{m}_j=\frac{1}{\gamma_j\dot{a}(\lambda_j)}, \quad j=1,\dots,n. \tag{5.80}$$

Notice that in view of the involutions (4.54), (4.62) the functions $k_\pm^{(0,2)}(x)$ and $\frac{1}{i}k_\pm^{(1)}(x)$ are real-valued.

We shall now outline a method for solving the inverse problem.

The input data consist of functions $r(\lambda), \tilde{r}(\lambda)$ and of a set of quantities $\{\lambda_j, m_j, \tilde{m}_j, j=1,\dots,n\}$ with the following properties.

I. *The functions $r(\lambda), \tilde{r}(\lambda)$ are of Schwartz type, vanish at $\lambda=0$ with all derivatives, and satisfy the involution*

$$r(-\lambda)=\bar{\tilde{r}}(\lambda), \quad \tilde{r}(-\lambda)=\bar{\tilde{r}}(\lambda) \tag{5.81}$$

and the relation

$$|r(\lambda)| = |\tilde{r}(\lambda)|. \tag{5.82}$$

II. *The relation*

$$\frac{\tilde{r}(\lambda)}{\bar{r}(\lambda)} = \frac{\bar{a}(\lambda)}{a(\lambda)}, \tag{5.83}$$

where

$$a(\lambda) = \prod_{j=1}^{n} \frac{\lambda - \lambda_j}{\lambda - \bar{\lambda}_j} \exp\left\{\frac{1}{2\pi i} \int_{-\infty}^{\infty} \frac{\log(1+|r(\mu)|^2)}{\lambda - \mu + i0} d\mu\right\}, \tag{5.84}$$

holds, and the pairwise distinct λ_j lie in the upper half-plane symmetrically relative to the imaginary axis: $\lambda_j = i\varkappa_j$, $\varkappa_j > 0$, $j = 1, \ldots, n_1$; $\lambda_{k+n_2} = -\bar{\lambda}_k$, $\operatorname{Im}\lambda_k$, $\operatorname{Re}\lambda_k > 0$, $k = n_1 + 1, \ldots, n_1 + n_2$; $n = n_1 + 2n_2$.

III. *The quantities m_j, $\tilde{m}_j$ satisfy*

$$m_j \tilde{m}_j = \frac{1}{\dot{a}^2(\lambda_j)}, \quad j = 1, \ldots, n, \tag{5.85}$$

and $m_j = -\bar{m}_j$, $\tilde{m}_j = -\bar{\tilde{m}}_j$, $j = 1, \ldots, n_1$; $m_{k+n_2} = -\bar{m}_k$, $\tilde{m}_{k+n_2} = -\bar{\tilde{m}}_k$, $k = n_1 + 1, \ldots, n_1 + n_2$.

Given these data, we construct the kernels $K^{(l)}_{\pm}(x)$, $l = 0, 1, 2$, and consider the systems of integral equations (5.71)-(5.72) and (5.76)-(5.77). *Then we claim that*

I′. *The systems* (5.71)-(5.72) *and* (5.76)-(5.77) *are uniquely solvable in the spaces $L_1^{(2\times 2)}(x, \infty)$ and $L_1^{(2\times 2)}(-\infty, x)$, respectively. Their solutions, the kernels $\Gamma^{(1,2)}_{\pm}(x, y)$, obey the involutions* (4.30)-(4.31) *and are of Schwartz type for $x, y \to \pm\infty$.*

II′. *The matrices $\tilde{T}_{\pm}(x, \lambda)$ constructed from $\Gamma^{(1,2)}_{\pm}(x, y)$ according to* (4.28)-(4.29) *satisfy the differential equations*

$$\frac{d\tilde{T}_{\pm}(x,\lambda)}{dx} = \frac{1}{4i}\left(\beta\theta_{\pm}(x)\sigma_3 + \frac{m\lambda}{4i}\sigma_2 - \frac{m}{4i\lambda}\Omega_{\pm}^{-2}(x)\sigma_2\Omega_{\pm}^2(x)\right)\tilde{T}_{\pm}(x,\lambda), \tag{5.86}$$

with

$$\frac{\beta}{4i}\theta_{\pm}(x)\sigma_3 = \pm(\sigma_2\Gamma^{(1)}_{\pm}(x,x)\sigma_2 - \Gamma^{(1)}_{\pm}(x,x)) \tag{5.87}$$

and

$$\Omega_{\pm}^{-2}(x)\sigma_2\Omega_{\pm}^2(x) = \left(\Gamma_{\pm}^{(2)}(x,x)\sigma_2 \pm \frac{m}{4i}I\right)\sigma_2\left(\Gamma_{\pm}^{(2)}(x,x)\sigma_2 \pm \frac{m}{4i}I\right)^{-1} \tag{5.88}$$

(*cf.* (4.35)–(4.36)).

III′. *The functions $\theta_{\pm}(x)$ are real-valued, the matrices $\Omega_{\pm}^2(x)$ are diagonal, unitary and unimodular, and*

$$\lim_{x\to\pm\infty} \theta_{\pm}(x) = 0, \qquad \lim_{x\to\pm\infty} \Omega_{\pm}^4(x) = I, \tag{5.89}$$

where the limiting values are attained in the sense of Schwartz.

IV′. *We have the relationship*

$$\theta_+(x) = \theta_-(x) = \theta(x), \qquad \Omega_+^4(x) = \Omega_-^4(x) = \Omega^4(x). \tag{5.90}$$

Normalizing $\Omega(x)$ by requiring that

$$\lim_{x\to-\infty} \Omega(x) = I \tag{5.91}$$

and setting

$$\Omega(x) = e^{\frac{i\beta\varphi(x)}{4}\sigma_3}, \qquad \pi(x) = \theta(x) - \frac{d\varphi(x)}{dx}, \tag{5.92}$$

we find that the matrix-functions

$$T_{\pm}(x,\lambda) = \Omega(x)\,\tilde{T}_{\pm}(x,\lambda) \tag{5.93}$$

satisfy the differential equation

$$\frac{dT_{\pm}(x,\lambda)}{dx} = U(x,\lambda)\,T_{\pm}(x,\lambda), \tag{5.94}$$

with a matrix $U(x,\lambda)$ of the form (4.1). The functions $\pi(x)$ and $\varphi(x)$ satisfy the boundary conditions (4.2) *with $Q \equiv n_1 \pmod 2$.*

V′. *The functions $a(\lambda)$ and $b(\lambda) = a(\lambda)\,r(\lambda)$ are the transition coefficients of the auxiliary linear problem* (5.94). *The discrete spectrum of the problem consists of the eigenvalues $\lambda_j, \bar{\lambda}_j$ with transition coefficients $\gamma_j, \bar{\gamma}_j$ where $\gamma_j = m_j\dot{a}(\lambda_j)$, $j = 1, \ldots, n$.*

Let us comment on the proof of these statements.

Assertion I′, properties (5.89) and relationship (5.90) are proved along the lines of § II.7, Part I. We shall therefore concentrate on the proof of assertion II′ and those properties of $\Omega_\pm(x)$ which are specific for the SG model.

Consider, for definiteness, the system (5.76)–(5.77); for shortness we shall drop the symbol $-$ in $\Gamma_-^{(1,2)}(x,y)$, $K_-^{(l)}(x,y)$, $\theta_-(x)$ and $\Omega_-(x)$. We will show that the matrices $\Gamma^{(1,2)}(x,y)$ satisfy (4.32)–(4.33), where $\theta(x)$ and $\Omega(x)$ are given respectively by (5.87) and (5.88), and the matrix $\Gamma^{(2)}(x,x)\sigma_2 - \frac{m}{4i} I$ occurring in (5.88) is nondegenerate. This will imply the differential equation (5.86).

For the proof we differentiate (5.76)–(5.77) with respect to x and y. Integrating by parts and using two formulae inferred from (5.78)–(5.79),

$$\frac{dK^{(1)}}{dx}(x) = \frac{m}{4i}\sigma_2 (K^{(0)}(x) - K^{(2)}(x)) \tag{5.95}$$

and

$$\frac{dK^{(0)}}{dx}(x) = \frac{m}{4i}\sigma_2 (K^{(-1)}(x) - K^{(1)}(x)), \tag{5.96}$$

where

$$K^{(-1)}(x) = k^{(-1)}(x)\sigma_1 \tag{5.97}$$

and $k^{(-1)}(x)$ is given by (5.79) with $l=-1$, we find

$$\begin{aligned}\frac{\partial\Gamma^{(1)}(x,y)}{\partial x} &+ \frac{m}{4i}\sigma_2 K^{(-1)}(x+y) + \Gamma^{(1)}(x,x)K^{(0)}(x+y)\\ &+ B(x)\sigma_2 K^{(1)}(x+y) + \int_{-\infty}^{x} \frac{\partial\Gamma^{(1)}(x,z)}{\partial x} K^{(0)}(z+y)\,dz\\ &+ \int_{-\infty}^{x} \frac{\partial\Gamma^{(2)}(x,z)}{\partial x} K^{(1)}(z+y)\,dz = 0,\end{aligned} \tag{5.98}$$

$$\begin{aligned}\frac{\partial\Gamma^{(2)}(x,y)}{\partial x} &+ \frac{m}{4i}\sigma_2 K^{(0)}(x+y) + \Gamma^{(1)}(x,x)K^{(1)}(x+y)\\ &+ B(x)\sigma_2 K^{(2)}(x+y) + \int_{-\infty}^{x} \frac{\partial\Gamma^{(1)}(x,z)}{\partial x} K^{(1)}(z+y)\,dz\\ &+ \int_{-\infty}^{x} \frac{\partial\Gamma^{(2)}(x,z)}{\partial x} K^{(2)}(z+y)\,dz = 0,\end{aligned} \tag{5.99}$$

where

$$B(x)=\Gamma^{(2)}(x,x)\sigma_2-\frac{m}{4i}I \tag{5.100}$$

and

$$\frac{\partial\Gamma^{(1)}(x,y)}{\partial y}+\frac{m}{4i}\sigma_2 K^{(-1)}(x+y)+\Gamma^{(1)}(x,x)K^{(0)}(x+y)$$
$$+B(x)\sigma_2 K^{(1)}(x+y)-\int_{-\infty}^{x}\frac{\partial\Gamma^{(1)}(x,z)}{\partial z}K^{(0)}(z+y)\,dz$$
$$-\int_{-\infty}^{x}\frac{\partial\Gamma^{(2)}(x,z)}{\partial z}K^{(1)}(z+y)\,dz=0, \tag{5.101}$$

$$\frac{\partial\Gamma^{(2)}(x,y)}{\partial y}+\frac{m}{4i}\sigma_2 K^{(0)}(x+y)+\Gamma^{(1)}(x,x)K^{(1)}(x+y)$$
$$+B(x)\sigma_2 K^{(2)}(x+y)-\int_{-\infty}^{x}\frac{\partial\Gamma^{(1)}(x,z)}{\partial z}K^{(1)}(z+y)\,dz$$
$$-\int_{-\infty}^{x}\frac{\partial\Gamma^{(2)}(x,z)}{\partial z}K^{(2)}(z+y)\,dz=0. \tag{5.102}$$

Multiply (5.101) by σ_2 both from the left and from the right and add (5.98). Using the fact that $K^{(l)}(x)$ anticommutes with σ_2 and the definition (5.87), we have

$$\frac{\partial\Gamma^{(1)}(x,y)}{\partial x}+\sigma_2\frac{\partial\Gamma^{(1)}(x,y)}{\partial y}\sigma_2-\frac{\beta}{4i}\theta(x)\sigma_3\Gamma^{(1)}(x,y)$$
$$+(B(x)\sigma_2-\sigma_2 B(x))K^{(1)}(x+y)$$
$$+\int_{-\infty}^{x}\left(\frac{\partial\Gamma^{(1)}(x,z)}{\partial x}+\sigma_2\frac{\partial\Gamma^{(1)}(x,z)}{\partial z}\sigma_2-\frac{\beta}{4i}\theta(x)\sigma_3\Gamma^{(1)}(x,z)\right)K^{(0)}(z+y)\,dz$$
$$+\int_{-\infty}^{x}\left(\frac{\partial\Gamma^{(2)}(x,z)}{\partial x}+\sigma_2\frac{\partial\Gamma^{(2)}(x,z)}{\partial z}\sigma_2-\frac{\beta}{4i}\theta(x)\sigma_3\Gamma^{(2)}(x,z)\right)K^{(1)}(z+y)\,dz=0, \tag{5.103}$$

where the matrix $\frac{\beta}{4i}\theta(x)\sigma_3 K^{(0)}(x+y)$ was reduced out by using (5.76). In order to eliminate also $(B(x)\sigma_2-\sigma_2 B(x))K^{(1)}(x+y)$ we exploit (5.77) and write it using (5.95) in the form

$$\Gamma^{(2)}(x,y)+K^{(1)}(x+y)+\int_{-\infty}^{x}\Gamma^{(1)}(x,z)K^{(1)}(z+y)\,dz$$

$$+\int_{-\infty}^{x}\Gamma^{(2)}(x,z)\left(K^{(0)}(z+y)-\frac{4i\sigma_2}{m}\frac{\partial}{\partial z}K^{(1)}(z+y)\right)dz=0. \tag{5.104}$$

Integrating here by parts we come down to

$$\frac{4i}{m}B(x)K^{(1)}(x+y)=\Gamma^{(2)}(x,y)+\int_{-\infty}^{x}\Gamma^{(1)}(x,z)K^{(1)}(z+y)\,dz$$

$$+\int_{-\infty}^{x}\Gamma^{(2)}(x,z)K^{(0)}(z+y)\,dz$$

$$+\frac{4i}{m}\int_{-\infty}^{x}\frac{\partial\Gamma^{(2)}(x,z)}{\partial z}\sigma_2K^{(1)}(z+y)\,dz. \tag{5.105}$$

Multiply (5.105) by $\frac{m}{4i}(\sigma_2-B(x)\sigma_2B^{-1}(x))$ from the left (it will be shown below that $B(x)$ is nondegenerate) and add (5.103). Taking into account (5.88) and (5.100), we obtain

$$\Phi^{(1)}(x,y)+\int_{-\infty}^{x}\Phi^{(1)}(x,z)K^{(0)}(z+y)\,dz$$

$$+\int_{-\infty}^{x}\Phi^{(2)}(x,z)K^{(1)}(z+y)\,dz=0, \tag{5.106}$$

where

$$\Phi^{(1)}(x,y)=\frac{\partial\Gamma^{(1)}(x,y)}{\partial x}+\sigma_2\frac{\partial\Gamma^{(2)}(x,y)}{\partial y}\sigma_2-\frac{\beta}{4i}\theta(x)\sigma_3\Gamma^{(1)}(x,y)$$

$$-\frac{m}{4i}(\sigma_2-\Omega^{-2}(x)\sigma_2\Omega^2(x))\Gamma^{(2)}(x,y), \tag{5.107}$$

$$\Phi^{(2)}(x,y)=\frac{\partial\Gamma^{(2)}(x,y)}{\partial x}+\Omega^{-2}(x)\sigma_2\Omega^2(x)\frac{\partial\Gamma^{(2)}(x,y)}{\partial y}\sigma_2$$

$$-\frac{\beta}{4i}\theta(x)\sigma_3\Gamma^{(2)}(x,y)$$

$$-\frac{m}{4i}(\sigma_2-\Omega^{-2}(x)\sigma_2\Omega^2(x))\Gamma^{(1)}(x,y). \tag{5.108}$$

Let us now consider equations (5.99) and (5.102). Multiply the latter by $B(x)\sigma_2 B^{-1}(x)$ from the left and by σ_2 from the right and add to the former. Reducing $\frac{m}{4i}(\sigma_2 - B(x)\sigma_2 B^{-1}(x))K^{(0)}(x+y)$ out of the resulting equation by means of (5.76), we find

$$\begin{aligned}
&\frac{\partial\Gamma^{(2)}(x,y)}{\partial x} + B(x)\sigma_2 B^{-1}(x)\frac{\partial\Gamma^{(2)}(x,y)}{\partial y}\sigma_2 \\
&-\frac{m}{4i}(\sigma_2 - B(x)\sigma_2 B^{-1}(x))\Gamma^{(1)}(x,y) \\
&+(\Gamma^{(1)}(x,x) - B(x)\sigma_2 B^{-1}(x)\Gamma^{(1)}(x,x)\sigma_2)K^{(1)}(x+y) \\
&+\int_{-\infty}^{x}\left(\frac{\partial\Gamma^{(1)}(x,z)}{\partial x} + B(x)\sigma_2 B^{-1}(x)\frac{\partial\Gamma^{(1)}(x,z)}{\partial z}\sigma_2\right. \\
&\left.-\frac{m}{4i}(\sigma_2 - B(x)\sigma_2 B^{-1}(x))\Gamma^{(2)}(x,z)\right)K^{(1)}(z+y)\,dz \\
&+\int_{-\infty}^{x}\left(\frac{\partial\Gamma^{(2)}(x,z)}{\partial x} + B(x)\sigma_2 B^{-1}(x)\frac{\partial\Gamma^{(2)}(x,z)}{\partial z}\sigma_2\right)K^{(2)}(z+y)\,dz \\
&-\frac{m}{4i}(\sigma_2 - B(x)\sigma_2 B^{-1}(x))\int_{-\infty}^{x}\Gamma^{(1)}(x,z)K^{(0)}(z+y)\,dz = 0. \qquad (5.109)
\end{aligned}$$

Next, we again exploit (5.95) which gives

$$\begin{aligned}
&\frac{m}{4i}\int_{-\infty}^{x}\Gamma^{(1)}(x,z)K^{(0)}(z+y)\,dz \\
&= \frac{m}{4i}\int_{-\infty}^{x}\Gamma^{(1)}(x,z)K^{(2)}(z+y)\,dz + \Gamma^{(1)}(x,x)\sigma_2 K^{(1)}(x+y) \\
&\quad -\int_{-\infty}^{x}\frac{\partial\Gamma^{(1)}(x,z)}{\partial z}\sigma_2 K^{(1)}(z+y)\,dz. \qquad (5.110)
\end{aligned}$$

Substituting this into (5.109), we obtain

$$\frac{\partial \Gamma^{(2)}(x,y)}{\partial x}+B(x)\sigma_2 B^{-1}(x)\frac{\partial \Gamma^{(2)}(x,y)}{\partial y}\sigma_2+\frac{\beta}{4i}\theta(x)\sigma_3 K^{(1)}(x+y)$$

$$-\frac{m}{4i}(\sigma_2-B(x)\sigma_2 B^{-1}(x))\Gamma^{(1)}(x,y)$$

$$+\int_{-\infty}^{x}\left(\frac{\partial \Gamma^{(1)}(x,z)}{\partial x}+\sigma_2\frac{\partial \Gamma^{(1)}(x,z)}{\partial z}\sigma_2-\frac{m}{4i}(\sigma_2-B(x)\sigma_2 B^{-1}(x))\right.$$

$$\left.\times\Gamma^{(2)}(x,z)\right)K^{(1)}(z+y)\,dz$$

$$+\int_{-\infty}^{x}\left(\frac{\partial \Gamma^{(2)}(x,z)}{\partial x}+B(x)\sigma_2 B^{-1}(x)\frac{\partial \Gamma^{(2)}(x,z)}{\partial z}\sigma_2\right.$$

$$\left.-\frac{m}{4i}(\sigma_2-B(x)\sigma_2 B^{-1}(x))\Gamma^{(1)}(x,z)\right)K^{(2)}(z+y)\,dz=0. \qquad (5.111)$$

Finally, eliminating the matrix $\frac{\beta}{4i}\theta(x)\sigma_3 K^{(1)}(x+y)$ with the help of (5.77) we come down to the equation

$$\Phi^{(2)}(x,y)+\int_{-\infty}^{x}\Phi^{(1)}(x,z)K^{(1)}(z+y)\,dz$$

$$+\int_{-\infty}^{x}\Phi^{(2)}(x,z)K^{(2)}(z+y)\,dz=0. \qquad (5.112)$$

We have thus shown that $\Phi^{(1)}(x,y)$ and $\Phi^{(2)}(x,y)$ satisfy (5.106) and (5.112), the homogeneous system of Gelfand-Levitan-Marchenko equations. But this system has only a trivial solution (the main point in the proof of the uniqueness and existence theorem for system (5.76)–(5.77)), hence

$$\Phi^{(1)}(x,y)=\Phi^{(2)}(x,y)=0 \qquad (5.113)$$

for all $y\leqslant x$. Therefore equations (4.32)–(4.33) hold.

To complete the proof of item II′, it suffices to show that $B(x)$ is non-degenerate. We will show that

$$\det B(x)=-\frac{m^2}{16} \qquad (5.114)$$

and

$$\Omega^2(x)=-\frac{m}{4i}B^{-1}(x). \qquad (5.115)$$

To prove (5.114), we shall first deduce it from the assumption that $\det B \neq 0$. In fact, if for some x_0 we have $\det B(x_0) \neq 0$, then (4.32) holds in a neighbourhood of x_0, so the matrix

$$\tilde{T}_-(x,\lambda) = E(x,\lambda) + \int_{-\infty}^{x} \Gamma^{(1)}(x,y)\,E(y,\lambda)\,dy + \frac{1}{\lambda}\int_{-\infty}^{x} \Gamma^{(2)}(x,y)\,E(y,\lambda)\,dy \tag{5.116}$$

satisfies the differential equation (5.86) and hence is unimodular. Using the relation

$$\frac{1}{\lambda}E(y,\lambda) = \lambda E(y,\lambda) - \frac{4i}{m}\sigma_2 \frac{dE(y,\lambda)}{dy} \tag{5.117}$$

and integrating by parts in the second integral in (5.116), we find the asymptotic behaviour

$$\tilde{T}_-(x,\lambda)E^{-1}(x,\lambda) = -\frac{4i}{m}B(x) + O(|\lambda|), \tag{5.118}$$

as $\lambda \to 0$. It follows that (5.114) holds for $x = x_0$. Also, it holds for $x \to -\infty$ since

$$\lim_{x \to -\infty} B(x) = -\frac{m}{4i}I. \tag{5.119}$$

By continuity, this implies the validity of (5.114) for all x. Relation (5.115) is a consequence of (5.114) and the normalization of $\Omega(x)$ as $x \to -\infty$.

The case of $\Gamma_+^{(1,2)}(x,y)$ is analyzed in a similar way.

This completes our discussion of the Gelfand-Levitan-Marchenko formulation.

3. Soliton solutions

Soliton solutions of the SG model, as was the case for the NS and HM models, correspond to

$$b(\lambda) = 0 \tag{5.120}$$

for all λ. For such data both the Riemann problem and the Gelfand-Levitan-Marchenko equations reduce to linear algebraic equations and can be solved in closed form. To describe soliton solutions we shall make use of the Riemann problem where one must set

$$G(x,\lambda)=I. \tag{5.121}$$

Consider first the simplest of cases when $n_1=1$, $n_2=0$. The data prescribed consist of a pure imaginary $\lambda_0=i\varkappa_0$, $\varkappa_0>0$, and a real $\gamma_0\neq 0$. The solution of the Riemann problem is given by

$$G_+(x,\lambda)=B(x,\lambda)\mathscr{E}^{-1}\Omega^{-1}(x), \tag{5.122}$$

$$G_-(x,\lambda)=\Omega(x)\mathscr{E}\,B^{-1}(x,\lambda), \tag{5.123}$$

where $B(x,\lambda)$ is the Blaschke-Potapov factor

$$B(x,\lambda)=I+\frac{\bar{\lambda}_0-\lambda_0}{\lambda-\bar{\lambda}_0}P(x), \tag{5.124}$$

$P(x)$ is an orthogonal projection operator,

$$P(x)=\frac{1}{1+\gamma_0^2(x)}\begin{pmatrix}\gamma_0^2(x) & \gamma_0(x)\\ \gamma_0(x) & 1\end{pmatrix}, \tag{5.125}$$

$$\gamma_0(x)=e^{-\frac{m}{2}\left(\varkappa_0+\frac{1}{\varkappa_0}\right)x}\gamma_0, \tag{5.126}$$

and $\Omega(x)$ is uniquely determined by the equation

$$\Omega^2(x)=-\mathscr{E}\sigma_3 B(x,0)\mathscr{E}^{-1} \tag{5.127}$$

and the requirement

$$\lim_{x\to-\infty}\Omega(x)=I. \tag{5.128}$$

The functions $\varphi(x)$, $\pi(x)$ can be found from (4.23) and (5.65) so that

$$\varphi(x)=-\frac{4}{\beta}\operatorname{arctg}\frac{1}{\gamma_0(x)}, \tag{5.129}$$

$$\pi(x)=-\frac{2m\left(\varkappa_0-\frac{1}{\varkappa_0}\right)\gamma_0(x)}{\beta(1+\gamma_0^2(x))}, \tag{5.130}$$

where the principal branch of arctg$x$ is taken: arctg$(\pm\infty)=\pm\frac{\pi}{2}$. For the topological charge Q this yields

$$Q=-\varepsilon_0, \qquad \varepsilon_0=\operatorname{sign}\gamma_0. \tag{5.131}$$

Assuming $\gamma_0(x)$ to be time-dependent according to (4.73), $\gamma_0(x,t)=e^{-\frac{m}{2}\left(\varkappa_0-\frac{1}{\varkappa_0}\right)t}\gamma_0(x)$, we have that $\varphi(x,t)=-\frac{4}{\beta}\operatorname{arctg}\frac{1}{\gamma_0(x,t)}$ satisfies the SG equation. Also,

$$\pi(x,t)=\frac{\partial\varphi}{\partial t}(x,t). \tag{5.132}$$

The expression for $\varphi(x,t)$ can be written as

$$\varphi(x,t)=-\frac{4\varepsilon_0}{\beta}\operatorname{arctg}\exp\left\{\frac{m(x-vt-x_0)}{\sqrt{1-v^2}}\right\}, \tag{5.133}$$

with

$$v=\frac{1-\varkappa_0^2}{1+\varkappa_0^2}, \qquad |v|<1, \tag{5.134}$$

$$x_0=\frac{\sqrt{1-v^2}}{m}\log|\gamma_0|. \tag{5.135}$$

Formula (5.133) suggests a natural interpretation of the above solution in terms of a relativistic particle of mass m and velocity v, whose center of inertia coordinate at $t=0$ is x_0. This solution corresponds to a *soliton of the SG model*. Besides the continuous parameters v and x_0, *the SG soliton has an important discrete characteristic, the topological charge* $Q=-\varepsilon_0$. Solutions of charge $Q=1$ are sometimes called (proper) solitons, and those of charge $Q=-1$ are called *antisolitons*.

The next simplest case occurs when $n_1=0$, $n_2=1$; the data consist of $\lambda_1=-\bar{\lambda}_2$, Im$\lambda_1$, Re$\lambda_1>0$, and $\gamma_1=\bar{\gamma}_2\neq 0$. The solution of the Riemann problem has the form

$$G_+(x,\lambda)=\Pi(x,\lambda)\mathscr{E}^{-1}\Omega^{-1}(x), \tag{5.136}$$

$$G_-(x,\lambda)=\Omega(x)\mathscr{E}\,\Pi^{-1}(x,\lambda). \tag{5.137}$$

Here $\Pi(x,\lambda)$ is a product of the Blaschke-Potapov factors,

$$\Pi(x,\lambda)=B_1(x,\lambda)B_2(x,\lambda) \tag{5.138}$$

with projection operators $P_1(x)$ and $P_2(x)$ determined by the equations

$$\Pi^*(x, \lambda_j)\,\xi_j = 0, \tag{5.139}$$

$$\xi_j(x) = \begin{pmatrix} \bar{\gamma}_j(x) \\ 1 \end{pmatrix}, \quad j = 1, 2, \tag{5.140}$$

and $\Omega(x)$ is recovered from

$$\Omega^2(x) = \mathscr{E}\,\Pi(x, 0)\,\mathscr{E}^{-1} \tag{5.141}$$

together with (5.128).

The solution of (5.139) in the general case was given in § II.5, Part I. Upon elementary transformations, the formulae cited there and (4.23) lead to

$$\varphi(x) = -\frac{4}{\beta}\operatorname{arctg}\frac{\lambda_1 - \bar{\lambda}_1}{\lambda_1 + \bar{\lambda}_1} \cdot \frac{\gamma_1(x) - \bar{\gamma}_1(x)}{1 + |\gamma_1(x)|^2}. \tag{5.142}$$

Introducing the time dependence by the replacement $\gamma_1(x) \to \exp\left\{\frac{mi}{2}\left(\lambda_1 + \frac{1}{\lambda_1}\right)t\right\}\gamma_1(x)$, we obtain a solution $\varphi(x, t)$ of the SG equation,

$$\varphi(x, t) = \frac{4}{\beta}\operatorname{arctg}\frac{\nu}{\zeta} \cdot \frac{\sin\left(\dfrac{m\omega_1(t - vx)}{\sqrt{1 - v^2}} + \varphi_0\right)}{\operatorname{ch}\left(\dfrac{m\omega_2(x - vt - x_0)}{\sqrt{1 - v^2}}\right)}, \tag{5.143}$$

with

$$\zeta = \operatorname{Re}\lambda_1, \quad \nu = \operatorname{Im}\lambda_1, \quad v = \frac{1 - |\lambda_1|^2}{1 + |\lambda_1|^2}, \quad \varphi_0 = \arg\gamma_1,$$
$$\omega_1 = \frac{\zeta}{|\lambda_1|}, \quad \omega_2 = \frac{\nu}{|\lambda_1|}, \quad x_0 = \frac{\sqrt{1 - v^2}}{m\omega_2}\log|\gamma_1|. \tag{5.144}$$

As usual, $\pi(x, t)$ is given by (5.132).

The function $\varphi(x, t)$ is specified by four real parameters and describes a particle-like solution of the SG equation with internal degrees of freedom. It is called a *double soliton or breather*. Along with translational motion of a relativistic particle of mass m, velocity v and initial center of inertia coordinate x_0, the breather oscillates both in space and time with frequencies

$\frac{mv\omega_1}{\sqrt{1-v^2}}$ and $\frac{m\omega_1}{\sqrt{1-v^2}}$, respectively. The parameter φ_0 plays the role of initial phase. In particular, if $v=0$, the breather is a t-periodic solution of the SG equation. *The breather has zero topological charge and can be interpreted as a relativistic bound state of a soliton and an antisoliton.*

If one compares the SG model to those discussed earlier, it should be noted that the SG soliton, as well as the soliton for the NS model in the case of finite density, has no internal degrees of freedom. On the other hand, the breather of the SG model is closer in its nature to the soliton of the NS model in the rapidly decreasing case, and to that of the HM model.

We shall now describe *the general n-soliton solution of the SG equation.* It is parametrized by n_1+n_2 distinct numbers $\lambda_j=-\bar{\lambda}_j=i\varkappa_j$, $\varkappa_j>0$, $j=1,\ldots,n_1$; $\lambda_{k+n_2}=-\bar{\lambda}_k$, $\operatorname{Im}\lambda_k$, $\operatorname{Re}\lambda_k>0$, $k=n_1+1,\ldots,n_1+n_2$, where $n=n_1+2n_2$, and by the nonzero quantities $\gamma_j=\bar{\gamma}_j$, $j=1,\ldots,n_1$; $\gamma_{k+n_2}=\bar{\gamma}_k$, $k=n_1+1,\ldots,n_1+n_2$. The solution $\varphi(x,t)$ is determined by

$$e^{\frac{i\beta\varphi(x,t)}{2}\sigma_3}=\mathscr{E}(-\sigma_3)^{n_1}\Pi(x,t,0)\mathscr{E}^{-1} \tag{5.145}$$

with the condition that

$$\lim_{x\to-\infty}\varphi(x,t)=0, \tag{5.146}$$

where $\Pi(x,t,\lambda)$ is an ordered product of the Blaschke-Potapov factors,

$$\Pi(x,t,\lambda)=\prod_{j=1}^{\overset{\frown}{n}} B_j(x,t,\lambda). \tag{5.147}$$

The projection operators $P_j(x,t)$ involved in $B_j(x,t,\lambda)$ are recovered from the system of equations

$$\Pi^*(x,t,\lambda_j)\,\xi_j=0, \tag{5.148}$$

where

$$\xi_j(x,t)=\begin{pmatrix}\bar{\gamma}_j(x,t)\\ 1\end{pmatrix} \tag{5.149}$$

and

$$\gamma_j(x,t)=e^{\frac{mi}{2}\left(\left(\lambda_j-\frac{1}{\lambda_j}\right)x+\left(\lambda_j+\frac{1}{\lambda_j}\right)t\right)}\gamma_j,\quad j=1,\ldots,n. \tag{5.150}$$

In generic position, the n-soliton solution, as $t\to\pm\infty$, breaks up into the sum of space-like separated solitons and breathers,

$$\varphi(x,t)=\sum_{j=1}^{n_1}\varphi_{sj}^{(\pm)}(x,t)+\sum_{k=n_1+1}^{n_1+n_2}\varphi_{bk}^{(\pm)}(x,t)+O(e^{-c|t|}), \tag{5.151}$$

with $c=\min\left\{\min_j \frac{m}{\sqrt{1-v_j^2}}\min_{i\neq j}|v_i-v_j|,\ \min_k \frac{m\omega_{2k}}{\sqrt{1-v_k^2}}\min_{k\neq l}|v_k-v_l|\right\}$. Here the $\varphi_{sj}^{(\pm)}(x,t)$ are solitons with parameters

$$v_j=\frac{1-\varkappa_j^2}{1+\varkappa_j^2},\quad x_{0j}^{(\pm)}=x_{0j}\pm\Delta x_{0j},\quad \varepsilon_j=-\operatorname{sign}\gamma_j, \tag{5.152}$$

where

$$x_{0j}=\frac{\sqrt{1-v_j^2}}{m}\log|\gamma_j| \tag{5.153}$$

and

$$\Delta x_{0j}=\frac{\sqrt{1-v_j^2}}{m}\left(\sum_{\substack{1\leq l\leq n\\ |\lambda_l|>|\lambda_j|}}\log\left|\frac{\lambda_j-\bar\lambda_l}{\lambda_j-\lambda_l}\right|-\sum_{\substack{1\leq l\leq n\\ |\lambda_l|<|\lambda_j|}}\log\left|\frac{\lambda_j-\bar\lambda_l}{\lambda_j-\lambda_l}\right|\right),\quad j=1,\ldots,n_1, \tag{5.154}$$

and the $\varphi_{bk}^{(\pm)}(x,t)$ are breathers with parameters

$$v_k=\frac{1-|\lambda_k|^2}{1+|\lambda_k|^2},\quad \omega_{1k}=\frac{\operatorname{Re}\lambda_k}{|\lambda_k|},\quad \omega_{2k}=\frac{\operatorname{Im}\lambda_k}{|\lambda_k|}, \tag{5.155}$$

$$x_{0k}^{(\pm)}=x_{0k}\pm\Delta x_{0k},\quad \varphi_{0k}^{(\pm)}=\varphi_{0k}\pm\Delta\varphi_{0k}, \tag{5.156}$$

where

$$x_{0k}=\frac{\sqrt{1-v_k^2}}{m\omega_{2k}}\log|\gamma_k|,\quad \varphi_{0k}=\arg\gamma_k \tag{5.157}$$

and

$$\Delta x_{0k}=\frac{\sqrt{1-v_k^2}}{m\omega_{2k}}\left(\sum_{\substack{1\leq l\leq n\\ |\lambda_l|>|\lambda_k|}}\log\left|\frac{\lambda_k-\bar\lambda_l}{\lambda_k-\lambda_l}\right|-\sum_{\substack{1\leq l\leq n\\ |\lambda_l|<|\lambda_k|}}\log\left|\frac{\lambda_k-\bar\lambda_l}{\lambda_k-\lambda_l}\right|\right), \tag{5.158}$$

$$\Delta\varphi_{0k}=\sum_{\substack{1\leq l\leq n\\ |\lambda_l|>|\lambda_k|}}\arg\frac{\lambda_k-\bar\lambda_l}{\lambda_k-\lambda_l}-\sum_{\substack{1\leq l\leq n\\ |\lambda_l|<|\lambda_k|}}\arg\frac{\lambda_k-\bar\lambda_l}{\lambda_k-\lambda_l},\quad k=n_1+1,\ldots,n_1+n_2. \tag{5.159}$$

Generic position means that all the velocities of solitons and breathers are distinct.

For the proof of (5.151)–(5.159) it suffices to use the results of § II.5, Part I, on the asymptotic behaviour of $\Pi(x, t, \lambda)$, as $t \to \pm\infty$, along the straight lines $x - vt = \text{const}$.

The above formulae show that soliton scattering theory for the SG model is factorized. In the next section this will be discussed from the Hamiltonian standpoint.

To conclude, we remark that, as it follows from (5.151), *the topological charge Q of the n-soliton solution is the sum of charges of the constituent solitons*,

$$Q = -\sum_{j=1}^{n_1} \varepsilon_j . \tag{5.160}$$

This relation is actually valid also when $b(\lambda) \neq 0$. Indeed, it can be shown that, for fixed $\lambda_j, \gamma_j, j = 1, \ldots, n$, the functions $\varphi(x)$ and $\pi(x)$ that solve the inverse problem depend on $b(\lambda)$ continuously. Since the topological charge Q is an integer, this implies the validity of (5.160) in the general case.

This completes our discussion of soliton dynamics and inverse problem techniques for the SG model.

§ 6. Hamiltonian Formulation of the SG Model

It will be shown here that the SG model in the rapidly decreasing case is a completely integrable Hamiltonian system. The proof will consist in an explicit construction of canonical action-angle variables. For that purpose we shall give an r-matrix expression for the Poisson brackets of the model, and use it to compute the Poisson brackets of the transition coefficients and of the discrete spectrum. An explicit expression for the local integrals of the model in terms of action-angle variables, a realization of the Poincare group generators and an interpretation of these results in terms of relativistic field theory will be given. To conclude the section, we shall discuss soliton scattering from the Hamiltonian viewpoint.

1. The fundamental Poisson brackets and the *r*-matrix

Consider the basic nonvanishing Poisson brackets of the SG model,

$$\{\pi(x), \varphi(y)\} = \delta(x-y) \tag{6.1}$$

and write them in the form

$$\left\{\pi(x), \sin\frac{\beta\varphi(y)}{2}\right\} = \frac{\beta}{2}\cos\frac{\beta\varphi(x)}{2}\,\delta(x-y), \tag{6.2}$$

$$\left\{\pi(x), \cos\frac{\beta\varphi(y)}{2}\right\} = -\frac{\beta}{2}\sin\frac{\beta\varphi(x)}{2}\,\delta(x-y). \tag{6.3}$$

The Poisson brackets of the entries of the matrix $U(x,\lambda)$ that enters into the auxiliary linear problem are then given by

$$\begin{aligned}\{U(x,\lambda) \underset{,}{\otimes} U(y,\mu)\} = \frac{m\beta^2}{32}\Bigg(&\left(\lambda+\frac{1}{\lambda}\right)\cos\frac{\beta\varphi(x)}{2}\cdot\sigma_1\otimes\sigma_3 \\ &-\left(\lambda-\frac{1}{\lambda}\right)\sin\frac{\beta\varphi(x)}{2}\cdot\sigma_2\otimes\sigma_3 \\ &-\left(\mu+\frac{1}{\mu}\right)\cos\frac{\beta\varphi(x)}{2}\cdot\sigma_3\otimes\sigma_1 \\ &+\left(\mu-\frac{1}{\mu}\right)\sin\frac{\beta\varphi(x)}{2}\cdot\sigma_3\otimes\sigma_2\Bigg)\delta(x-y). \end{aligned} \tag{6.4}$$

Our objective is to express the right hand side of (6.4) as the commutator

$$[r(\lambda,\mu), U(x,\lambda)\otimes I + I\otimes U(x,\mu)]\,\delta(x-y). \tag{6.5}$$

The right hand side of (6.4) does not contain the functions $\pi(x)$ and $\pi(y)$, so the explicit form of $U(x,\lambda)$ implies the following relation for $r(\lambda,\mu)$, which ensures that (6.5) is independent of π,

$$[r(\lambda,\mu), \sigma_3\otimes I + I\otimes\sigma_3] = 0. \tag{6.6}$$

Guided by (6.6), we shall look for an r-matrix of the form

$$r(\lambda,\mu) = f(\lambda,\mu)(I\otimes I - \sigma_3\otimes\sigma_3) + g(\lambda,\mu)(\sigma_1\otimes\sigma_1 + \sigma_2\otimes\sigma_2). \tag{6.7}$$

Using the commutation relations for the Pauli matrices, we derive a system of four equations for two unknown functions $f(\lambda,\mu)$ and $g(\lambda,\mu)$,

$$\begin{aligned} &\left(\lambda+\frac{1}{\lambda}\right)f(\lambda,\mu) + \left(\mu+\frac{1}{\mu}\right)g(\lambda,\mu) = \frac{\beta^2}{16}\left(\lambda-\frac{1}{\lambda}\right), \\ &\left(\mu+\frac{1}{\mu}\right)f(\lambda,\mu) + \left(\lambda+\frac{1}{\lambda}\right)g(\lambda,\mu) = -\frac{\beta^2}{16}\left(\mu-\frac{1}{\mu}\right), \\ &\left(\lambda-\frac{1}{\lambda}\right)f(\lambda,\mu) + \left(\mu-\frac{1}{\mu}\right)g(\lambda,\mu) = \frac{\beta^2}{16}\left(\lambda+\frac{1}{\lambda}\right), \\ &\left(\mu-\frac{1}{\mu}\right)f(\lambda,\mu) + \left(\lambda-\frac{1}{\lambda}\right)g(\lambda,\mu) = -\frac{\beta^2}{16}\left(\mu+\frac{1}{\mu}\right). \end{aligned} \tag{6.8}$$

The system (6.8) has a unique solution given by

$$f(\lambda,\mu)=\frac{\gamma}{2}\,\frac{\lambda^2+\mu^2}{\lambda^2-\mu^2},\qquad g(\lambda,\mu)=-\frac{\gamma\lambda\mu}{\lambda^2-\mu^2},\tag{6.9}$$

with

$$\gamma=\frac{\beta^2}{8}.\tag{6.10}$$

In terms of the variables

$$\alpha=\log\lambda,\qquad \beta=\log\mu\tag{6.11}$$

we have

$$f=\frac{\gamma}{2}\,\frac{\operatorname{ch}(\alpha-\beta)}{\operatorname{sh}(\alpha-\beta)},\qquad g=-\frac{\gamma}{2\operatorname{sh}(\alpha-\beta)},\tag{6.12}$$

so that

$$r(\lambda,\mu)=r(\alpha-\beta).\tag{6.13}$$

In matrix notation, and using (6.11), (6.13) we can write (6.7) as

$$r(\alpha)=\frac{\gamma}{\operatorname{sh}\alpha}\begin{pmatrix}0&0&0&0\\0&\operatorname{ch}\alpha&-1&0\\0&-1&\operatorname{ch}\alpha&0\\0&0&0&0\end{pmatrix}.\tag{6.14}$$

So we have derived the fundamental Poisson brackets for the SG model

$$\{U(x,\lambda)\overset{\otimes}{,}U(y,\mu)\}=[r(\lambda,\mu),\,U(x,\lambda)\otimes I+I\otimes U(x,\mu)]\,\delta(x-y).\tag{6.15}$$

This yields the Poisson brackets for the transition matrix

$$\{T(x,y,\lambda)\overset{\otimes}{,}T(x,y,\mu)\}=[r(\lambda,\mu),\,T(x,y,\lambda)\otimes T(x,y,\mu)]\tag{6.16}$$

for $y<x$.

2. The Poisson brackets of the transition coefficients and the discrete spectrum

Setting in (6.16) $y\to\pm\infty$, $x\to+\infty$ as prescribed by (4.18) and (4.49), we find the following expressions for the Poisson brackets of the Jost solutions

$T_\pm(x,\lambda)$ and of the reduced monodromy matrix $T(\lambda)$:

$$\{T_\pm(x,\lambda) \underset{,}{\otimes} T_\pm(x,\mu)\} = \mp r(\lambda,\mu)\, T_\pm(x,\lambda) \otimes T_\pm(x,\mu) \pm T_\pm(x,\lambda) \otimes T_\pm(x,\mu)\, r_\pm(\lambda,\mu), \tag{6.17}$$

$$\{T_+(x,\lambda) \underset{,}{\otimes} T_-(x,\mu)\} = 0 \tag{6.18}$$

and

$$\{T(\lambda) \underset{,}{\otimes} T(\mu)\} = r_+(\lambda,\mu)\, T(\lambda) \otimes T(\mu) - T(\lambda) \otimes T(\mu)\, r_-(\lambda,\mu), \tag{6.19}$$

where

$$r_\pm(\lambda,\mu) = \frac{\gamma}{2}\begin{pmatrix} \frac{\lambda-\mu}{\lambda+\mu} & 0 & 0 & 0 \\ 0 & \text{p.v.}\,\frac{\lambda+\mu}{\lambda-\mu} & \mp\pi i(\lambda+\mu)\delta(\lambda-\mu) & 0 \\ 0 & \pm\pi i(\lambda+\mu)\delta(\lambda-\mu) & \text{p.v.}\,\frac{\lambda+\mu}{\lambda-\mu} & 0 \\ 0 & 0 & 0 & \frac{\lambda-\mu}{\lambda+\mu} \end{pmatrix}. \tag{6.20}$$

Here in view of the involutions (4.20) and (4.51) we assume that $\lambda,\mu>0$; in the derivation we have used the relation

$$\lim_{y\to\pm\infty} \text{p.v.}\, \frac{e^{\frac{mi}{2}\left(\lambda-\frac{1}{\lambda}-\mu+\frac{1}{\mu}\right)y}}{\lambda-\mu} = \pm\pi i\delta(\lambda-\mu), \tag{6.21}$$

which holds for such λ and μ.

From (6.17)–(6.20) we obtain the following expressions for the Poisson brackets of the transition coefficients and the discrete spectrum for $\lambda,\mu>0$:

$$\{a(\lambda), a(\mu)\} = \{a(\lambda), \bar{a}(\mu)\} = 0, \tag{6.22}$$

$$\{b(\lambda), b(\mu)\} = 0, \tag{6.23}$$

$$\{b(\lambda), \bar{b}(\mu)\} = -2\pi i\gamma\lambda\, |a(\lambda)|^2\, \delta(\lambda-\mu), \tag{6.24}$$

$$\{a(\lambda), b(\mu)\} = \frac{2\gamma\lambda\mu}{(\lambda-\mu+i0)(\lambda+\mu)}\, a(\lambda) b(\mu), \tag{6.25}$$

$$\{a(\lambda), \bar{b}(\mu)\} = -\frac{2\gamma\lambda\mu}{(\lambda-\mu+i0)(\lambda+\mu)} a(\lambda)\bar{b}(\mu) \tag{6.26}$$

and

$$\{b(\lambda), \lambda_j\} = \{\bar{b}(\lambda), \lambda_j\} = 0, \tag{6.27}$$

$$\{b(\lambda), \gamma_j\} = \{\bar{b}(\lambda), \gamma_j\} = 0, \tag{6.28}$$

$$\{a(\lambda), \gamma_j\} = \frac{2\gamma\lambda\lambda_j}{\lambda^2-\lambda_j^2} a(\lambda)\gamma_j, \tag{6.29}$$

$$\{a(\lambda), \bar{\gamma}_j\} = -\frac{2\gamma\lambda\bar{\lambda}_j}{\lambda^2-\bar{\lambda}_j^2} a(\lambda)\bar{\gamma}_j, \tag{6.30}$$

and also

$$\{\lambda_j, \lambda_l\} = \{\lambda_j, \bar{\lambda}_l\} = 0, \tag{6.31}$$

$$\{\gamma_j, \gamma_l\} = \{\gamma_j, \bar{\gamma}_l\} = 0, \tag{6.32}$$

$$\{\gamma_j, \lambda_l\} = \gamma\gamma_j\lambda_l\delta_{jl}, \quad j, l = 1, \ldots, n. \tag{6.33}$$

Thus the continuous spectrum variables are in involution with the discrete spectrum ones, and the nonvanishing Poisson brackets of the inverse problem data ($b(\lambda), \bar{b}(\lambda), \lambda > 0; \lambda_j, \bar{\lambda}_j, \gamma_j, \bar{\gamma}_j, j = 1, \ldots, n$) *are given by* (6.24) *and* (6.33).

From (6.22) it follows that $\log a(\lambda)$ is a generating function for *integrals of the motion in involution*. In particular,

$$\{I_{2l+1}, I_{2m+1}\} = 0, \tag{6.34}$$

where I_{2k+1}, $k = -\infty, \ldots, \infty$, are the local integrals for the SG model defined in § 4.

3. Canonical variables of action-angle type

The formulae of the preceeding subsection show that *the variables*

$$\varrho(\lambda) = -\frac{1}{2\pi\gamma\lambda} \log(1-|b(\lambda)|^2), \quad \varphi(\lambda) = -\arg b(\lambda), \tag{6.35}$$

where $\lambda \geqslant 0$ *and*

$$p_j = -\frac{1}{\gamma} \log \varkappa_j, \quad q_j = \log|\gamma_j|, \quad j = 1, \ldots, n_1; \tag{6.36}$$

$$\xi_k = -\frac{2}{\gamma}\log|\lambda_k|, \qquad \eta_k = \log|\gamma_k|, \tag{6.37}$$

$$\varrho_k = \frac{2}{\gamma}\arg\lambda_k, \qquad \varphi_k = \arg\gamma_k, \qquad k = n_1+1, \ldots, n_1+n_2, \tag{6.38}$$

form a canonical family. The nonvanishing Poisson brackets of these variables have the form

$$\{\varrho(\lambda), \varphi(\mu)\} = \delta(\lambda-\mu), \qquad \lambda, \mu \geqslant 0 \tag{6.39}$$

$$\{p_i, q_j\} = \delta_{ij}, \qquad \{\xi_k, \eta_l\} = \{\varrho_k, \varphi_l\} = \delta_{kl}, \tag{6.40}$$

where $i, j = 1, \ldots, n_1$; $k, l = n_1+1, \ldots, n_1+n_2$.

The range of the variables $\varrho(\lambda)$ and ϱ_k is $0 \leqslant \varrho(\lambda) < \infty$ and $0 \leqslant \varrho_k < \dfrac{\pi}{\gamma}$ respectively; these play the role of action variables conjugate to the angle variables $\varphi(\lambda)$ and φ_k, $0 \leqslant \varphi(\lambda)$, $\varphi_k < 2\pi$. Observe that the variable $\varrho(\lambda)$ is nonsingular by virtue of the condition (A) and the relation $b(0) = 0$. The variables p_j, q_j and ξ_k, η_k range over the whole real line.

Let us emphasize that, unlike the cases considered earlier, the variables (6.35)–(6.38) do not give a complete parametrization of the n-soliton subspace of the phase space of the model. To complete the description, one must in addition specify the quantities $\varepsilon_j = \pm 1$, $j = 1, \ldots, n_1$, the topological charges of solitons. So, the submanifold of phase space containing n_1 solitons and n_2 breathers has 2^{n_1} connected components.

Thus the mapping $\mathscr{F}$ discussed in §§ 4–5 defines a canonical transformation. It linearizes the dynamics of the SG model. In fact, the local integrals I_{2l+1} depend only on the action variables,

$$\begin{aligned} \operatorname{sign}(2l+1) I_{2l+1} = {} & -2\gamma \int_0^\infty \varrho(\lambda)\lambda^{2l+1}\, d\lambda \\ & + \frac{(-1)^{l-1}2}{2l+1} \sum_{j=1}^{n_1} e^{-(2l+1)\gamma p_j} \\ & - \frac{4}{2l+1} \sum_{k=n_1+1}^{n_1+n_2} e^{-\frac{(2l+1)\gamma\xi_k}{2}} \sin\frac{2l+1}{2}\gamma\varrho_k, \end{aligned} \tag{6.41}$$

$l = -\infty, \ldots, \infty$. Hence all the *higher SG equations*

$$\frac{\partial \pi}{\partial t} = \{I, \pi\}, \quad \frac{\partial \varphi}{\partial t} = \{I, \varphi\}, \tag{6.42}$$

where

$$I = \sum_l \operatorname{sign}(2l+1) c_{2l+1} I_{2l+1}, \tag{6.43}$$

are completely integrable Hamiltonian systems, and their time evolution is given by

$$\begin{aligned} b(\lambda, t) &= e^{-2i\gamma I(\lambda)t} b(\lambda, 0), \quad \lambda_j(t) = \lambda_j(0), \\ \gamma_j(t) &= e^{-2i\gamma I(\lambda_j)t} \gamma_j(0), \quad j = 1, \ldots, n, \end{aligned} \tag{6.44}$$

where

$$I(\lambda) = \sum_l \operatorname{sign}(2l+1) c_{2l+1} \lambda^{2l+1}. \tag{6.45}$$

In particular, setting $c_{-1} = -c_1 = \frac{m}{4\gamma}$ and $c_{2l+1} = 0$ for other l, we find I to be the Hamiltonian of the SG model, and (6.44)–(6.45) turn into the familiar expressions (4.73).

We now proceed to interprete the independent modes of the SG model in terms of relativistic field theory. For that purpose it is advantageous to introduce another canonical set of variables in place of (6.35)–(6.38), so as to make manifest the Lorentz-covariant nature of excitations in the model. Namely, we set

$$k(\lambda) = \frac{m}{2}\left(\frac{1}{\lambda} - \lambda\right), \quad -\infty < k < \infty, \tag{6.46}$$

$$\varphi(k) = \varphi(\lambda(k)), \quad \varrho(k) = -\frac{d\lambda(k)}{dk} \varrho(\lambda(k)), \tag{6.47}$$

where $\lambda(k)$ is the inverse function for $k(\lambda)$,

$$\lambda(k) = \frac{1}{m}\left(\sqrt{k^2 + m^2} - k\right) \tag{6.48}$$

and

$$P_{sj} = \frac{m}{\gamma} \operatorname{sh} \gamma p_j, \quad Q_{sj} = \frac{q_j}{m \operatorname{ch} \gamma p_j}, \quad j = 1, \ldots, n_1; \tag{6.49}$$

$$P_{bk}=\frac{2m}{\gamma}\operatorname{sh}\frac{\gamma\xi_k}{2}\sin\frac{\gamma}{2}\varrho_k,\quad Q_{bk}=\frac{\eta_k}{m\operatorname{ch}\frac{\gamma\xi_k}{2}\sin\frac{\gamma}{2}\varrho_k},\tag{6.50}$$

where $k=n_1+1,\ldots,n_1+n_2$. Obviously, the variables $\varrho(k)$, $\varphi(k)$, P_{sj}, Q_{sj}, P_{bk}, Q_{bk} and ϱ_k, φ_k are also canonical. Using the expressions (4.99)–(4.100), (6.41) and (6.46)–(6.50), for the momentum P and the Hamiltonian H we obtain

$$P=\int_{-\infty}^{\infty}k\varrho(k)\,dk+\sum_{j=1}^{n_1}P_{sj}+\sum_{k=n_1+1}^{n_1+n_2}P_{bk}\tag{6.51}$$

and

$$H=\int_{-\infty}^{\infty}\sqrt{k^2+m^2}\,\varrho(k)\,dk+\sum_{j=1}^{n_1}\sqrt{P_{sj}^2+M_s^2}+\sum_{k=n_1+1}^{n_1+n_2}\sqrt{P_{bk}^2+M_{bk}^2},\tag{6.52}$$

with

$$M_s=\frac{m}{\gamma},\quad M_{bk}=\frac{2m}{\gamma}\sin\frac{\gamma}{2}\varrho_k.\tag{6.53}$$

The above formulae are sums over independent modes and allow for a clear field-theoretic interpretation.

The first summands in (6.51)–(6.52) are interpreted in terms of a wave packet of continuous spectrum modes with density $\varrho(k)$. A single mode of index k describes a particle with momentum and energy given by

$$p=k,\quad h=\sqrt{k^2+m^2},\tag{6.54}$$

related by the relativistic dispersion law

$$h^2=p^2+m^2.\tag{6.55}$$

This particle has zero topological charge. In a different way, these modes describe a neutral relativistic particle of mass m.

The second terms are the contribution from solitons associated with charged relativistic particles of mass M_s (and topological charge $Q=\pm 1$).

The third terms in these formulae correspond to breathers. The latter are associated with a neutral relativistic particle having internal degrees of freedom. Its mass M_b depends on the generalized momentum of internal motion ϱ as is shown by (6.53), and varies from zero to twice the mass of the soliton. Such a particle may be interpreted as a relativistic bound state of a soliton and an antisoliton.

Thus the excitation spectrum of the SG model is a fairly rich one and describes several kinds of particles. The ordinary perturbation theory arguments would associate to our model only particles of the first type that correspond to the linearized SG equation in the vicinity of $\varphi=0$,

$$\frac{\partial^2\varphi}{\partial t^2}-\frac{\partial^2\varphi}{\partial x^2}+m^2\varphi=0, \tag{6.56}$$

i.e. to the Klein-Gordon equation.

The appearance of solitons, antisolitons and their bound states in the excitation spectrum is entirely due to the specific, and in a way unique, form of the nonlinear interaction. In the linear limit, as $\beta\to 0$, solitons and breathers go over into solutions with infinite energy and disappear from the excitation spectrum.

To conclude this subsection we note that when the zeros λ_k accumulate towards the real line, breathers go into continuous spectrum modes. More precisely, under the assumption that

$$\lambda_k=\mu_k+\frac{i\mu_k\gamma\varrho(\mu_k)}{n_2}, \qquad k=n_1+1,\ldots,n_1+n_2, \tag{6.57}$$

where the real numbers μ_k fill in the positive line uniformly as $n_2\to\infty$, the third terms in the local integrals I_{2l+1} (6.41) go into the first terms which correspond to the continuous spectrum.

4. Realization of the Lie algebra of the Poincaré group in terms of action-angle variables

In the last subsection the generators of translations in x and t, i.e. the momentum P and the Hamiltonian H, were expressed through canonical action-angle variables. Here we shall derive an expression for the Lorentz boost generator K (see § I.1). For that purpose we will evaluate all variational derivatives of K with respect to the variables (6.35)–(6.38). To do so it clearly suffices to compute the Poisson brackets of K with the transition coefficients and discrete spectrum quantities.

For any functional F we set

$$\delta F=\{K,F\}, \tag{6.58}$$

thus defining a variation along K. We have (see § I.1)

$$\delta\varphi(x)=x\pi(x), \tag{6.59}$$

$$\delta\pi(x)=x\left(\frac{\partial^2\varphi}{\partial x^2}-\frac{m^2}{\beta}\sin\beta\varphi\right)+\frac{\partial\varphi}{\partial x}. \tag{6.60}$$

Recalling the explicit form of U and V in the zero curvature representation of § I.1, we find

$$\delta U(x,\lambda)=x\frac{\partial V}{\partial x}-x[U,V]+\frac{\beta}{4i}\frac{\partial\varphi}{\partial x}\sigma_3. \tag{6.61}$$

Let us now calculate the variation $\delta T(x,y,\lambda)$ of the transition matrix of the auxiliary linear problem for the SG model. Using (4.1) and (6.61) we have

$$\begin{aligned}\frac{\partial}{\partial x}\delta T&=U(x,\lambda)\delta T+\delta U(x,\lambda)T\\&=U\delta T+x\frac{\partial V}{\partial x}T-xUVT+xVUT+\frac{\beta}{4i}\frac{\partial\varphi}{\partial x}\sigma_3 T\\&=U\delta T+x\frac{\partial V}{\partial x}T+xV\frac{\partial T}{\partial x}-xUVT+\frac{\beta}{4i}\frac{\partial\varphi}{\partial x}\sigma_3 T\\&=U\delta T+x\frac{\partial}{\partial x}(VT)-xUVT+\frac{\beta}{4i}\frac{\partial\varphi}{\partial x}\sigma_3 T,\end{aligned} \tag{6.62}$$

or

$$\frac{\partial}{\partial x}(\delta T-xVT)=U(\delta T-xVT)-\left(V-\frac{\beta}{4i}\frac{\partial\varphi}{\partial x}\sigma_3\right)T. \tag{6.63}$$

Then, differentiating (4.1) with respect to λ and using again the explicit form of U and V, we get

$$\frac{\partial^2}{\partial x\partial\lambda}T=U\frac{\partial T}{\partial\lambda}+\frac{\partial U}{\partial\lambda}T=U\frac{\partial T}{\partial\lambda}+\frac{1}{\lambda}\left(V-\frac{\beta}{4i}\frac{\partial\varphi}{\partial x}\sigma_3\right)T, \tag{6.64}$$

so that (6.63) can be rewritten as

$$\frac{\partial}{\partial x}\left(\delta T-xVT+\lambda\frac{\partial T}{\partial\lambda}\right)=U(x,\lambda)\left(\delta T-xVT+\lambda\frac{\partial T}{\partial\lambda}\right). \tag{6.65}$$

This coincides with the equation of the auxiliary linear problem for our model, hence

$$\delta T - x\,V(x,\lambda)\,T + \lambda\,\frac{\partial T}{\partial \lambda} = TC(y,\lambda). \tag{6.66}$$

Using the boundary condition $T(x,y,\lambda)|_{x=y} = I$ we obtain

$$C(y,\lambda) = -y\,V(y,\lambda), \tag{6.67}$$

so the final expression for δT is

$$\delta T(x,y,\lambda) = x\,V(x,\lambda)\,T(x,y,\lambda) - y\,T(x,y,\lambda)\,V(y,\lambda) - \lambda\,\frac{\partial T}{\partial \lambda}(x,y,\lambda). \tag{6.68}$$

Taking the appropriate limits we get the expressions for the variations of the Jost solutions

$$\delta T_{\pm}(x,\lambda) = x\,V(x,\lambda)\,T_{\pm}(x,\lambda) - \lambda\,\frac{\partial T_{\pm}(x,\lambda)}{\partial \lambda} \tag{6.69}$$

and of the reduced monodromy matrix

$$\delta T(\lambda) = -\lambda\,\frac{\partial T}{\partial \lambda}(\lambda). \tag{6.70}$$

This gives

$$\delta a(\lambda) = -\lambda\,\frac{da}{d\lambda}(\lambda), \qquad \delta b(\lambda) = -\lambda\,\frac{db}{d\lambda}(\lambda) \tag{6.71}$$

and

$$\delta\lambda_j = \lambda_j, \qquad \delta\gamma_j = 0, \qquad j = 1, \ldots, n, \tag{6.72}$$

which implies

$$\delta\varphi(\lambda) = -\lambda\,\frac{d\varphi}{d\lambda}(\lambda) = \frac{\delta K}{\delta\varrho(\lambda)}, \tag{6.73}$$

$$\delta\varrho(\lambda) = -\varrho(\lambda) - \lambda\,\frac{d\varrho(\lambda)}{d\lambda} = -\frac{\delta K}{\delta\varphi(\lambda)}, \tag{6.74}$$

$$\delta p_j = -\frac{1}{\gamma} = -\frac{\partial K}{\partial q_j}, \tag{6.75}$$

$$\delta q_j = 0 = \frac{\partial K}{\partial p_j}, \qquad j = 1, \ldots, n_1, \tag{6.76}$$

and

$$\delta \xi_k = -\frac{2}{\gamma} = -\frac{\partial K}{\partial \eta_k}, \tag{6.77}$$

$$\delta \eta_k = 0 = \frac{\partial K}{\partial \xi_k}, \tag{6.78}$$

$$\delta \varrho_k = 0 = -\frac{\partial K}{\partial \varphi_k}, \tag{6.79}$$

$$\delta \varphi_k = 0 = \frac{\partial K}{\partial \varrho_k}, \quad k = n_1+1, \ldots, n_1+n_2. \tag{6.80}$$

Here we recalled at the last stage that (6.58) is a Poisson bracket and took into account the canonical nature of the variables $\varrho(\lambda)$, $\varphi(\lambda)$, p_j, q_j, ϱ_k, φ_k, ξ_k, η_k.

Integrating these formulae we come down to the desired expression for K in terms of canonical action-angle variables,

$$K = -\int_0^\infty \lambda \frac{d\varphi(\lambda)}{d\lambda} \varrho(\lambda)\, d\lambda + \frac{1}{\gamma} \sum_{j=1}^{n_1} q_j + \frac{2}{\gamma} \sum_{k=n_1+1}^{n_1+n_2} \eta_k. \tag{6.81}$$

In the variables (6.46)–(6.50) the expression for K takes a manifestly relativistically-covariant form,

$$K = \int_{-\infty}^{\infty} \sqrt{k^2+m^2} \frac{d\varphi(k)}{dk} \varrho(k)\, dk + \sum_{j=1}^{n_1} \sqrt{P_{sj}^2 + M_s^2}\, Q_{sj}$$
$$+ \sum_{k=n_1+1}^{n_1+n_2} \sqrt{P_{bk}^2 + M_{bk}^2}\, Q_{bk}, \tag{6.82}$$

where K is expressed as a sum over independent modes.

5. Soliton scattering from the Hamiltonian point of view

Here the n-soliton solution of the SG model will be labelled by a set of variables $\{p_j, q_j, \varepsilon_j, j=1, \ldots, n_1;\ \varrho_k, \xi_k, \varphi_k, \eta_k, k=n_1+1, \ldots, n_1+n_2\}$ that are simply related to $\{v_j, x_{0j}, \varepsilon_j,\ j=1, \ldots, n_1;\ v_k, \omega_{1k}, \omega_{2k}, x_{0k}, \varphi_{0k}, k=n_1+1, \ldots, n_1+n_2\}$ (compare (5.152)–(5.153), (5.155), (5.157) with (6.36)–(6.38)). In generic position, as $t \to \pm\infty$, it breaks up into the sum of spatially separated solitons and breathers with parameters $p_j^{(\pm)}, q_j^{(\pm)}, \varepsilon_j$ and $\varrho_k^{(\pm)}, \xi_k^{(\pm)}, \varphi_k^{(\pm)}, \eta_k^{(\pm)}$, respectively, where

$$p_j^{(+)}=p_j^{(-)}=p_j, \qquad \xi_k^{(+)}=\xi_k^{(-)}=\xi_k, \qquad \varrho_k^{(+)}=\varrho_k^{(-)}=\varrho_k, \tag{6.83}$$

$$q_j^{(\pm)}=q_j\pm\Delta q_j, \qquad \eta_k^{(\pm)}=\eta_k\pm\Delta\eta_k, \qquad \varphi_k^{(\pm)}=\varphi_k\pm\Delta\varphi_k, \tag{6.84}$$

and

$$\Delta q_j = \sum_{|\lambda_l|>|\lambda_j|} \log\left|\frac{\lambda_j-\bar{\lambda}_l}{\lambda_j-\lambda_l}\right| - \sum_{|\lambda_l|<|\lambda_j|} \log\left|\frac{\lambda_j-\bar{\lambda}_l}{\lambda_j-\lambda_l}\right|, \tag{6.85}$$

$$\Delta \eta_k = \sum_{|\lambda_l|>|\lambda_k|} \log\left|\frac{\lambda_k-\bar{\lambda}_l}{\lambda_k-\lambda_l}\right| - \sum_{|\lambda_l|<|\lambda_k|} \log\left|\frac{\lambda_k-\bar{\lambda}_l}{\lambda_k-\lambda_l}\right|, \tag{6.86}$$

$$\Delta \varphi_k = \sum_{|\lambda_l|>|\lambda_k|} \arg\frac{\lambda_k-\bar{\lambda}_l}{\lambda_k-\lambda_l} - \sum_{|\lambda_l|<|\lambda_k|} \arg\frac{\lambda_k-\bar{\lambda}_l}{\lambda_k-\lambda_l}, \tag{6.87}$$

$j=1,\ldots,n_1$, $k=n_1+1,\ldots,n_1+n_2$; the quantities λ_l range over the whole set $\lambda_1,\ldots,\lambda_n$ with the above restrictions. Here

$$\begin{aligned} &\lambda_j=e^{-\gamma p_j}, \qquad j=1,\ldots,n_1, \\ &\lambda_k=-\bar{\lambda}_{k+n_2}=e^{-\frac{\gamma}{2}(\xi_k-i\varrho_k)}, \qquad k=n_1+1,\ldots,n_1+n_2 \end{aligned} \tag{6.88}$$

(see Subsection 3 above and § 5).

The transformations $W_\pm$,

$$\begin{aligned} &W_\pm:\{p_j, q_j, \varepsilon_j, j=1,\ldots,n_1;\ \xi_k, \varrho_k, \eta_k, \varphi_k, k=n_1+1,\ldots,n_1+n_2\} \\ &\to\{p_j^{(\pm)}, q_j^{(\pm)}, \varepsilon_j, j=1,\ldots,n_1;\ \xi_k^{(\pm)}, \varrho_k^{(\pm)}, \eta_k^{(\pm)}, \varphi_k^{(\pm)}, k=n_1+1,\ldots,n_1+n_2\}, \end{aligned} \tag{6.89}$$

given by (6.83)–(6.87) are canonical. Indeed, the verification of the relations

$$\begin{aligned} &\frac{\partial\Delta q_j}{\partial p_l}=\frac{\partial\Delta q_l}{\partial p_j}, \qquad \frac{\partial\Delta q_j}{\partial\varrho_k}=\frac{\partial\Delta\varphi_k}{\partial p_j}, \qquad \frac{\partial\Delta q_j}{\partial\xi_k}=\frac{\partial\Delta\eta_k}{\partial p_j}, \\ &\frac{\partial\Delta\eta_k}{\partial\xi_m}=\frac{\partial\Delta\eta_m}{\partial\xi_k}, \qquad \frac{\partial\Delta\eta_k}{\partial\varrho_m}=\frac{\partial\Delta\varphi_m}{\partial\xi_k}, \qquad \frac{\partial\Delta\varphi_k}{\partial\varrho_m}=\frac{\partial\Delta\varphi_m}{\partial\varrho_k}, \end{aligned} \tag{6.90}$$

$$j,l=1,\ldots,n_1;\ k,m=n_1+1,\ldots,n_1+n_2,$$

is straightforward. Hence $W_\pm$ are defined by the generating functions $\pm K(p_1,\ldots,p_{n_1};\xi_{n_1+1},\ldots,\xi_{n_1+n_2},\varrho_{n_1+1},\ldots,\varrho_{n_1+n_2})$:

$$q_j^{(\pm)} = q_j \pm \frac{\partial K}{\partial p_j}, \quad \eta_k^{(\pm)} = \eta_k \pm \frac{\partial K}{\partial \xi_k}, \quad \varphi_k^{(\pm)} = \varphi_k \pm \frac{\partial K}{\partial \varrho_k},$$
$$j = 1, \ldots, n_1; \; k = n_1 + 1, \ldots, n_1 + n_2. \tag{6.91}$$

The scattering transformation S,

$$\begin{aligned} S: \{ & p_j^{(-)}, q_j^{(-)}, \varepsilon_j; \xi_k^{(-)}, \varrho_k^{(-)}, \eta_k^{(-)}, \varphi_k^{(-)} \} \\ & \rightarrow \{ p_j^{(+)}, q_j^{(+)}, \varepsilon_j; \xi_k^{(+)}, \varrho_k^{(+)}, \eta_k^{(+)}, \varphi_k^{(+)} \}, \end{aligned} \tag{6.92}$$

can be written as

$$S = W_+ W_-^{-1} \tag{6.93}$$

and is obviously canonical with the generating function (the classical S-matrix)

$$\begin{aligned} & S(p_1, \ldots, p_{n_1}; \xi_{n_1+1}, \ldots, \xi_{n_1+n_2}, \varrho_{n_1+1}, \ldots, \varrho_{n_1+n_2}) \\ & = 2K(p_1, \ldots, p_{n_1}, \xi_{n_1+1}, \ldots, \xi_{n_1+n_2}, \varrho_{n_1+1}, \ldots, \varrho_{n_1+n_2}). \end{aligned} \tag{6.94}$$

Since scattering factorizes, K can be expressed as

$$\begin{aligned} & K(p_1, \ldots, p_{n_1}; \xi_{n_1+1}, \ldots, \xi_{n_1+n_2}, \varrho_{n_1+1}, \ldots, \varrho_{n_1+n_2}) \\ & = \sum_{p_j > p_l} K_{ss}(p_j, p_l) + \sum_{2p_j > \xi_k} K_{sb}(p_j, \xi_k, \varrho_k) \\ & \quad + \sum_{\xi_k > 2p_l} K_{bs}(\xi_k, \varrho_k, p_l) + \sum_{\xi_k > \xi_m} K_{bb}(\xi_k, \varrho_k, \xi_m, \varrho_m). \end{aligned} \tag{6.95}$$

Here K_{ss}, K_{sb}, K_{bs} and K_{bb} are the generating functions for the soliton-soliton, soliton-breather, breather-soliton and breather-breather scattering, respectively.

Let us first compute K_{ss}. Setting $n_1 = 2$, $n_2 = 0$ in (6.87) and (6.88) we have, for $p_1 > p_2$,

$$\Delta q_1 = -\Delta q_2 = \log \operatorname{cth} \frac{\gamma}{2}(p_1 - p_2), \tag{6.96}$$

so that

$$K_{ss}(p_1, p_2) = K_{ss}(p_1 - p_2), \tag{6.97}$$

with

$$\frac{dK_{ss}(p)}{dp} = \log \operatorname{cth} \frac{\gamma}{2} p \tag{6.98}$$

for $p > 0$. We cannot solve (6.98) in elementary functions, but we can easily express $K_{ss}(p)$ in the form

$$K_{ss}(p) = \frac{i}{2\gamma} \int_0^{\pi} \log \frac{e^{\gamma p} e^{-i\theta} + 1}{e^{\gamma p} + e^{-i\theta}} d\theta - \frac{\pi^2}{4\gamma}, \tag{6.99}$$

where the constant of integration is chosen from the natural requirement

$$\lim_{p \to +\infty} K_{ss}(p) = 0. \tag{6.100}$$

In a similar manner, taking $n_1 = n_2 = 1$ we get

$$K_{sb}(p, \xi, \varrho) = K_{ss}\left(p - \frac{\pi i}{2\gamma} - \frac{\xi}{2} + \frac{i\varrho}{2}\right) + K_{ss}\left(p + \frac{\pi i}{2\gamma} - \frac{\xi}{2} - \frac{i\varrho}{2}\right), \tag{6.101}$$

$$K_{bs}(\xi, \varrho, p) = K_{ss}\left(\frac{\xi}{2} + \frac{i\varrho}{2} - p - \frac{\pi i}{2\gamma}\right) + K_{ss}\left(\frac{\xi}{2} - \frac{i\varrho}{2} - p + \frac{\pi i}{2\gamma}\right). \tag{6.102}$$

Finally, taking $n_1 = 0$, $n_2 = 2$ we have

$$\begin{aligned} K_{bb}(\xi_1, \varrho_1, \xi_2, \varrho_2) &= K_{sb}\left(\frac{\xi_1 - i\varrho_1}{2} + \frac{\pi i}{2\gamma}, \xi_2, \varrho_2\right) \\ &\quad + K_{sb}\left(\frac{\xi_1 + i\varrho_1}{2} - \frac{\pi i}{2\gamma}, \xi_2, \varrho_2\right) \\ &= K_{ss}\left(\frac{\xi_1 - \xi_2 - i\varrho_1 + i\varrho_2}{2}\right) + K_{ss}\left(\frac{\xi_1 - \xi_2 + i\varrho_1 - i\varrho_2}{2}\right) \\ &\quad + K_{ss}\left(\frac{\xi_1 - \xi_2 + i\varrho_1 + i\varrho_2}{2} - \frac{\pi i}{\gamma}\right) \\ &\quad + K_{ss}\left(\frac{\xi_1 - \xi_2 - i\varrho_1 - i\varrho_2}{2} + \frac{\pi i}{\gamma}\right). \end{aligned} \tag{6.103}$$

Expressions (6.102)–(6.103) confirm that the breather is a bound state of a soliton and an antisoliton. Indeed, they agree with the fact that the breather may be obtained from a two-soliton solution of zero topological charge labelled by $\lambda_1 = i\varkappa_1$, $\lambda_2 = i\varkappa_2$, $\gamma_1 = \bar{\gamma}_1$, $\gamma_2 = \bar{\gamma}_2$, $\gamma_1\gamma_2 < 0$ by analytic continuation $\lambda_1 = -\bar{\lambda}_2$, $\gamma_1 = \bar{\gamma}_2$.

Of course, soliton scattering may also be described in terms of the parameters P_{sj}, Q_{sj}, P_{bk}, Q_{bk}, ϱ_k, φ_k. We have

$$Q_{sj}^{(+)} = Q_{sj}^{(-)} + \frac{\partial S}{\partial P_{sj}}, \quad j=1,\ldots,n_1;$$

$$Q_{bk}^{(+)} = Q_{bk}^{(-)} + \frac{\partial S}{\partial P_{bk}}, \quad \varphi_k^{(+)} = \varphi_k^{(-)} + \frac{\partial S}{\partial \varrho_k}, \tag{6.104}$$

$$k = n_1+1, \ldots, n_1+n_2,$$

where the generating function S results from (6.94)–(6.95) by a change of variables (6.49)–(6.50).

This concludes our description of the SG model in the laboratory coordinates x, t,

$$\frac{\partial^2 \varphi}{\partial t^2} - \frac{\partial^2 \varphi}{\partial x^2} + \frac{m^2}{\beta} \sin \beta\varphi = 0. \tag{6.105}$$

It represents a unique example of completely integrable model of relativistic field theory with a rich excitation spectrum and a highly nontrivial factorized scattering theory.

§ 7. The SG Model in Light-Cone Coordinates

The SG equation is often considered in the light-cone coordinates

$$\xi = \frac{t+x}{2}, \quad \eta = \frac{t-x}{2}, \tag{7.1}$$

where it takes the form

$$\frac{\partial^2 \chi}{\partial \xi \partial \eta} + \frac{m^2}{\beta} \sin \beta\chi = 0. \tag{7.2}$$

Equations (7.2) and (6.105) are locally equivalent and their solutions go into one another under the change of variables

$$\chi(\xi,\eta) = \varphi(\xi-\eta, \xi+\eta), \quad \varphi(x,t) = \chi\left(\frac{t+x}{2}, \frac{t-x}{2}\right). \tag{7.3}$$

A less trivial problem, however, is to describe the relationship between the classes of solutions that correspond to various boundary conditions. Here we shall identify the class of solutions $\chi(\xi, \eta)$ associated with the rapidly decreasing boundary conditions for $\varphi(x, t)$.

A solution of (7.2) is naturally parametrized by initial data on one of the characteristics, say, on $\eta=0$,

$$\chi(\xi)=\chi(\xi, \eta)|_{\eta=0}. \tag{7.4}$$

The rapidly decreasing boundary conditions $\left(\bmod \frac{2\pi}{\beta}\right)$ for $\chi(\xi)$

$$\lim_{\xi\to-\infty} \chi(\xi)=0, \qquad \lim_{\xi\to+\infty} \chi(\xi)=\frac{2\pi Q}{\beta}, \tag{7.5}$$

where the boundary values are attained in the sense of Schwartz and Q is an integer, allow us to introduce the topological charge

$$Q=\frac{\beta}{2\pi}\int_{-\infty}^{\infty} \frac{d\chi(\xi)}{d\xi}\, d\xi. \tag{7.6}$$

As will be seen later on, *the initial data $\chi(\xi)$ induced by a solution $\varphi(x, t)$ under the rapidly decreasing boundary conditions satisfy an infinite sequence of constraints in addition to (7.5).* These imply, in particular, the decay of all derivatives of $\chi(\xi, \eta)$ with respect to η at $\eta=0$, as $|\xi|\to\infty$,

$$\lim_{|\xi|\to\infty} \left.\frac{\partial^n \chi(\xi, \eta)}{\partial \eta^n}\right|_{\eta=0}=0. \tag{7.7}$$

These relations can be written as

$$Q_n=\int_{-\infty}^{\infty} F_n(\chi, \xi)\, d\xi=0, \qquad n=1, 2, \ldots, \tag{7.8}$$

where F_n is defined implicitly by

$$F_n(\chi, \xi)=\left.\frac{\partial^{n-1}}{\partial \eta^{n-1}} \sin\beta\chi\right|_{\eta=0}. \tag{7.9}$$

The right hand side of (7.9) may be expressed in terms of the initial data $\chi(\xi)$ by using

$$\frac{\partial}{\partial \xi} \frac{\partial^k \chi}{\partial \eta^k} = -\frac{m^2}{\beta} \frac{\partial^{k-1}}{\partial \eta^{k-1}} \sin \beta \chi, \tag{7.10}$$

$k=1, \ldots, n-1$, and integrating successively over ξ. In particular,

$$F_1 = \sin \beta \chi(\xi), \quad F_2 = -m^2 \cos \beta \chi(\xi) \int_{-\infty}^{\xi} \sin \beta \chi(\xi') d\xi'. \tag{7.11}$$

Similar constraints arise for the higher SG equations in light-cone coordinates. We will not write them down explicitly in terms of the initial data $\chi(\xi)$. Later on they will be described in terms of the auxiliary linear problem data.

Notice that in the linear limit

$$\frac{\partial^2 \chi}{\partial \xi \partial \eta} + m^2 \chi = 0 \tag{7.12}$$

the corresponding constraints take the form

$$\int_{-\infty}^{\infty} \xi^n \chi(\xi) d\xi = 0, \quad n = 0, 1, \ldots, \tag{7.13}$$

so that the Fourier transform of $\chi(\xi)$ vanishes at the origin with all derivatives.

Let us now outline the Hamiltonian picture associated with the SG equation in light-cone coordinates and parametrized by $\chi(\xi)$. The Poisson structure is formally defined by the Poisson brackets

$$\{\chi(\xi), \chi(\xi')\} = \tfrac{1}{4} \operatorname{sign}(\xi - \xi'), \tag{7.14}$$

and the phase space consists of functions $\chi(\xi)$ satisfying (7.5) and the above constraints. The Hamiltonian of the model is

$$H = \frac{2m^2}{\beta^2} \int_{-\infty}^{\infty} (1 - \cos \beta \chi(\xi)) d\xi \tag{7.15}$$

and the equations of motion

$$\frac{\partial \chi}{\partial \eta} = \{H, \chi\} \tag{7.16}$$

coincide with (7.2). The role of momentum is played by

$$P = -\int_{-\infty}^{\infty} \left(\frac{d\chi(\xi)}{d\xi}\right)^2 d\xi \,. \tag{7.17}$$

The presence of constraints has no effect on the equations of motion (7.16). It can be verified that the constraints are in involution with both the Hamiltonian H and the momentum P on the level surface of the constraints. We will see below that the Poisson structure described in this rather implicit way is in fact induced by the standard Poisson structure on $\pi(x), \varphi(x)$.

We shall now state the main propositions of this section.

I. *Let $\chi(\xi, \eta)$ be a solution of (7.2) satisfying (7.5) and the above constraints. Then (7.3) gives a solutions $\varphi(x, t)$ of* (6.105) *in the class of rapidly decreasing initial data*

$$\varphi(x) = \chi\left(\frac{x}{2}, -\frac{x}{2}\right), \tag{7.18}$$

$$\pi(x) = \frac{1}{2}\left(\frac{\partial}{\partial\xi} + \frac{\partial}{\partial\eta}\right)\chi(\xi, \eta)\bigg|_{\xi = -\eta = \frac{x}{2}} . \tag{7.19}$$

The topological charges of χ and φ are equal.

II. *Let $\varphi(x, t)$ be a solution of* (6.105) *under the rapidly decreasing boundary conditions. Then* (7.3) *gives a solution $\chi(\xi, \eta)$ of* (7.2) *with the initial data*

$$\chi(\xi) = \varphi(\xi, \xi), \tag{7.20}$$

satisfying (7.5) *and the constraints. The topological charges of these solutions coincide.*

These two assertions mean that the two classes of solutions of the SG equation coincide, one parametrized by the initial data $\pi(x), \varphi(x)$ in laboratory coordinates, the other parametrized by $\chi(\xi)$ in light-cone coordinates.

III. *For the classes of solutions indicated above the Hamiltonians and momenta are related by*

$$H(\chi) = P(\pi, \varphi) + H(\pi, \varphi), \tag{7.21}$$

$$P(\chi) = P(\pi, \varphi) - H(\pi, \varphi). \tag{7.22}$$

IV. *The Poisson structures* (6.1) *and* (7.14) *are equivalent. On the solutions of the SG equation we have*

$$\int_{-\infty}^{\infty} d\chi'(\xi) \wedge d\chi(\xi) d\xi = \int_{-\infty}^{\infty} d\pi(x) \wedge d\varphi(x) dx \,, \tag{7.23}$$

the prime indicating differentiation with respect to ξ.

The proof of assertions I–IV will use the inverse scattering formalism. Let us give the necessary formulations for the case of (7.2).

1) The zero curvature representation results from (I.1.30)–(I.1.31) upon a change of variables (7.1). The corresponding matrices U_χ and V_χ have the form

$$U_\chi(\xi,\eta,\lambda)=V_{\pi,\varphi}(x,t,\lambda)+U_{\pi,\varphi}(x,t,\lambda)$$
$$=\frac{\beta}{4i}\frac{\partial\chi}{\partial\xi}\sigma_3+\frac{m\lambda}{2i}e^{\frac{i\beta\chi(\xi)}{2}\sigma_3}\sigma_2, \tag{7.24}$$

$$V_\chi(\xi,\eta,\lambda)=V_{\pi,\varphi}(x,t,\lambda)-U_{\pi,\varphi}(x,t,\lambda)$$
$$=-\frac{\beta}{4i}\frac{\partial\chi}{\partial\eta}\sigma_3+\frac{m}{2i\lambda}e^{-\frac{i\beta\chi(\xi)}{2}\sigma_3}\sigma_2. \tag{7.25}$$

The auxiliary linear problem

$$\frac{dF}{d\xi}=U_\chi(\xi,\lambda)F \tag{7.26}$$

by means of a gauge transformation

$$F(\xi,\lambda)=e^{\frac{i\beta\chi(\xi)}{4}\sigma_3}\tilde{F}(\xi,\lambda) \tag{7.27}$$

is reduced to

$$\frac{d\tilde{F}}{d\xi}=\left(\frac{m\lambda}{2i}\sigma_2+\frac{\beta}{2i}\frac{d\chi}{d\xi}\sigma_3\right)\tilde{F}. \tag{7.28}$$

Choosing a new basis in $\mathbb{C}^2$ which leads to a change of the Pauli matrices,

$$\sigma_2\to\sigma_3,\quad \sigma_3\to-\sigma_2,\quad \sigma_1\to\sigma_1, \tag{7.29}$$

we are left with the matrix $U(\xi,\lambda)$ of the form

$$U(\xi,\lambda)=\frac{m\lambda}{2i}\sigma_3-\frac{\beta}{2i}\frac{d\chi(\xi)}{d\xi}\sigma_2. \tag{7.30}$$

This matrix coincides literally with its counterpart in the auxiliary linear problem for the NS equation,

$$U(\xi,\lambda)=U^{\mathrm{NS}}(\xi,m\lambda), \tag{7.31}$$

where $\varkappa=-\frac{\beta^2}{4}<0$ and

$$\psi(\xi)=-\bar{\psi}(\xi)=i\frac{d\chi}{d\xi}(\xi). \tag{7.32}$$

(see § I.2, Part I). The last condition means that there is an additional involution

$$U(\xi,-\lambda)=\sigma_2\,U(\xi,\lambda)\sigma_2. \tag{7.33}$$

2) Along with the general properties of Chapter I, Part I, the transition matrix $T(\xi,\xi',\lambda)$, the Jost solutions $T_\pm(\xi,\lambda)$ and the reduced monodromy matrix $T(\lambda)$ satisfy an additional involution implied by (7.33). For $T(\lambda)$, it is given by

$$T(-\lambda)=\bar{T}(\lambda). \tag{7.34}$$

Thus the transition coefficients for the continuous spectrum, $a(\lambda)$ and $b(\lambda)$, satisfy a supplementary condition,

$$a(\lambda)=\bar{a}(-\lambda), \quad b(\lambda)=\bar{b}(-\lambda), \tag{7.35}$$

typical for the SG model in laboratory coordinates (see (4.54)).

In a similar way, it is shown that the discrete spectrum $\lambda_j, \bar{\lambda}_j$ and the associated transition coefficients $\gamma_j, \bar{\gamma}_j$, $j=1,\ldots,n$, have all the properties listed in § 4.

3) The boundary conditions (7.5) lead to

$$a(0)=(-1)^{Q}, \quad b(0)=0. \tag{7.36}$$

The constraints $Q_n=0$ and their analogues engendered by the higher SG equations imply that all derivatives of $b(\lambda)$ vanish as $\lambda\to 0$:

$$\left.\frac{d^n b(\lambda)}{d\lambda^n}\right|_{\lambda=0}=0, \quad n=0,1,\ldots. \tag{7.37}$$

The last assertion cannot be proven here since we have not written all the constraints explicitly. Instead, we shall regard (7.37) as the definition of a complete set of constraints.

With this understood, there is an asymptotic expansion

$$\log a(\lambda)=i\sum_{n=0}^{\infty} I_{-n}\lambda^n, \quad I_{-2n}=0, \quad n>0, \tag{7.38}$$

as $\lambda\to 0$, where

$$I_0 \equiv \pi Q(\mathrm{mod}\, 2\pi), \tag{7.39}$$

$$I_{-1}=\frac{\beta^2}{4m}H, \tag{7.40}$$

and the I_{-n} for $n>1$ generate the higher SG equations. These integrals of the motion are nonlocal.

For $|\lambda|\to\infty$, we have the asymptotic expansion

$$\log a(\lambda)=i\sum_{n=1}^{\infty}\frac{I_n}{\lambda^n}, \quad I_{2n}=0, \tag{7.41}$$

that gives a sequence of local integrals of the motion I_n. Their densities are polynomials in $\chi(\xi)$ and its derivatives with respect to ξ. In particular,

$$I_1=\frac{\beta^2}{4m}P. \tag{7.42}$$

4) The time dependence of the transition coefficients is given by

$$\begin{gathered} a(\lambda,\eta)=a(\lambda,0), \quad b(\lambda,\eta)=e^{-\frac{mi}{\lambda}\eta}b(\lambda,0), \\ \lambda_j(\eta)=\lambda_j(0), \quad \gamma_j(\eta)=e^{\frac{mi}{\lambda_j}\eta}\gamma_j(0), \quad j=1,\ldots,n. \end{gathered} \tag{7.43}$$

The dynamics induced by $\frac{4m}{\beta^2}I_{-l}$, $l\equiv 1(\mathrm{mod}\,2)$, is described by similar formulae where $\exp\left\{\frac{mi}{\lambda}\eta\right\}$ $\left(\text{respectively } \exp\left\{\frac{mi}{\lambda_j}\eta\right\}\right)$ is replaced by $\exp\left\{\frac{mi}{\lambda^l}\eta\right\}$ $\left(\text{respectively by } \exp\left\{\frac{mi}{\lambda_j^l}\eta\right\}\right)$. We point out that (7.37) is left invariant by the dynamics.

5) The inverse problem for our model is a special case of the inverse problem for the NS model considered in Chapter II of Part I. The additional involution (7.35) implies that $\frac{d\chi}{d\xi}(\xi)$ is real-valued.

The results listed above enable us to prove assertions I–IV. We start with assertion I. Consider the auxiliary linear problem (7.28) with the coefficient $\frac{\partial\chi}{\partial\xi}(\xi,\eta)$ for fixed η. Let $\tilde{T}_\pm(\xi,\eta,\lambda)$ be the corresponding Jost solutions. By the zero curvature condition, the matrices

$$F_\pm(\xi,\eta,\lambda)=e^{\frac{i\beta\chi(\xi)}{4}\sigma_3}\,\tilde{T}_\pm(\xi,\eta,\lambda)\,e^{-\frac{mi}{2\lambda}\eta\sigma_3} \tag{7.44}$$

satisfy

$$\frac{\partial F_\pm}{\partial\xi}=U_\chi(\xi,\eta,\lambda)F_\pm\,, \tag{7.45}$$

$$\frac{\partial F_\pm}{\partial\eta}=V_\chi(\xi,\eta,\lambda)F_\pm\,. \tag{7.46}$$

For $\xi\to\pm\infty$, $\eta=\text{const}$ they have the asymptotic behaviour

$$\lim_{\xi\to-\infty}F_-(\xi,\eta,\lambda)\exp\left\{\frac{mi}{2}\left(\lambda\xi+\frac{\eta}{\lambda}\right)\sigma_3\right\}=\mathscr{E}\,, \tag{7.47}$$

$$\lim_{\xi\to+\infty}F_+(\xi,\eta,\lambda)\exp\left\{\frac{mi}{2}\left(\lambda\xi+\frac{\eta}{\lambda}\right)\sigma_3\right\}
=\begin{cases}(-1)^{Q/2}\mathscr{E}\,, & Q \text{ even}\,,\\ (-1)^{(Q-1)/2}i\sigma_3\mathscr{E}\,, & Q\text{ odd}\,,\end{cases} \tag{7.48}$$

where the matrix $\mathscr{E}$ is defined in (5.21) and realizes the automorphism (7.29).

These asymptotic expressions also hold along the spacelike straight lines $\eta=c\xi$, $c<0$, as $\xi\to\pm\infty$. This follows from

$$\lim_{\xi\to\pm\infty}\int_{-\infty}^{\infty}e^{\frac{mi}{2}\left(\lambda+\frac{c}{\lambda}\right)\xi}f(\lambda)\,d\lambda=0\,, \tag{7.49}$$

where $f(\lambda)$ lies in Schwartz space and vanishes with all derivatives at $\lambda=0$. Indeed, these are precisely the expressions that enter into the kernels of the Wiener-Hopf or the Gelfand-Levitan-Marchenko equations (see §§ II.2–II.4 of Part I), and hence the latter decay along spacelike directions. Therefore the corresponding solutions, i.e. the kernels of the integral representations for the Jost solutions, are also decaying, which ensures the asymptotic expressions (7.47)–(7.48).

Letting $\eta = -\xi + O(1)$ we find that the matrices

$$\tilde{F}_\pm(x,t,\lambda) = F_\pm\left(\frac{t+x}{2}, \frac{t-x}{2}, \lambda\right) \tag{7.50}$$

have the following asymptotic behaviour as $x \to \pm\infty$, $t = \text{const}$:

$$\lim_{x\to-\infty} \tilde{F}_-(x,t,\lambda) \exp\left\{\frac{mi}{4}\left(\left(\lambda - \frac{1}{\lambda}\right)x + \left(\lambda + \frac{1}{\lambda}\right)t\right)\sigma_3\right\} = \mathscr{E}, \tag{7.51}$$

$$\lim_{x\to+\infty} \tilde{F}_+(x,t,\lambda) \exp\left\{\frac{mi}{4}\left(\left(\lambda - \frac{1}{\lambda}\right)x + \left(\lambda + \frac{1}{\lambda}\right)t\right)\sigma_3\right\}$$
$$= \begin{cases} (-1)^{Q/2}\mathscr{E}, & Q \text{ even}, \\ (-1)^{(Q-1)/2} i\sigma_3 \mathscr{E}, & Q \text{ odd}. \end{cases} \tag{7.52}$$

The corresponding matrices U and V,

$$U(x,t,\lambda) = \frac{\partial \tilde{F}_\pm}{\partial x}\tilde{F}_\pm^{-1}, \quad V(x,t,\lambda) = \frac{\partial \tilde{F}_\pm}{\partial t}\tilde{F}_\pm^{-1} \tag{7.53}$$

are representable in the form (I.1.30)–(I.1.31) with $\varphi(x,t)$ given by (7.3) and $\pi(x,t) = \frac{\partial\varphi}{\partial t}(x,t)$. The asymptotic formulae (7.51)–(7.52) yield

$$\lim_{x\to-\infty} U(x,t,\lambda) = \frac{m}{4i}\left(\lambda - \frac{1}{\lambda}\right)\sigma_2,$$
$$\lim_{x\to-\infty} V(x,t,\lambda) = \frac{m}{4i}\left(\lambda + \frac{1}{\lambda}\right)\sigma_2, \tag{7.54}$$

$$\lim_{x\to+\infty} U(x,t,\lambda) = \frac{(-1)^Q m}{4i}\left(\lambda - \frac{1}{\lambda}\right)\sigma_2,$$
$$\lim_{x\to+\infty} V(x,t,\lambda) = \frac{(-1)^Q m}{4i}\left(\lambda + \frac{1}{\lambda}\right)\sigma_2, \tag{7.55}$$

which implies that the initial data (7.18)–(7.19) satisfy the rapidly decreasing boundary conditions. The matrices

$$T_\pm(x,t,\lambda) = \tilde{F}_\pm(x,t,\lambda)\, e^{-\frac{mi}{4}\left(\lambda + \frac{1}{\lambda}\right)t\sigma_3} \tag{7.56}$$

are the Jost solutions of the auxiliary linear problem (4.1) with the same transition coefficients and discrete spectrum as for (7.28).

This proves assertion I. Assertion II is proved in a similar way.

To prove assertion III, consider a 1-form on $\mathbb{R}^2$

$$\omega = (h+p)\,dx - \left(h + p - \frac{2m^2}{\beta^2}(1-\cos\beta\varphi)\right)dt, \tag{7.57}$$

where

$$h = \frac{\pi^2}{2} + \frac{1}{2}\left(\frac{\partial\varphi}{\partial x}\right)^2 + \frac{m^2}{\beta^2}(1-\cos\beta\varphi), \tag{7.58}$$

$$p = -\pi\frac{\partial\varphi}{\partial x}. \tag{7.59}$$

The form ω is closed on solutions of the SG equation in view of the energy-momentum conservation law

$$\frac{\partial h}{\partial t} = -\frac{\partial p}{\partial x}, \tag{7.60}$$

$$\frac{\partial p}{\partial t} = -\frac{\partial}{\partial x}\left(h - \frac{2m^2}{\beta^2}(1-\cos\beta\varphi)\right). \tag{7.61}$$

Let us integrate ω around a closed path l in the x, t plane formed by two right triangles shown in Fig. 3.

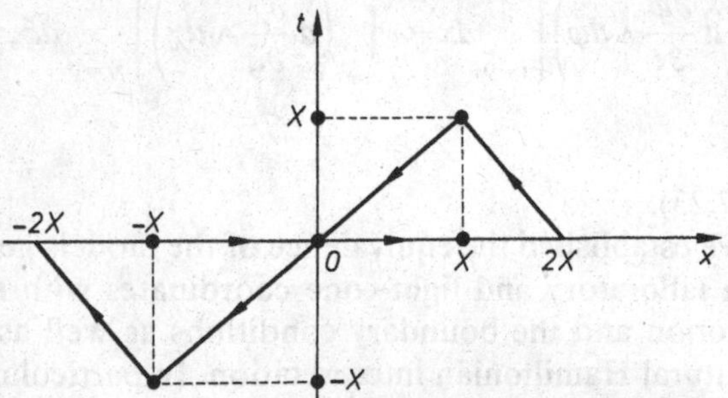

Fig. 3

We have

$$0=\int_{l}\omega=\int_{-2X}^{2X}(h+p)\Big|_{t=0}dx-\frac{2m^2}{\beta^2}\int_{-X}^{X}(1-\cos\beta\chi(\xi))\Big|_{\eta=0}d\xi$$
$$+\int_{-X}^{0}\left(\frac{\partial\chi}{\partial\eta}\right)^2\Big|_{\xi=X}d\eta+\int_{0}^{X}\left(\frac{\partial\chi}{\partial\eta}\right)^2\Big|_{\xi=-X}d\eta, \tag{7.62}$$

where we have used that $\frac{\partial\chi}{\partial\eta}=\frac{\partial\varphi}{\partial t}-\frac{\partial\varphi}{\partial x}$. The rapid decay of $\frac{\partial\chi}{\partial\eta}$, as $\xi\to\pm\infty$, implies that the last two terms in (7.62) vanish, as $X\to\infty$, so that in the limit we recover (7.21). Equation (7.22) is proved in a similar manner.

To prove assertion IV, we consider the evolution of the variation $d\varphi$ of φ. It is governed by the linearized equation

$$\frac{\partial^2 d\varphi}{\partial t^2}-\frac{\partial^2 d\varphi}{\partial x^2}+m^2\cos\beta\varphi\, d\varphi=0, \tag{7.63}$$

which yields that the 1-form θ

$$\theta=\left(d\frac{\partial\varphi}{\partial t}\wedge d\varphi\right)dx+\left(d\frac{\partial\varphi}{\partial x}\wedge d\varphi\right)dt \tag{7.64}$$

is closed on $\mathbb{R}^2$. Integration around the contour l gives

$$\int_{-\infty}^{\infty}\left(d\frac{\partial\varphi}{\partial t}\wedge d\varphi\right)\Big|_{t=0}dx=\int_{-\infty}^{\infty}\left(d\frac{\partial\chi}{\partial\xi}\wedge d\chi\right)\Big|_{\eta=0}d\xi, \tag{7.65}$$

which proves (7.23).

Thus we have established the equivalence of the models generated by the SG equation in laboratory and light-cone coordinates with regard to their equations of motion and the boundary conditions as well as in the framework of their natural Hamiltonian interpretation. In particular, both models are completely integrable and are described by the same set of canonical action-angle variables.

§ 8. The Landau-Lifshitz Equation as a Universal Integrable Model with Two-Dimensional Auxiliary Space

The LL model is described by the equations of motion

$$\frac{\partial \vec{S}}{\partial t} = \vec{S} \wedge \frac{\partial^2 \vec{S}}{\partial x^2} + \vec{S} \wedge J\vec{S}, \tag{8.1}$$

with J a diagonal matrix, $J = \operatorname{diag}(J_1, J_2, J_3)$, $J_1 < J_2 < J_3$. The corresponding matrices U and V in the zero curvature representation are given by

$$U(x, t, \lambda) = \frac{1}{i} \sum_{a=1}^{3} u_a(\lambda) S_a \sigma_a, \tag{8.2}$$

$$V(x, t, \lambda) = 2 i u_1(\lambda) u_2(\lambda) u_3(\lambda) \sum_{a=1}^{3} \frac{S_a}{u_a(\lambda)} \sigma_a + \frac{1}{i} \sum_{a,b,c=1}^{3} u_a(\lambda) \varepsilon_{abc} S_b \frac{\partial S_c}{\partial x} \sigma_a, \tag{8.3}$$

where

$$u_1(\lambda) = \varrho \frac{1}{\operatorname{sn}(\lambda, k)}, \quad u_2(\lambda) = \varrho \frac{\operatorname{dn}(\lambda, k)}{\operatorname{sn}(\lambda, k)}, \quad u_3(\lambda) = \varrho \frac{\operatorname{cn}(\lambda, k)}{\operatorname{sn}(\lambda, k)} \tag{8.4}$$

and

$$\varrho = \frac{1}{2} \sqrt{J_3 - J_1} > 0, \quad 0 < k = \sqrt{\frac{J_2 - J_1}{J_3 - J_1}} < 1 \tag{8.5}$$

(see § I.1).

In contrast to the previously discussed NS, HM, and SG models, where the spectral parameter λ ranges over the whole complex plane $\mathbb{C}$, the natural range of λ for the LL model is an elliptic curve, the torus $E = \mathbb{C}/\Gamma$ where Γ is a lattice with generators $4K$ and $4iK'$. Here K and K' are the complete elliptic integrals of moduli k and $k' = \sqrt{1-k^2}$, respectively. As a result, the investigation of the direct and inverse scattering problems for the auxiliary linear equation of the LL model proves to be more complicated technically, and will not be reported here. We will content ourselves with a few remarks of kinematical nature. Specifically, we will show that the LL model admits an r-matrix formulation and describe the limiting procedures that lead to the HM, NS, and SG models.

We start with the Hamiltonian formulation. Consider the basic Poisson brackets for the LL model

$$\{S_a(x), S_b(y)\} = -\varepsilon_{abc} S_c(x)\delta(x-y) \tag{8.6}$$

and write them in terms of $U(x,\lambda)$

$$\{U(x,\lambda) \underset{,}{\otimes} U(y,\mu)\} = \sum_{a,b,c=1}^{3} \varepsilon_{abc} u_a(\lambda) u_b(\mu) S_c(x)(\sigma_a \otimes \sigma_b)\delta(x-y). \tag{8.7}$$

Our nearest goal is to express the right hand side of (8.7) as a commutator of the form

$$[r(\lambda,\mu), U(x,\lambda) \otimes I + I \otimes U(x,\mu)]\delta(x-y).$$

To do so we shall use the addition theorems for the Jacobi elliptic functions. In terms of the $u_a(\lambda)$, these yield

$$u_a(\lambda) u_b(\mu) = u_a(\lambda-\mu) u_c(\mu) - u_b(\lambda-\mu) u_c(\lambda), \tag{8.8}$$

where the triple (a,b,c) is a cyclic permutation of the indices 1, 2, 3. From (8.7)-(8.8) we can immediately derive *the fundamental Poisson brackets for the LL model*

$$\{U(x,\lambda) \underset{,}{\otimes} U(y,\mu)\} = [r(\lambda-\mu), U(x,\lambda) \otimes I + I \otimes U(x,\mu)]\delta(x-y), \tag{8.9}$$

where the matrix $r(\lambda)$ has the form

$$r(\lambda) = -\frac{1}{2}\sum_{a=1}^{3} u_a(\lambda)\sigma_a \otimes \sigma_a. \tag{8.10}$$

It is easily seen from (8.8) that $r(\lambda)$ satisfies the equations

$$r(-\lambda) = -Pr(\lambda)P \tag{8.11}$$

and

$$[r_{12}(\lambda-\mu), r_{13}(\lambda) + r_{23}(\mu)] + [r_{13}(\lambda), r_{23}(\mu)] = 0 \tag{8.12}$$

(see § III.1, Part I) which ensure that the Poisson brackets defined by (8.9) are skew-symmetric and satisfy the Jacobi identity. Actually, (8.9) is the general solution of (8.11)-(8.12) for the case of 4×4 matrices. This will be discussed in more detail in Chapter IV where the phase space of the matrices $U(x,\lambda)$ will be interpreted as the simplest orbit of a suitable infinite-dimensional Lie algebra.

Let us now verify that the previously discussed integrable models with two-dimensional auxiliary space and minimal pole divisor can be obtained from the LL model by appropriate limiting procedures.

The simplest case is when $k \to 0$, so that $K \to \infty$, $K' \to \frac{\pi}{2}$ and the elliptic curve E degenerates into a rational curve – the cylinder $\mathbb{C}/\frac{\pi}{2}\mathbb{Z}$. The Jacobi elliptic functions then turn into trigonometric functions,

$$\operatorname{sn}(\lambda, k) \to \sin\lambda, \quad \operatorname{cn}(\lambda, k) \to \cos\lambda, \quad \operatorname{dn}(\lambda, k) \to 1. \tag{8.13}$$

The result is the partially anisotropic HM model, a special case of the LL model that corresponds to $J_1 = J_2 < J_3$. The zero curvature representation and the r-matrix for this model result from (8.3)–(8.4) and (8.10) upon the replacement (8.13). Setting $\lambda = i\alpha$ and $\varrho = i\gamma$ in the r-matrix thus obtained, we come down to the expression (6.14) for the r-matrix of the SG model, up to an irrelevant term proportional to $I \otimes I$.

The completely isotropic LL model – the HM model – is obtained in the limit as $\varrho \to 0$ so that $J_2 \to J_3$. Replacing λ by $\dfrac{2\varrho}{\lambda}$ in the corresponding formulae for the partially anisotropic HM model and letting $\varrho \to 0$, we get the zero curvature representation (I.1.14) and the r-matrix (3.9) for the HM model.

The LL model is thus the most general magnet model admitting an r-matrix formulation.

We will now show that the SG and NS models are limiting cases of the LL model as well. Hence, the latter is indeed a universal model for integrable systems with two-dimensional auxiliary space. It was not chosen to be the basic model of the book only because its analysis is technically more complicated than that for the NS model.

When deriving the SG and NS models, besides degenerating the elliptic curve E we must also contract the phase space of the LL model. This is made possible by the so far unexploited freedom in describing the phase space, namely: instead of a sphere of radius 1 in $\mathbb{R}^3$ which is the domain of $\vec{S}(x)$, we may consider a sphere of arbitrary radius $R > 0$; in addition, we can introduce an arbitrary "coupling constant" $\eta > 0$ into the Poisson structure (8.6),

$$\{S_a(x), S_b(y)\} = -\eta\varepsilon_{abc} S_c(x)\delta(x-y). \tag{8.14}$$

Of course, for R and η fixed, this arbitrariness can be reduced out by a dilatation of $\vec{S}$ and a change of the independent variable x. However, in the limits $R \to \infty$ or $\eta \to 0$ considered below, such a replacement is unnatural.

We shall thus assume that $\vec{S}(x)$ belongs to the sphere of radius R,

$$\vec{S}^2(x) = R^2, \tag{8.15}$$

and satisfies

$$\frac{\partial \vec{S}}{\partial t} = \frac{1}{R^2}\vec{S} \wedge \frac{\partial^2 \vec{S}}{\partial x^2} + \vec{S} \wedge J\vec{S} \tag{8.16}$$

and that its Poisson brackets are given by (8.14). The r-matrix in (8.9) differs from (8.10) by the factor η, and the matrix V in the zero curvature representation must be modified: the second term in (8.3) must be divided by R^2.

Let us consider a passage to the SG model; we start with the equations of motion. Make a change of variables $\vec{S}(x,t) \rightarrow \pi(x,t), \varphi(x,t)$ in (8.16) according to

$$S_1 = -\frac{\beta\pi}{2}, \quad S_2 = \sqrt{R^2 - \frac{\beta^2\pi^2}{4}} \sin\frac{\beta\varphi}{2}, \quad S_3 = \sqrt{R^2 - \frac{\beta^2\pi^2}{4}} \cos\frac{\beta\varphi}{2}, \tag{8.17}$$

where $\beta > 0$, and take the parameters J_1, J_2, J_3 to be

$$J_2 = J_1 + 1, \quad J_3 = J_2 + \frac{m^2}{R^2}. \tag{8.18}$$

It is easily seen that the new variables allow us to take the limit as $R \rightarrow \infty$ in (8.16) and recover the SG equation

$$\frac{\partial^2 \varphi}{\partial t^2} - \frac{\partial^2 \varphi}{\partial x^2} + \frac{m^2}{\beta} \sin\beta\varphi = 0 \tag{8.19}$$

and $\pi = \dfrac{\partial \varphi}{\partial t}$.

To carry out a passage to the limit as $R \rightarrow \infty$ in the zero curvature representation, it is convenient to shift the spectral parameter, $\lambda = \alpha + K$, so that

$$\tilde{u}_1(\alpha) = u_1(\alpha + K) = \varrho \frac{\mathrm{dn}(\alpha, k)}{\mathrm{cn}(\alpha, k)}, \tag{8.20}$$

$$\tilde{u}_2(\alpha) = u_2(\alpha + K) = \varrho \frac{k'}{\mathrm{cn}(\alpha, k)}, \tag{8.21}$$

$$\tilde{u}_3(\alpha) = u_3(\alpha + K) = -\varrho k' \frac{\mathrm{sn}(\alpha, k)}{\mathrm{cn}(\alpha, k)}. \tag{8.22}$$

Using

$$\mathrm{sn}(\alpha, 1) = \mathrm{th}\,\alpha, \quad \mathrm{cn}(\alpha, 1) = \mathrm{dn}(\alpha, 1) = \frac{1}{\mathrm{ch}\,\alpha} \tag{8.23}$$

and the relations resulting from (8.5), (8.18)

$$\varrho = \frac{1}{2} + O\left(\frac{1}{R^2}\right), \quad k' = \frac{m}{R} + O\left(\frac{1}{R^3}\right), \tag{8.24}$$

we find the following asymptotic expressions for the coefficients $\tilde{u}_a(\alpha)$, as $R \to \infty$ $(k \to 1)$,

$$\tilde{u}_1(\alpha) = \frac{1}{2} + O\left(\frac{1}{R^2}\right), \tag{8.25}$$

$$\tilde{u}_2(\alpha) = \frac{m}{2R} \operatorname{ch} \alpha + O\left(\frac{1}{R^3}\right), \tag{8.26}$$

$$\tilde{u}_3(\alpha) = -\frac{m}{2R} \operatorname{sh} \alpha + O\left(\frac{1}{R^3}\right). \tag{8.27}$$

Now substitute these formulae and the expressions for S_a (8.17) into (8.2) and the modified formula (8.3) and take the limit, as $R \to \infty$. We then arrive at a zero curvature representation for the SG model, which upon an automorphism of the Pauli matrices,

$$\sigma_1 \to -\sigma_3, \quad \sigma_2 \to \sigma_1, \quad \sigma_3 \to -\sigma_2, \tag{8.28}$$

and a change $\lambda = e^\alpha$ coincides with (I.1.30)-(I.1.32).

The standard Poisson structure for the SG model results from the Poisson brackets (8.14) with $\eta = \frac{\beta^2}{4}$ by making a change of variables (8.17) and taking the limit as $R \to \infty$. As noted above, the factor $\frac{\beta^2}{4}$ appears also in the r-matrix; using (8.23) and (8.28), we can derive the r-matrix for the SG model (up to an irrelevant term proportional to $I \otimes I$).

Let us now discuss the passage to the NS model. We set $R = 1$ and consider the limit as $\eta \to 0$. Let

$$(S_1 + iS_2)(x, t) = \sqrt{2\eta} \cdot e^{\frac{2i\varkappa t}{\eta}} \psi(x, t), \tag{8.29}$$

$$S_3(x, t) = \sqrt{1 - 2\eta |\psi(x, t)|^2}, \tag{8.30}$$

$$J_1 = J_2, \quad J_3 = J_1 - \frac{2\varkappa}{\eta}, \tag{8.31}$$

where $\varkappa > 0$ is a new parameter. Inserting these formulae into (8.1) and taking the limit, as $\eta \to 0$, we find that $\psi(x, t)$ satisfies the NS equation

$$i \frac{\partial \psi}{\partial t} = - \frac{\partial^2 \psi}{\partial x^2} + 2\varkappa |\psi|^2 \psi . \tag{8.32}$$

To evaluate the limit in the zero curvature representation for the partially anisotropic HM model, it is convenient to set $\lambda = \alpha + \frac{\pi}{2}$ so that

$$\tilde{u}_1(\alpha) = \tilde{u}_2(\alpha) = u_1(\alpha + \tfrac{\pi}{2}) = u_2(\alpha + \tfrac{\pi}{2}) = \frac{\sqrt{-\dfrac{\varkappa}{2\eta}}}{\cos \alpha}, \tag{8.33}$$

$$\tilde{u}_3(\alpha) = u_3(\alpha + \tfrac{\pi}{2}) = -\sqrt{-\frac{\varkappa}{2\eta}} \cdot \operatorname{tg} \alpha . \tag{8.34}$$

Setting once again $\alpha = -\sqrt{\dfrac{\eta}{-2\varkappa}}\,\lambda$, we insert (8.29)–(8.30) and (8.33)–(8.34) into (8.2)–(8.3), perform a gauge transformation with matrix $\exp \dfrac{i\varkappa t}{\eta} \sigma_3$, and take the limit as $\eta \to 0$. It is not hard to see that we come down to the zero curvature representation for the NS model stated in § I.2 of Part I.

The standard Poisson structure for the NS model results from the Poisson brackets (8.14) by using (8.29)–(8.30) and letting $\eta \to 0$. The same limit gives the r-matrix for the NS model as in § III.1, Part I (up to an irrelevant unit summand).

The geometric meaning of the above contractions of phase space for the LL model is obvious. For fixed x, the dynamical variables of the LL model belong to the sphere $\mathbf{S}^2$, and the Poisson brackets (8.6) are induced by the symplectic structure, the area form on $\mathbf{S}^2$. The corresponding phase spaces for the SG and NS models are the cylinder $\mathbf{S}^1 \times \mathbb{R}^1$ and the plane $\mathbb{R}^2$, respectively. These manifolds are contractions of the sphere $\mathbf{S}^2$ when one stretches a strip along a particular meridian or a spherical cap centered at the north pole, respectively. The area form on $\mathbf{S}^1 \times \mathbb{R}^1$ or $\mathbb{R}^2$ is what defines the symplectic structures of the SG or NS models.

This concludes our discussion of concrete continuous integrable models. In Chapter IV we shall resume their general investigation and classification from the Lie-algebraic point of view.

9. Notes and References

1. The inverse scattering method for the HM model was developed in [T 1977]. The complete integrability of the model in the rapidly decreasing case was established in [ZT 1979], [F 1980]; in particular, in [F 1980] canonical action-angle variables were constructed. The relationship of the Poisson structures for the HM and NS models under the gauge transformation was observed in [KR 1978].

2. The inverse scattering method for the SG model in laboratory coordinates was formulated in [TF 1974], [ZTF 1974]. In [TF 1974], [TF 1975] the model was shown to be completely integrable, canonical action-angle variables were constructed and an interpretation of the excitation spectrum in terms of relativistic field theory was proposed. An expression for the boost generator K in terms of action-angle variables was given in [AM 1983].

3. The SG model in light-cone variables was integrated via the inverse scattering method in [AKNS 1973], [T 1974]. The relation between the SG models in laboratory and light-cone coordinates was analyzed in [TF 1976], [KN 1978]. The constraints (7.8) and the equivalence of the Hamiltonian pictures for these models were established in [TF 1976].

4. The SG equation in light-cone coordinates written as an evolution equation with respect to η,

$$\frac{\partial \chi}{\partial \eta} = -\frac{m^2}{\beta} \int_{-\infty}^{\xi} \sin\beta\chi(\xi')\,d\xi', \tag{9.1}$$

is nonlocal. The class of rapidly decreasing initial data is not preserved by the dynamics. A complete system of constraints in § 7 specifies a subset of initial data invariant under the dynamics of all the higher SG equations

$$\frac{\partial \chi}{\partial \eta} = -\frac{2m}{\beta^2} \int_{-\infty}^{\xi} \frac{\delta I_{-n}}{\delta \chi(\xi')}\,d\xi', \tag{9.2}$$

$n = 1, 2, \ldots$. These constraints are given by a procedure analogous to that of § 7.

5. The densities $J_n^{(1)}$ of the integrals of the SG model in laboratory coordinates evaluated on solutions of the equations of motion satisfy

$$\frac{\partial J_n^{(1)}}{\partial t} = \frac{\partial J_n^{(0)}}{\partial x}, \tag{9.3}$$

$-\infty<n<\infty$, $n\equiv 1\,(\mathrm{mod}\,2)$. Here $J_n^{(0)}(x,t)$ is a polynomial in $\varphi(x,t)$, $\pi(x,t)$ and their derivatives with respect to x easily determined from the zero curvature representation (see, for instance, [VV 1985]). From (9.3) we have

$$\frac{\partial}{\partial\eta}\left(J_n^{(0)}+J_n^{(1)}\right)=\frac{\partial}{\partial\xi}\left(J_n^{(0)}-J_n^{(1)}\right) \tag{9.4}$$

hence (see § 7)

$$\int_{-\infty}^{\infty} J_n^{(1)}|_{t=0}\,dx=\int_{-\infty}^{\infty}\left(J_n^{(0)}+J_n^{(1)}\right)|_{\eta=0}\,d\xi. \tag{9.5}$$

The expressions $J_n^{(+)}(\chi)=J_n^{(0)}(\pi,\varphi)+J_n^{(1)}(\pi,\varphi)$ are local in χ and $\dfrac{\partial\chi}{\partial\eta}$; the nonlocality of the integrals $J_n^{(+)}(\chi)$ of the SG model in light-cone coordinates for $n<-1$ is due to the necessity to reduce $\dfrac{\partial\chi}{\partial\eta}$ out of $J_n^{(0)}(\pi,\varphi)+J_n^{(1)}(\pi,\varphi)$ by using (9.1).

6. The relationship described in § 7 between the Hamiltonian pictures of the SG models in laboratory and light-cone coordinates is a fairly general one; in particular, it holds for the principal chiral field model and the n-field model.

7. The equation

$$\frac{\partial^2\varphi}{\partial t^2}-\frac{\partial^2\varphi}{\partial x^2}+\frac{m^2}{\beta}\,\mathrm{sh}\,\beta\varphi=0 \tag{9.6}$$

in the rapidly decreasing case in analyzed in exactly the same way as the SG equation after replacing $\beta\to i\beta$ in the auxiliary linear problem (4.1). The latter then becomes a self-adjoint problem and the coefficient $a(\lambda)$ has no zeros. Hence the model described by (9.6) has no solitons. In this sense, the relationship between (9.6) and the SG model is the same as between the NS models in the rapidly decreasing case for $\varkappa>0$ and $\varkappa<0$, respectively.

8. In some sense, the analogue for solitons of the SG model is provided by singular solutions of (9.6). For the general approach to singular solutions and their particle-like interpretation, see the survey [PP 1985]. A version of the inverse scattering method for constructing singular solutions of nonlinear equations is presented in [APP 1982], [APP 1983], [A 1984].

9. Asymptotic expressions for solutions of the SG equation in laboratory coordinates, as $t\to\pm\infty$, and in light-cone coordinates, as $\eta\to\pm\infty$, were derived in [ZM 1975], [IP 1982]. The corresponding asymptotic formulae for the two-dimensional Toda lattice including in particular (9.6) were obtained in [No 1984].

10. Just as for the NS model, a hierarchy of Poisson structures can be defined for the HM and SG models. The Λ-operators that generate the hierarchy are given in [M 1978], [GY 1984], and [K 1984], respectively.

11. The zero curvature representation for the LL model was found in [S 1979] and [BR 1981]. In [S 1979] the auxiliary linear problem for the rapidly decreasing boundary conditions was studied and action-angle variables were introduced. We point out that it was in [S 1979] in the framework of the LL model that the notion of (classical) r-matrix appeared for the first time.

12. For the notation and information concerning the Jacobi elliptic functions, see, for instance, [WW 1927].

13. The inverse problem for the LL model was formulated and studied as a matrix Riemann problem on an elliptic curve in [Mi 1982], [R 1984] with a description of the n-soliton solutions.

The paper [R 1984] also treated the matrix Riemann problem on a general algebraic curve (compact Riemann surface).

14. A construction of soliton solutions for the LL equation using the dressing procedure was given in [B 1983], [Bo 1983], [BBI 1984a].

15. Finite gap solutions of the SG equation, both in laboratory and light-cone coordinates, were obtained in [I 1976], [KK 1976]. A solution of the reality problem in terms of algebraic geometry was given in [C 1980], and in terms of explicit theta-functional formulae in [BE 1982] (the case of two gaps) and in [DN 1982] (the general case). For the general reality problem in finite-gap integration theory, see [N 1984].

16. A general algebraic-geometric description of finite-gap solutions of the LL equation is presented in [C 1983]. An explicit construction which allows to express these solutions in terms of theta functions was given in [BBI 1983], [BBI 1984b] for the partially anisotropic HM model, and in [BB 1983], [B 1985] for the LL equation. From the point of view of finite-gap integration, the LL equation is distinguished by the fact that the relevant algebraic curve Γ is a two-sheeted covering of the elliptic curve rather than that of the complex plane, as was the case for the models with rational dependence on the spectral parameter. As a result, the final expressions for finite-gap solutions involve the theta functions of Prym rather than those of Riemann [DJK 1983], [B 1985].

17. For the construction of finite-gap solutions of the general zero curvature equation with rational dependence on the spectral parameter, see, in addition to the papers mentioned above, the surveys [K 1977], [K 1983], [D 1983].

18. The LL equation is connected with integrable systems of classical mechanics. In [V 1983] if was shown that stationary (i.e. t-independent) solutions of the LL equation solve the Neumann problem of the motion of a particle on a two-dimensional sphere, whereas solutions that depend only on the combination $x-vt$ correspond to Clebsch's integrable case of the

motion of a rigid body in fluid. Explicit formulae for these solutions in terms of Prym theta functions are given in [B 1985].

19. The complete integrability of the models of Chapter II: the KdV model, the vector NS model, the N-wave model and the $SO(N)$-SG model (for the characteristic case $N=3$) under the rapidly decreasing boundary conditions was proved in [ZF 1971], [Ku 1982], [M 1976] and [Bu 1978], respectively, where the canonical action-angle variables were determined.

20. In the vector NS model, solitons possess a polarization which in general changes in the course of interaction [M 1973]. Nevertheless, soliton scattering theory remains factorized as before [Ku 1976], [Ku 1979].

21. The models discussed in § I.2 admit an r-matrix formulation

$$\{U(x,\lambda) \underset{,}{\otimes} U(y,\mu)\} = [r(\lambda,\mu), U(x,\lambda) \otimes I + I \otimes U(x,\mu)]\delta(x-y), \tag{9.7}$$

where for the vector NS model with n colours and the $N=\dfrac{n(n-1)}{2}$ wave model one has $r(\lambda,\mu)=r(\lambda-\mu)$ with

$$r(\lambda) = -\varkappa \frac{P}{\lambda} \tag{9.8}$$

and

$$r(\lambda) = \frac{P}{\lambda} \tag{9.9}$$

respectively, where P is the permutation matrix in $\mathbb{C}^n \otimes \mathbb{C}^n$ (see [KS 1980]). For the two-dimensional Toda model one has

$$r(\lambda,\mu) = -\frac{\lambda^n+\mu^n}{2(\lambda^n-\mu^n)} \sum_{i=1}^{n-1} H_i \otimes H_i - \sum_{\alpha} \frac{\lambda^{p(\alpha)}\mu^{n-p(\alpha)}}{\lambda^n-\mu^n} E_\alpha \otimes E_{-\alpha} \tag{9.10}$$

(see [Ku 1981]). Here α ranges over all roots of the Lie algebra A_{n-1}, $p(\alpha)=1,\ldots,n-1$ is the height (mod n) of the root α, and H_i, E_α are the Cartan-Weyl basis vectors for A_{n-1}.

22. The basic Poisson brackets (I.3.15) for the KdV model contain the derivative of the delta function (the non-ultralocal case), so that the Poisson brackets $\{U(x,\lambda) \underset{,}{\otimes} U(y,\mu)\}$ cannot be expressed in the form (9.7). The latter is equally true of the SG model in light-cone coordinates. However, the Poisson brackets of the transition matrices for these models do have the familiar representation

$$\{T(x,y,\lambda) \underset{,}{\otimes} T(x,y,\mu)\} = [r(\lambda,\mu), T(x,y,\lambda) \otimes T(x,y,\mu)], \quad y \leqslant x \tag{9.11}$$

(see [Ts 1981]) where for the KdV model

$$r(\lambda, \mu) = \frac{2}{\lambda^2 - \mu^2} P, \tag{9.12}$$

and the r-matrices for the SG model in laboratory and light-cone coordinates coincide.

References

[A 1984] Arkadiev, V. A.: Application of the inverse scattering method to singular solutions of nonlinear equations. III. Teor. Mat. Fiz. *58*, 38–49 (1984) [Russian]

[AKNS 1973] Ablowitz, M. J., Kaup, D. J., Newell, A. C., Segur, H.: Method for solving the Sine-Gordon equation. Phys. Rev. Lett. *30*, 1262–1264 (1973)

[AM 1983] Alonso Martínez, L.: Group-theoretical analysis of the Sine-Gordon equation as a relativistic dynamical system. J. Math. Phys. *24*, 982–989 (1983)

[APP 1982, 1983] Arkadiev, V. A., Pogrebkov, A. K., Polivanov, M. K.: Application of the inverse scattering method to singular solutions of nonlinear equations. I. Teor. Mat. Fiz. *53*, 163–180 (1982). II. Teor. Mat. Fiz. *54*, 23–37 (1983) [Russian]; English transl. in Theor. Math. Phys. *53*, 151–162 (1982), *54*, 12–32 (1983)

[B 1983] Bobenko, A. I.: The Landau-Lifshitz equation. Dressing up of the solutions and elementary excitations. In: Differential geometry, Lie groups and mechanics. V. Zap. Nauchn. Semin. LOMI *123*, 58–66 (1983) [Russian]

[B 1985] Bobenko, A. I.: Real algebraic-geometric solutions of the Landau-Lifshitz equation in terms of Prym theta-functions. Funk. Anal. Prilož. *19* (1), 6–19 (1985) [Russian]

[BB 1983] Bikbaev, R. F., Bobenko, A. I.: On finite-gap integration of the Landau-Lifshitz equation. XYZ case. Preprint LOMI E-8-83. Leningrad, 1983

[BBI 1983] Bikbaev, R. F., Bobenko, A. I., Its, A. R.: Finite-gap integration of the Landau-Lifshitz equation. Dokl. Akad. Nauk SSSR *272*, 1293–1298 (1983) [Russian]; English transl. in Sov. Math. Dokl. *28*, 512–516 (1983)

[BBI 1984a, b] Bikbaev, R. F., Bobenko, A. I., Its, A. R.: The Landau-Lifshitz equation. The theory of exact solutions. (Part I, II). Preprints Don PTI-4-6(81), Don PTI-4-7(82), Donetsk 1984 [Russian]

[BE 1982] Belokolos, E. D., Enolsky, V. Z.: Solution in elliptic functions of the nonlinear partial differential equations integrable by the inverse scattering method. Usp. Mat. Nauk *37* (4), 89 (1982) [Russian]

[Bo 1983] Borisov, A. B.: Multisoliton solutions of the nonisotropic magnet equations. Fizika Materialov i Metallovedenie *55* (2), 230–234 (1983) [Russian]

[BR 1981] Borovik, A. E., Robuk, V. N.: Linear pseudopotentials and conservation laws for the Landau-Lifshitz equation describing the nonlinear dynamics of a ferromagnet with uniaxial anisotropy. Teor. Mat. Fiz. *46*, 371–381 (1981) [Russian]; English transl. in Theor. Math. Phys. *46*, 242–248 (1981)

[Bu 1978] Budagov, A. S.: A completely integrable model of classical field theory with nontrivial particle interaction in two space-time dimensions. In: Problems in quantum field theory and statistical physics. I. Zap. Nauchn. Semin. LOMI *77*, 24–56 (1978) [Russian]

[C 1980] Cherednik, I. V.: Reality conditions in finite-gap integration. Dokl. Akad. Nauk SSR *252*, 1104–1108 (1980) [Russian]; English transl. in Sov. Phys. Dokl. *25*, 450–452 (1980)

[C 1983] Cherednik, I. V.: Integrable differential equations and coverings of elliptic curves. Izv. Akad. Nauk SSSR, Ser. Mat. *47*, 384–406 (1983) [Russian]; English transl. in Math. USSR - Izv. *22*, 357–377 (1984)

[D 1983] Dubrovin, B. A.: Matrix finite-gap operators. Itogi Nauki Tekh., Ser. Sovrem. Probl. Mat. *23*, 33–78 Moscow, VINITI 1983 [Russian]

[DJK 1983] Date, E., Jimbo, M., Kashiwara, M.: Landau-Lifshitz equation: solitons, quasi-periodic solutions and infinite dimensional Lie algebras. J. Phys. *A 16*, 221–236 (1983)

[DN 1982] Dubrovin, B. A., Natanson, S. M.: Real two-gap solutions of the sine-Gordon equation. Funk. Anal. Prilož. *16* (1), 27–43 (1982) [Russian]; English transl. in Funct. Anal. Appl. *16*, 21–33 (1982)

[F 1980] Fogedby, H. C.: Solitons and magnons in the classical Heisenberg chain. J. Phys. *A 13*, 1467–1499 (1980)

[GY 1984] Gerdjikov, V. S., Yanovski, A. B.: Gauge covariant formulation of the generating operator. 1. The Zakharov-Shabat system. Phys. Lett. *103A*, 232–236 (1984)

[I 1976] Its, A. R.: Finite-gap solutions of the sine-Gordon equation. For an exposition see Matveev V. B.: Abelian functions and solutions. Preprint No 373, Inst. Fiz. Teor. Univ. Wroclaw (1976)

[IP 1982] Its, A. R., Petrov, V. E.: “Isomonodromic” solutions of the sine-Gordon equation and the long time asymptotics of its rapidly decaying solutions. Dokl. Akad. Nauk SSSR *265*, 1302–1306 (1982) [Russian]; English transl. in Sov. Math. Dokl. *26*, 244–247 (1983)

[K 1977] Krichever, I. M.: Methods of algebraic geometry in the theory of nonlinear equations. Usp. Mat. Nauk *32* (6), 183–208 (1977) [Russian]; English transl. in Russian Math. Surveys *32* (6), 185–213 (1977)

[K 1983] Krichever, I. M.: Nonlinear equations and elliptic curves. Itogi Nauki Tekh., Ser. Sovrem. Probl. Math. *23*, 79–136 Moscow, VINITI 1983 [Russian]

[K 1984] Kaup, D.: The squared eigenstates of the sine-Gordon eigenvalue problem. J. Math. Phys. *25*, 2467–2471 (1984)

[KK 1976] Kosel, V. A., Kotlyarov, V. P.: Almost-periodic solutions of the equations $u_{tt}-u_{xx}+\sin u=0$. Dokl. Akad. Nauk. Ukr. SSR, Ser. A. *10*, 878–881 (1976) [Russian]

[KN 1978] Kaup, D. J., Newell, A. C.: The Goursat and Cauchy problems for the Sine-Gordon equation. SIAM J. Appl. Math. *34*, 37–54 (1978)

[KR 1978] Kulish, P. P., Reyman, A. G.: A hierarchy of symplectic forms for the Schrödinger and the Dirac equations on the line. In: Problems in quantum field theory and statistical physics. I. Zapiski Nauchn. Semin. LOMI *77*, 134–147 (1978) [Russian]; English transl. in J. Sov. Math. *22*, 1627–1637 (1983)

[KS 1980] Kulish, P. P., Sklyanin, E. K.: Solutions of the Yang-Baxter equation. In: Differential geometry, Lie groups and mechanics. III. Zapiski Nauchn. Semin. LOMI *95*, 129–160 (1980) [Russian]; English transl. in J. Sov. Math. *19* (5) 1596–1620 (1982)

[Ku 1976] Kulish, P. P.: Factorization of the classical and quantum *S*-matrix and conservation laws. Teor. Mat. Fiz. *26*, 198–205 (1976) [Russian]; English transl. in Theor. Math. Phys. *26*, 132–137 (1976)

[Ku 1979] Kulish, P. P.: Scattering of solitons with internal degrees of freedom. In: Problems in high energy physics and quantum field theory. II International seminar, Protvino, 463–470 (1979) [Russian]

[Ku 1981] Kulish, P. P.: Quantum difference nonlinear Schrödinger equation. Lett. Math. Phys. *5*, 191–197 (1981)

[Ku 1982] Kulish, P. P.: Action-angle variables for the multi-component nonlinear Schrödinger equation. In: Boundary value problems of mathematical physics and related topics of function theory. 14. Zap. Nauchn. Semin. LOMI *115*, 126–136 (1982) [Russian]

[M 1973] Manakov, S. V.: On the theory of two-dimensional stationary self-focusing of electro-magnetic waves. Zh. Exp. Teor. Fiz. *65*, 506–516 (1973) [Russian]; English transl. in Sov. Phys. JETP *38*, 248–253 (1974)

[M 1976] Manakov, S. V.: Example of a completely integrable nonlinear wave field with nontrivial dynamics (Lee model). Teor. Mat. Fiz. *28*, 172–179 (1976) [Russian]; English transl. in Theor. Math. Phys. *28*, 709–714 (1976)

[M 1978] Magri, F.: A simple model of the integrable Hamiltonian equation. J. Math. Phys. *19*, 1156–1162 (1978)

[Mi 1982] Mikhailov, A. V.: The Landau-Lifshitz equation and the Riemann boundary problem on a torus. Phys. Lett. *92A*, 51–55 (1982)

[N 1984] Novikov, S. P.: Algebraic-topological approach to reality problems. Real action variables in the theory of finite-gap solutions of the sine-Gordon equation. In: Differential geometry, Lie groups and mechanics. VI. Zapiski Nauchn. Semin. LOMI *133*, 177–196 (1984) [Russian]

[No 1984] Novokshenov, V. Yu.: Asymptotics as $t\to\infty$ of the solution of the Cauchy problem for the two-dimensional generalization of the Toda lattice. Izv. Akad. Nauk SSSR, Ser. Mat. *48*, 372–410 (1984) [Russian]

[PP 1985] Pogrebkov, A. K., Polivanov, M, K.: Singular solutions of the Liouville and Sinh-Gordon equations. Sov. Sci. Rev., Ser. C. Review in Math. Physics *5*, 120–169 (1985)

[R 1984] Rodin, Yu. L.: The Riemann boundary problem on Riemann surfaces and the inverse scattering problem for the Landau-Lifshitz equation. Physica *D 11*, 90–108 (1984)

[S 1979] Sklyanin, E. K.: On complete integrability of the Landau-Lifshitz equation. Preprint LOMI E-3-79, Leningrad, 1979

[T 1974] Takhtajan, L. A.: Exact theory of propagation of ultra-short optical pulses in two-level media. Zh. Exp. Teor. Fiz. *66*, 476–489 (1974) [Russian]

[T 1977] Takhtajan, L. A.: Integration of the continuous Heisenberg spin chain through the inverse scattering method. Phys. Lett. *64A*, 235–237 (1977)

[TF 1974] Takhtajan, L. A., Faddeev, L. D.: Essentially nonlinear one-dimensional model of classical field theory. Teor. Mat. Fiz. *21* (2), 160–174 (1974) [Russian]; English transl. in Theor. Math. Phys. *21*, 1046–1057 (1974)

[TF 1975] Takhtajan, L. A., Faddeev, L. D.: Essentially nonlinear one-dimensional model of classical field theory. (Addition). Teor. Mat. Fiz. *22*, 143 (1975) [Russian]; English transl. in Theor. Math. Phys. *22*, 100 (1975)

[TF 1976] Takhtajan, L. A., Faddeev, L. D.: The Hamiltonian system connected with the equation $u_{\xi\eta}+\sin u=0$. Trudy. Mat. Inst. Steklov *142*, 254–266 (1976) [Russian]; English transl. in Proc. Steklov Inst. Math. *3*, 277–289 (1979)

[Ts 1981] Tsiplyaev, S. A.: Commutation relations of the transition matrix in the classical and quantum inverse scattering methods. (The local case). Teor. Mat. Fiz. *48*, 24–33 (1981) [Russian]; English transl. in Theor. Math. Phys. *48*, 580–586 (1982)

[V 1983] Veselov, A. P.: The Landau-Lifshitz equation and integrable systems of classical mechanics. Dokl. Akad. Nauk SSSR *270*, 1094–1097 (1983) [Russian]; English transl. in Sov. Phys. Dokl. *28*, 458–459 (1983)

[VV 1985] Vladimirov, V. S., Volovich, I. V.: Local and nonlocal currents for nonlinear equations. Teor. Mat. Fiz. *62*, 3–29 (1985) [Russian]

[WW 1927] Whittaker, E. T., Watson, G. N.: A Course of Modern Analysis. Cambridge, University Press 1927

[ZF 1971] Zakharov, V. E., Faddeev, L. D.: Korteweg-de Vries equation, a completely integrable Hamiltonian system. Funk. Anal. Prilož. *5* (4), 18–27 (1971) [Russian]; English transl. in Funct. Anal. Appl. *5*, 280–287 (1971)

[ZM 1975] Zakharov, V. E., Manakov, S. V.: The theory of resonant interaction of wave packets in nonlinear media. Zh. Eksp. Teor. Fiz. *69*, 1654–1673 (1975) [Russian]; English transl. in Soviet Phys. JETP *42*, 842–850 (1976)

[ZT 1979] Zakharov, V. E., Takhtajan, L. A.: Equivalence of the nonlinear Schrödinger equation and the Heisenberg ferromagnet equation. Teoret. Mat. Fiz. *38* (1), 26–35 (1979) [Russian]; English transl. in Theor. Math. Phys. *38*, 17–23 (1979)

[ZTF 1974] Zakharov, V. E., Takhtajan, L. A., Faddeev, L. D.: A complete description of the solution of the sine-Gordon equation. Dokl. Akad. Nauk SSSR *219*, 1334–1337 (1974) [Russian]; English transl. in Sov. Phys. Dokl. *19*, 824–826 (1975)

Chapter III
Fundamental Models on the Lattice

Here we shall give a complete list of results pertaining to the Toda model, a fundamental model on the lattice. We will show that the r-matrix approach applies to this case and may be used to prove the complete integrability of the model in the quasi-periodic case. For the rapidly decreasing boundary conditions we will analyze the mapping $\mathscr{F}$ from the initial data of the auxiliary linear problem to the transition coefficients and outline a method for solving the inverse problem, i.e. for constructing $\mathscr{F}^{-1}$. On the basis of the r-matrix approach it will be shown that $\mathscr{F}$ is a canonical transformation to action-angle type variables establishing the complete integrability of the Toda model in the rapidly decreasing case. We shall also define a lattice version of the LL model, the most general integrable lattice system with two-dimensional auxiliary space.

§ 1. Complete Integrability of the Toda Model in the Quasi-Periodic Case

The equations of motion for the model are

$$\frac{d^2 q_n}{dt^2} = e^{q_{n+1}-q_n} - e^{q_n - q_{n-1}}, \quad n=1,\ldots,N, \tag{1.1}$$

where

$$q_{N+n} = q_n + c. \tag{1.2}$$

This is a Hamiltonian system on the phase space $\mathscr{M} = \mathbb{R}^{2N}$ with coordinates $(p_1,\ldots,p_N, q_1,\ldots,q_N)$, endowed with the standard Poisson structure

$$\{p_i, p_j\} = \{q_i, q_j\} = 0, \quad \{p_i, q_j\} = \delta_{ij}, \quad i,j=1,\ldots,N. \tag{1.3}$$

The Hamiltonian is

$$H=\sum_{n=1}^{N}\left(\frac{1}{2}p_n^2+e^{q_{n+1}-q_n}\right) \tag{1.4}$$

(see § I.2).

We will show that *the model is completely integrable* in the sense of classical mechanics with finitely many degrees of freedom. By the Liouville-Arnold theorem, one need only to produce a set of N involutive integrals of the motion I_n

$$\{H,I_n\}=0, \quad \{I_n,I_m\}=0, \quad n,m=1,\ldots,N, \tag{1.5}$$

that are functionally independent,

$$\operatorname{rank}\left(\frac{\partial I_m}{\partial p_n},\frac{\partial I_m}{\partial q_n}\right)=N \tag{1.6}$$

on a dense subset of $\mathcal{M}$. The left hand side of (1.6) is an $N\times 2N$ matrix composed of the first derivatives of the I_m.

For the proof we consider the auxiliary linear problem for the Toda model,

$$F_{n+1}=L_n(\lambda)F_n, \tag{1.7}$$

where

$$L_n(\lambda)=\begin{pmatrix} p_n+\lambda & e^{q_n}\\ -e^{-q_n} & 0\end{pmatrix} \tag{1.8}$$

(see § I.2) and apply the r-matrix approach. *A natural analogue of the fundamental Poisson brackets of Chapter II is given by*

$$\{L_n(\lambda)\underset{,}{\otimes}L_m(\mu)\}=[r(\lambda,\mu),L_n(\lambda)\otimes L_n(\mu)]\,\delta_{nm}. \tag{1.9}$$

Indeed, $L_n(\lambda)$ may be regarded as a transition matrix (over one lattice step), and so its Poisson brackets should be modelled on the corresponding expressions for $T(x,y,\lambda)$.

To calculate the r-matrix we express $L_n(\lambda)$ as

$$L_n(\lambda)=(p_n+\lambda)\sigma+\operatorname{sh}q_n\sigma_1+i\operatorname{ch}q_n\sigma_2, \tag{1.10}$$

with

$$\sigma=\frac{I+\sigma_3}{2}. \tag{1.11}$$

From

$$\{p_n, \operatorname{sh} q_m\} = \operatorname{ch} q_n \delta_{nm}, \qquad \{p_n, \operatorname{ch} q_m\} = \operatorname{sh} q_n \delta_{nm} \tag{1.12}$$

we find

$$\begin{aligned}\{L_n(\lambda) \overset{\otimes}{,} L_m(\mu)\} = (i \operatorname{sh} q_n (\sigma \otimes \sigma_2 - \sigma_2 \otimes \sigma) \\ + \operatorname{ch} q_n (\sigma \otimes \sigma_1 - \sigma_1 \otimes \sigma)) \delta_{nm},\end{aligned} \tag{1.13}$$

so that the left hand side of (1.9) is linear in $\operatorname{sh} q_n$ and $\operatorname{ch} q_n$ and does not depend on λ, μ or p_n. In the product $L_n(\lambda) \otimes L_n(\mu)$ the terms linear in $\operatorname{sh} q_n$ and $\operatorname{ch} q_n$ have the form

$$\operatorname{sh} q_n (\lambda \sigma \otimes \sigma_1 + \mu \sigma_1 \otimes \sigma) + i \operatorname{ch} q_n (\lambda \sigma \otimes \sigma_2 + \mu \sigma_2 \otimes \sigma)$$

and the remaining terms commute with the permutation matrix P. We shall therefore look for an r-matrix of the form

$$r(\lambda, \mu) = f(\lambda, \mu) P, \tag{1.14}$$

where $f(\lambda, \mu)$ is an unknown function. We have

$$\begin{aligned}[P, \lambda \sigma \otimes \sigma_1 + \mu \sigma_1 \otimes \sigma] &= (\lambda - \mu) P (\sigma \otimes \sigma_1 - \sigma_1 \otimes \sigma) \\ &= i(\lambda - \mu)(\sigma \otimes \sigma_2 - \sigma_2 \otimes \sigma),\end{aligned} \tag{1.15}$$

$$\begin{aligned}[P, \lambda \sigma \otimes \sigma_2 + \mu \sigma_2 \otimes \sigma] &= (\lambda - \mu) P (\sigma \otimes \sigma_2 - \sigma_2 \otimes \sigma) \\ &= i(\lambda - \mu)(\sigma_1 \otimes \sigma - \sigma \otimes \sigma_1),\end{aligned} \tag{1.16}$$

where we have used the expression

$$P = \frac{1}{2}\left(I \otimes I + \sum_{a=1}^{3} \sigma_a \otimes \sigma_a\right) \tag{1.17}$$

and the multiplication formulae for the Pauli matrices. It follows that (1.9) will hold if $f(\lambda, \mu)$ is chosen to be

$$f(\lambda, \mu) = \frac{1}{\lambda - \mu}. \tag{1.18}$$

As a result, $L_n(\lambda)$ obeys the fundamental lattice Poisson brackets (1.9) with the r-matrix

$$r(\lambda, \mu) = r(\lambda - \mu) = \frac{P}{\lambda - \mu} \tag{1.19}$$

that already occurred for the NS model in Part I.

Introducing the monodromy matrix

$$T_N(\lambda)=\prod_{n=1}^{\overset{\curvearrowleft}{N}} L_n(\lambda), \tag{1.20}$$

we derive from (1.9) the corresponding Poisson brackets

$$\{T_N(\lambda) \overset{\otimes}{,} T_N(\mu)\}=[r(\lambda-\mu), T_N(\lambda)\otimes T_N(\mu)]. \tag{1.21}$$

As was already observed in § I.7, under the periodic boundary conditions the trace of the monodromy matrix is a generating function for integrals of the motion. In the quasi-periodic case

$$L_{N+1}(\lambda)=Q(c)L_1(\lambda)Q^{-1}(c), \tag{1.22}$$

where

$$Q(c)=\exp\frac{c\sigma_3}{2}, \tag{1.23}$$

a similar role is played by the function

$$F_N(\lambda)=\operatorname{tr} T_N(\lambda)Q^{-1}(c) \tag{1.24}$$

(cf. the NS model under the quasi-periodic boundary conditions in § I.2, Part I), which is a polynomial in λ of degree N,

$$F_N(\lambda)=e^{-\frac{c}{2}}\lambda^N+\sum_{n=1}^{N} I_n\lambda^{N-n}, \tag{1.25}$$

the coefficients I_n in turn being polynomials in p_j and $e^{\pm q_j}$. In particular, we have

$$I_1=e^{-\frac{c}{2}}\sum_{n=1}^{N} p_n, \tag{1.26}$$

$$\begin{aligned} I_2&=e^{-\frac{c}{2}}\left(\sum_{1\leq k<n\leq N} p_k p_n-\sum_{n=1}^{N-1} e^{q_{n+1}-q_n}-e^{q_1+c-q_N}\right)\\ &=e^{-\frac{c}{2}}\left(\sum_{1\leq k<n\leq N} p_k p_n-\sum_{n=1}^{N} e^{q_{n+1}-q_n}\right), \end{aligned} \tag{1.27}$$

so that

$$H = \frac{e^c}{2} I_1^2 - e^{\frac{c}{2}} I_2. \tag{1.28}$$

Since $r(\lambda)$ commutes with $Q(c) \otimes Q(c)$, (1.21) yields

$$\{F_N(\lambda), F_N(\mu)\} = 0, \tag{1.29}$$

hence $I_1, \ldots, I_N$ is an involutive family of integrals of the motion which contains the Hamiltonian of the model.

To conclude the proof of complete integrability of the Toda model it only remains to verify that the integrals I_n are functionally independent. Obviously,

$$I_n = e^{-\frac{c}{2}} S_n(p_1, \ldots, p_N) + I_n', \tag{1.30}$$

where $S_n(p_1, \ldots, p_N)$ is the n-th elementary symmetric function and I_n' is a polynomial in $p_1, \ldots, p_N$ of degree not greater than $n-1$. Hence (1.6) holds for large p_n, and since everything is polynomial, it holds in the whole phase space $\mathcal{M}$ with the exception of an algebraic subvariety (relative to the coordinates p_n, e^{q_n}) of dimension less than N.

An explicit description of action-angle variables requires recourse to methods of algebraic geometry, which are not our concern in this book.

§ 2. The Auxiliary Linear Problem for the Toda Model in the Rapidly Decreasing Case

Here we shall introduce the principal characteristics of the auxiliary linear problem

$$F_{n+1} = L_n(\lambda) F_n \tag{2.1}$$

in the rapidly decreasing case

$$\lim_{n \to -\infty} q_n = 0, \qquad \lim_{n \to +\infty} q_n = c, \qquad \lim_{|n| \to \infty} p_n = 0. \tag{2.2}$$

We assume that the limiting values in (2.2) are attained sufficiently fast: the quantities q_n, $q_n - c$, p_n and their differencies of all orders decrease faster than any power of $|n|^{-1}$, as $|n| \to \infty$ (the lattice analogue of Schwartz's conditions).

1. The transition matrix and Jost solutions

The transition matrix $T(n, m, \lambda)$ is defined to be the solution of (2.1) with the initial condition

$$T(n, m, \lambda)|_{n=m} = I; \tag{2.3}$$

for $n > m$ it is given by

$$T(n, m, \lambda) = \overset{\curvearrowleft}{\prod_{k=m}^{n-1}} L_k(\lambda) \tag{2.4}$$

and for $n < m$

$$T(n, m, \lambda) = T^{-1}(m, n, \lambda) = \overset{\curvearrowright}{\prod_{k=n}^{m-1}} L_k^{-1}(\lambda). \tag{2.5}$$

The matrix $T(n, m, \lambda)$ is unimodular and is a polynomial in λ of degree $|n-m|$; it obeys the involution

$$\bar{T}(n, m, \lambda) = T(n, m, \bar{\lambda}). \tag{2.6}$$

As $n \to \pm\infty$, the auxiliary linear problem (2.1) simplifies and becomes

$$E_{n+1} = L_{\pm}(\lambda) E_n, \tag{2.7}$$

where

$$L_{-}(\lambda) = L(\lambda) = \begin{pmatrix} \lambda & 1 \\ -1 & 0 \end{pmatrix}, \tag{2.8}$$

$$L_{+}(\lambda) = Q(c) L(\lambda) Q^{-1}(c). \tag{2.9}$$

For $\lambda \neq 2$, $L(\lambda)$ can be reduced to diagonal form

$$L(\lambda) = U(\lambda) \begin{pmatrix} \dfrac{1}{z(\lambda)} & 0 \\ 0 & z(\lambda) \end{pmatrix} U^{-1}(\lambda), \tag{2.10}$$

with

$$U(\lambda) = \begin{pmatrix} 1 & -z(\lambda) \\ -z(\lambda) & 1 \end{pmatrix}, \tag{2.11}$$

and $z(\lambda)$ is determined from

$$z+\frac{1}{z}=\lambda, \tag{2.12}$$

so that

$$z(\lambda)=\frac{\lambda+\sqrt{\lambda^2-4}}{2}. \tag{2.13}$$

The function $z(\lambda)$ is the analogue of $k(\lambda)$ for the NS model in the finite density case (see § I.8 of Part I) and is well defined on the Riemann surface of the function $\sqrt{\lambda^2-4}$. It is often advantageous to use z instead of the spectral parameter λ; in that case $F(z)$ will stand for $F(\lambda(z))$ for any function $F(\lambda)$.

The solution of (2.7) is given by

$$E_n^{(-)}(z)=E_n(z)=U(z)\begin{pmatrix} z^{-n} & 0 \\ 0 & z^n \end{pmatrix} \tag{2.14}$$

and

$$E_n^{(+)}(z)=Q(c)\,E_n(z). \tag{2.15}$$

The matrix $E_n(z)$ obeys the involutions

$$\overline{E_n}(z)=E_n(\bar{z}), \tag{2.16}$$

$$E_n\left(\frac{1}{\bar{z}}\right)=-\frac{1}{z}E_n(z)\sigma_1 \tag{2.17}$$

and the relation

$$\det E_n(z)=1-z^2. \tag{2.18}$$

On the circle $|z|=1$ the entries of $E_n(z)$ are bounded for all n, which corresponds to *the continuous spectrum* of the auxiliary linear problem (2.7). In terms of λ, the continuous spectrum fills in the interval $-2\leqslant\lambda\leqslant 2$. The matrix $E_n(z)$ degenerates at $z=\pm 1$ so that (2.7) has *virtual levels* on the boundary of the spectrum (cf. §§ I.8–I.9 of Part I). The interior and exterior of the unit circle relative to the variable z play a similar role to the upper and lower half-planes of the variable $k(\lambda)$ for the NS model in the finite density case. The analytic properties of $E_n(z)$ are similar to those of $E_\varrho(x,k)$ in § I.8, Part I.

The Jost solutions $T_\pm(n,z)$ for $|z|=1$ are defined to be the limits

$$T_{\pm}(n,z) = \lim_{m\to\pm\infty} T(n,m,z)\,E_m^{(\pm)}(z). \tag{2.19}$$

Alternatively, they can be identified as solutions of (2.1) with the asymptotic conditions

$$T_{\pm}(n,z) = E_n^{(\pm)}(z) + O(1) \tag{2.20}$$

as $n \to \pm\infty$.

The matrices $T_{\pm}(n,z)$ for $|z|=1$ obey the involutions

$$\overline{T_{\pm}}(n,z) = T_{\pm}(n,\bar{z}), \tag{2.21}$$

$$\overline{T_{\pm}}(n,z) = -\frac{1}{z}\,T_{\pm}(n,z)\,\sigma_1 \tag{2.22}$$

and the relation

$$\det T_{\pm}(n,z) = 1 - z^2. \tag{2.23}$$

Their analytic properties are as follows: the columns $T_{-}^{(1)}(n,z)$ and $T_{+}^{(2)}(n,z)$ can be analytically continued inside the unit circle, $|z| \leqslant 1$, whereas the columns $T_{+}^{(1)}(n,z)$ and $T_{-}^{(2)}(n,z)$ can be analytically continued outside it, $|z| \geqslant 1$, with the following asymptotic behaviour:

$$z^n\,T_{-}^{(1)}(n,z) = \begin{pmatrix}1\\0\end{pmatrix} + O(|z|), \tag{2.24}$$

$$|z| \leqslant 1$$

$$z^{-n}\,T_{+}^{(2)}(n,z) = \begin{pmatrix}0\\ e^{-\frac{c}{2}}\end{pmatrix} + O(|z|) \tag{2.25}$$

as $z \to 0$ and

$$z^n\,T_{+}^{(1)}(n,z) = -e^{-\frac{c}{2}}\,z\begin{pmatrix}0\\1\end{pmatrix} + O(1), \tag{2.26}$$

$$|z| \geqslant 1$$

$$z^{-n}\,T_{-}^{(2)}(n,z) = -z\begin{pmatrix}1\\0\end{pmatrix} + O(1) \tag{2.27}$$

as $|z| \to \infty$.

To prove the existence of the Jost solutions and to study their properties it is convenient to make a gauge transformation

$$F_n = \Omega_n \tilde{F}_n, \tag{2.28}$$

where

$$\Omega_n = \begin{pmatrix} e^{\frac{q_n}{2}} & 0 \\ 0 & -e^{-\frac{q_{n-1}}{2}} \end{pmatrix}, \tag{2.29}$$

which carries the inverse linear problem (2.1) into

$$\tilde{F}_{n+1} = \tilde{L}_n(\lambda)\tilde{F}_n, \tag{2.30}$$

with

$$\tilde{L}_n(\lambda) = \Omega_{n+1}^{-1} L_n(\lambda)\Omega_n = \begin{pmatrix} e^{\frac{q_n - q_{n+1}}{2}}(p_n + \lambda) & -e^{-\frac{q_{n+1} - 2q_n + q_{n-1}}{2}} \\ 1 & 0 \end{pmatrix}. \tag{2.31}$$

Setting

$$\tilde{F}_n = \begin{pmatrix} f_n \\ g_n \end{pmatrix} \tag{2.32}$$

we have $g_{n+1} = f_n$ and

$$c_{n+1} f_{n+1} - p_n f_n + c_n f_{n-1} = \lambda f_n, \tag{2.33}$$

with

$$c_n = e^{\frac{q_n - q_{n-1}}{2}}. \tag{2.34}$$

Thus the auxiliary linear problem (2.1) is equivalent to the eigenvalue problem (2.33) for an infinite Jacobi matrix $\mathscr{L}$,

$$\mathscr{L}_{nm} = c_n \delta_{n,m+1} - p_n \delta_{nm} + c_{n+1}\delta_{n+1,m}. \tag{2.35}$$

Let us show that this problem, for $|z| = 1$, has solutions $\psi_\pm(n, z)$ with the asymptotic behaviour

$$\psi_\pm(n, z) = z^n + o(1) \tag{2.36}$$

as $n \to \pm\infty$ $\left(\text{remind that } \lambda = z + \dfrac{1}{z}\right)$. We look for the solutions of the form

$$\psi_+(n, z) = z^n + \sum_{m=n}^{\infty} \Gamma(n, m) z^m \tag{2.37}$$

and

$$\psi_-(n,z)=z^n+\sum_{m=-\infty}^{n}\tilde{\Gamma}(n,m)z^m\,, \tag{2.38}$$

where

$$\lim_{n,m\to\infty}\Gamma(n,m)=\lim_{n,m\to-\infty}\tilde{\Gamma}(n,m)=0\,. \tag{2.39}$$

Consider, for definiteness, (2.37) and substitute it into (2.33). Collecting the coefficients of a given power of z we deduce

$$c_n(1+\Gamma(n-1,n-1))=1+\Gamma(n,n)\,, \tag{2.40}$$

$$c_n\Gamma(n-1,n)-p_n(1+\Gamma(n,n))=\Gamma(n,n+1) \tag{2.41}$$

and

$$\begin{aligned}\Gamma(n,m+1)+\Gamma(n,m-1)=c_{n+1}(\delta_{m-n,1}+\Gamma(n+1,m))\\ -p_n\Gamma(n,m)+c_n\Gamma(n-1,m)\end{aligned} \tag{2.42}$$

for $m>n$. In the class of kernels $\Gamma(n,m)$ satisfying (2.39), equations (2.40)–(2.42) are uniquely solvable. In fact, (2.40) allows to determine $\Gamma(n,n)$ whereas (2.41) gives $\Gamma(n,n+1)$ for all n, so that (2.42), a second order finite difference equation, has a unique solution in the region $m>n$. The limiting values in (2.39) are attained in the sense of Schwartz. This establishes the existence of the solution $\psi_+(n,z)$.

The existence of $\psi_-(n,z)$ is proved in a similar manner.

In terms of $\psi_\pm(n,z)$, the Jost solutions $T_\pm(n,z)$ are

$$T_\pm(n,z)=\Omega_n\begin{pmatrix}\psi_\pm\left(n,\dfrac{1}{z}\right) & -z\psi_\pm(n,z)\\ \psi_\pm\left(n-1,\dfrac{1}{z}\right) & -z\psi_\pm(n-1,z)\end{pmatrix} \tag{2.43}$$

and clearly satisfy the above requirements.

2. The reduced monodromy matrix and transition coefficients

The reduced monodromy matrix $T(z)$ is defined for $|z|=1$, $z\neq\pm1$, as a ratio of the Jost solutions,

$$T(z)=T_+^{-1}(n,z)\,T_-(n,z)\,, \tag{2.44}$$

and can be expressed as the limit

$$T(z) = \lim_{\substack{n \to \infty \\ m \to -\infty}} E_n^{-1}(z) Q^{-1}(c) T(n, m, z) E_m(z). \tag{2.45}$$

$T(z)$ is a unimodular matrix satisfying

$$\bar{T}(z) = \sigma_1 T(z) \sigma_1, \tag{2.46}$$

$$\bar{T}(z) = T(\bar{z}) \tag{2.47}$$

and can be written as

$$T(z) = \begin{pmatrix} a(z) & \bar{b}(z) \\ b(z) & \bar{a}(z) \end{pmatrix}, \tag{2.48}$$

where $a(z)$ and $b(z)$ are *the transition coefficients for the continuous spectrum*. These are defined for $|z| = 1$, $z \neq \pm 1$, satisfy the normalization condition

$$|a(z)|^2 - |b(z)|^2 = 1 \tag{2.49}$$

and are symmetric,

$$\bar{a}(z) = a(\bar{z}), \quad \bar{b}(z) = b(\bar{z}). \tag{2.50}$$

For the coefficient $a(z)$ we have

$$a(z) = \frac{1}{1-z^2} \det(T_-^{(1)}(n, z), T_+^{(2)}(n, z)), \tag{2.51}$$

which shows that it has an analytic continuation into the unit disk $|z| < 1$ and

$$a(0) = e^{-\frac{c}{2}}. \tag{2.52}$$

A similar expression for $b(z)$,

$$b(z) = \frac{1}{1-z^2} \det(T_-^{(1)}(n, z), T_+^{(1)}(n, z)), \tag{2.53}$$

shows that in general it has no continuation off the circle $|z| = 1$. Such a continuation is possible if there exists $N > 0$ such that $c_n = 1$, $p_n = 0$ for $n > N$.

We shall now discuss the alternatives for the behaviour of $a(z)$ and $b(z)$ in the vicinity of $z=\pm 1$. If the columns $T_{-}^{(1)}(n,z)$ and $T_{+}^{(2)}(n,z)$ are linearly independent at $z=1$ or $z=-1$, then $a(z)$ *is singular and has the expansion*

$$a(z)=\frac{a_{\pm}}{z\mp 1}+O(1), \tag{2.54}$$

with $a_{\pm}$ *nonzero and real* (cf. the NS model in the case of finite density in § I.9, Part I). This is precisely what happens in a generic situation. In the special situation when $T_{-}^{(1)}(n,z)$ and $T_{+}^{(2)}(n,z)$ become linearly dependent at $z=1$ or $z=-1$, the coefficients a_{+} or a_{-} or both vanish, and $a(z)$ is nonsingular near the corresponding points. In that case $z=1$ or $z=-1$ or both values *are virtual levels*. They are located on the boundary, $\lambda=\pm 2$, of the continuous spectrum of the auxiliary linear problem.

The coefficient $b(z)$ *is either singular or regular near* $z=\pm 1$ *simultaneously with* $a(z)$. Indeed, we have

$$T_{+}^{(1)}(n,z)|_{z=\pm 1}=\mp T_{+}^{(2)}(n,z)|_{z=\pm 1}, \tag{2.55}$$

so that if a_{+} or a_{-} does not vanish, then

$$b(z)=\mp\frac{a_{\pm}}{z\mp 1}+O(1). \tag{2.56}$$

In particular, under this assumption we have

$$\lim_{z\to\pm 1}\frac{b(z)}{a(z)}=\mp 1. \tag{2.57}$$

(cf. the corresponding formulae in § I.9, Part I).

In view of the normalization condition, the zeros of $a(z)$ may only lie inside the circle $|z|=1$ and their number N is finite. If $a(z_j)=0$, then

$$T_{-}^{(1)}(n,z_j)=\gamma_j T_{+}^{(2)}(n,z_j), \qquad \gamma_j\neq 0 \tag{2.58}$$

and

$$\psi_{-}\left(n,\frac{1}{z_j}\right)=-z_j\gamma_j\psi_{+}(n,z_j), \qquad j=1,\dots,N. \tag{2.59}$$

Thus $\lambda_j = z_j + \frac{1}{z_j}$ are *the discrete eigenvalues* of the self-adjoint operator $\mathscr{L}$, hence λ_j and consequently z_j are real, $-1 < z_j < 1$, $z_j \neq 0$, $j = 1, \ldots, N$. The associated *transition coefficients for the discrete spectrum* γ_j are also real.

Let us show that *the zeros z_j are simple*. From (2.43) we have

$$a(z) = -\frac{c_n z}{1-z^2}\left(\psi_+(n,z)\psi_-\left(n-1,\frac{1}{z}\right) - \psi_+(n-1,z)\psi_-\left(n,\frac{1}{z}\right)\right). \tag{2.60}$$

Differentiating this with respect to z and setting $z = z_j$ we find

$$\begin{aligned}\dot{a}(z_j) = \frac{c_n z_j}{1-z_j^2}&\left(\dot{\psi}_+(n-1,z_j)\psi_-\left(n,\frac{1}{z_j}\right) - \dot{\psi}_+(n,z_j)\psi_-\left(n-1,\frac{1}{z_j}\right)\right.\\ &\left.-\frac{1}{z_j^2}\psi_+(n-1,z_j)\dot{\psi}_-\left(n,\frac{1}{z_j}\right) + \frac{1}{z_j^2}\psi_+(n,z_j)\dot{\psi}_-\left(n-1,\frac{1}{z_j}\right)\right),\end{aligned} \tag{2.61}$$

where the dot indicates differentiation with respect to z. From (2.33) and

$$c_{n+1}\dot{f}_{n+1} - p_n\dot{f}_n + c_n\dot{f}_{n-1} = \left(z+\frac{1}{z}\right)\dot{f}_n + \left(1-\frac{1}{z^2}\right)f_n \tag{2.62}$$

we deduce that the quantities

$$\phi_+(n,z) = c_n\left(\dot{\psi}_+(n,z)\psi_-\left(n-1,\frac{1}{z}\right) - \dot{\psi}_+(n-1,z)\psi_-\left(n,\frac{1}{z}\right)\right) \tag{2.63}$$

and

$$\phi_-(n,z) = \frac{c_n}{z^2}\left(\psi_+(n-1,z)\dot{\psi}_-\left(n,\frac{1}{z}\right) - \psi_+(n,z)\dot{\psi}_-\left(n-1,\frac{1}{z}\right)\right) \tag{2.64}$$

satisfy

$$\phi_\pm(n+1,z) = \phi_\pm(n,z) \pm \left(1-\frac{1}{z^2}\right)\psi_+(n,z)\psi_-\left(n,\frac{1}{z}\right). \tag{2.65}$$

Setting $z = z_j$ and using (2.59) we obtain

$$\phi_+(n,z_j) = \frac{\gamma_j(z_j^2-1)}{z_j}\sum_{k=n}^{\infty}\psi_+^2(k,z_j) \tag{2.66}$$

and

$$\phi_-(n,z_j)=\frac{\gamma_j(z_j^2-1)}{z_j}\sum_{k=-\infty}^{n-1}\psi_+^2(k,z_j), \tag{2.67}$$

so that

$$\dot a(z_j)=\gamma_j\sum_{n=-\infty}^{\infty}\psi_+^2(n,z_j)\neq 0. \tag{2.68}$$

This equation also shows that

$$\operatorname{sign}\gamma_j=\operatorname{sign}\dot a(z_j),\qquad j=1,\ldots,N \tag{2.69}$$

(cf. the corresponding arguments in § I.9, Part I).

The function $a(z)$ is uniquely determined by the coefficient $b(z)$ and the zeros $z_1,\ldots,z_N$. To derive the corresponding dispersion relation consider Schwarz's formula

$$f(z)=\operatorname{Im}f(0)+\frac{1}{2\pi i}\int_{|\zeta|=1}\operatorname{Re}f(\zeta)\,\frac{\zeta+z}{\zeta-z}\,\frac{d\zeta}{\zeta}, \tag{2.70}$$

where $f(z)$ is analytic in the disk $|z|\leqslant 1$, and apply this to

$$f(z)=\log\prod_{j=1}^{N}\operatorname{sign}z_j\,\frac{zz_j-1}{z-z_j}\,a(z), \tag{2.71}$$

where the principal branch of the logarithm is taken. Using $a(0)>0$ and the normalization condition we find

$$a(z)=\prod_{j=1}^{N}\operatorname{sign}z_j\,\frac{z-z_j}{zz_j-1}\exp\left\{\frac{1}{4\pi i}\int_{|\zeta|=1}\log(1+|b(\zeta)|^2)\,\frac{\zeta+z}{\zeta-z}\,\frac{d\zeta}{\zeta}\right\}. \tag{2.72}$$

Taking account of (2.50), we obtain the final expression for $a(z)$

$$a(z)=\prod_{j=1}^{N}\operatorname{sign}z_j\,\frac{z-z_j}{zz_j-1}\exp\left\{\frac{1}{2\pi i}\int_{C}\log(1+|b(\zeta)|^2)\,\frac{1-z^2}{(1-z\zeta)(\zeta-z)}\,d\zeta\right\}, \tag{2.73}$$

where C is the semi-circle $|\zeta|=1$, $0\leqslant\arg\zeta\leqslant\pi$.

The data $b(z)$, z_j and c are not independent. Firstly, from (2.52) it follows that

$$e^{-\frac{c}{2}}=\prod_{j=1}^{N}|z_j|\exp\left\{\frac{1}{2\pi i}\int_C\log(1+|b(\zeta)|^2)\frac{d\zeta}{\zeta}\right\}. \tag{2.74}$$

This relation will be called the *condition* (*c*). *Secondly, in a generic situation when*

$$b(z)=\frac{b_\pm}{z\mp1}+O(1) \tag{2.75}$$

near $z=\pm1$, *we have*

$$\operatorname{sign}b_\pm=\prod_{j=1}^{N}(\mp\operatorname{sign}z_j) \tag{2.76}$$

(cf. the condition (θ) and the conditions for the determination of signs in § I.9, Part I).

To derive (2.76) we shall study the asymptotic behaviour of $a(z)$ as $z\to\pm1$, $|z|<1$ by using the dispersion relation (2.73). The dominant contribution into (2.73) comes from the singular term (2.75) and has the form

$$I_\pm=\frac{1}{2\pi i}\int_{C_\pm}\log\frac{|b_\pm|^2}{|\zeta\mp1|^2}\,\frac{1-z^2}{(1-z\zeta)(\zeta-z)}\,\frac{d\zeta}{\zeta}, \tag{2.77}$$

where $C_\pm$ are small neighbourhoods of $\zeta=\pm1$ on C. We have

$$\begin{aligned}I_+&=\frac{1}{\pi i}\int_{C_+}\log\frac{|b_+|\cdot|\zeta+1|}{2|\zeta-1|}\cdot\frac{1-z^2}{(1-z\zeta)(\zeta-z)}\,\frac{d\zeta}{\zeta}+O(|z-1|)\\&=\frac{1}{2\pi i}\int_{|\zeta|=1}\log\left|\frac{b_+}{2}\,\frac{\zeta+1}{\zeta-1}\right|\cdot\frac{\zeta+z}{\zeta-z}\,\frac{d\zeta}{\zeta}+O(|z-1|)\\&=\log\left(-\frac{|b_+|}{2}\cdot\frac{z+1}{z-1}\right)+O(|z-1|),\end{aligned} \tag{2.78}$$

where the last equality made use of Schwarz's formula. This yields

$$a(z)=-\frac{|b_+|\prod_{j=1}^{N}(-\operatorname{sign}z_j)}{z-1}+O(1) \tag{2.79}$$

as $z \to 1$. Comparing this with (2.57) we arrive at (2.76) for the sign $+$.

The second formula in (2.76) is proved in a similar way.

Let us emphasize that, as for the NS model in the finite density case, complications in the analytic properties of the transition coefficients are caused by the fact that the continuous spectrum of the auxiliary linear problem has boundary points $\lambda = \pm 2$.

This completes our analysis of the mapping $\mathscr{F}: (p_n, q_n) \to (b(z), \bar{b}(z), z_j, \gamma_j, j=1, \ldots, N)$ from the initial data of the Toda model to the characteristics of the auxiliary linear problem (2.1).

3. Time evolution of the transition coefficients

We shall consider the evolution of the transition coefficients when $p_n(t)$ and $q_n(t)$ satisfy the Toda model equations of motion. Using the zero curvature representation (see § I.2) we obtain

$$\frac{dT}{dt}(n, m, z) = V_n(z)\, T(n, m, z) - T(n, m, z)\, V_m(z), \tag{2.80}$$

with

$$V_n(z) = \begin{pmatrix} 0 & -e^{q_n} \\ e^{q_{n-1}} & z + \dfrac{1}{z} \end{pmatrix}. \tag{2.81}$$

Letting in (2.80) $n \to \infty$, $m \to -\infty$ according to the definitions (2.19), (2.45) and using

$$\begin{aligned} &\lim_{n \to \pm\infty} (E_n^{(\pm)}(z))^{-1}\, V_n(z)\, E_n^{(\pm)}(z) \\ &= \lim_{n \to \pm\infty} (E_n^{(\pm)}(z))^{-1}\, L_{\pm}(z)\, E_n^{(\pm)}(z) = V(z), \end{aligned} \tag{2.82}$$

with

$$V(z) = \begin{pmatrix} z & 0 \\ 0 & \dfrac{1}{z} \end{pmatrix}, \tag{2.83}$$

we derive the evolution equations for the Jost solutions

$$\frac{dT_{\pm}(n, z)}{dt} = V_n(z)\, T_{\pm}(n, z) - T_{\pm}(n, z)\, V(z) \tag{2.84}$$

and for the reduced monodromy matrix,

$$\frac{dT}{dt}(z) = \frac{1}{2}\left(z - \frac{1}{z}\right)[\sigma_3, T(z)]. \tag{2.85}$$

These lead to the following *explicit time dependence of the transition coefficients*:

$$a(z,t) = a(z,0), \quad b(z,t) = e^{-\left(z-\frac{1}{z}\right)t} b(z,0), \tag{2.86}$$

$$z_j(t) = z_j(0), \quad \gamma_j(t) = e^{-\left(z_j-\frac{1}{z_j}\right)t} \gamma_j(0), \quad j = 1, \ldots, N. \tag{2.87}$$

As in the cases considered earlier, the coefficient $a(z)$ is a generating function for integrals of the motion. We close this section by describing a family of local integrals of the motion. The latter are understood to have the form

$$F = \sum_{n=-\infty}^{\infty} f_n, \tag{2.88}$$

where f_n is a polynomial in p_n, c_n and their higher differencies.

4. Local integrals of the motion

We will show that *the expansion of* $\log a(z)$ *into a Taylor series at* $z = 0$,

$$\log a(z) = -\frac{c}{2} + \sum_{n=1}^{\infty} I_n z^n, \tag{2.89}$$

gives a sequence of local integrals of the motion for the Toda model including its Hamiltonian. In the previous examples of continuous models we were dealing with the asymptotic expansion of $\log a(\lambda)$ near the points $\lambda = \infty$ or $\lambda = 0$ where $b(\lambda)$ was rapidly decreasing. This enabled us to start with the asymptotic expansion of the transition matrix $T(x, y, \lambda)$ and then let $x \to +\infty$, $y \to -\infty$. For the Toda model, $b(z)$ in general is not defined near $z = 0$, so this method does not apply. We will outline another method for computing the coefficients I_n based on a direct analysis of the auxiliary linear problem (2.33) in the rapidly decreasing case.

Consider (2.60) for $|z| < 1$ and let $n \to +\infty$. Taking account of (2.36) we have

$$a(z)=\frac{z}{1-z^2}\lim_{n\to+\infty}(z^{n-1}\varphi(n,z)-z^n\varphi(n-1,z)),\tag{2.90}$$

where we have set $\varphi(n,z)=\psi_-\left(n,\frac{1}{z}\right)$. For z small, $\varphi(n,z)$ can be expressed as

$$\varphi(n,z)=z^{-n}\prod_{k=-\infty}^{n}\frac{\chi(k,z)}{c_k}.\tag{2.91}$$

Substituting this expression into (2.90) and using (2.2) and (2.34) yields

$$a(z)=\frac{z}{1-z^2}\lim_{n\to+\infty}\left(\frac{1}{z}\prod_{k=-\infty}^{n}\frac{\chi(k,z)}{c_k}-z\prod_{k=-\infty}^{n-1}\frac{\chi(k,z)}{c_k}\right)$$

$$=\prod_{n=-\infty}^{\infty}\frac{\chi(n,z)}{c_n}=e^{-\frac{c}{2}}\prod_{n=-\infty}^{\infty}\chi(n,z).\tag{2.92}$$

We shall now present a procedure for computing $\chi(n,z)$. Substituting (2.91) into (2.33) gives the equation

$$\chi(n,z)(\chi(n+1,z)-1-zp_n-z^2)=-z^2c_n^2,\tag{2.93}$$

which has a solution of the form

$$\chi(n,z)=\sum_{m=0}^{\infty}\chi(n,m)z^m,\tag{2.94}$$

where

$$\chi(n,0)=1,\quad \chi(n,1)=p_{n-1},\tag{2.95}$$

$$\chi(n,2)=1-c_{n-1}^2\tag{2.96}$$

and for $m>2$

$$\chi(n,m)=c_{n-1}^2\chi(n-1,m-2)-\sum_{k=3}^{m-1}\chi(n,k)\chi(n-1,m-k).\tag{2.97}$$

Formulae (2.92) and (2.94)-(2.97) allow us to express the I_m in terms of the p_n and c_n. In particular, we have

$$I_1=P=\sum_{n=-\infty}^{\infty}p_n\tag{2.98}$$

and

$$I_2 = -H = -\sum_{n=-\infty}^{\infty}\left(\frac{1}{2}p_n^2 + c_n^2 - 1\right). \tag{2.99}$$

By means of the dispersion relation (2.73), I_n can be expressed in terms of the transition coefficients and the discrete spectrum of the auxiliary linear problem. The corresponding *trace identities* are

$$I_n = \frac{1}{2\pi i}\int_C \log(1+|b(\zeta)|^2)(\zeta^n + \zeta^{-n})\frac{d\zeta}{\zeta} + \frac{1}{n}\sum_{j=1}^{N}(z_j^n - z_j^{-n}), \qquad n = 1, 2, \ldots \tag{2.100}$$

In § 4 we shall discuss whether the functionals I_n belong the algebra of observables on the phase space of our model.

This completes the analysis of the auxiliary linear problem and the mapping $\mathscr{F}$ for the Toda model.

§ 3. The Inverse Problem and Soliton Dynamics for the Toda Model in the Rapidly Decreasing Case

In this section we shall describe the mapping $\mathscr{F}^{-1}$, i.e. solve the inverse problem of reconstructing the p_n and q_n from the transition coefficients and the discrete spectrum. As before, we may take two routes, the matrix Riemann problem or the Gelfand-Levitan-Marchenko formulation. The presence of a boundary in the continuous spectrum of the auxiliary linear problem and the ensuing constraints on the transition coefficients and the discrete spectrum (the condition (c) etc.) lead to complications in the first approach (cf. the NS model in the case of finite density in § II.6, Part I). We shall therefore only deal with the Gelfand-Levitan-Marchenko formulation. At the end of this section it will be used to describe soliton dynamics for the Toda model.

1. The Gelfand-Levitan-Marchenko formulation

The formulation is based on the relationship between the Jost solutions for $|z|=1$,

$$T_-(n,z) = T_+(n,z)\,T(z), \tag{3.1}$$

which in terms of $\psi_\pm(n,z)$ is written as

$$\frac{1}{a(z)}\psi_-\left(n,\frac{1}{z}\right)=\psi_+\left(n,\frac{1}{z}\right)+r(z)\psi_+(n,z) \tag{3.2}$$

and

$$\frac{1}{a(z)}\psi_+(n,z)=\psi_-(n,z)+\tilde{r}(z)\psi_-\left(n,\frac{1}{z}\right), \tag{3.3}$$

where

$$r(z)=-z\frac{b(z)}{a(z)}, \quad \tilde{r}(z)=\frac{\bar{b}(z)}{za(z)}. \tag{3.4}$$

Let us consider, for definiteness, (3.2) and make the following operations. Insert (2.37)–(2.38) into (3.2), multiply by $\frac{1}{2\pi i}z^{m-1}$, $m \geqslant n$, and integrate over the circle $|z|=1$. Using (2.52), (2.59) and Cauchy's formula, we deduce

$$\delta_{n,m}+\Gamma(n,m)+K(n+m)+\sum_{l=n}^{\infty}\Gamma(n,l)K(l+m)=e^{\frac{c}{2}}\delta_{n,m}(1+\tilde{\Gamma}(n,n)), \tag{3.5}$$

where

$$K(n)=\frac{1}{2\pi i}\int_{|z|=1} r(z)z^n\frac{dz}{z}+\sum_{j=1}^{N}m_j z_j^n, \tag{3.6}$$

and

$$m_j=\frac{\gamma_j}{\dot{a}(z_j)}, \quad j=1,\ldots,N, \tag{3.7}$$

the dot indicating differentiation with respect to z.

In contrast to the previous examples of continuous models, (3.5) contains an additional term on the right hand side induced by the residue of $\frac{1}{a(z)}\psi_-(n,z)z^{m-1}$ at $z=0$. It can be expressed through $\Gamma(n,n)$ as follows. Consider (2.40); from (2.1), (2.34) and (2.39) we derive

$$1+\Gamma(n,n)=e^{\frac{q_n-c}{2}}. \tag{3.8}$$

An analogous equation for $\tilde{\Gamma}(n,n)$,

$$c_{n+1}(1+\tilde{\Gamma}(n+1,n+1))=1+\tilde{\Gamma}(n,n) \tag{3.9}$$

yields

$$1+\tilde{\Gamma}(n,n)=e^{-\frac{q_n}{2}}, \tag{3.10}$$

hence

$$e^{\frac{c}{2}}(1+\tilde{\Gamma}(n,n))=(1+\Gamma(n,n))^{-1}. \tag{3.11}$$

As a result, (3.5) becomes

$$\frac{\delta_{n,m}}{1+\Gamma(n,n)}=\delta_{n,m}+\Gamma(n,m)+K(n+m)+\sum_{l=n}^{\infty}\Gamma(n,l)K(l+m). \tag{3.12}$$

Unfortunately, this is a nonlinear equation for $\Gamma(n,m)$.

To reduce (3.12) to a linear equation, set

$$X(n,m)=\frac{\Gamma(n,m)}{1+\Gamma(n,n)}, \quad m>n, \tag{3.13}$$

and multiply (3.12) by $(1+\Gamma(n,n))^{-1}$. For $m>n$ we obtain a linear equation,

$$X(n,m)+K(n+m)+\sum_{l=n+1}^{\infty}X(n,l)K(l+m)=0, \tag{3.14}$$

and for $m=n$

$$\frac{1}{(1+\Gamma(n,n))^2}=1+K(2n)+\sum_{l=n+1}^{\infty}X(n,l)K(l+n). \tag{3.15}$$

Equation (3.14) is precisely the Gelfand-Levitan-Marchenko equation from the right, and (3.15) allows to recover $\Gamma(n,n)$ from $X(n,m)$.

In a similar manner, (3.3) yields the Gelfand-Levitan-Marchenko equation from the left:

$$\tilde{X}(n,m)+\tilde{K}(n+m)+\sum_{l=-\infty}^{n-1}\tilde{X}(n,l)\tilde{K}(l+m)=0, \quad n>m, \tag{3.16}$$

where

$$\tilde{K}(n)=\frac{1}{2\pi i}\int_{|z|=1}\tilde{r}(z)z^{-n}\frac{dz}{z}+\sum_{j=1}^{N}\tilde{m}_j z_j^{-n-2}, \tag{3.17}$$

$$\tilde{m}_j = \frac{1}{\gamma_j \dot{a}(z_j)}, \qquad j=1, \ldots, N, \tag{3.18}$$

and

$$\tilde{X}(n, m) = \frac{\tilde{\Gamma}(n, m)}{1+\tilde{\Gamma}(n, n)}, \tag{3.19}$$

with the relation

$$\frac{1}{(1+\tilde{\Gamma}(n, n))^2} = 1+\tilde{K}(2n)+ \sum_{l=-\infty}^{n-1} \tilde{X}(n, l)\tilde{K}(l+n). \tag{3.20}$$

Let us now outline a procedure for solving the inverse problem.

The input data consist of functions $r(z), \tilde{r}(z)$ and of a set of real numbers $m_j, \tilde{m}_j, z_j, j=1, \ldots, N$; c with the following properties.

I. *$r(z), \tilde{r}(z)$ are smooth functions on the circle $|z|=1$ and obey the involution*

$$\bar{r}(z)=r(\bar{z}), \qquad \bar{\tilde{r}}(z)=\tilde{r}(\bar{z}) \tag{3.21}$$

and the relation

$$|r(z)|=|\tilde{r}(z)| \leqslant 1, \tag{3.22}$$

where equality in the estimate can only be attained at $z=\pm 1$, in which case

$$r(\pm 1)=-\tilde{r}(\pm 1)=1. \tag{3.23}$$

II. *The pairwise distinct numbers $z_j \neq 0$ lie in the interval $-1<z_j<1$, while m_j and $\tilde{m}_j$ are positive, $j=1, \ldots, N$.*

III. *The condition (c) holds,*

$$e^{-\frac{c}{2}} = \prod_{j=1}^{N} |z_j| \exp\left\{\frac{i}{4\pi} \int_{|\zeta|=1} \log(1-|r(\zeta)|^2) \frac{d\zeta}{\zeta}\right\}. \tag{3.24}$$

IV. *The relations*

$$\frac{\tilde{r}(z)}{\bar{r}(z)} = -\frac{\bar{a}(z)}{a(z)} \tag{3.25}$$

hold and

$$m_j \tilde{m}_j = \frac{1}{\dot{a}^2(z_j)}, \quad j=1, \ldots, N, \tag{3.26}$$

where

$$a(z) = \prod_{j=1}^{N} \operatorname{sign} z_j \frac{z-z_j}{z z_j - 1} \times \exp\left\{\frac{1}{4\pi i} \int_{|\zeta|=1} \log(1-|r(\zeta)|^2) \frac{z+\zeta}{z-\zeta} \frac{d\zeta}{\zeta}\right\}. \tag{3.27}$$

Given these data, construct the kernels $K(n)$, $\tilde{K}(n)$ and consider (3.14), (3.16). We claim that

I′. *Equations* (3.14), (3.16) *are uniquely solvable in the spaces* $l_1(n+1, \infty)$ *and* $l_1(-\infty, n-1)$, *respectively. Their solutions* $X(n, m)$ *and* $\tilde{X}(n, m)$ *are rapidly decreasing as* $n, m \to +\infty$ *or* $n, m \to -\infty$, *respectively.*

II′. *The right hand sides of* (3.15) *and* (3.20) *are positive, and hence* $1+\Gamma(n, n)$ *and* $1+\tilde{\Gamma}(n, n)$ *can be taken positive.*

III′. *Let*

$$\Gamma(n, m) = (1+\Gamma(n, n)) X(n, m) \tag{3.28}$$

and

$$\tilde{\Gamma}(n, m) = (1+\tilde{\Gamma}(n, n)) \tilde{X}(n, m). \tag{3.29}$$

Then the functions $\psi_\pm(n, z)$ *defined by* (2.37)–(2.38) *satisfy*

$$c_{n+1}^{(\pm)} \psi_\pm(n+1, z) - p_n^{(\pm)} \psi_\pm(n, z) + c_n^{(\pm)} \psi_\pm(n-1, z) = \left(z + \frac{1}{z}\right) \psi_\pm(n, z), \tag{3.30}$$

with $c_n^{(\pm)}$ *positive,*

$$c_n^{(+)} = \frac{1+\Gamma(n, n)}{1+\Gamma(n-1, n-1)}, \quad c_n^{(-)} = \frac{1+\tilde{\Gamma}(n-1, n-1)}{1+\tilde{\Gamma}(n, n)} \tag{3.31}$$

and

$$p_n^{(+)} = \frac{c_n^{(+)} \Gamma(n-1, n) - \Gamma(n, n+1)}{1+\Gamma(n, n)}, \tag{3.32}$$

$$p_n^{(-)} = \frac{c_n^{(-)} \tilde{\Gamma}(n+1, n) - \tilde{\Gamma}(n, n-1)}{1+\tilde{\Gamma}(n, n)}. \tag{3.33}$$

IV′. *The relations*

$$\lim_{n\to\pm\infty} c_n^{(\pm)}=1, \qquad \lim_{n\to\pm\infty} p_n^{(\pm)}=0 \tag{3.34}$$

hold, where the limiting values are attained in the sense of Schwartz.

V′. *The relations*

$$p_n^{(+)}=p_n^{(-)}=p_n, \qquad c_n^{(+)}=c_n^{(-)}=e^{\frac{q_n-q_{n-1}}{2}} \tag{3.35}$$

hold, so that

$$\lim_{n\to-\infty} q_n=0, \qquad \lim_{n\to+\infty} q_n=c, \qquad \lim_{|n|\to\infty} p_n=0, \tag{3.36}$$

where the limiting values are attained in the sense of Schwartz.

VI′. *The functions* $a(z)$ *and* $b(z)=-\dfrac{a(z)r(z)}{z}$ *are the transition coefficients for the auxiliary linear problem*

$$F_{n+1}=L_n(\lambda)F_n, \tag{3.37}$$

where

$$L_n(\lambda)=\begin{pmatrix} p_n+\lambda & e^{q_n} \\ -e^{-q_n} & 0 \end{pmatrix}. \tag{3.38}$$

Its discrete spectrum consists of the eigenvalues $\lambda_j=z_j+\dfrac{1}{z_j}$ *with transition coefficients* $\gamma_j=m_j\dot{a}(z_j)$, $j=1,\ldots,N$.

We will not give the proof of these assertions since it is a straightforward lattice transcription of the argument of § II.7, Part I. In conclusion, we only note that the Gelfand-Levitan-Marchenko formalism can be used to show that if the time dependence of the inverse problem data is given by (2.86)–(2.87), then the reconstructed $p_n(t)$ and $q_n(t)$ satisfy the Toda equations.

2. Soliton solutions

Soliton solutions of the Toda model correspond to

$$b(z)=0 \tag{3.39}$$

for all z on the circle $|z|=1$. In this case the requirements on the data $\{c, z_j, m_j, \tilde{m}_j, j=1,\ldots,N\}$ simplify and amount to the following.

I. *The quantities* $z_j\neq 0$ *lie in the interval* $-1<z_j<1$ *and are pairwise distinct.*

II. *The condition* (c) *holds,*

$$e^{-c}=\prod_{j=1}^{N} z_j^2 . \tag{3.40}$$

III. *The quantities m_j, $\tilde{m}_j$ are positive and related by*

$$m_j \tilde{m}_j = \frac{1}{\dot{a}^2(z_j)}, \quad j=1, \ldots, N, \tag{3.41}$$

where

$$a(z)=\prod_{j=1}^{N} \operatorname{sign} z_j \frac{z-z_j}{z z_j - 1} . \tag{3.42}$$

For such data the Gelfand-Levitan-Marchenko equations (3.14)-(3.16) *reduce to linear algebraic equations and can be solved in closed form.*

Consider first the case $N=1$. The kernel $K(n)$ of (3.14) has the form

$$K(n)=m_1 z_1^n \tag{3.43}$$

and is one-dimensional. Setting

$$X(n, m)=X(n) m_1 z_1^m \tag{3.44}$$

we find from (3.14)

$$X(n)+z_1^n+X(n) m_1 \sum_{l=n+1}^{\infty} z_1^{2l}=0, \tag{3.45}$$

so that

$$X(n)=-\frac{z_1^n}{1+|\gamma_1| z_1^{2n+2}}, \tag{3.46}$$

where we have used

$$m_1=-\operatorname{sign} z_1 \gamma_1 (1-z_1^2)=|\gamma_1|(1-z_1^2). \tag{3.47}$$

Substituting (3.44) and (3.46) into (3.15) gives

$$\frac{1}{(1+\Gamma(n, n))^2}=1+m_1 z_1^{2n}-\frac{m_1^2 z_1^{2n}}{1+|\gamma_1| z_1^{2n+2}} \sum_{l=n+1}^{\infty} z_1^{2l}=\frac{1+|\gamma_1| z_1^{2n}}{1+|\gamma_1| z_1^{2n+2}} . \tag{3.48}$$

Now from (3.8) we find

$$e^{q_n}=e^c\,\frac{1+|\gamma_1|z_1^{2n+2}}{1+|\gamma_1|z_1^{2n}}.\tag{3.49}$$

(Remind that in this case $e^{-c}=z_1^2$.)

The time dependence is introduced by replacing γ_1 with $\gamma_1(t)$,

$$\gamma_1(t)=e^{-\left(z_1-\frac{1}{z_1}\right)t}\gamma_1.\tag{3.50}$$

If we denote

$$z_1=\varepsilon e^{-\alpha_1},\quad \alpha_1>0,\quad \varepsilon=\pm 1,\tag{3.51}$$

the solutions $q_n(t)$ and $p_n(t)$ of the equations of motion for the Toda model are finally given by

$$q_n(t)=c+\log\frac{1+\exp\{-2\alpha_1(n+1-v_1t+n_{01})\}}{1+\exp\{-2\alpha_1(n-v_1t+n_{01})\}}\tag{3.52}$$

and

$$p_n(t)=\frac{dq_n}{dt}(t),\tag{3.53}$$

with

$$v_1=\varepsilon_1\frac{\operatorname{sh}\alpha_1}{\alpha_1},\qquad n_{01}=-\frac{1}{2\alpha_1}\log|\gamma_1|.\tag{3.54}$$

The solution (3.52) represents a wave propagating along the lattice with velocity v_1, $|v_1|>1$, whose center of inertia position at $t=0$ is n_{01}. By the general definition of Part I, it should be called a *soliton for the Toda model.* The soliton is characterized by two real parameters, v_1 and n_{01}.

Let us now consider the general case of arbitrary N. As before, the kernel $K(n+m)$ is degenerate,

$$K(n+m)=\sum_{j=1}^{N}\sqrt{m_j}\,z_j^n\sqrt{m_j}\,z_j^m,\tag{3.55}$$

where $\sqrt{m_j}>0$; we look for a solution of (3.14) of the form

$$X(n,m)=\sum_{j=1}^{N}X_j(n)\sqrt{m_j}\,z_j^m.\tag{3.56}$$

Substituting (3.56) into (3.14) yields a system of equations

$$M(n)X(n) = -Y(n), \tag{3.57}$$

where $X(n)$ is a column-vector with entries $X_j(n)$ and $Y(n)$ is one with entries $\sqrt{m_j}\, z_j^n$, $j=1,\ldots,N$, and $M(n)$ is an $N\times N$ matrix with entries given by

$$M(n)_{ij} = \delta_{ij} + \frac{\sqrt{m_i m_j}\,(z_i z_j)^{n+1}}{1 - z_i z_j}, \tag{3.58}$$

$i,j=1,\ldots,N$.

From (3.56)–(3.57) we deduce

$$X(n,m) = -Y^\tau(n)\,M^{-1}(n)\,Y(m). \tag{3.59}$$

Substituting this into (3.15) gives

$$\begin{aligned}\frac{1}{(1+\Gamma(n,n))^2} &= 1 + Y^\tau(n)\,Y(n) + Y^\tau(n)(M(n)-I)X(n)\\ &= 1 - Y^\tau(n)X(n) = 1 + Y^\tau(n)\,M^{-1}(n)\,Y(n).\end{aligned} \tag{3.60}$$

The last formula can be simplified. Notice that (3.58) yields

$$M(n-1) - M(n) = Y(n)\,Y^\tau(n), \tag{3.61}$$

or

$$M(n-1)\,M^{-1}(n) = I + Y(n)\,Y^\tau(n)\,M^{-1}(n). \tag{3.62}$$

The matrix $B(n) = Y(n)\,Y^\tau(n)\,M^{-1}(n)$ is one-dimensional and

$$B^2(n) = \alpha(n)\,B(n), \qquad \alpha(n) = Y^\tau(n)\,M^{-1}(n)\,Y(n). \tag{3.63}$$

By comparing (3.60) and (3.62)–(3.63) it follows that

$$(1+\Gamma(n,n))^2 = \frac{\det M(n)}{\det M(n-1)}. \tag{3.64}$$

Introducing the time dependence by

$$\gamma_j(t) = e^{-\left(z_j - \frac{1}{z_j}\right)t}\,\gamma_j, \qquad j=1,\ldots,N, \tag{3.65}$$

we derive from (3.64) *an expression for the N-soliton solution of the Toda model,*

$$q_n(t)=c+\log\frac{\det M(n,t)}{\det M(n-1,t)}. \tag{3.66}$$

The expression for $p_n(t)$ is given by (3.53) as usual.

As in the earlier examples, *the N-soliton solution describes a scattering process of N solitons. Specifically, for large $|t|$ the solution $q_n(t)$ can be expressed as the sum of one-soliton solutions,*

$$q_n(t)=\sum_{j=1}^{N} q_n^{(+j)}(t)+O(e^{-at}) \tag{3.67}$$

for $t\to+\infty$ and

$$q_n(t)=\sum_{j=1}^{N} q_n^{(-j)}(t)+O(e^{at}) \tag{3.68}$$

for $t\to-\infty$. Here $a=\min\alpha_j \min\limits_{i\neq j}|v_i-v_j|$, and $q_n^{(\pm j)}(t)$ are solitons with parameters $c_j, v_j, n_{0j}^{(\pm)}$:

$$q_n^{(\pm j)}(t)=q_{c_j}(n-v_jt+n_{0j}^{(\pm)}), \tag{3.69}$$

where

$$c_j=-\log z_j^2, \qquad v_j=2\operatorname{sign} z_j\frac{\operatorname{sh}\frac{c_j}{2}}{c_j} \tag{3.70}$$

and

$$n_{0j}^{(+)}=n_{0j}+\frac{1}{c_j}\left(\sum_{v_k<v_j}\log\left|\frac{1-z_jz_k}{z_j-z_k}\right|-\sum_{v_k>v_j}\log\left|\frac{1-z_jz_k}{z_j-z_k}\right|\right), \tag{3.71}$$

$$n_{0j}^{(-)}=n_{0j}-\frac{1}{c_j}\left(\sum_{v_k<v_j}\log\left|\frac{1-z_jz_k}{z_j-z_k}\right|-\sum_{v_k>v_j}\log\left|\frac{1-z_jz_k}{z_j-z_k}\right|\right),$$

$$n_{0j}=\frac{1}{c_j}\log|\gamma_j|, \qquad j=1,\ldots,N. \tag{3.72}$$

The proof of these formulae is based on computations essentially analogous to those of § II.8, Part I.

As for the NS model in the finite density case, the N-soliton solution $q_n(t)$ with parameter c breaks up into solitons $q_n^{(\pm j)}(t)$ with distinct parameters c_j. Thus, only solitons with c_j distinct interact. The relation

$$c = \sum_{j=1}^{N} c_j \tag{3.73}$$

can be thought of as a conservation law. The interpretation of (3.67)–(3.72) in terms of scattering theory is similar to that for the previous examples.

This concludes our discussion of the inverse problem techniques and soliton dynamics for the Toda model.

§ 4. Complete Integrability of the Toda Model in the Rapidly Decreasing Case

In this section we shall consider the mapping $\mathscr{F}$ from the standpoint of canonical transformations in phase space. We shall see that, as in the finite density case of the NS model, the programme of constructing canonical action-angle variables for the Toda model reveals some interesting peculiarities connected with the presence of boundary points in the continuous spectrum of the auxiliary linear problem. We will demonstrate their effect on the Hamiltonian interpretation of soliton scattering theory.

1. The Poisson structure and the algebra of observables

The phase space $\mathscr{M}_c$ of the Toda model is parametrized by coordinates p_n, q_n subject to the rapidly decreasing boundary conditions

$$\lim_{n\to-\infty} q_n = 0, \qquad \lim_{n\to+\infty} q_n = c, \qquad \lim_{|n|\to\infty} p_n = 0. \tag{4.1}$$

The Poisson structure on $\mathscr{M}_c$ is given by the formal Poisson brackets

$$\{p_n, p_m\} = \{q_n, q_m\} = 0, \qquad \{p_n, q_m\} = \delta_{nm}. \tag{4.2}$$

The algebra of observables is composed of admissible functionals $F(p_n, q_n)$. A functional $F(p_n, q_n)$ is admissible if the induced Hamiltonian flow leaves $\mathscr{M}_c$ invariant. In particular, such a functional must satisfy

$$\lim_{|n|\to\infty} \frac{\partial F}{\partial p_n} = \lim_{|n|\to\infty} \frac{\partial F}{\partial q_n} = 0. \tag{4.3}$$

A simplest example of an inadmissible functional is

$$P = \sum_{n=-\infty}^{\infty} p_n, \tag{4.4}$$

which is the first coefficient in the expansion of $\log a(z)$ into a Taylor series at $z=0$ (see Subsection 4 of § 2). Its flow shifts all the q_n simultaneously and so violates the boundary conditions (4.1).

There is the following analogy with the NS model in the finite density case: the quantity c, as well as the phase θ, stands for the index of the phase space $\mathscr{M}_c$ and is related to the transition coefficients and the discrete spectrum by the condition (c). The functional P is analogous to N_ϱ in the finite density case of the NS model. This model has taught us that care is needed when studying the formal Poisson brackets of the transition coefficients on the boundary of the discrete spectrum. In what follows, we shall pay special attention to selecting admissible observables out of the family of local integrals of the motion I_n produced by the trace identities.

2. The Poisson brackets of transition coefficients and discrete spectrum

Consider the Poisson brackets for the transition matrix $T(n,m,z)$ that follow from the fundamental Poisson brackets (1.9):

$$\{T(n,m,z) \otimes T(n,m,z')\} = [r(z,z'), T(n,m,z) \otimes T(n,m,z')], \quad m<n, \tag{4.5}$$

where

$$r(z,z') = r(\lambda(z) - \lambda(z')), \tag{4.6}$$

and $r(\lambda)$ is given by (1.19), and let $n \to +\infty$, $m \to \pm\infty$ according to the definitions (2.19), (2.45). As a result, we obtain the following expressions for the Poisson brackets of the Jost solutions $T_\pm(n,z)$ and of the reduced monodromy matrix $T(z)$:

$$\begin{aligned}&\{T_\pm(n,z) \underset{,}{\otimes} T_\pm(n,z')\}\\ &= \mp r(z,z') T_\pm(n,z) \otimes T_\pm(n,z') \pm T_\pm(n,z) \otimes T_\pm(n,z') r_\pm(z,z'),\end{aligned} \tag{4.7}$$

$$\{T_+(n,z) \underset{,}{\otimes} T_-(n,z)\} = 0 \tag{4.8}$$

and

$$\{T(z) \underset{,}{\otimes} T(z')\} = r_+(z,z') T(z) \otimes T(z') - T(z) \otimes T(z') r_-(z,z'). \tag{4.9}$$

Here

$$r_{\pm}(z,z') = \begin{pmatrix} \text{p.v.}\,\dfrac{zz'\alpha(z,z')}{(z-z')(zz'-1)} & 0 & 0 & 0 \\ 0 & \text{p.v.}\,\dfrac{zz'\beta(z,z')}{(z-z')(zz'-1)} & \mp\pi i\,\dfrac{\delta(zz'^{-1})z}{1-z^2} & 0 \\ 0 & \pm\pi i\,\dfrac{\delta(zz'^{-1})z}{1-z^2} & \text{p.v.}\,\dfrac{zz'\beta(z,z')}{(z-z')(zz'-1)} & 0 \\ 0 & 0 & 0 & \text{p.v.}\,\dfrac{zz'\alpha(z,z')}{(z-z')(zz'-1)} \end{pmatrix} \tag{4.10}$$

and

$$\alpha(z,z') = \frac{(zz'-1)^2}{(1-z^2)(1-z'^2)}, \quad \beta(z,z') = -\frac{(z-z')^2}{(1-z^2)(1-z'^2)}, \tag{4.11}$$

so that

$$\alpha(z,z') + \beta(z,z') = 1, \tag{4.12}$$

and in view of the involutions (2.21) and (2.47), we assume that $|z| = |z'| = 1$, $\operatorname{Im} z, \operatorname{Im} z' > 0$ where $z, z' \neq \pm 1$; the delta function $\delta(zz'^{-1})$ is defined in a natural way,

$$\int_{|z'|=1} \delta(zz'^{-1})\, f(z')\, \frac{dz'}{z'} = f(z). \tag{4.13}$$

When deriving (4.10)–(4.11) we have also made use of

$$\lim_{n\to\pm\infty} \text{p.v.}\, \frac{(zz'^{-1})^n}{1-zz'^{-1}} = \mp\pi i\,\delta(zz'^{-1}), \tag{4.14}$$

where $|z| = |z'| = 1$.

The Sochocki-Plemelj formula

$$\frac{1}{z - z' e^{-0}} = \lim_{\substack{\tilde{z}\to z' \\ |\tilde{z}|<1}} \frac{1}{z-\tilde{z}} = \text{p.v.}\,\frac{1}{z-z'} + \pi i\,\frac{\delta(zz'^{-1})}{z} \tag{4.15}$$

together with (4.7)–(4.11) leads to following Poisson brackets of the transition coefficients and the discrete spectrum:

$$\{a(z), a(z')\} = \{a(z), \bar{a}(z')\} = 0, \tag{4.16}$$

$$\{b(z), b(z')\} = 0, \tag{4.17}$$

$$\{b(z), \bar{b}(z')\} = 2\pi i \frac{z|a(z)|^2}{1-z^2}\,\delta(zz'^{-1}), \tag{4.18}$$

$$\{a(z), b(z')\} = \frac{zz'((1-zz')^2+(z-z')^2)a(z)b(z')}{(ze^{-0}-z')(1-zz')(1-z^2)(1-z'^2)}, \tag{4.19}$$

$$\{a(z), \bar{b}(z')\} = \frac{zz'((1-zz')^2+(z-z')^2)a(z)\bar{b}(z')}{(ze^{-0}-z')(1-zz')(1-z^2)(1-z'^2)} \tag{4.20}$$

and

$$\{a(z), \gamma_j\} = \frac{zz_j((1-zz_j)^2+(z-z_j)^2)a(z)\gamma_j}{(z-z_j)(1-zz_j)(1-z^2)(1-z_j^2)}, \tag{4.21}$$

$$\{b(z), z_j\} = \{b(z), \gamma_j\} = 0, \tag{4.22}$$

$$\{z_i, z_j\} = \{\gamma_i, \gamma_j\} = 0, \tag{4.23}$$

$$\{z_i, \gamma_j\} = -\frac{z_i^2}{1-z_i^2}\gamma_j\delta_{ij}, \qquad i,j=1,\ldots,N. \tag{4.24}$$

Due to the analyticity of $a(z)$, (4.19)–(4.21) remain valid for $|z|<1$ as well.

As in the case of the NS model, this gives a set of independent variables with simple Poisson brackets. Namely, consider (4.19) and (4.21) for $|z|<1$, let $|z|\to 1$ and split off the imaginary and real part, respectively. Then for $|z|=|z'|=1$, $\operatorname{Im} z \geqslant 0$, $\operatorname{Im} z' > 0$, we get

$$\{\log|a(z)|, \arg b(z')\} = -\frac{\pi\delta(zz'^{-1})z}{1-z^2} + \frac{\pi z'}{1-z'^2}(\delta(z)+\delta(-z)) \tag{4.25}$$

and

$$\{\log|a(z)|, \log|\gamma_j|\} = \frac{\pi i z_j}{1-z_j^2}(\delta(z)+\delta(-z)). \tag{4.26}$$

The terms containing $\delta(\pm z)$ result from the singular denominator $(1-z^2)^{-1}$ in (4.19) and (4.21); the delta function $\delta(\pm z)$ is defined by

$$\int_C \delta(\pm z) f(z)\frac{dz}{z} = \frac{1}{2} f(\pm 1), \tag{4.27}$$

where C is the semi-circle $|z|=1$, $0 \leqslant \arg z \leqslant \pi$. (Cf. the analogous formulae in § III.9, Part I.)

Let us introduce a set of variables

$$\varrho(\theta)=\frac{\sin\theta}{\pi}\log(1+|b(e^{i\theta})|^2), \qquad \varphi(\theta)=-\arg b(e^{i\theta}), \qquad 0\leqslant\theta\leqslant\pi, \tag{4.28}$$

$$\tilde{p}_j=\lambda_j=z_j+\frac{1}{z_j}, \qquad \tilde{q}_j=\log|\gamma_j|, \quad j=1,\ldots,N, \tag{4.29}$$

with the following ranges

$$0\leqslant\varrho(\theta)<\infty, \qquad 0\leqslant\varphi(\theta)<2\pi, \tag{4.30}$$

$$|\tilde{p}_j|>2, \qquad -\infty<\tilde{q}_j<\infty. \tag{4.31}$$

Using (4.24)–(4.26) we see that the nonvanishing Poisson brackets of these variables are

$$\{\varrho(\theta),\varphi(\theta')\}=\delta(\theta-\theta')-\frac{\sin\theta}{\sin\theta'}(\delta(\theta)+\delta(\theta-\pi)), \tag{4.32}$$

$$\{\varrho(\theta),\tilde{q}_j\}=-\frac{2\sin\theta z_j}{z_j^2-1}(\delta(\theta)+\delta(\theta-\pi)), \tag{4.33}$$

$$\{\tilde{p}_i,\tilde{q}_j\}=\delta_{ij}, \qquad i,j=1,\ldots,N. \tag{4.34}$$

These Poisson brackets would have canonical form if the right hand sides of (4.32)–(4.33) did not contain terms proportional to $\sin\theta(\delta(\theta)+\delta(\theta-\pi))$. *These additional terms should be interpreted in the same spirit as in § III.9, Part I. They must be taken into account every time we are dealing with functionals of the form*

$$F(\varrho)=\int_0^\pi\frac{\varrho(\theta)}{\sin\theta}f(\theta)\,d\theta, \tag{4.35}$$

where $f(\theta)$ is smooth for $0\leqslant\theta\leqslant\pi$, $f(0)=f(\pi)\neq 0$. We shall encounter such functionals in the next subsection.

3. Hamiltonian dynamics and integrals of the motion in terms of the variables $\varrho(\theta)$, $\varphi(\theta)$, $\tilde{p}_j$, $\tilde{q}_j$

The variables introduced above may be regarded as coordinates on the phase space $\mathcal{M}_c$ in whose terms the Poisson structure (4.2) takes the form (4.32)–(4.34). *These variables, however, are not completely independent.* Specifically, we have the condition (c)

$$c = -\int_0^\pi \frac{\varrho(\theta)}{\sin\theta}\,d\theta - \sum_{j=1}^N \log z_j^2. \qquad (4.36)$$

Besides, in general position $\frac{\varrho(\theta)}{\sin\theta}$ has a singularity of the type $\log\frac{1}{|\sin\theta|}$ as $\theta\to 0,\pi$, whereas $\varphi(0)$ and $\varphi(\pi)$ are fixed and equal to 0 or π according to (2.76). If $\theta=0$ or $\theta=\pi$ or both are virtual levels, $\frac{\varrho(\theta)}{\sin\theta}$ is finite at these points and $\varphi(\theta)$ takes the value 0 or π.

As an illustration *let us show that although at first sight the right hand side of* (4.36) *depends on the dynamical variables* $\varrho(\theta)$ *and* $\tilde{p}_j$, *it is actually in involution with* $\varphi(\theta)$ *and* $\tilde{q}_j$ (cf. § III.9, Part I). Indeed, from (4.32)–(4.34) we have

$$\{c,\varphi(\theta)\} = -\frac{1}{\sin\theta} + \frac{1}{\sin\theta}\int_0^\pi \frac{\sin\theta'}{\sin\theta'}(\delta(\theta')+\delta(\theta'-\pi))\,d\theta' = 0 \qquad (4.37)$$

and

$$\{c,\tilde{q}_j\} = \frac{2z_j}{z_j^2-1} - \{\log z_j^2, \tilde{q}_j\} = 0. \qquad (4.38)$$

We will now show that *the introduction of the new variables trivializes the dynamics of the Toda model.* The Hamiltonian H and the equations of motion can be written as

$$H = -\int_0^\pi \frac{\cos 2\theta}{\sin\theta}\,\varrho(\theta)\,d\theta + \frac{1}{2}\sum_{j=1}^N\left(\frac{1}{z_j^2} - z_j^2\right), \qquad (4.39)$$

$$\frac{\partial\varrho(\theta)}{\partial t} = \{H,\varrho(\theta)\} = 0, \qquad \frac{d\tilde{p}_j}{dt} = \{H,\tilde{p}_j\} = 0, \qquad (4.40)$$

$$\frac{\partial\varphi(\theta)}{\partial t} = \{H,\varphi(\theta)\} = -\frac{\cos 2\theta}{\sin\theta} + \frac{1}{\sin\theta} = 2\sin\theta, \qquad (4.41)$$

$$\frac{d\tilde{q}_j}{dt} = \{H,\tilde{q}_j\} = \frac{z_j^4+1}{z_j(1-z_j^2)} - \frac{2z_j}{1-z_j^2} = -\left(z_j - \frac{1}{z_j}\right) \qquad (4.42)$$

and their solution is a trivial matter. The result is equivalent to (2.86)–(2.87).

We emphasize that if the additional terms in the Poisson brackets (4.32)–(4.34) *were neglected, the time dependence of the transition coefficients would be incorrect.*

The trace identities (2.100) yield expressions for the local integrals I_n,

$$I_n = \int_0^{\pi} \frac{\cos n\theta}{\sin\theta} \varrho(\theta)\, d\theta + \frac{1}{n} \sum_{j=1}^{N} (z_j^n - z_j^{-n}), \tag{4.43}$$

so that these depend only on the variables $\varrho(\theta)$ and $\tilde{p}_j$. *These expressions show that the functionals I_{2n+1} are inadmissible.* In fact, the equations of motion

$$\frac{\partial\varphi(\theta)}{\partial t} = \{I_{2n+1}, \varphi(\theta)\} \tag{4.44}$$

have the form

$$\frac{\partial\varphi(\theta)}{\partial t} = \frac{\cos(2n+1)\theta}{\sin\theta} \tag{4.45}$$

(an additional term in the Poisson bracket (4.32) gives no contribution into (4.44)). The solution

$$\varphi(\theta, t) = \varphi(\theta, 0) + \frac{\cos(2n+1)\theta}{\sin\theta}\, t \tag{4.46}$$

for $t > 0$ is singular at $\theta = 0$ and $\theta = \pi$ and hence the dynamics induced by I_{2n+1} does not preserve the phase space $\mathscr{M}_c$. In particular, this shows once again that $P = -I_1$ is inadmissible.

The functionals I_{2n} are admissible and correspond to observables on the phase space $\mathscr{M}_c$. The induced equations of motion in the variables $\varrho(\theta)$, $\varphi(\theta)$, $\tilde{p}_j$, $\tilde{q}_j$ are

$$\frac{\partial\varphi(\theta)}{\partial t} = \{I_{2n}, \varphi(\theta)\} = \frac{\cos 2n\theta - 1}{\sin\theta} = -\frac{2\sin^2 n\theta}{\sin\theta}, \tag{4.47}$$

$$\frac{d\tilde{q}_j}{dt} = \{I_{2n}, \tilde{q}_j\} = \frac{(z_j^{2n} + z_j^{-2n})}{z_j - z_j^{-1}} - \frac{2}{z_j - z_j^{-1}} = \frac{(z_j^n - z_j^{-n})^2}{z_j - z_j^{-1}} \tag{4.48}$$

(now there is a contribution from the additional terms in (4.32)-(4.33)). The time evolution of the transition coefficients is given by

$$b(z,t)=\exp\left\{\frac{(z^n-z^{-n})^2}{z-z^{-1}}\,t\right\}b(z,0), \tag{4.49}$$

$$\gamma_j(t)=\exp\left\{\frac{(z_j^n-z_j^{-n})^2}{z_j-z_j^{-1}}\,t\right\}\gamma_j(0), \quad j=1,\ldots,N. \tag{4.50}$$

For $n=1$ this gives (upon reversing the sign of time) the familiar expressions (2.86)-(2.87).

One should not be misled into thinking that the other "half" of the local integrals are inadmissible. In fact, the quantities

$$\tilde{I}_n=I_n-I_{n-2}, \quad n>1, \tag{4.51}$$

with $I_0=-c$ are already admissible. They may be expressed as

$$\tilde{I}_n=\int_0^\pi \frac{\cos n\theta-\cos(n-2)\theta}{\sin\theta}\,\varrho(\theta)\,d\theta+\sum_{j=1}^N\left(\frac{z_j^n-z_j^{-n}}{n}-\frac{z_j^{n-2}-z_j^{2-n}}{n-2}\right), \tag{4.52}$$

and the integrand $\dfrac{\cos n\theta-\cos(n-2)\theta}{\sin\theta}$ is nonsingular at $\theta=0$ and $\theta=\pi$. *Hence the functionals $\tilde{I}_n$ correspond to observables on the phase space $\mathcal{M}_c$, and when writing down the induced equations of motion one may neglect the additional terms in* (4.32)-(4.33). A similar regularization was made for the NS model in the finite density case. The only quantity that does not admit this kind of regularization is P (cf. § III.9, Part I).

Hamilton's equations of motion

$$\frac{dp_n}{dt}=\{\tilde{I}_l,p_n\}, \quad \frac{dq_n}{dt}=\{\tilde{I}_l,q_n\}, \tag{4.53}$$

$n=-\infty,\ldots,\infty$, are naturally called *the higher Toda equations. All of them are exactly solvable.*

The above results imply that the Toda model and all its higher analogues are completely integrable Hamiltonian systems. The variables $\varrho(\theta)$, $\varphi(\theta)$, $\tilde{p}_j$, and $\tilde{q}_j$ are effectively their action-angle variables.

The regularized integrals of the motion $\tilde{I}_l$ display separation of modes in a natural manner. Thus for $\tilde{H}=-\tilde{I}_2=H-c$ we have

$$\tilde{H}=2\int_0^\infty \sin\theta\,\varrho(\theta)\,d\theta+\frac{1}{2}\sum_{j=1}^N(z_j^{-2}-z_j^2+2\log z_j^2), \tag{4.54}$$

which can be interpreted as a sum over independent modes. The continuous spectrum mode with index θ has positive energy given by

$$h(\theta)=2\sin\theta, \quad 0\leqslant\theta\leqslant\pi, \tag{4.55}$$

and the discrete spectrum mode (soliton) also has positive energy

$$h(z)=\tfrac{1}{2}z^{-2}-\tfrac{1}{2}z^2+\log z^2, \quad -1<z<1. \tag{4.56}$$

4. Soliton dynamics

The Poisson brackets (4.32)–(4.33) show that in general *soliton dynamics cannot be decoupled from the continuous spectrum modes dynamics in a Hamiltonian manner.* In other words, the constraint $\varrho(\theta)=0$ is inconsistent with these Poisson brackets. Nevertheless (cf. the NS model in the finite density case in § III.9, Part I), the equations of motion generated by the regularized functionals $\tilde{I}_l$ have an independent Hamiltonian formulation in the N-soliton submanifold of the phase space. Namely, on the phase space with coordinates $\tilde{p}_j, \tilde{q}_j, j=1,\ldots,N$, subject to $|\tilde{p}_j|>2$ endowed with the Poisson structure

$$\{\tilde{p}_i, \tilde{q}_j\}=\delta_{ij}, \quad i,j=1,\ldots,N, \tag{4.57}$$

the Hamiltonians

$$\tilde{I}_l^{(\mathrm{sol})}=\tilde{I}_l\big|_{\varrho(\theta)=0} \tag{4.58}$$

induce an evolution that coincides with soliton dynamics governed by the higher Toda equations.

Just as in the finite density case of the NS model, soliton scattering given by (3.67)–(3.72) *is not described by a canonical transformation* if the asymptotic variables $\tilde{p}_j, \tilde{q}_j^{(\pm)}=\tilde{q}_j\pm\Delta\tilde{q}_j$, where

$$\Delta\tilde{q}_j=\sum_{v_k<v_j}\log\left|\frac{1-z_jz_k}{z_j-z_k}\right|-\sum_{v_k>v_j}\log\left|\frac{1-z_jz_k}{z_j-z_k}\right|, \tag{4.59}$$

are supposed to have the same Poisson brackets as $\tilde{p}_j, \tilde{q}_j$.

In fact, for the two-soliton scattering we have

$$\Delta\tilde{q}_1=-\Delta\tilde{q}_2=\log\frac{|\tilde{p}_1\tilde{p}_2+\sqrt{\tilde{p}_1^2-4}\sqrt{\tilde{p}_2^2-4}-4|}{2(\tilde{p}_1-\tilde{p}_2)} \tag{4.60}$$

for $\tilde{p}_1>\tilde{p}_2$, and this is clearly not a function of the difference $\tilde{p}_1-\tilde{p}_2$ only.

This means, of course, that the a priori hypothesis that the $\tilde{p}_j, \tilde{q}_j^{(\pm)}$ form a canonical set is false. The correct choice of canonical asymptotic variables for soliton dynamics (and also for continuous spectrum modes) requires a separate analysis and is not our concern here.

The example of the Toda model shows that the inverse scattering method for lattice models is no less efficient than for continuous ones. The fundamental Poisson brackets (1.9) for $L_n(\lambda)$ are of crucial importance for the Hamiltonian interpretation of the method. This ends our description of the Toda model.

§ 5. The Lattice LL Model as a Universal Integrable System with Two-Dimensional Auxiliary Space

In § II.8 we have seen that the LL model is in some sense universal among integrable models with two-dimensional phase space for fixed x, which admit a zero curvature representation with two-dimensional auxiliary space. In particular, the SG, NS and HM models were interpreted as its limiting cases. Here we shall introduce a lattice analogue of the LL model, the LLL model, and consider the corresponding limiting cases. In this way, besides the LHM and LNS_1 models described in § I.2 we shall obtain a natural lattice analogue of the SG model - the LSG model.

As was observed in § I.2, the easiest thing to define when passing from continuous to discrete models is the matrix $L_n(\lambda)$ of the zero curvature representation. It is a more direct descendant of its continuous counterpart, the matrix $U(x,\lambda)$ of the auxiliary linear problem, than other entities such as $V_n(\lambda)$ and the associated equations of motion, or the Poisson structure and the Hamiltonian. We shall therefore proceed as follows: first, guided by natural requirements we shall define $L_n(\lambda)$ and then describe the LLL model itself.

The principal condition is that $L_n(\lambda)$ should satisfy the fundamental lattice Poisson brackets

$$\{L_n(\lambda) \underset{,}{\otimes} L_m(\mu)\} = [r(\lambda-\mu), L_n(\lambda) \otimes L_n(\mu)]\,\delta_{nm}. \tag{5.1}$$

The significance of these relations was illustrated above by the Toda model. We shall take $r(\lambda)$ to be the r-matrix of the LL model

$$r(\lambda) = -\frac{1}{2}\sum_{a=1}^{3} u_a(\lambda)\sigma_a \otimes \sigma_a, \tag{5.2}$$

where

$$u_1(\lambda)=\varrho\frac{1}{\operatorname{sn}(\lambda,k)},\quad u_2(\lambda)=\varrho\frac{\operatorname{dn}(\lambda,k)}{\operatorname{sn}(\lambda,k)},\quad u_3(\lambda)=\varrho\frac{\operatorname{cn}(\lambda,k)}{\operatorname{sn}(\lambda,k)},\tag{5.3}$$

and

$$\varrho=\frac{1}{2}\sqrt{J_3-J_1},\quad 0<k=\sqrt{\frac{J_2-J_1}{J_3-J_1}}<1\tag{5.4}$$

with $J_1<J_2<J_3$ (see § II.8). This is quite natural since (5.1) may be interpreted as the Poisson brackets of the one step transition matrix to the next lattice site, i.e. for a small interval Δ in the corresponding continuous model (see § III.1 of Part I and § 1).

Using this analogy, we can approximately write $L_n(\lambda)$ as

$$\begin{aligned}L_n(\lambda)&=I+\int_{\Delta_n}U(x,\lambda)\,dx+O(\Delta^2)\\&=I+\frac{1}{i}\sum_{a=1}^{3}u_a(\lambda)\int_{\Delta_n}S_a(x)\,dx+O(\Delta^2)\end{aligned}\tag{5.5}$$

(see the expression (II.8.2) for $U(x,\lambda)$). Terms of order $O(\Delta^2)$ in (5.5) are not specified by the initial continuous model. The discussion of the LHM and LNS_1 models in § I.2 shows that these terms are determined from the zero curvature representation. We will presently see that they are also uniquely determined by the fundamental Poisson brackets (5.1). As suggested by (5.5), it is natural to look for $L_n(\lambda)$ in the form

$$L_n(\lambda)=\mathscr{S}_0^{(n)}I+\frac{1}{i}\sum_{a=1}^{3}u_a(\lambda)\mathscr{S}_a^{(n)}\sigma_a,\tag{5.6}$$

where $\mathscr{S}_\alpha^{(n)}$, $\alpha=0,1,2,3$, are some new dynamical variables. To recover the LL model in the continuum limit, these must have the asymptotic behaviour

$$\mathscr{S}_0^{(n)}=1+O(\Delta^2),\quad \mathscr{S}_a^{(n)}=\Delta S_a(x)+O(\Delta^3),\tag{5.7}$$

with $\Delta n=x$, $\Delta\to 0$, $S_1^2(x)+S_2^2(x)+S_3^2(x)=1$.

Remarkably, the fundamental Poisson brackets (5.1) *with the r-matrix* (5.2)-(5.3) *are satisfied for* $L_n(\lambda)$ *of the form* (5.6) *if* $\mathscr{S}_0^{(n)}$, $\mathscr{S}_a^{(n)}$ *have the following Poisson brackets*

$$\{\mathscr{S}_a^{(n)},\mathscr{S}_0^{(m)}\}=J_{bc}\mathscr{S}_b^{(n)}\mathscr{S}_c^{(n)}\delta_{nm}\tag{5.8}$$

and

$$\{\mathcal{S}_a^{(n)}, \mathcal{S}_b^{(m)}\} = -\mathcal{S}_0^{(n)}\mathcal{S}_c^{(n)}\delta_{nm}. \tag{5.9}$$

Here and below (a, b, c) is a cyclic permutation of the indices 1, 2, 3, and we have set

$$J_{bc} = \tfrac{1}{4}(J_c - J_b). \tag{5.10}$$

To derive (5.8)-(5.9) one should use (II.8.8) and the identities

$$u_a(\lambda-\mu)u_b(\lambda)u_a(\mu) - u_b(\lambda-\mu)u_a(\lambda)u_b(\mu) = J_{ab}u_c(\lambda), \tag{5.11}$$

which are a consequence of addition theorems for the Jacobi elliptic functions. They can also be verified directly by comparing the poles in λ on the right and on the left of (5.11) and using the Liouville theorem.

Let us discuss the Poisson brackets (5.8)-(5.9).

1. These Poisson brackets are ultralocal: the variables $\mathcal{S}_a^{(n)}$ that belong to different sites are in involution. Hence we can first consider (5.8)-(5.9) in one site (suppressing the dependence on n) as the Poisson brackets on $\mathbb{R}^4$

$$\{\mathcal{S}_a, \mathcal{S}_0\} = J_{bc}\mathcal{S}_b\mathcal{S}_c, \tag{5.12}$$

$$\{\mathcal{S}_a, \mathcal{S}_b\} = -\mathcal{S}_0\mathcal{S}_c. \tag{5.13}$$

2. The Jacobi identity for the Poisson brackets (5.8)-(5.9) and (5.12)-(5.13) is ensured by the equation (II.8.12) for $r(\lambda)$. However, it can easily be verified directly by making use of the obvious relation

$$J_{12} + J_{23} + J_{31} = 0. \tag{5.14}$$

3. Unlike the Lie-Poisson brackets that occur for the HM and LHM models (see §§ I.1-I.2), *the Poisson brackets* (5.12)-(5.13) *are quadratic in the generators* $\mathcal{S}_0, \mathcal{S}_1, \mathcal{S}_2, \mathcal{S}_3$. In a natural sense, they are a deformation of the Poisson brackets for the LHM model. In particular, in the continuum limit (5.7) they go over into the Lie-Poisson brackets for the HM model.

4. The Poisson structure (5.12)-(5.13) is degenerate. *Its annihilator is generated by two polynomials,*

$$\mathcal{C}_0 = \sum_{a=1}^{3} \mathcal{S}_a^2 \tag{5.15}$$

and

$$\mathcal{C}_1 = \mathcal{S}_0^2 - \frac{1}{4}\sum_{a=1}^{3} J_a\mathcal{S}_a^2. \tag{5.16}$$

The equations

$$\mathscr{C}_0 = c_0, \qquad \mathscr{C}_1 = c_1, \tag{5.17}$$

where c_0 and c_1 are real, define a symplectic submanifold $\Gamma = \Gamma(J_a, c_0, c_1)$ in $\mathbb{R}^4$.

5. The manifold Γ is in general disconnected. Under the condition

$$c_1 > -\frac{J_1}{4} c_0 \tag{5.18}$$

Γ is homeomorphic to a disjoint union of two spheres $\mathbf{S}^2$. The additional requirement $\mathscr{S}_0 > 0$ selects one of them; the corresponding phase space will be denoted by Γ_0. If

$$-\frac{J_3}{4} c_0 < c_1 < -\frac{J_2}{4} c_0, \tag{5.19}$$

Γ is homeomorphic to the union of two spheres as before. However, if $-\frac{J_2}{4} c_0 < c_1 < -\frac{J_1}{4} c_0$, Γ is connected and homeomorphic to the torus $\mathbf{T}^2 = \mathbf{S}^1 \times \mathbf{S}^1$; this type of phase space will occur later on when describing the LSG model. If $c_1 < -\frac{J_3}{4} c_0$, (5.17) has no solution in $\mathbb{R}^4$.

6. Let us return to the lattice Poisson brackets (5.8)–(5.9). Their natural domain is the product of N copies of $\mathbb{R}^4$, N being the number of lattice sites. The phase space $\mathscr{M}$ of the LLL model will be the product of the Γ_0's where c_0 and c_1 do not depend on the index n, so that the model is homogeneous in space. In the continuum limit, with

$$c_0 = \Delta^2, \qquad c_1 = 1, \tag{5.20}$$

$\mathscr{M}$ goes into the phase space of the LL model.

We thus have defined the phase space $\mathscr{M}$ of the LLL model and the matrix $L_n(\lambda)$ of the corresponding auxiliary linear problem

$$F_{n+1} = L_n(\lambda) F_n. \tag{5.21}$$

The latter leads to the monodromy matrix

$$T_N(\lambda) = \prod_{n=1}^{\overset{\curvearrowleft}{N}} L_n(\lambda), \tag{5.22}$$

whose Poisson brackets have the same form as for $L_n(\lambda)$,

$$\{T_N(\lambda) \underset{,}{\otimes} T_N(\mu)\} = [r(\lambda-\mu), T_N(\lambda) \otimes T_N(\mu)]. \tag{5.23}$$

It follows that the functions

$$F_N(\lambda) = \operatorname{tr} T_N(\lambda) \tag{5.24}$$

form an involutive family of observables on $\mathscr{M}$,

$$\{F_N(\lambda), F_N(\mu)\} = 0. \tag{5.25}$$

The choice of the family (5.24) corresponds to the periodic boundary conditions

$$\mathscr{S}_\alpha^{(n+N)} = \mathscr{S}_\alpha^{(n)}, \quad \alpha = 0, 1, 2, 3. \tag{5.26}$$

Let us show that this family contains local observables representable as the sum over lattice sites

$$G_k = \sum_{n=1}^{N} g(\mathscr{S}_\alpha^{(n)}, \ldots, \mathscr{S}_\alpha^{(n+k)}), \tag{5.27}$$

with $k < N$. We shall say that G_k describes the interaction of $k+1$ nearest neighbours on the lattice. In particular, H will describe the interaction of two nearest neighbours.

To define it we proceed as follows. Observe that the expression (5.24) for $F_N(\lambda)$ simplifies if $\lambda = \lambda_0$, where λ_0 is a point at which $L_n(\lambda)$ degenerates. In fact, we have

$$L_n(\lambda_0) = \alpha_n \beta_n^\tau \tag{5.28}$$

where α_n and β_n are column-vectors and τ indicates transposition. It follows that

$$F_N(\lambda_0) = \prod_{n=1}^{N} \beta_{n+1}^\tau \alpha_n, \quad \beta_{N+1} = \beta_1, \tag{5.29}$$

so that $\log F_N(\lambda_0)$ is a local observable describing a nearest neighbour interaction. Unfortunately, this quantity is complex in general. To define a real-valued observable one should use two involutions satisfied by $L_n(\lambda)$,

$$\overline{L_n}(\lambda) = \sigma_2 L_n(\bar{\lambda}) \sigma_2 \tag{5.30}$$

and

$$L_n(-\lambda) = \sigma_2 L_n^\tau(\lambda) \sigma_2 \tag{5.31}$$

(these are immediate consequences of (5.3) and (5.6)). The first one implies $\overline{F_N(\lambda)} = F_N(\bar{\lambda})$, so that

$$H = \log \frac{|F_N(\lambda_0)|^2}{2} \tag{5.32}$$

also belongs to the involutive family generated by $F_N(\lambda)$. The second involution allows us to calculate H explicitly.

Indeed, from

$$\det L_n(\lambda_0) = c_1 + c_0\left(u_1^2(\lambda_0) + \frac{J_1}{4}\right) = 0 \tag{5.33}$$

and (5.18) it follows that λ_0 can be taken pure imaginary, which yields

$$L_n(\bar{\lambda}_0) = L_n(-\lambda_0) = \sigma_2 \beta_n \alpha_n^\tau \sigma_2, \tag{5.34}$$

hence

$$F_N(\bar{\lambda}_0) = F_N(-\lambda_0) = \prod_{n=1}^{N} \alpha_{n+1}^\tau \beta_n, \qquad \alpha_{N+1} = \alpha_1. \tag{5.35}$$

This implies

$$H = \sum_{n=1}^{N} \log \frac{\beta_{n+1}^\tau \alpha_n \alpha_{n+1}^\tau \beta_n}{2} = \sum_{n=1}^{N} \log h(\mathscr{S}_\alpha^{(n)}, \mathscr{S}_\alpha^{(n+1)}), \tag{5.36}$$

with

$$h(\mathscr{S}_\alpha^{(n)}, \mathscr{S}_\alpha^{(n+1)}) = \frac{1}{2}\operatorname{tr} L_{n+1}(\lambda_0) L_n(\lambda_0)$$
$$= \mathscr{S}_0^{(n)} \mathscr{S}_0^{(n+1)} + \sum_{a=1}^{3}\left(\frac{c_1}{c_0} + \frac{J_a}{4}\right) \mathscr{S}_a^{(n)} \mathscr{S}_a^{(n+1)}. \tag{5.37}$$

To derive the last equation we have made use of (5.6), (5.15)–(5.17) and (5.33).

The quantity H is what we shall take to be the Hamiltonian of the LLL model. The corresponding equations of motion

$$\frac{d\mathscr{S}_\alpha^{(n)}}{dt} = \{H, \mathscr{S}_\alpha^{(n)}\}, \qquad \alpha = 0, 1, 2, 3, \tag{5.38}$$

are not so very instructive and will not be stated here explicitly. We will rather discuss their general properties.

1) *In the continuum limit defined by (5.7) and (5.20), H goes into the Hamiltonian of the LL model (see § I.1)*

$$-2H+2N\log 2=\frac{\Delta}{2}\int\left(\left(\frac{d\vec{S}}{dx}\right)^2-J(\vec{S})\right)dx+O(\Delta^2) \tag{5.39}$$

and (5.38) gives the LL equation.

2) *The LLL model is a completely integrable Hamiltonian system.* Indeed, a family of $N-1$ independent integrals of the motion in involution comprising H may be produced as follows:

$$I_k=\frac{d^k}{d\lambda^k}\log|F_N(\lambda)|^2\bigg|_{\lambda=\lambda_0}, \quad k=0,\ldots,N-2. \tag{5.40}$$

The quantities I_k are local and describe the interaction of $k+2$ nearest neighbours. The missing integral can be taken in the form $\arg F_N(\lambda_0)$.

3) *The equations of motion (5.38) are representable as a zero curvature condition,*

$$\frac{dL_n}{dt}(\lambda)=V_{n+1}(\lambda)L_n(\lambda)-L_n(\lambda)V_n(\lambda). \tag{5.41}$$

In fact, arguing in complete analogy to § III.3, we find

$$\{\log F_N(\mu), L_n(\lambda)\}=V_{n+1}(\lambda,\mu)L_n(\lambda)-L_n(\lambda)V_n(\lambda,\mu), \tag{5.42}$$

where

$$V_n(\lambda,\mu)=\frac{1}{F_N(\mu)}\operatorname{tr}_1\left(\left(\prod_{k=n}^{\overset{\curvearrowleft}{N}} L_k(\mu)\otimes I\right)r(\mu-\lambda)\left(\prod_{k=1}^{\overset{\curvearrowleft}{n-1}} L_k(\mu)\otimes I\right)\right) \tag{5.43}$$

and we have used the notation tr_1 explained there. It follows from (5.32) that $\{H, L_n(\lambda)\}$ coincides with the right hand side of (5.41) where

$$V_n(\lambda)=\tfrac{1}{2}(V_n(\lambda,\lambda_0)+V_n(\lambda,\bar{\lambda}_0)). \tag{5.44}$$

Thus the equations of motion

$$\frac{dL_n}{dt}(\lambda)=\{H, L_n(\lambda)\} \tag{5.45}$$

are representable in the form (5.41).

The expression for $V_n(\lambda)$ can be simplified by using (5.29), (5.35) and (5.37). We have

$$V_n(\lambda)=\frac{\operatorname{tr}_1(\alpha_{n-1}\beta_n^\tau\otimes I)r(\lambda_0-\lambda)}{2\beta_n^\tau\alpha_{n-1}}-\frac{\operatorname{tr}_1(\sigma_2\beta_{n-1}\alpha_n^\tau\sigma_2\otimes I)r(\lambda_0+\lambda)}{2\alpha_n^\tau\beta_{n-1}}$$
$$=-\frac{1}{h(S_\alpha^{(n-1)},S_\alpha^{(n)})}\operatorname{tr}_1((L_{n-1}(\lambda_0)L_n(\lambda_0)\otimes I)r(\lambda-\lambda_0)$$
$$+(L_{n-1}(-\lambda_0)L_n(-\lambda_0)\otimes I)r(\lambda+\lambda_0)). \tag{5.46}$$

The last formula shows that $V_n(\lambda)$ depends only on the two nearest neighbours.

We emphasize that, as in the continuum case, *the fundamental Poisson brackets can replace the zero curvature representation.* This is yet another demonstration of the utility and universality of the notion of r-matrix.

This concludes our description of the LLL model. We shall now consider its limiting cases obtained by degenerating the elliptic curve E (see § II.8).

The simplest limit corresponds to $k\to 0$ so that $J_1=J_2<J_3$. The corresponding $L_n(\lambda)$ becomes

$$L_n(\lambda)=\mathscr{S}_0^{(n)}I+\frac{\varrho}{i\sin\lambda}(\mathscr{S}_1^{(n)}\sigma_1+\mathscr{S}_2^{(n)}\sigma_2+\cos\lambda\,\mathscr{S}_3^{(n)}\sigma_3), \tag{5.47}$$

where the variables $\mathscr{S}_\alpha^{(n)}$ satisfy the Poisson brackets (5.8)–(5.9) with $J_{12}=0$, $J_{13}=J_{23}=\varrho^2$. In this case there is an explicit expression for the $\mathscr{S}_\alpha$ (in each site) in terms of the usual variables S_1, S_2, S_3 on a sphere of radius R in $\mathbb{R}^3$

$$S_1^2+S_2^2+S_3^2=R^2 \tag{5.48}$$

with the Lie-Poisson brackets

$$\{S_a,S_b\}=-S_c. \tag{5.49}$$

Namely, we set

$$\mathscr{S}_0=\operatorname{ch}(\varrho S_3),\qquad \mathscr{S}_3=\frac{1}{\varrho}\operatorname{sh}(\varrho S_3),$$
$$\mathscr{S}_1=\frac{1}{\varrho}F(S_3)S_1,\qquad \mathscr{S}_2=\frac{1}{\varrho}F(S_3)S_2, \tag{5.50}$$

where

$$F(x)=\sqrt{\frac{\operatorname{sh}^2\varrho R-\operatorname{sh}^2\varrho x}{R^2-x^2}}. \tag{5.51}$$

The variables $\mathcal{S}_\alpha$ then have the Poisson brackets (5.12)–(5.13); the invariants $\mathcal{C}_0$ and $\mathcal{C}_1$ take the values

$$c_0 = \frac{\mathrm{sh}^2 \varrho R}{\varrho^2}, \qquad c_1 = 1 - \frac{J_1}{4} c_0 . \tag{5.52}$$

Substituting (5.50)–(5.51) into (5.47) gives the matrix $L_n(\lambda)$ for a model which for obvious reasons will be called *the partially anisotropic LHM model*; its r-matrix results from (5.2)–(5.3) in the limit as $k \to 0$ and is given by

$$r(\lambda) = -\frac{\varrho}{2\sin\lambda}(\sigma_1 \otimes \sigma_1 + \sigma_2 \otimes \sigma_2 + \cos\lambda\, \sigma_3 \otimes \sigma_3). \tag{5.53}$$

The same r-matrix serves for the partially anisotropic HM model (see § I.8), which is a continuum limit of our model upon the naive replacement

$$S_a^{(n)} = \Delta S_a(x), \qquad R = \Delta . \tag{5.54}$$

The partially anisotropic LHM model admits further degeneration. Specifically, in the limit as $\varrho \to 0$ $\left(\text{and replacing } \lambda \text{ by } \frac{2\varrho}{\lambda}\right)$ we come down to the isotropic case $J_1 = J_2 = J_3$ that corresponds to the LHM model of § I.2. The associated r-matrix results from (5.53) in this limit and coincides with the r-matrix for the HM model in § II.3. As was explained in § I.2, this also gives the LNS_1 model.

Let us now describe a lattice analogue of the SG model, the LSG model. It is in essence just another real form of the partially anisotropic LHM model considered above. More precisely, we interchange the roles of J_1, J_2, J_3 and assume that $J_1 = J_2 > J_3$, whereas (5.3)–(5.4) (with $k = 0$) and the form of $L_n(\lambda)$ (5.47) are left intact. The constraints (5.19) become

$$-\frac{J_1}{4} c_0 < c_1 < -\frac{J_3}{4} c_0 \tag{5.55}$$

and the phase space of the model in one lattice site is homeomorphic to the torus T^2. The variables $\mathcal{S}_\alpha$ are expressed as functions of the canonical variables π and φ on the torus,

$$\{\pi, \varphi\} = 1 . \tag{5.56}$$

Namely, let

$$\mathcal{S}_0 = s\cos\frac{\beta\varphi}{2}, \qquad \mathcal{S}_3 = \frac{s}{\gamma}\sin\frac{\beta\varphi}{2},$$
$$\mathcal{S}_1 = -\frac{f(\varphi)}{\gamma}\sin\frac{\beta\pi}{4}, \quad \mathcal{S}_2 = -\frac{f(\varphi)}{\gamma}\cos\frac{\beta\pi}{4}, \tag{5.57}$$

where

$$f(x) = \sqrt{1+\frac{s^2\cos\beta x}{2}}, \tag{5.58}$$

with $\gamma = \frac{\beta^2}{8} > 0$ and $s > 0$ arbitrary constants. Then the $\mathcal{S}_a$ have the Poisson brackets (5.12)–(5.13) with parameters

$$J_{12} = 0, \quad J_{13} = J_{23} = -\gamma^2 \tag{5.59}$$

where $J_1 = J_2 = 4\gamma^2$, $J_3 = 0$, and the invariants $\mathscr{C}_0$ and $\mathscr{C}_1$ take the values

$$c_0 = \frac{s^2+2}{2\gamma^2}, \quad c_1 = \frac{s^2-2}{2}. \tag{5.60}$$

Consider now the matrix $L_n^{SG}(\alpha)$ of the form

$$L_n^{SG}(\alpha) = \frac{1}{i}\operatorname{sh}\alpha\,\sigma_2 L_n(i\alpha), \tag{5.61}$$

where $L_n(\lambda)$ is given by (5.47) with $\varrho = i\gamma$. Expressing $\mathcal{S}_\alpha^{(n)}$ through π_n and φ_n according to (5.57)–(5.58) we find

$$L_n^{SG}(\alpha) = f(\varphi_n)\cos\frac{\beta\pi_n}{4}I + \frac{1}{i}f(\varphi_n)\sin\frac{\beta\pi_n}{4}\sigma_3$$
$$+\frac{s}{i}\left(\operatorname{ch}\alpha\sin\frac{\beta\varphi_n}{2}\sigma_1 + \operatorname{sh}\alpha\cos\frac{\beta\varphi_n}{2}\sigma_2\right). \tag{5.62}$$

The matrix $L_n^{SG}(\alpha)$ satisfies the fundamental Poisson brackets (5.1) with the r-matrix given by (5.53) for $\lambda = i\alpha$, $\varrho = i\gamma$. The matrix $r(\alpha)$ coincides (up to an irrelevant summand proportional to $I \otimes I$) with the r-matrix for the SG model of § II.6.

The Hamiltonian of the LSG model is

$$H^{\mathrm{LSG}} = \sum_{n=1}^{N} \log\left(\frac{2s^2\gamma^2}{s^2+2}(-\mathcal{S}_1^{(n)}\mathcal{S}_1^{(n+1)} + \mathcal{S}_2^{(n)}\mathcal{S}_2^{(n+1)}) + \frac{2-s^2}{2+s^2}\gamma^2\mathcal{S}_3^{(n)}\mathcal{S}_3^{(n+1)} + \mathcal{S}_0^{(n)}\mathcal{S}_0^{(n+1)}\right), \tag{5.63}$$

where the $\mathcal{S}_\alpha^{(n)}$ should be replaced by their expressions (5.57)-(5.58) in terms of the π_n, φ_n; H results from (5.36)-(5.37) by taking account of (5.60)-(5.61). Notice that the opposite sign of $\mathcal{S}_1^{(n)}\mathcal{S}_1^{(n+1)}$ and $\mathcal{S}_3^{(n)}\mathcal{S}_3^{(n+1)}$ as compared to (5.37) agrees with (5.61) which may be interpreted as the operation of alternating the sign,

$$\mathcal{S}_1^{(n)} \to (-1)^n\mathcal{S}_1^{(n)}, \qquad \mathcal{S}_3^{(n)} \to (-1)^n\mathcal{S}_3^{(n)} \tag{5.64}$$

(cf. the argument of § I.2).

The auxiliary linear problem (upon the replacement $\lambda = e^\alpha$), the Hamiltonian H, and other characteristics of the LSG model in the continuum limit

$$\pi_n = \Delta\pi(x), \qquad \varphi_n = \varphi(x), \qquad s = \frac{m\Delta}{2} \tag{5.65}$$

turn into the corresponding entities for the SG model, see § I.1. This is the reason for referring to the above completely integrable lattice model as the lattice SG model.

We point out that although the LSG and the partially anisotropic LHM models are essentially equivalent on the lattice, their continuous counterparts lie far apart, since they result from different continuum limits.

The list of models generated by the LLL model is in no way exhausted by the above examples. We may consider the higher analogues of the LLL model with the Hamiltonians I_k, their contractions, and also other values of the parameters J_a and of the invariants $\mathscr{C}_0$, $\mathscr{C}_1$. We may, moreover, vary the structure of $L_n(\lambda)$ by replacing

$$L_n(\lambda) \to A\,L_n(\lambda), \tag{5.66}$$

where $A \otimes A$ commutes with the r-matrix. In this way one can derive the Toda model as well.

Still, we have chosen the Toda model to be our basic example of a lattice model because its investigation is technically simpler. At the same time, it gives a fairly satisfactory illustration of the main features of the inverse scattering formalism for lattice models.

§ 6. Notes and References

1. The complete integrability of the Toda model in the periodic case was established by S. V. Manakov [M 1974a] and H. Flaschka [F 1974a, b], who made use of the Lax representation with the matrix $\mathscr{L}$ given by (2.35),

$$\mathscr{L}_{nm} = c_n \delta_{n,m+1} - p_n \delta_{nm} + c_m \delta_{n,m-1}, \tag{6.1}$$

with $\delta_{n+N,m} = \delta_{n,N+m} = \delta_{nm}$. The functions $\operatorname{tr} \mathscr{L}^k$, $k=1, \ldots, N$, form an involutive family on the phase space of the model, and $H = \frac{1}{2} \operatorname{tr} \mathscr{L}^2$. The matrix $L_n(\lambda)$ of the form (1.8) was introduced in [TF 1979].

2. The general solution of the periodic Toda model in terms of theta functions was derived in [K 1978]. The corresponding canonical action-angle variables were introduced in [FM 1976].

3. The auxiliary linear problem (2.33) (for $N = \infty$),

$$\mathscr{L} f = \lambda f, \tag{6.2}$$

and the related inverse problem were investigated in [M 1974a], [F 1974a, b] (without discussing the peculiarities connected with the boundary of the continuous spectrum and the condition (c)); see also the books [ZMNP 1980] and [T 1981]. The latter contains various physical applications of the Toda model.

4. For $p_n = 0$, (6.2) turns into the auxiliary linear problem for the Volterra model introduced in § I.2,

$$c_{n+1} f_{n+1} + c_n f_{n-1} = \lambda f_n \tag{6.3}$$

where $c_n = \sqrt{u_n}$ (see [M 1974a]). In that case the transition coefficients $a(z)$ and $b(z)$ obey an additional involution

$$a(z) = a(-z), \qquad b(z) = -b(-z). \tag{6.4}$$

Equation (6.3) is a lattice analogue of the one-dimensional Schrödinger equation and was studied for that reason in [CK 1973].

5. In the continuum limit, the Toda equations of motion go over into the equation of a nonlinear string,

$$\frac{\partial^2 u}{\partial t^2} = \frac{\partial^2 u}{\partial x^2} + \frac{\partial^2 u^2}{\partial x^2} + \frac{\partial^4 u}{\partial x^4}, \tag{6.5}$$

which can be solved by the inverse scattering method as well (see [Z 1973]), while the Volterra equations of motion go over into the KdV equation (see

[M 1974a], [ZMNP 1980]). Also, the zero curvature representations for lattice models go over into their continuous counterparts.

6. The action-angle variables of § 4 (but ignoring the additional terms in the Poisson brackets (4.32)–(4.33)) were introduced in [M 1974b] (see also [E 1981]) which from the very beginning used the Hamiltonian $\tilde{H}$.

7. It would be interesting to describe the topology of the phase space $\mathcal{M}_c$ in terms of the variables $\varrho(\theta)$, $\varphi(\theta)$, $\tilde{p}_j$, $\tilde{q}_j$ and the associated algebra of observables.

8. For models whose continuous spectrum has a boundary, the correct choice of canonical asymptotic variables for soliton dynamics (and of continuous spectrum modes) presents a nontrivial problem (cf. the NS model in the case of finite density). For the KdV equation, the problem was solved in [BFT 1986]. The method of this paper can in principle be applied to the Toda model and the NS model in the case of finite density.

9. As in the case of the NS and KdV models, a hierarchy of Poisson structures can be defined for the Toda model, starting with the Poisson brackets (1.3). The second Poisson structure for the Toda model was defined in [A 1979]; in terms of the variables p_n, $u_n = e^{q_n - q_{n-1}}$ it is given by

$$\{p_n, u_m\} = -p_n u_m (\delta_{nm} - \delta_{n-1,m}),$$

$$\{p_n, p_m\} = u_m^2 \delta_{n,m+1} - u_n^2 \delta_{n,m-1}, \tag{6.6}$$

$$\{u_n, u_m\} = -u_n u_m (\delta_{n+1,m} - \delta_{n-1,m}).$$

Unlike the Poisson structure (1.3), the Poisson brackets (6.6) have a nontrivial restriction to the submanifold $p_n = 0$. The resulting Poisson structure

$$\{u_n, u_m\} = -u_n u_m (\delta_{n+1,m} - \delta_{n-1,m}) \tag{6.7}$$

gives rise to the equations of motion for the Volterra model

$$\frac{du_n}{dt} = u_n (u_{n+1} - u_{n-1}), \tag{6.8}$$

if the Hamiltonian is set to be

$$H = \sum_n u_n. \tag{6.9}$$

In the continuum limit $u_n \to 1 - \Delta u(x)$, the Poisson brackets (6.7) go over into the Poisson brackets (I.3.15) for the KdV model.

The Poisson brackets (I.2.18) for the Volterra model result from the third Poisson structure for the Toda model by restricting to the submanifold

$p_n = 0$. In the continuum limit, as $u_n \to 1 - \Delta^2 u(x)$, this gives the second Poisson structure for the KdV model (see § III.10 of Part I).

10. The LLL model and the quadratic algebra of Poisson brackets were introduced by E. K. Sklyanin [S 1982]. The rapidly decreasing case of the model was investigated in [V 1985] where action-angle variables were also described.

11. The LSG model was stated in [IK 1981 a, b]. In [T 1982] the rapidly decreasing case of the model was shown to be completely integrable and action-angle variables were reported.

12. In [T 1982] and [V 1985] it was shown that the action-angle variables for the LLL and LSG models coincide with their analogues for the LL and SG models, and it was pointed out that this fact is due the coincidence of the associated r-matrices.

13. The scheme for constructing the local Hamiltonians for lattice models outlined above was developed in [IK 1982 a, b]. We note that the Hamiltonian defined in [IK 1982 a, b] and considered in [T 1982] differs from the one discussed in § 5.

14. Our derivation of the lattice analogue of the LL model was based on the construction of the matrix $L_n(\lambda)$ of the auxiliary linear problem which satisfies the fundamental Poisson brackets. We believe this to be the most fruitful principle for deriving integrable lattice analogues of continuous models.

References

[A 1979] Adler, M.: On a trace functional for formal pseudodifferential operators and symplectic structure of the Korteweg-de Vries type equations. Invent. Math. *50*, 219–248 (1979)

[BFT 1986] Buslaev, V. S., Faddeev, L. D., Takhtajan, L. A.: Scattering theory for the Korteweg-de Vries equation and its Hamiltonian interpretation. Physica *D 18*, 255–266 (1986)

[CK 1973] Case, K. M., Kac, M.: A discrete version of the inverse scattering problem. J. Math. Phys. *14*, 594–603 (1973)

[E 1981] Eilenberger, G.: Solitons. Mathematical method for Physicists. Berlin, Springer 1981

[F 1974 a] Flaschka, H.: The Toda lattice. I. Existence of integrals. Phys. Rev. *B 9*, 1924–1925 (1974)

[F 1974 b] Flaschka, H.: On the Toda lattice. II. Inverse scattering solution. Prog. Theor. Phys. *51*, 703–716 (1974)

[FM 1976] Flaschka, H., McLaughlin, D.: Canonically conjugate variables for the Korteweg-de Vries equation and the Toda lattice with periodic boundary conditions. Prog. Theor. Phys. *55*, 438–456 (1976)

[IK 1981 a] Izergin, A. G., Korepin, V. E.: The lattice sine-Gordon model. Vestn. Leningr. Univ. Ser. Fiz. Khim. *22*, 84–87 (1981) [Russian]

[IK 1981 b] Izergin, A. G., Korepin, V. E.: The lattice quantum Sine-Gordon model. Lett. Math. Phys. *5*, 199–205 (1981)

[IK 1982a] Isergin, A. G., Korepin, V. E.: Lattice regularizations of two-dimensional quantum field models. In: Problems in quantum field theory and statistical physics. 3. Zap. Nauchn. Semin. LOMI *120*, 75–91 (1982) [Russian]

[IK 1982b] Isergin, A. G., Korepin, V. E.: Lattice versions of quantum field theory models in two dimensions. Nucl. Phys. *B205*, 401–413 (1982)

[K 1978] Krichever, I. M.: Algebraic curves and nonlinear difference equations. Usp. Mat. Nauk *33* (4), 215–216 (1978) [Russian]

[M 1974a] Manakov, S. V.: Complete integrability and stochastization of discrete dynamical systems. Zh. Exp. Teor. Fiz. *67*, 543–555 (1974) [Russian]; English transl. in Sov. Phys. JETP *40*, 269–274 (1975)

[M 1974b] Manakov, S. V.: Inverse scattering method as applied to problems of the physics of waves in nonlinear media. Candidate thesis, Chernogolovka 1974 [Russian]

[S 1982] Sklyanin, E. K.: Algebraic structures connected with the Yang-Baxter equation. Funk. Anal. Prilož. *16* (4), 27–34 (1982) [Russian]; English transl. in Funct. Anal. Appl. *16*, 263–270 (1982)

[T 1981] Toda, M.: Theory of non-linear lattices. Springer series in solid state physics 20. Berlin–Heidelberg–New York, Springer 1981

[T 1982] Tarasov, V. O.: Classical version of the lattice sine-Gordon model. In: Problems in quantum field theory and statistical physics. 3. Zap. Nauchn. Semin. LOMI *120*, 173–187 (1982) [Russian]

[TF 1979] Takhtajan, L. A., Faddeev, L. D.: The quantum inverse problem method and the XYZ Heisenberg model. Usp. Mat. Nauk *34* (5), 13–63 (1979) [Russian]; English transl. in Russian Math. Surveys *34* (5), 11–68 (1979)

[V 1985] Volkov, A. Yu.: A discrete version of the Landau-Lifshitz equation. In: Problems in quantum field theory and statistical physics. 5. Zap. Nauchn. Semin. LOMI *145*, 62–71 (1985) [Russian]

[Z 1973] Zakharov, V. E.: On stochastization of one-dimensional chains of nonlinear oscillators. Zh. Exp. Teor. Fiz. *65*, 219–225 (1973) [Russian]; English transl. in Sov. Phys. JETP *38*, 108–110 (1974)

[ZMNP 1980] Zakharov, V. E., Manakov, S. V., Novikov, S. P., Pitaievski, L. P.: Theory of Solitons. The Inverse Problem Method. Moscow, Nauka 1980 [Russian]; English transl.: New York, Plenum 1984

Chapter IV
Lie-Algebraic Approach to the Classification and Analysis of Integrable Models

In this chapter we shall summarize and generalize our experience in describing integrable models gained from the study of particular examples. The principal entities of the inverse scattering method and its Hamiltonian interpretation were the auxiliary linear problem operator $L=\frac{d}{dx}-U(x,\lambda)$ and the fundamental Poisson brackets for $U(x,\lambda)$ involving the r-matrix. Similar objects were introduced for lattice models. We will show that these notions have a simple geometric interpretation.

We shall define a natural partition of the integrable models into three families: rational, trigonometric and elliptic, according to the dependence of $U(x,\lambda)$ and $r(\lambda)$ on the spectral parameter λ. We shall interprete the fundamental Poisson brackets for the rational family in terms of an infinite-dimensional Lie algebra associated with the current algebra. The trigonometric and elliptic families will be obtained by averaging over a one- or two-dimensional lattice in the complex plane of the spectral variable λ. There are also similar families for lattice models; we shall discuss the associated fundamental Poisson brackets. Concentrating our attention on the rational case, we shall also discuss the dynamics of integrable models from a general point of view. This will lead to a natural geometric interpretation of the Riemann problem. We shall also give a Lie-algebraic interpretation of the hierarchy of Poisson structures and the associated Λ-operator.

§ 1. Fundamental Poisson Brackets Generated by the Current Algebra

We recall the definition of the standard Poisson structure associated with an arbitrary connected Lie group G, $\dim G=n$. Let $\mathfrak{g}$ be its Lie algebra and let X_a, $a=1,\ldots,n$, be a set of generators of $\mathfrak{g}$ with structure constants C_{ab}^{c},

$$[X_a, X_b] = C^c_{ab} X_c. \tag{1.1}$$

From here on we adopt the convention on summation over repeated indices. The linear space $\mathfrak{g}^*$ dual to $\mathfrak{g}$ has natural coordinates u_a: if $\xi = \xi^a X_a$ lies in $\mathfrak{g}$, then $u(\xi) = (u, \xi) = u_a \xi^a$. In the algebra $\mathscr{A}$ of smooth functions $f(u)$ on $\mathfrak{g}^*$ we define a bracket $\{,\}: \mathscr{A} \times \mathscr{A} \to \mathscr{A}$ by

$$\{f_1, f_2\}(u) = -C^c_{ab} \frac{\partial f_1}{\partial u_a} \frac{\partial f_2}{\partial u_b} u_c, \tag{1.2}$$

which is obviously skew-symmetric and satisfies the Jacobi identity by virtue of the Jacobi identity for the commutator (1.1). Thus (1.2) defines a Poisson structure on the phase space $\mathfrak{g}^*$. The Poisson bracket (1.2) for the coordinates u_a takes the form

$$\{u_a, u_b\} = -C^c_{ab} u_c \tag{1.3}$$

and is called the *Lie-Poisson bracket.*

In general, the Poisson structure (1.2) is degenerate. Its annihilator coincides with the *algebra of Casimir functions* $I(\mathfrak{g})$ which consists of functions $f(u)$ invariant under the coadjoint action $u \to \mathrm{Ad}^* g \cdot u$ of G on $\mathfrak{g}^*$ given by

$$\mathrm{Ad}^* g \cdot u(\xi) = u(g^{-1} \xi g). \tag{1.4}$$

The restriction of the Poisson bracket (1.2) to any orbit of this action is nondegenerate, so that a Poisson submanifold in $\mathfrak{g}^*$ is a union of orbits.

These definitions actually involve only the Lie algebra $\mathfrak{g}$ of G and its coadjoint action $u \to \mathrm{ad}^* \eta \cdot u$ given by

$$\mathrm{ad}^* \eta \cdot u(\xi) = u([\xi, \eta]). \tag{1.5}$$

The orbits of the Ad^* action of G on $\mathfrak{g}^*$ are integral manifolds for the distribution spanned by the vector fields $\mathrm{ad}^* \eta$ for all η in $\mathfrak{g}$. Hence we may speak of the Lie-Poisson structure induced by a Lie bracket.

In the study of integrable systems we shall deal with infinite-dimensional Lie algebras. All the above definitions carry over to this case in a natural manner.

Let us consider the *current algebra* $C(\mathfrak{g})$ *associated with the Lie algebra* $\mathfrak{g}$. It consists of formal Laurent series $\xi(\lambda)$ in the variable λ,

$$\xi(\lambda) = \sum_{k > -\infty}^{\infty} \xi_k \lambda^k, \tag{1.6}$$

with ξ_k in $\mathfrak{g}$ and the symbol $k \gg -\infty$ indicating that the series in powers of λ^{-1} truncates. The commutator in $C(\mathfrak{g})$ is defined in the obvious way

$$[\xi(\lambda), \eta(\lambda)] = \sum_{k \gg -\infty}^{\infty} \sum_{i+j=k} [\xi_i, \eta_j] \lambda^k . \tag{1.7}$$

A set of generators of $C(\mathfrak{g})$ is formed by the elements

$$X_{a,k} = X_a \lambda^k , \quad a = 1, \dots, n; \quad k = -\infty, \dots, \infty , \tag{1.8}$$

where the X_a are the generators of $\mathfrak{g}$ with structure constants C^c_{ab}. Their commutator clearly is

$$[X_{a,k}, X_{b,l}] = C^c_{ab} X_{c,k+l} . \tag{1.9}$$

Let $u_{a,k}$ denote the coordinates of an element u of the dual space $C^*(\mathfrak{g})$; in agreement with (1.6) we assume that $u_{a,k} = 0$ for large positive k. The corresponding pairing is given by

$$u(\xi) = (u, \xi) = \sum_k u_{a,k} \xi^a_k , \tag{1.10}$$

where the sum over k is always finite. The Lie-Poisson bracket of the coordinates $u_{a,k}$ has the form

$$\{u_{a,k}, u_{b,l}\} = -C^c_{ab} u_{c,k+l} . \tag{1.11}$$

It will be convenient to introduce a generating function $u_a(\lambda)$ for the coordinates $u_{a,k}$ of a point u in $C^*(\mathfrak{g})$ as a formal Laurent series

$$u_a(\lambda) = \sum_{k=-\infty}^{k \ll \infty} u_{a,k} \lambda^{-k-1} . \tag{1.12}$$

The pairing (1.10) is given by a neat formula

$$u(\xi) = \operatorname{Res} u_a(\lambda) \xi^a(\lambda) , \tag{1.13}$$

where the symbol Res indicates the coefficient of λ^{-1} in a Laurent series.

The variable λ determines a *gradation* of $C(\mathfrak{g})$,

$$C(\mathfrak{g}) = \sum_{k \gg -\infty}^{\infty} \mathfrak{g} \lambda^k = \sum_{k \gg -\infty}^{\infty} C_k , \tag{1.14}$$

where

$$[C_k, C_l] \subset C_{k+l}, \tag{1.15}$$

which, in particular, allows us to decompose $C(\mathfrak{g})$ into a linear sum of two subalgebras,

$$C(\mathfrak{g}) = C_+(\mathfrak{g}) + C_-(\mathfrak{g}), \tag{1.16}$$

where

$$C_+(\mathfrak{g}) = \sum_{k=0}^{\infty} C_k, \quad C_-(\mathfrak{g}) = \sum_{k > -\infty}^{k=-1} C_k. \tag{1.17}$$

There is a similar decomposition for $C^*(\mathfrak{g})$,

$$C^*(\mathfrak{g}) = C^*_+(\mathfrak{g}) + C^*_-(\mathfrak{g}), \tag{1.18}$$

which in terms of the generating function $u_a(\lambda)$ becomes

$$u_a(\lambda) = u_a^+(\lambda) + u_a^-(\lambda), \tag{1.19}$$

with

$$u_a^+(\lambda) = \sum_{k=-\infty}^{k=-1} u_{a,k}\lambda^{-k-1}, \quad u_a^-(\lambda) = \sum_{k=0}^{k<\infty} u_{a,k}\lambda^{-k-1}. \tag{1.20}$$

The subspaces $C^*_\pm(\mathfrak{g})$ are orthogonal to $C_\pm(\mathfrak{g})$ relative to the pairing (1.10), and $C^*_\pm(\mathfrak{g}) = (C_\mp(\mathfrak{g}))^*$. The Lie-Poisson bracket (1.11) has a natural restriction to these subspaces.

The resulting Poisson structure on $C^*_\pm(\mathfrak{g})$ has an elegant expression in terms of the generating functions $u_a^\pm(\lambda)$. Namely, multiply both sides of (1.11) by $\lambda^{-k-1}\mu^{-l-1}$ and sum over $k, l<0$ and $k, l \geqslant 0$. Then

$$\{u_a^\pm(\lambda), u_b^\pm(\mu)\} = \mp C_{ab}^c \frac{u_c^\pm(\lambda) - u_c^\pm(\mu)}{\lambda - \mu}. \tag{1.21}$$

The corresponding expression for the Poisson bracket $\{u_a^+(\lambda), u_b^-(\mu)\}$ is not so elegant. Fortunately, we shall not need it because we are going to define a new Poisson structure on the phase space $C^*(\mathfrak{g})$. To do so, we use the decomposition (1.16) to define on the vector space $C(\mathfrak{g})$ a new structure of Lie algebra with commutator $[,]_0$ by setting

$$[\xi_+, \eta_+]_0 = [\xi_+, \eta_+], \quad [\xi_-, \eta_-]_0 = -[\xi_-, \eta_-] \tag{1.22}$$

and

$$[\xi_+, \eta_-]_0 = 0, \tag{1.23}$$

where $\xi=\xi_+ + \xi_-$, $\eta=\eta_+ + \eta_-$ are elements of $C(\mathfrak{g})$. By introducing the operator

$$R=\tfrac{1}{2}(P_+ - P_-), \tag{1.24}$$

where $P_\pm$ are the projection operators onto $C_\pm(\mathfrak{g})$, $P_+P_-=P_-P_+=0$, (1.22)–(1.23) may be combined into a single formula

$$[\xi,\eta]_0=[R\xi,\eta]+[\xi,R\eta]. \tag{1.25}$$

The infinite-dimensional Lie algebra with the commutator $[\,,]_0$ defined above will be denoted by $C_0(\mathfrak{g})$. This algebra will in fact play a key role in the classification of integrable models.

The corresponding Lie-Poisson brackets $\{\,,\}_0$ on the phase space $C^*(\mathfrak{g})$ are given by (1.21) without the $\pm$ sign on the right hand side, and

$$\{u_a^+(\lambda), u_b^-(\mu)\}=0. \tag{1.26}$$

We shall now unite (1.21) and (1.26) into a single elegant formula under the assumption that $\mathfrak{g}$ has a nondegenerate symmetric bilinear form $\langle\,,\rangle$ invariant under the adjoint action. For example, if $\mathfrak{g}$ is semi-simple, the Killing form may be taken as $\langle\,,\rangle$.

Consider a nondegenerate matrix K with entries

$$K_{ab}=\langle X_a, X_b\rangle. \tag{1.27}$$

(If $\mathfrak{g}$ is semi-simple and represented as a matrix algebra, we may assume $K_{ab}=\operatorname{tr} X_a X_b$). Let K^{ab} denote the entries of the inverse matrix K^{-1}. Let an element Π of $\mathfrak{g}\otimes\mathfrak{g}$ and elements A^a of $\mathfrak{g}$ be defined by

$$\Pi=K^{ab}X_a\otimes X_b, \tag{1.28}$$

$$A^a=K^{ab}X_b. \tag{1.29}$$

Then we have the relations

$$[\Pi, A^c\otimes I]=-[\Pi, I\otimes A^c]=C^c_{ab}A^a\otimes A^b, \tag{1.30}$$

where the symbols $A\otimes I$ and $I\otimes A$ denote the natural imbeddings of A into $\mathfrak{g}\otimes\mathfrak{g}$. To derive these relations one must use, besides (1.1), the skew-symmetry of the structure constants: the tensor $C^{abc}=K^{aa'}K^{bb'}C^c_{a'b'}$ is totally skew-symmetric, which follows from the invariance of $\langle\,,\rangle$.

Using Π and A^a we can define an element $r(\lambda)$ of $\mathfrak{g}\otimes\mathfrak{g}$,

$$r(\lambda)=\frac{\Pi}{\lambda}, \tag{1.31}$$

and a formal Laurent series $U(\lambda)$ with coefficients in $\mathfrak{g}^* \otimes \mathfrak{g}$,

$$U(\lambda)=u_a(\lambda)A^a. \tag{1.32}$$

The Lie-Poisson brackets (1.21)–(1.26) associated with $C_0(\mathfrak{g})$ can be written in these terms as a single formula

$$\{U(\lambda) \underset{,}{\otimes} U(\mu)\}_0=[r(\lambda-\mu), U(\lambda)\otimes I+I\otimes U(\mu)], \tag{1.33}$$

where on the right hand side we have used the natural notation $\{\underset{,}{\otimes}\}_0$ (see § III.1, Part I). Formula (1.33) results from (1.21) upon multiplying by $A^a \otimes A^b$ and using (1.30). From here up to § 4 we shall only deal with the Poisson brackets $\{,\}_0$, where for notational simplicity the index 0 will be dropped.

It is instructive to compare the Lie-Poisson brackets for the Lie algebra $C_0(\mathfrak{g})$ given by (1.33) with the fundamental Poisson brackets for the continuous models of § III.1, Part I, and of §§ II.3, II.6, II.8. These expressions are practically identical; a formal difference is that (1.33) does not contain the spatial variable x. The dependence on x can easily be assimilated in our treatment by considering a direct product of the algebras $C(\mathfrak{g})$ over all x. In a more formal way, one should use the current algebra $\mathscr{C}((\mathfrak{g}))$ that consists of Laurent series $\xi(\lambda, x)$ with coefficients depending on x and satisfying certain boundary conditions (e.g., periodic or rapidly decreasing). The algebra $\mathscr{C}((\mathfrak{g}))$ has generators $X_{a,k}(x)$ with the commutator

$$[X_{a,k}(x), X_{b,l}(y)]=C^c_{ab}X_{c,k+l}(x)\delta(x-y). \tag{1.34}$$

Reproducing the above arguments in the context of $\mathscr{C}((\mathfrak{g}))$, we come down to the *Lie-Poisson brackets on the phase space* $\mathscr{C}^*((\mathfrak{g}))$

$$\{U(x,\lambda) \underset{,}{\otimes} U(y,\mu)\}=[r(\lambda-\mu), U(x,\lambda)\otimes I+I\otimes U(x,\mu)]\delta(x-y), \tag{1.35}$$

which have exactly the same structure as the fundamental Poisson brackets for continuous models.

Nevertheless, (1.35) has a different content than, say, (II.3.8). Thus, the latter deals with a particular matrix $U(x,\lambda)$ in the auxiliary space, which is a rational function of the spectral parameter λ, whereas (1.35) involves a formal Laurent series with coefficients in a given Lie algebra $\mathfrak{g}$. The agreement is reached by noticing that the fundamental Poisson brackets for a particular model are a realization of the Poisson brackets (1.35) in a particular matrix representation of $\mathfrak{g}$ (the representation space playing the role of auxiliary space), with a further restriction to an orbit of the associated algebra

$\mathscr{C}_0((\mathfrak{g}))$ in the phase space $\mathscr{C}^*((\mathfrak{g}))$. Here, for shortness, we speak of the orbits of a Lie algebra having in mind the abovementioned integral submanifolds for the distribution induced by the coadjoint action. The orbits in question are products over x of the orbits of the coadjoint action of the Lie algebra $C_0(\mathfrak{g})$ in $C^*(\mathfrak{g})$. Spatial homogeneity requires these orbits to be the same for all x. We shall therefore suppress the dependence on x in what follows.

For the applications to integrable models, we are mostly interested in finite-dimensional orbits of $C_0(\mathfrak{g})$. (Notice that generic orbits have infinite dimension.) To determine them, it is convenient to specify finite-dimensional Poisson submanifolds of $C^*(\mathfrak{g})$ by imposing constraints on the coordinates $u_{a,k}$ (or on their generating functions $u_a(\lambda)$, $U(\lambda)$) invariant under the coadjoint action of $C_0(\mathfrak{g})$.

Clearly, the simplest example of such a Poisson submanifold is the subspace $C^*_{N,M}$ of $C^*(\mathfrak{g})$ defined by

$$u_{a,k}=0 \tag{1.36}$$

if $k\geqslant N$ or $k\leqslant -M-1$, with $N, M\geqslant 0$. The action of $C_0(\mathfrak{g})$ on $C^*_{N,M}$ reduces to the action of a finite-dimensional Lie algebra $C_{N,M}(\mathfrak{g})$ with generators $X_{a,k}$, $-M\leqslant k<N$, and the commutator

$$[X_{a,k}, X_{b,l}]=\begin{cases} C^c_{ab}X_{c,k+l} & \text{if } k,l\geqslant 0,\ k+l<N, \\ -C^c_{ab}X_{c,k+l} & \text{if } k,l<0,\ -M-1<k+l, \\ 0 & \text{otherwise}. \end{cases} \tag{1.37}$$

The orbits of this algebra in $C^*_{N,M}$ are the required phase spaces associated with integrable models. In more detail, coordinates in $C^*_{N,M}$ are given by the set $\{u_{a,k},\ k=-M,\dots,N-1\}$; their Lie-Poisson brackets are

$$\{u_{a,k}, u_{b,l}\}=\begin{cases} -C^c_{ab}u_{c,k+l} & \text{if } k,l\geqslant 0,\ k+l<N, \\ C^c_{ab}u_{c,k+l} & \text{if } k,l<0,\ -M-1<k+l, \\ 0 & \text{otherwise}, \end{cases} \tag{1.38}$$

and Poisson submanifolds are unions of orbits of the Lie algebra $C_{N,M}(\mathfrak{g})$. The problem of specifying such orbits is a finite-dimensional one and can be solved by traditional methods, for instance, by fixing the values of Casimir functions.

The generating function for the $u_{a,k}$ (or $u_{a,k}(x)$ upon restoring the x-dependence) is now a rational function of λ,

$$U(\lambda)=\sum_{k=-M}^{N-1} u_{a,k}A^a\lambda^{-k-1}, \tag{1.39}$$

and if a representation of $\mathfrak{g}$ is chosen, it is a matrix in the representation space. This is what gives the matrix $U(x,\lambda)$ of the auxiliary linear problem for the integrable models to be discussed.

Let us consider several examples.

1. $N=1, M=0$.

The corresponding $U(\lambda)$ has the form

$$U(\lambda)=\frac{U_0}{\lambda}=\frac{S_a A^a}{\lambda}, \tag{1.40}$$

where the S_a are dynamical variables on $C^*_{1,0}=\mathfrak{g}^*$ with the Poisson brackets

$$\{S_a, S_b\}=-C^c_{ab}S_c. \tag{1.41}$$

In the simplest case $\mathfrak{g}=su(2)$, there are three dynamical variables S_a, $a=1,2,3$. The fundamental representation of $su(2)$ with generators $X_a=\frac{1}{2i}\sigma_a$, structure constants $C_{abc}=\varepsilon_{abc}$, and matrices $A^a=i\sigma_a$ leads to

$$U(\lambda)=\frac{iS_a\sigma_a}{\lambda}, \tag{1.42}$$

$$r(\lambda)=\frac{1}{2\lambda}\sigma_a\otimes\sigma_a. \tag{1.43}$$

The corresponding orbits are determined by

$$S_1^2+S_2^2+S_3^2=\text{const}. \tag{1.44}$$

The dynamical variables S_a satisfying (1.44) with the Poisson brackets (1.41) were used in the description of the phase space for the HM model in § I.1. The matrix $U(x,\lambda)$ (1.42) and the r-matrix (1.43) turn into their counterparts from § II.3 upon a change $\lambda\to-\frac{2}{\lambda}$ if one recalls that the permutation matrix P in $\mathbb{C}^2\otimes\mathbb{C}^2$ is given by

$$P=\tfrac{1}{2}(I\otimes I+\sigma_a\otimes\sigma_a) \tag{1.45}$$

and drops the irrelevant summand proportional to $I\otimes I$.

For an arbitrary semi-simple Lie algebra $\mathfrak{g}$ the above example leads to an integrable generalization of the HM model, the $\mathfrak{g}$-invariant magnet.

2. $N=0, M=2$.

The corresponding $U(\lambda)$ has the form

$$U(\lambda)=U_{-1}+\lambda U_{-2}=Q_a A^a+\lambda J_a A^a, \tag{1.46}$$

where the Q_a, J_a are dynamical variables on the phase space $C^*_{0,2}$ with the Poisson brackets

$$\{Q_a, Q_b\}=C^c_{ab} J_c, \quad \{Q_a, J_b\}=\{J_a, J_b\}=0. \tag{1.47}$$

We consider the simplest cases $\mathfrak{g}=su(2)$ or $\mathfrak{g}=su(1,1)$ in the fundamental representation with generators $X_a=\frac{1}{2i}\sigma_a$, $a=1,2,3$, and $X_1=\frac{1}{2}\sigma_1$, $X_2=\frac{1}{2}\sigma_2$, $X_3=\frac{1}{2i}\sigma_3$ respectively. The orbit in question is specified by

$$J_1=J_2=0, \quad J_3=\frac{\varepsilon}{2}, \tag{1.48}$$

$$Q_3=0, \quad Q_1+iQ_2=\psi, \tag{1.49}$$

with $\varepsilon=-1$ for $\mathfrak{g}=su(2)$ and $\varepsilon=1$ for $\mathfrak{g}=su(1,1)$; the only nonvanishing Poisson bracket is

$$\{\psi, \bar{\psi}\}=i. \tag{1.50}$$

As a result, $U(\lambda)$ can be written as

$$U(\lambda)=-\frac{\varepsilon\lambda}{2i}\sigma_3+\sqrt{\varepsilon}\begin{pmatrix}0 & \bar{\psi}\\ \psi & 0\end{pmatrix}, \tag{1.51}$$

and after the replacement $\lambda\to-\varepsilon\lambda$ turns into the matrix $U(x,\lambda)$ for the NS model (see § I.2, Part I) with $\varkappa=\varepsilon$. The matrix

$$r(\lambda)=\frac{1}{\lambda}\left(P-\frac{I\otimes I}{2}\right), \tag{1.52}$$

is carried by this replacement into the r-matrix for the NS model of § III.1, Part I (up to an irrelevant unit summand).

Other Lie algebras $\mathfrak{g}$ will give vector and matrix generalization of the NS model.

Thus the general scheme outlined above not only includes the two principal models of the book, but also provides their nontrivial generalizations. We have seen that the NS and HM models are indeed the simplest ones in an infinite sequence of examples: the Lie algebra $\mathfrak{g}$, the integers N and $M\geqslant 0$ and the orbit of the Lie algebra $C_{N,M}(\mathfrak{g})$ may be taken arbitrarily. The auxiliary linear problem with matrix $U(x,\lambda)$ of the form (1.39) and the r-matrix

(1.31) give rise to a zero curvature representation for the corresponding Hamilton equations of motion. A Lie-algebraic interpretation of the associated Hamiltonians and a scheme for solving the equations of motion will be given in § 4.

Here we point out that the above examples do not exhaust all relevant finite-dimensional phase spaces (at fixed x). Let us exhibit another family of interesting examples. To start with, we notice that the Poisson brackets (1.41) may be derived from (1.33) by inserting $U(\lambda)$ as given by (1.40). It does not matter that the pole of $U(\lambda)$ is at $\lambda=0$; the replacement

$$U(\lambda)=\frac{S_a A^a}{\lambda-c} \tag{1.53}$$

leads to the same result. Such a $U(\lambda)$ belongs to $C_+^*(\mathfrak{g})$, and the corresponding coefficients $u_{a,k}$ are nonzero for all $k\leqslant 0$ and are related by

$$u_{a,k-1}=\frac{1}{c}\,u_{a,k}\,, \tag{1.54}$$

which follows from expanding $(\lambda-c)^{-1}$ into a geometric progression. The latter relations are invariant under the coadjoint action of $C_0(\mathfrak{g})$. The same is true for

$$U(\lambda)=\sum_{i=1}^{N}\sum_{k=0}^{n_i}\frac{S_{a,k}^{(i)}A^a}{(\lambda-c_i)^{k_i+1}}\,. \tag{1.55}$$

So, generating functions of the type (1.55) form a Poisson submanifold in $C^*(\mathfrak{g})$ parametrized by the coordinates $S_{a,k}^{(i)}$. In these coordinates the Poisson brackets (1.33) become

$$\{S_{a,k}^{(i)},S_{b,l}^{(j)}\}=\begin{cases}-C_{ab}^{c}\,\delta^{ij}S_{c,k+l}^{(i)} & \text{if } k+l\leqslant n_i\,,\\ 0 & \text{otherwise}\end{cases} \tag{1.56}$$

and are just the Lie-Poisson brackets of a finite-dimensional Lie algebra, the direct sum of the algebras $C_{n_i+1,0}(\mathfrak{g})$ over all poles.

We have already come across the matrices $U(x,\lambda)$ of the form (1.55) in §§ I.6–I.7 when studying the general solution of the zero curvature equation. Here they have appeared as a result of general Lie-algebraic considerations, and the related integrable models have been endowed with a natural Poisson structure.

Thus, in this section we outlined a general construction of the matrices $U(x,\lambda)$ satisfying the fundamental Poisson brackets with the r-matrix (1.31) and explained their geometric origin. By repeating the reasoning of Part I, with every $U(x, \lambda)$ one can associate a family of integrable models. Specifically, consider the auxiliary linear problem

$$\frac{dF}{dx} = U(x, \lambda) F \tag{1.57}$$

and its monodromy matrix

$$T(\lambda) = \overset{\curvearrowleft}{\exp} \int_{-L}^{L} U(x, \lambda)\, dx \tag{1.58}$$

(where for definiteness the periodic boundary conditions are assumed). The functionals $\operatorname{tr} T(\lambda)$ and other algebraic invariants of $T(\lambda)$ form an involutive family, and Hamilton's equations of motion induced by any functional of this family admit a zero curvature representation. The geometric meaning of these constructions will be clarified in § 4.

§ 2. Trigonometric and Elliptic r-Matrices and the Related Fundamental Poisson Brackets

In the preceding section, for any semi-simple Lie algebra we introduced an *r*-matrix,

$$r(\lambda) = \frac{\Pi}{\lambda} = \frac{K^{ab} X_a \otimes X_b}{\lambda}. \tag{2.1}$$

(Here we take the liberty of using the term *r*-matrix also for an element $r(\lambda)$ of $\mathfrak{g} \otimes \mathfrak{g}$). This is an extension of the *r*-matrix for the NS and HM models (see § III.1 of Part I and § II.3) that corresponds to the Lie algebra $\mathfrak{g} = su(2)$ and has the form $\frac{P}{\lambda}$ where P is the permutation matrix in $\mathbb{C}^2 \otimes \mathbb{C}^2$. Still, in other cases such as the SG or LL model (see § II.6 and § II.8) we encounter more complicated *r*-matrices which depend on the spectral parameter λ through trigonometric or elliptic functions, respectively. It is natural to call (2.1) a *rational* *r*-matrix and consider the *r*-matrices of the SG or LL model as examples of *trigonometric* or *elliptic* *r*-matrices. In § 1 we saw that rational *r*-matrices determine the structure of the Lie algebra $C_0(\mathfrak{g})$. Now the problem is to describe trigonometric and elliptic *r*-matrices and find their geometric interpretation, which will be our concern in this section.

We begin by constructing a vast family of such r-matrices. The basic functional equations are

$$r_{12}(-\lambda) = -r_{21}(\lambda) \tag{2.2}$$

and

$$[r_{12}(\lambda-\mu), r_{13}(\lambda)+r_{23}(\mu)]+[r_{13}(\lambda), r_{23}(\mu)]=0, \tag{2.3}$$

which ensure the skew-symmetry and the Jacobi identity for the fundamental Poisson brackets (see § III.1, Part I). The subscripts 12, 21, 13, 23 indicate a specific imbedding of r from $\mathfrak{g}\otimes\mathfrak{g}$ into $\mathfrak{g}\otimes\mathfrak{g}\otimes\mathfrak{g}$ (cf. the analogous matrix notation in § III.1, Part I). Obviously, these relations hold for the r-matrix (2.1), with (2.3) being equivalent to the Jacobi identity for the structure constants C^c_{ab}.

The remarkable property of (2.3) is that *it allows averaging over a lattice in the complex λ-plane.* More precisely, let θ be an automorphism of a semisimple Lie algebra $\mathfrak{g}$ of finite order q, $\theta^q = I$, and let $\Lambda_1 = \{n\omega, -\infty < n < \infty\}$ be a one-dimensional lattice in $\mathbb{C}$ with generator ω. Let the action of the additive group of translations Λ_1 on the r-matrix (2.1) be defined by

$$r(\lambda) \to r^{(n)}(\lambda) = (\theta^n \otimes I)\, r(\lambda - n\omega) = (I \otimes \theta^{-n})\, r(\lambda - n\omega). \tag{2.4}$$

The last equation in (2.4) reflects the invariance of the r-matrix (2.1) under the diagonal action of θ,

$$(\theta \otimes \theta)\, r(\lambda) = r(\lambda), \tag{2.5}$$

which is an obvious consequence of

$$(\theta \otimes \theta)\Pi = \Pi. \tag{2.6}$$

Let

$$r^{\Lambda_1}(\lambda) = \sum_{n=-\infty}^{\infty} r^{(n)}(\lambda) \tag{2.7}$$

be the result of averaging the r-matrix (2.1) over Λ_1. The averaged r-matrix, $r^{\Lambda_1}(\lambda)$, is quasi-periodic

$$r^{\Lambda_1}(\lambda+\omega) = (\theta \otimes I)\, r^{\Lambda_1}(\lambda) = (I \otimes \theta^{-1})\, r^{\Lambda_1}(\lambda), \tag{2.8}$$

satisfies (2.2) and at first sight also seems to satisfy (2.3). In fact, replace λ in (2.3) by $\lambda - n\omega$, μ by $\mu - m\omega$, and act on the right hand side by the automorphism $\theta^n \otimes \theta^m \otimes I$. Using (2.5), we find

$$[r_{12}^{(n-m)}(\lambda-\mu), r_{13}^{(n)}(\lambda)+r_{23}^{(m)}(\mu)]+[r_{13}^{(n)}(\lambda), r_{23}^{(m)}(\mu)]=0, \tag{2.9}$$

so the relation (2.3) for $r^{\Lambda_1}(\lambda)$ results from summing over n and m.

However, the argument is too naive and in general incorrect. The point is that the series (2.7) converges only as a principal value series,

$$\text{p.v.} \sum_{n=-\infty}^{\infty} = \lim_{N\to\infty} \sum_{n=-N}^{N}, \tag{2.10}$$

so that one cannot replace summation over n and m by summation over $n-m$ and m or over $n-m$ and n.

Let us see what the requirements on θ are in order that (2.7) *do satisfy* (2.3). Using

$$\text{p.v.} \sum_{n=-\infty}^{\infty} \frac{1}{\lambda-n\omega} = \frac{\pi}{\omega} \operatorname{ctg} \frac{\pi\lambda}{\omega}, \tag{2.11}$$

we find for (2.7) the expression

$$r^{\Lambda_1}(\lambda) = \frac{\pi}{q\omega} \sum_{k=0}^{q-1} \operatorname{ctg} \frac{\pi(\lambda-k\omega)}{q\omega} (\theta^k \otimes I)\Pi, \tag{2.12}$$

so that $r^{\Lambda_1}(\lambda)$ is indeed quasi-periodic in the sense of (2.8).

We now proceed with (2.3) and denote its left hand side by $\Phi(\lambda,\mu)$:

$$\Phi(\lambda,\mu)=[r_{12}^{\Lambda_1}(\lambda-\mu), r_{13}^{\Lambda_1}(\lambda)+r_{23}^{\Lambda_1}(\mu)]+[r_{13}^{\Lambda_1}(\lambda), r_{23}^{\Lambda_1}(\mu)]. \tag{2.13}$$

Consider $\Phi(\lambda,\mu)$ as a function of λ for μ fixed, $\mu \not\equiv 0 (\operatorname{mod} \Lambda_1)$. It satisfies the quasi-periodicity condition

$$\Phi(\lambda+\omega,\mu)=(\theta\otimes I\otimes I)\Phi(\lambda,\mu) \tag{2.14}$$

and may have only simple poles at $\lambda\equiv\mu(\operatorname{mod}\Lambda_1)$ and $\lambda\equiv 0(\operatorname{mod}\Lambda_1)$. We will show that $\Phi(\lambda,\mu)$ is an entire function of λ. In fact, the residue at $\lambda=\mu$ is $[\Pi_{12}, r_{13}^{\Lambda_1}(\mu)+r_{23}^{\Lambda_1}(\mu)]$ and vanishes in view of

$$[\Pi, A\otimes I+I\otimes A]=0 \tag{2.15}$$

(see (1.30)). The case $\lambda=0$ is treated in a similar way; here one should also make use of (2.2). Next, $\Phi(\lambda,\mu)$ is bounded, so that the Liouville theorem yields

$$\Phi(\lambda,\mu)=\Phi(\pm i\infty,\mu)=-\frac{\pi^2}{\omega^2}[\mathscr{P}_{12},\mathscr{P}_{23}]\pm\frac{\pi i}{\omega}[\mathscr{P}_{12}+\mathscr{P}_{13},r_{23}^{\Lambda_1}(\mu)], \tag{2.16}$$

with

$$\mathscr{P}=\frac{1}{q}\sum_{k=0}^{q-1}(\theta^k\otimes I)\Pi. \tag{2.17}$$

This implies

$$[\mathscr{P}_{12}+\mathscr{P}_{13},r_{23}^{\Lambda_1}(\mu)]=0 \tag{2.18}$$

and

$$\Phi(\lambda,\mu)=-\frac{\pi^2}{\omega^2}[\mathscr{P}_{12},\mathscr{P}_{23}]. \tag{2.19}$$

Thus we have shown that (2.12) satisfies (2.3) if

$$[\mathscr{P}_{12},\mathscr{P}_{23}]=0. \tag{2.20}$$

This is the desired necessary condition on θ. As is easily seen, this is equivalent to

$$[\tilde{X}_a,\tilde{X}_b]=0 \tag{2.21}$$

for any generators X_a, X_b, where $\tilde{X}$ denotes the average $\tilde{X}=\frac{1}{q}(I+\theta+\ldots+\theta^{q-1})X$ invariant under θ. *Condition* (2.21) *amounts to saying that the subalgebra* $\mathfrak{h}$ *of fixed points of* θ *is abelian.*

We thus obtain a new family of r-matrices of the type (2.12) parametrized by a one-dimensional lattice Λ_1 and an automorphism θ of finite order whose fixed subalgebra is abelian. Formula (2.12) shows that it is natural to refer to such r-matrices as *trigonometric*.

Another family of r-matrices is obtained by averaging over a two-dimensional lattice $\Lambda_2=\left\{n_1\omega_1+n_2\omega_2;\ \operatorname{Im}\frac{\omega_2}{\omega_1}>0,\ n_1,n_2=-\infty,\ldots,\infty\right\}$. Let θ_1 and θ_2 be automorphisms of $\mathfrak{g}$ of order q_1 and q_2, respectively, and let $r^{\Lambda_2}(\lambda)$ be the averaged r-matrix

$$r^{\Lambda_2}(\lambda)=\sum_{n_1=-\infty}^{\infty}\sum_{n_2=-\infty}^{\infty}(\theta_1^{n_1}\theta_2^{n_2}\otimes I)r(\lambda-n_1\omega_1-n_2\omega_2). \tag{2.22}$$

In order for the "naive proof" of (2.3) based on (2.9) to be valid, these automorphisms must commute. However, the series (2.22) should be regarded in the sense of (2.10). By repeating the above derivation of (2.3) which uses the Liouville theorem it can be shown that θ_1 and θ_2 must have no fixed points in common. Such pairs of automorphisms exist only for Lie algebras of type A_{n-1}, i. e. for $\mathfrak{g}=su(n)$, and then $q_1=q_2=n$. In the fundamental (vector) representation of $su(n)$ we have, up to an inner automorphism,

$$\theta_i \xi = T_i \xi T_i^{-1}, \qquad i=1,2, \tag{2.23}$$

where T_1 and T_2 are $n\times n$ matrices with entries

$$(T_1)_{kl}=\zeta^k \delta_{kl}, \qquad (T_2)_{kl}=\delta_{k+1,l} \tag{2.24}$$

and ξ is a primitive n-th root of unity, $k, l=1, \ldots, n$, $\delta_{k+n,l}=\delta_{k,l+n}=\delta_{k,l}$.

So the corresponding r-matrix may be fully characterized as a meromorphic matrix function with values in the complexification of $su(n)\times su(n)$, i. e. in $sl(n,\mathbb{C})\times sl(n,\mathbb{C})$, satisfying the quasi-periodicity conditions

$$r^{\wedge_2}(\lambda+\omega_i)=(T_i\otimes I)r^{\wedge_2}(\lambda)(T_i^{-1}\otimes I), \qquad i=1,2 \tag{2.25}$$

and the requirement

$$r^{\wedge_2}(\lambda)=\frac{\Pi}{\lambda}+O(1) \tag{2.26}$$

as $\lambda\to 0$. Its matrix entries are elliptic functions with period lattice $n\Lambda_2$ and simple poles at the points of Λ_2. The r-matrices of this kind are naturally called *elliptic*.

We shall consider the simplest examples that correspond to the Lie algebra $su(2)$ in the fundamental representation. Let the generator of the one-dimensional lattice Λ_1 be $\omega=\pi$ and let an automorphism θ be defined by

$$\theta\xi=\sigma_3\xi\sigma_3. \tag{2.27}$$

As follows from (2.12), the associated trigonometric r-matrix is

$$r(\lambda)=\frac{1}{2\sin\lambda}(\sigma_1\otimes\sigma_1+\sigma_2\otimes\sigma_2+\cos\lambda\,\sigma_3\otimes\sigma_3). \tag{2.28}$$

This r-matrix has already occurred in the study of the partially anisotropic HM model and the SG model (see § II.8 and § II.6; in the latter case λ must be replaced by $i\alpha$).

Let the generators of the two-dimensional lattice Λ_2 be $\omega_1=2K$, $\omega_2=2iK'$ where K and K' are the complete elliptic integrals of moduli k and $k'=\sqrt{1-k^2}$, respectively; we set

$$T_1=\sigma_3, \quad T_2=\sigma_1. \tag{2.29}$$

The associated elliptic r-matrix is

$$r(\lambda)=\frac{1}{2\,\mathrm{sn}(\lambda,k)}(\sigma_1\otimes\sigma_1+\mathrm{dn}(\lambda,k)\sigma_2\otimes\sigma_2+\mathrm{cn}(\lambda,k)\sigma_3\otimes\sigma_3), \tag{2.30}$$

which can be verified by summing the series (2.22). Of course, relations (2.25)–(2.26) follow immediately from the expression for $r(\lambda)$. This r-matrix occurred in the description of the LL model (see § II.8).

We thus have shown that all the r-matrices that appear in specific models are contained in one of the three families described above: the rational r-matrices having a Lie-algebraic interpretation and trigonometric and elliptic ones resulting from rational r-matrices by averaging.

Each of these r-matrices gives rise to the fundamental Poisson brackets

$$\{U(\lambda)\underset{,}{\otimes} U(\mu)\}=[r(\lambda-\mu), U(\lambda)\otimes I+I\otimes U(\mu)], \tag{2.31}$$

where we have again suppressed the x-dependence. The element $U(\lambda)$ of $\mathfrak{g}$ must satisfy the quasi-periodicity conditions

$$U(\lambda+\omega)=\theta\, U(\lambda) \tag{2.32}$$

in the trigonometric case and

$$U(\lambda+\omega_i)=\theta_i\, U(\lambda), \quad i=1,2, \tag{2.33}$$

in the elliptic case. *A natural method for constructing such a $U(\lambda)$ is to apply the averaging procedure to elements $U(\lambda)$ which belong to a finite dimensional phase space in the rational case.* A large variety of examples of this kind was exhibited in § 1. However, the requirement of convergence of the corresponding series imposes additional constraints on the choice of a rational $U(\lambda)$.

The most representative example is provided by

$$U(\lambda)=\sum_{i=1}^{N}\sum_{k=0}^{n_i}\frac{S_{a,k}^{(i)}A^a}{(\lambda-c_i)^{k+1}}. \tag{2.34}$$

Setting for such $U(\lambda)$

$$U^{\Lambda}(\lambda)=\sum_{n=-\infty}^{\infty}\theta^n\, U(\lambda-n\omega) \tag{2.35}$$

in the trigonometric case and

$$U^{\Lambda_2}(\lambda)=\sum_{n_1=-\infty}^{\infty}\sum_{n_2=-\infty}^{\infty}\theta_1^{n_1}\theta_2^{n_2}U(\lambda-n_1\omega_1-n_2\omega_2) \tag{2.36}$$

in the elliptic case (with the summation convention (2.10)) we find that both $U^{\Lambda_1}(\lambda)$ and $U^{\Lambda_2}(\lambda)$ satisfy the fundamental Poisson brackets with the r-matrices $r^{\Lambda_1}(\lambda)$ and $r^{\Lambda_2}(\lambda)$ respectively. In fact, setting

$$U^{(n)}(\lambda)=\theta^n\, U(\lambda-n\omega) \tag{2.37}$$

we get from (2.31)

$$\{U^{(n)}(\lambda)\underset{,}{\otimes} U^{(m)}(\mu)\}=[r^{(n-m)}(\lambda-\mu),\, U^{(n)}(\lambda)\otimes I+I\otimes U^{(m)}(\mu)], \tag{2.38}$$

and summing over n and m we conclude that $U^{\Lambda_1}(\lambda)$ satisfies the fundamental Poisson brackets with the r-matrix $r^{\Lambda_1}(\lambda)$. For a rigorous proof one should compare the poles of the left and right hand sides of the fundamental Poisson brackets and make use of the Liouville theorem. The elliptic case is analyzed in a similar manner.

The simplest illustration of this construction is the matrix $U^{\mathrm{LL}}(\lambda)$ for the LL model obtained by averaging the corresponding matrix $U^{\mathrm{HM}}(\lambda)$ for the HM model,

$$U^{\mathrm{HM}}(\lambda)=\frac{iS_a\sigma_a}{\lambda} \tag{2.39}$$

(see (1.42)). We have

$$\begin{aligned}
U^{\mathrm{LL}}(\lambda) &= \sum_{n_1=-\infty}^{\infty}\sum_{n_2=-\infty}^{\infty}\sigma_3^{n_1}\sigma_1^{n_2}U^{\mathrm{HM}}(\lambda-2n_1K-2in_2K')\sigma_1^{n_2}\sigma_3^{n_1}\\
&= i\sum_{n_1=-\infty}^{\infty}\sum_{n_2=-\infty}^{\infty}\frac{(-1)^{n_1}}{\lambda-2n_1K-2in_2K'}S_1\sigma_1\\
&\quad +i\sum_{n_1=-\infty}^{\infty}\sum_{n_2=-\infty}^{\infty}\frac{(-1)^{n_1+n_2}}{\lambda-2n_1K-2in_2K'}S_2\sigma_2\\
&\quad +i\sum_{n_1=-\infty}^{\infty}\sum_{n_2=-\infty}^{\infty}\frac{(-1)^{n_2}}{\lambda-2n_1K-2in_2K'}S_3\sigma_3\\
&= \frac{i}{\operatorname{sn}(\lambda,k)}(S_1\sigma_1+\operatorname{dn}(\lambda,k)S_2\sigma_2+\operatorname{cn}(\lambda,k)S_3\sigma_3).
\end{aligned} \tag{2.40}$$

In the trigonometric case we obviously obtain the matrix $U(\lambda)$ of the partially anisotropic HM model.

The matrix $U(\alpha)$ of the SG model

$$U(\alpha)=\operatorname{ch}\alpha\, S_1\sigma_1+\operatorname{sh}\alpha\, S_2\sigma_2+S_3\sigma_3, \tag{2.41}$$

where

$$S_1=\frac{m}{2i}\sin\frac{\beta\varphi}{2}, \quad S_2=\frac{m}{2i}\cos\frac{\beta\varphi}{2}, \quad S_3=\frac{\beta\pi}{4i}, \tag{2.42}$$

is an example of a quasi-periodic matrix $U(\alpha)$ (with $\omega=i\pi$ and θ given by (2.27)) satisfying the fundamental Poisson brackets with the r-matrix (2.28) (with $\lambda=i\alpha$), which cannot be obtained by a straightforward averaging procedure. It can be derived, however, by contracting the matrix $U^{\Lambda_1}(\lambda)$, which in turn is obtained by averaging a double-pole matrix $U(\lambda)$,

$$U(\lambda)=\frac{S_a^{(1)}\sigma_a}{\lambda-c}+\frac{S_a^{(2)}\sigma_a}{\lambda+c}. \tag{2.43}$$

We may therefore claim that the averaging procedure enables us to build up a classification of continuous models with finite number of degrees of freedom for fixed x, described by trigonometric or elliptic r-matrices, if contraction of phase space is also allowed.

We thus have a scheme for constructing the matrix $U(x,\lambda)$ of the auxiliary linear problem satisfying the fundamental Poisson brackets with an r-matrix from the rational, trigonometric, or elliptic families. In the rational case $U(x,\lambda)$ ranges over an orbit (of finite dimension for fixed x) of the coadjoint action of the Lie algebra $\mathscr{C}((\mathfrak{g}))$. Hence the above scheme may be regarded as a method for classifying the associated integrable models. In the trigonometric and elliptic cases we have outlined an averaging procedure which enables us to produce a large family of quasi-periodic matrices $U(x,\lambda)$.

In the next section we shall extend the discussion to lattice models.

§ 3. Fundamental Poisson Brackets on the Lattice

Our analysis of lattice models in Chapter III was based on the auxiliary linear problem

$$F_{n+1}=L_n(\lambda)F_n \tag{3.1}$$

with the matrix $L_n(\lambda)$ satisfying the fundamental Poisson brackets

$$\{L_n(\lambda) \otimes L_m(\mu)\} = [r(\lambda-\mu), L_n(\lambda) \otimes L_n(\mu)]\, \delta_{nm}. \tag{3.2}$$

For (3.2) to be compatible, the r-matrix should satisfy (2.2)–(2.3). These fundamental lattice Poisson brackets have no simple Lie-algebraic interpretation. Still, the corresponding r-matrices have been already described in connection with continuous models. We shall therefore discuss here only the choice of finite-dimensional phase spaces for fixed n, and the associated $L_n(\lambda)$. The index n will be suppressed in what follows.

Let us first consider rational r-matrices and concentrate for simplicity on the Lie algebra $\mathfrak{g} = su(N)$ in the fundamental representation. The corresponding r-matrix (2.1) (up to an irrelevant unit summand) is given by

$$r(\lambda) = \frac{P}{\lambda}, \tag{3.3}$$

where P is the permutation matrix in $\mathbb{C}^N \otimes \mathbb{C}^N$. The simplest matrix satisfying (3.2) is

$$L(\lambda) = I + \frac{S_a A^a}{\lambda}, \tag{3.4}$$

where the dynamical variables S_a have the Lie-Poisson brackets of the Lie algebra $su(N)$,

$$\{S_a, S_b\} = -C^c_{ab} S_c, \tag{3.5}$$

and may be restricted to an orbit. The verification of (3.2) is straightforward: in addition to (1.30) one should use

$$[P, A \otimes A] = 0. \tag{3.6}$$

Notice that (3.4) is the naive lattice version of the operator L of the continuous auxiliary linear problem,

$$L = \frac{d}{dx} - \frac{S_a \sigma_a}{\lambda}, \tag{3.7}$$

which appeared in the HM model for $\mathfrak{g} = su(2)$ (see § 1). Thus (3.4) provides a lattice version of the continuous $\mathfrak{g}$-invariant magnet for the case $\mathfrak{g} = su(N)$.

Since (3.2) is multiplicative, a product $L(\lambda)$ of the simplest matrices,

$$L(\lambda) = L^{(1)}(\lambda + c_1) \dots L^{(m)}(\lambda + c_m), \tag{3.8}$$

also satisfies the fundamental Poisson brackets (3.2). Of course, the dynamical variables $S_a^{(i)}$ entering into $L^{(i)}(\lambda)$ are in involution for distinct i. Expression (3.8) is the lattice analogue of the multi-pole matrix $U(\lambda)$ in (1.55) with simple poles. Conversely, by taking various continuum limits of (3.8) one can obtain a rational matrix $U(\lambda)$ of the general type, in particular with multiple poles.

We thus have constructed a large set of matrices $L(\lambda)$ that describe integrable lattice models and are associated with rational r-matrices for the Lie algebra $\mathfrak{g}=su(N)$. The construction of $L(\lambda)$ satisfying the fundamental Poisson brackets (3.2) with rational r-matrices for other classical series of Lie algebras is more difficult and will not concern us here.

We shall now turn to the fundamental Poisson brackets (3.2) with trigonometric and elliptic r-matrices. Here we can only suggest to look for suitable combinations generalizing (3.4) and satisfying the quasi-periodicity conditions

$$L(\lambda+\omega_i)=\theta_i L(\lambda), \qquad i=1,2. \tag{3.9}$$

In the case of $\mathfrak{g}=su(2)$ and the fundamental representation, we already know the simplest matrix $L(\lambda)$: this is the one for the LLL model (see § III.5). Its trigonometric degenerate case gives the matrix $L(\lambda)$ for the partially anisotropic LHM model, or for the LSG model if one takes another real form. Similar expressions exist for $\mathfrak{g}=su(N)$ for any N, but we will not write them explicitly. Knowing the simplest $L(\lambda)$'s we may construct more complicated ones using (3.8).

Notice that the averaging procedure of § 2 introduced for continuous models can also be extented to the lattice case. Let $L(\lambda)$ satisfy the fundamental Poisson brackets (3.2) with a rational r-matrix. Setting

$$L^{(n)}(\lambda)=\theta^n L(\lambda-n\omega), \tag{3.10}$$

we can write (3.2) as

$$\{L^{(n)}(\lambda) \underset{,}{\otimes} L^{(m)}(\mu)\}=[r^{(n-m)}(\lambda-\mu), L^{(n)}(\lambda)\otimes L^{(m)}(\mu)], \tag{3.11}$$

where $r^{(n-m)}(\lambda)$ was defined in (2.4). Taking a formal product of (3.11) over all n and m we come down to relation (3.2) that involves the matrix

$$L^{\wedge}(\lambda)=\prod_{n=-\infty}^{\infty} L^{(n)}(\lambda) \tag{3.12}$$

and the r-matrix $r^{\wedge}(\lambda)$ given by (2.7). This method, however, needs substantial justification which is beyond the purposes of the book.

These few remarks about lattice models is all we wanted to say here. The classes of $L_n(\lambda)$ displayed above include the models discussed in the book and allow for interesting generalizations.

Thus, in this and the two previous sections, the auxiliary linear problems for integrable models and the corresponding fundamental Poisson brackets were discussed from a general standpoint. The most complete geometric interpretation was obtained for continuous models with a rational r-matrix. The associated fundamental Poisson brackets were shown to be generated by a special infinite-dimensional Lie algebra $\mathscr{C}((\mathfrak{g}))$; the corresponding continuous models have the highest degree of symmetry. Continuous models with trigonometric and elliptic r-matrices have a partially broken symmetry and were described in less detail. In particular, the analogue of the algebra $\mathscr{C}_0((\mathfrak{g}))$ was not defined. Finally, when dealing with lattice models, we in fact restricted our analysis to suitable substitutions. Nevertheless, at least for the rational r-matrix associated with $su(N)$, these substitutions in the continuum limit give the whole class of the matrices $U(x, \lambda)$ described in § 1. The role of the Lie algebra $\mathscr{C}_0((\mathfrak{g}))$ is transferred in lattice models to an object which is rather a Lie group, but a detailed discussion of this topic cannot be persued here. This brings us to the end of the exposition of our classification scheme for integrable models.

§ 4. Geometric Interpretation of the Zero Curvature Representation and the Riemann Problem Method

The Riemann problem of factorization of matrix-valued functions played a significant role in our book. Firstly, it was used to solve the inverse problem – the problem of inverting the mapping $\mathscr{F}$ (see §§ II.1-3, II.6 of Part I and §§ II.2, II.5) and so was a constituent part of the method for solving the initial value problem for integrable nonlinear equations. In particular, the zero curvature representation for these equations was a consequence of the Riemann problem formalism. Secondly, in § I.6 the Riemann problem was used to outline a method for constructing a large class of special solutions of the general zero curvature equation, the dressing procedure. In this section the method for solving the initial value problem for integrable nonlinear equations and the dressing procedure will be discussed from a general standpoint. More specifically, we shall outline a geometric scheme which gives rise to Hamilton's equations possessing a rich set of integrals of the motion in involution and admitting a zero curvature representation. The dressing procedure is incorporated in the scheme in a natural way. The key role in the scheme is played by the infinite-dimensional Lie algebra $\mathscr{C}((\mathfrak{g}))$ with two commutators $[\,,]$ and $[\,,]_0$, and its central extension.

The construction will be divided into several steps. First, in Subsection 1 we shall consider a model situation, starting with a finite-dimensional Lie algebra $\mathfrak{g}$. This will serve to introduce and illustrate the geometric techniques for constructing integrable equations and their solution via a factorization problem in the Lie group G. Then in Subsection 2 $\mathfrak{g}$ will be replaced by the infinite-dimensional Lie algebra $\mathscr{C}(\mathfrak{g})$ consisting of functions of x with values in $\mathfrak{g}$, and its central extension. This will give rise to the zero curvature representation and the monodromy matrix of the auxiliary linear problem in a natural way. The final scheme of Subsection 3 will result from replacing $\mathfrak{g}$ in Subsection 2 by the current algebra $C(\mathfrak{g})$ thus leading to the Lie algebra $\mathscr{C}((\mathfrak{g}))$. So the variables x and λ are brought into action in reverse order compared to § 1. The abstract factorization problem of Subsection 1 will become in Subsection 3 the traditional Riemann problem of analytic factorization of matrix-functions.

1. The factorization problem as a method for constructing integrable Hamiltonian equations and their solutions

Let G be a finite-dimensional Lie group such that its Lie algebra $\mathfrak{g}$ can be split into a linear sum of two subalgebras

$$\mathfrak{g}=\mathfrak{g}_{+}+\mathfrak{g}_{-}. \tag{4.1}$$

Let $P_{\pm}$ be the corresponding projection operators,

$$P_{\pm}\mathfrak{g}_{\pm}=\mathfrak{g}_{\pm}, \qquad P_{\pm}\mathfrak{g}_{\mp}=0. \tag{4.2}$$

Let

$$R=\tfrac{1}{2}(P_{+}-P_{-}) \tag{4.3}$$

and define a second Lie structure on $\mathfrak{g}$ with the commutator given by

$$[\xi,\eta]_0=[R\xi,\eta]+[\xi,R\eta], \tag{4.4}$$

where $[,]$ is the original commutator in $\mathfrak{g}$ (cf. § 1). On the phase space $\mathfrak{g}^*$ we consider the Lie-Poisson brackets $\{,\}$ and $\{,\}_0$ associated with the Lie brackets $[,]$ and $[,]_0$, respectively. Let $I(\mathfrak{g})$ denote the annihilator of the Poisson structure $\{,\}$, the algebra of Casimir functions consisting of invariants of the coadjoint action Ad^* of G on $\mathfrak{g}^*$: for $f(u)$ in $I(\mathfrak{g})$ we have

$$f(\mathrm{Ad}^*g\cdot u)=f(u) \tag{4.5}$$

for all g in G.

The remarkable property of the Poisson structure $\{,\}_0$ is that $I(\mathfrak{g})$ *is an involutive algebra with respect to* $\{,\}_0$.

To prove this we shall use an invariant definition of the Lie-Poisson bracket given by (1.2). For any function $f(u)$ on $\mathfrak{g}^*$ let $\nabla f(u)$ denote its gradient which is an element of $\mathfrak{g}$ defined by

$$\nabla f(u) = \frac{\partial f(u)}{\partial u_a} X_a . \tag{4.6}$$

With this notation, (1.2) can be written as

$$\{f_1, f_2\}(u) = -(u, [\nabla f_1(u), \nabla f_2(u)]) \tag{4.7}$$

where u is a point of $\mathfrak{g}^*$ at which the Poisson bracket is evaluated, and $(,)$ is the pairing between $\mathfrak{g}$ and $\mathfrak{g}^*$ (see § 1). In particular, setting $f_1(u) = u_a \xi^a$ and $f_2(u) = f(u)$ with $f(u)$ in $I(\mathfrak{g})$, we find

$$(u, [\nabla f(u), \xi]) = 0 \tag{4.8}$$

for any element $\xi = \xi_a X^a$ of $\mathfrak{g}$. Now the involutive property of the algebra $I(\mathfrak{g})$ is obvious: for any functions $f_1(u)$, $f_2(u)$ in $I(\mathfrak{g})$ we have

$$\{f_1, f_2\}_0(u) = -(u, [R\nabla f_1(u), \nabla f_2(u)]) - (u, [\nabla f_1(u), R\nabla f_2(u)]) = 0, \tag{4.9}$$

since every term on the right hand side vanishes in view of (4.8).

We thus have obtained a large family of functions which are in involution with respect to the Poisson bracket $\{,\}_0$. It is natural to look at Hamilton's equations of motion defined by these functions. They have the form

$$\frac{du_a}{dt} = \{f, u_a\}_0(u), \tag{4.10}$$

where a function f from $I(\mathfrak{g})$ is the Hamiltonian. In view of (4.8) these equations may be written as

$$\frac{du_a}{dt} = -(u, [R\nabla f(u), \nabla u_a]) = (\mathrm{ad}^*(R\nabla f(u)) \cdot u)_a , \tag{4.11}$$

or

$$\left(\frac{du}{dt}, \xi\right) = -(u, [R\nabla f(u), \xi]) \tag{4.12}$$

for all ξ in $\mathfrak{g}$. In the semisimple case this equation can be given a more elegant form: for a point U in $\mathfrak{g}$,

$$U = u_a A^a, \tag{4.13}$$

(cf. (1.32)) it becomes

$$\frac{dU}{dt} = [R\,\nabla f(u), U]. \tag{4.14}$$

The representation of (4.10) in the form (4.12) or (4.14) is of crucial importance. We shall presently see that it leads to a method for solving the initial value problem $u_a(t)|_{t=0} = u_a^0$ for the nonlinear equation (4.10) in terms of a factorization problem in the Lie group G.

Let $G_\pm$ be the subgroups of G associated with the Lie subalgebras $\mathfrak{g}_\pm$. For any g in G sufficiently close to I there is a factorization

$$g = g_+ g_-, \tag{4.15}$$

with $g_\pm$ in $G_\pm$; the decomposition is unique if $g_\pm$ are also supposed to be close to the unit element of G. *The factorization problem (4.15) is the abstract analogue of the Riemann problem.*

We will show how (4.15) allows us to solve the nonlinear equation (4.10). *Consider a one-parameter subgroup of G given by $g(t) = \exp\{-t\,\nabla f(u^0)\}$ and the associated family of factorization problems (for t small enough)*

$$g(t) = g_+(t)\, g_-(t), \tag{4.16}$$

with $g_\pm(t)|_{t=0} = I$. The solution of the equations of motion (4.10) with initial condition u^0 is then given by

$$u(t) = \mathrm{Ad}^* g_+^{-1}(t) \cdot u^0 = \mathrm{Ad}^* g_-(t) u^0, \tag{4.17}$$

or

$$(u(t), \xi) = (u^0, g_+(t)\,\xi\, g_+^{-1}(t)) = (u^0, g_-^{-1}(t)\,\xi\, g_-(t)) \tag{4.18}$$

for all ξ in $\mathfrak{g}$.

The two representations for $u(t)$ in (4.17) or (4.18) coincide due to the relation

$$\mathrm{Ad}^* g(t) \cdot u^0 = u^0, \tag{4.19}$$

whose infinitesimal version is given by (4.8).

For the proof let us differentiate (4.16) with respect to t and write the result as

$$g_+^{-1}(t)\frac{dg}{dt}(t)g^{-1}(t)g_+(t)=g_+^{-1}(t)\frac{dg_+}{dt}(t)+\frac{dg_-}{dt}(t)g_-^{-1}(t). \tag{4.20}$$

Recalling that $\frac{dg}{dt}(t)g^{-1}(t)=-\nabla f(u^0)$ and setting

$$\xi_+(t)=g_+^{-1}\frac{dg_+}{dt}(t), \quad \xi_-(t)=\frac{dg_-}{dt}(t)g_-^{-1}(t), \tag{4.21}$$

we find

$$-g_+^{-1}(t)\nabla f(u^0)g_+(t)=\xi_+(t)+\xi_-(t). \tag{4.22}$$

It is not hard to see that the left hand side of this equation equals $-\nabla f(u(t))$. To show this one must use that $f(u)$ is Ad*-invariant, hence

$$f(u)=f(\mathrm{Ad}^* g_+^{-1}(t)u), \tag{4.23}$$

then differentiate this equation with respect to u and use (4.17). In other words, the gradient of an Ad*-invariant function "transforms by conjugation". As a result, we get the decomposition

$$-\nabla f(u(t))=\xi_+(t)+\xi_-(t), \tag{4.24}$$

so that

$$\xi_\pm(t)=-P_\pm(\nabla f(u(t))). \tag{4.25}$$

Now, differentiating (4.18) with respect to t we find

$$\begin{aligned}\left(\frac{du(t)}{dt},\xi\right)&=\left(u^0,\frac{dg_+}{dt}(t)\,\xi\,g_+^{-1}(t)\right)\\&\quad-\left(u^0,g_+(t)\,\xi\,g_+^{-1}(t)\frac{dg_+}{dt}(t)g_+^{-1}(t)\right)\\&=(u(t),[\xi_+,\xi])\end{aligned} \tag{4.26}$$

and

$$\left(\frac{du(t)}{dt},\xi\right)=-(u(t),[\xi_-,\xi]). \tag{4.27}$$

Since

$$\tfrac{1}{2}(\xi_+(t)-\xi_-(t))=-R\,\nabla f(u(t)), \tag{4.28}$$

by virtue of (4.26)–(4.27) we conclude that $u(t)$ satisfies (4.12) or (4.10).

Thus we have shown that the solution of the nonlinear equation (4.10) reduces to constructing a one-parameter subgroup $g(t)$ and solving the factorization problem (4.16). Both these problems are linear. In other words, in a model finite-dimensional situation we have explained the linearization procedure for Hamilton's equations of special form (4.10). As we have seen, it already contains the principal aspects of the inverse scattering method, i.e. the Hamiltonian structure of the equations of motion, the existence of a set of integrals of the motion in involution, and a method for solving the initial value problem by means of the factorization problem. The scheme is based, in essence, on a single formula that describes the splitting of the original Lie algebra into the sum of two its subalgebras.

2. A central extension of the Lie algebra $\mathscr{C}(\mathfrak{g})$ and the zero curvature equation

Consider the Lie algebra $\mathscr{C}(\mathfrak{g})$ of functions $\xi(x)$ with values in a finite dimensional Lie algebra $\mathfrak{g}$. We shall assume, for definiteness, that $\xi(x)$ satisfies the periodic boundary conditions

$$\xi(x+2L)=\xi(x). \tag{4.29}$$

The algebra $\mathscr{C}(\mathfrak{g})$ is spanned by the generators $X_a(x)$, $-L\leqslant x<L$, with the commutator

$$[X_a(x), X_b(y)]=C^c_{ab}X_c(x)\delta(x-y) \tag{4.30}$$

(cf. (1.34)).

We shall suppose that $\mathfrak{g}$ has an invariant bilinear form $\langle,\rangle$ and introduce a *central extension* $\tilde{\mathscr{C}}(\mathfrak{g})$ *of the Lie algebra* $\mathscr{C}(\mathfrak{g})$ *defined by the Maurer-Cartan 2-cocycle*

$$\omega(\xi,\eta)=\int\limits_{-L}^{L}\left\langle \xi(x), \frac{d\eta}{dx}(x)\right\rangle dx. \tag{4.31}$$

The elements of $\tilde{\mathscr{C}}(\mathfrak{g})$ are representend by pairs $\tilde{\xi}=(\xi(x),\sigma)$ where $\xi(x)$ lies in $\mathscr{C}(\mathfrak{g})$ and σ is a complex number; the commutator is

$$[\tilde{\xi},\tilde{\eta}]=([\xi(x),\eta(x)],\omega(\xi,\eta)). \tag{4.32}$$

The Jacobi identity for (4.32) holds in view of the cocycle property

$$\omega([\xi,\eta],\zeta)+\omega([\zeta,\xi],\eta)+\omega([\eta,\zeta],\xi)=0, \tag{4.33}$$

which follows from (4.31) by integrating by parts and using the invariance of $\langle,\rangle$. Elements of the form $(0,\sigma)$ constitute the center of $\tilde{\mathscr{C}}(\mathfrak{g})$.

The generators of $\tilde{\mathscr{C}}(\mathfrak{g})$ are the $\tilde{X}_a(x)$ and I; their commutation relations are

$$[\tilde{X}_a(x), \tilde{X}_b(y)] = C^c_{ab}\tilde{X}_c(x)\delta(x-y) + K_{ab}\delta'(x-y)I, \tag{4.34}$$

$$[\tilde{X}_a(x), I] = 0 \tag{4.35}$$

(cf. (4.30)) where $\delta'(x-y)$ indicates the derivative of the delta function $\delta(x-y)$ with respect to the argument. Here

$$K_{ab} = \langle X_a, X_b \rangle, \tag{4.36}$$

where the X_a are the generators of $\mathfrak{g}$.

The dual space $\tilde{\mathscr{C}}^*(\mathfrak{g})$ of the Lie algebra $\tilde{\mathscr{C}}(\mathfrak{g})$ is formed by elements $\tilde{u}$ with coordinates $(u_a(x), c)$; the corresponding pairing is given by

$$\tilde{u}(\tilde{\xi}) = (\tilde{u}, \tilde{\xi}) = \int_{-L}^{L} u_a(x)\xi^a(x)\,dx + c\sigma. \tag{4.37}$$

The coadjoint action of the center of $\tilde{\mathscr{C}}(\mathfrak{g})$ is trivial, so that the $\widetilde{\mathrm{ad}}^*$-action of $\tilde{\mathscr{C}}(\mathfrak{g})$ reduces to the action of $\mathscr{C}(\mathfrak{g})$ which will be denoted by the same symbol. By definition, we have

$$\widetilde{\mathrm{ad}}^*\xi\cdot\tilde{u}(x) = \left(C^c_{ab}u_c(x)\xi^b(x) + cK_{ab}\frac{d\xi^b}{dx}, 0\right), \tag{4.38}$$

where $\xi(x) = \xi^a(x)X_a$. If K_{ab} is nondegenerate, which will be assumed from here on, (4.38) can be written in an elegant manner

$$(\widetilde{\mathrm{ad}}^*\xi\cdot U)(x) = c\frac{d\xi(x)}{dx} + [\xi(x), U(x)], \tag{4.39}$$

where $U(x)$ is a $\mathfrak{g}$-valued function associated with $\tilde{u} = (u_a(x), c)$ according to

$$U(x) = u_a(x)A^a, \qquad A^a = K^{ab}X_b \tag{4.40}$$

(cf. (4.14)). The last formula defines an identification of the dual space $\tilde{\mathscr{C}}^*(\mathfrak{g})$ with $\mathscr{C}(\mathfrak{g})$.

The action $\widetilde{\mathrm{ad}}^*$ of the Lie algebra $\mathscr{C}(\mathfrak{g})$ lifts to the action of the Lie group $\mathscr{C}(G)$ consisting of periodic functions $g(x)$ with values in the Lie group G:

$$(\widetilde{\mathrm{Ad}}^*g\cdot U)(x) = c\frac{dg(x)}{dx}g^{-1}(x) + g(x)U(x)g^{-1}(x), \tag{4.41}$$

which is an extension of the usual action Ad* (by conjugation).

It is appropriate to compare the last formula with the gauge transformation introduced in § I.2 of Part I. The comparison shows that the element $\tilde{u}=(u_a(x), c)$ is conveniently associated with the differential operator

$$L=c\frac{d}{dx}-U(x). \tag{4.42}$$

Then the $\widetilde{\mathrm{Ad}}^*$*-action of* $\mathscr{C}(G)$ *can be expressed as*

$$\widetilde{\mathrm{Ad}}^* g\cdot L=g(x)Lg^{-1}(x), \tag{4.43}$$

where the right hand side is understood to be the composition of multiplication operators by the functions $g(x)$ and $g^{-1}(x)$ with the differential operator L. The monodromy "matrix"

$$T(\tilde{u})=\widehat{\exp}\frac{1}{c}\int_{-L}^{L} U(x)\,dx \tag{4.44}$$

is a G-valued functional on $\mathscr{C}^*(\mathfrak{g})$; its transformation under $\widetilde{\mathrm{Ad}}^*$ is given by $T\to g(L)Tg^{-1}(L)$ (where we have used that $g(x)$ is periodic). *Hence the invariants of the finite-dimensional action* Ad *of* G

$$\mathrm{Ad}\,g\cdot T=gTg^{-1} \tag{4.45}$$

are invariant under the action $\widetilde{\mathrm{Ad}}^*$ *of* $\mathscr{C}(G)$ *and generate the algebra* $I(\tilde{\mathscr{C}}(\mathfrak{g}))$ *of Casimir functions of* $\tilde{\mathscr{C}}(\mathfrak{g})$. In fact, if the monodromy matrices $T(\tilde{u}_1)$ and $T(\tilde{u}_2)$ are conjugate to one another in G,

$$T(\tilde{u}_2)=g\,T(\tilde{u}_1)g^{-1}, \tag{4.46}$$

then

$$g(x)=F_2(x)gF_1^{-1}(x), \tag{4.47}$$

where

$$L_1F_1(x)=0, \qquad L_2F_2(x)=0 \tag{4.48}$$

and

$$F_1(x)|_{x=-L}=F_2(x)|_{x=-L}=I, \tag{4.49}$$

is a periodic function and so belongs to $\mathscr{C}(G)$, and

$$\widetilde{\mathrm{Ad}}^* g\cdot U_1(x)=U_2(x). \tag{4.50}$$

Suppose now that the Lie algebra $\mathfrak{g}$ admits the splitting (4.1). Then in the reasoning of Subsection 1 $\mathfrak{g}$ may be replaced by $\mathscr{C}(\mathfrak{g})$. More precisely, in accord with (4.1) we split $\mathscr{C}(\mathfrak{g})$ into a linear sum of two subalgebras,

$$\mathscr{C}(\mathfrak{g}) = \mathscr{C}_+(\mathfrak{g}) + \mathscr{C}_-(\mathfrak{g}), \tag{4.51}$$

where $\mathscr{C}_\pm(\mathfrak{g}) = \mathscr{C}(\mathfrak{g}_\pm)$, and define a second Lie bracket on $\mathscr{C}(\mathfrak{g})$ by

$$\begin{aligned}&[(\xi(x), \sigma), (\eta(x), \tau)]_0 \\ &= ([\xi_+(x), \eta_+(x)] - [\xi_-(x), \eta_-(x)], \omega(\xi_+, \eta_+) - \omega(\xi_-, \eta_-)),\end{aligned} \tag{4.52}$$

where $\xi(x) = \xi_+(x) + \xi_-(x)$, $\eta(x) = \eta_+(x) + \eta_-(x)$. The corresponding Lie algebra will be denoted by $\mathscr{C}_0(\mathfrak{g})$. It is obtained from the Lie algebra $\mathscr{C}_0(\mathfrak{g})$ with the commutator

$$[\xi(x), \eta(x)]_0 = [R\,\xi(x), \eta(x)] + [\xi(x), R\eta(x)] \tag{4.53}$$

(cf. (4.4)), where the operator R is determined by the splitting (4.51) according to (4.3), by means of a central extension with the 2-cocycle

$$\omega_0(\xi, \eta) = \omega(R\xi, \eta) + \omega(\xi, R\eta). \tag{4.54}$$

Now we define the Lie-Poisson brackets $\{,\}_0$ on the phase space $\mathscr{C}^*(\mathfrak{g})$ and consider Hamilton's equations of motion

$$\frac{\partial \tilde{u}}{\partial t} = \{f, \tilde{u}\}_0, \tag{4.55}$$

with $\tilde{u} = (u_a(x), c)$ and $f(\tilde{u})$ lying in $I(\mathscr{C}(\mathfrak{g}))$. Comparing the general formula (4.11) of Subsection 1 with (4.39) we find that in terms of $U(x) = u_a(x)A^a$ equation (4.55) becomes

$$\frac{\partial U(x)}{\partial t} = c\,\frac{\partial V(x)}{\partial x} + [V(x), U(x)], \tag{4.56}$$

$$\frac{dc}{dt} = 0, \tag{4.57}$$

with

$$V(x) = R\,\nabla f(u) \tag{4.58}$$

and

$$\nabla f(u)=\frac{\delta f}{\delta u_a(x)}X_a\,. \tag{4.59}$$

We see that the phase space $\mathscr{C}^*(\mathfrak{g})$ *stratifies into the Poisson submanifolds* $c=\text{const}$. *On the reduced phase space* $\mathscr{C}^*(\mathfrak{g})$ *with* $c=1$ *Hamilton's equation (4.55) turns into the zero curvature equation. Its integrals of the motion are given by elements of the Casimir algebra* $I(\tilde{\mathscr{C}}(\mathfrak{g}))$, *i.e. by invariants of the monodromy matrix (4.44) (for* $c=1$). *The functional dimension of* $I(\mathscr{C}(\mathfrak{g}))$ *is equal to the dimension of the Cartan subalgebra of* $\mathfrak{g}$.

This closes our description of the general scheme which gives rise to Hamilton's equations possessing integrals of the motion in involution and admitting a zero curvature representation. However, the previously discussed zero curvature representations connected with integrable models involve the matrices U and V that depend not only on x and t but also on the spectral parameter λ. This is a specialization of the general scheme for the case when the current algebra $C(\mathfrak{g})$ (see § 1) is taken in place of the Lie algebra $\mathfrak{g}$.

3. Realization of the general scheme for the Lie algebra $\mathscr{C}((\mathfrak{g}))$; the Riemann problem and a family of Poisson structures

Let us replace the Lie algebra $\mathfrak{g}$ in the discussion of Subsection 2 by the current algebra $C(\mathfrak{g})$ with the splitting (see § 1)

$$C(\mathfrak{g})=C_+(\mathfrak{g})+C_-(\mathfrak{g})\,. \tag{4.60}$$

To be able to define the corresponding current group, we have to modify the definition given in § 1. Specifically, we shall suppose $C(\mathfrak{g})$ to consist of functions $\xi(\lambda)$ with values in $\mathfrak{g}$ which are analytic in $\mathbb{C}\backslash\{0\}$. The subalgebras $C_+(\mathfrak{g})$ and $C_-(\mathfrak{g})$ will then consist of functions $\xi_+(\lambda)$ and $\xi_-(\lambda)$ analytic in $\mathbb{C}$ or $(\mathbb{C}\backslash\{0\})\cup\{\infty\}$, respectively, with $\xi_-(\infty)=0$. (Other definitions of $C(\mathfrak{g})$ and $C_\pm(\mathfrak{g})$ are also possible, one of them will be encountered in § 5.) The Lie groups $C(G)$ and $C_\pm(G)$ consist of G-valued functions $g(\lambda)$ and $g_\pm(\lambda)$ analytic in their respective domains with the normalization condition $g_-(\infty)=I$. The factorization problem

$$g(\lambda)=g_+(\lambda)g_-(\lambda) \tag{4.61}$$

in the Lie group $C(G)$ may be interpreted as a Riemann problem for a contour which separates the points $\lambda=0$ and $\lambda=\infty$.

Every invariant bilinear form $\langle\,,\rangle$ *on* $\mathfrak{g}$ *gives rise to an infinite family of invariant forms* $\langle\,,\rangle_p$ *on the current algebra* $C(\mathfrak{g})$ *given by*

$$\langle\xi,\eta\rangle_p=\operatorname{Res}\lambda^p\langle\xi(\lambda),\eta(\lambda)\rangle,\quad p=-\infty,\dots,\infty\,. \tag{4.62}$$

Each such form gives rise to a 2-cocycle on the Lie algebra $\mathscr{C}((\mathfrak{g})) = \mathscr{C}(C(\mathfrak{g}))$,

$$\omega_p(\xi, \eta) = \int_{-L}^{L} \operatorname{Res} \lambda^p \left\langle \xi(\lambda, x), \frac{d}{dx} \eta(\lambda, x) \right\rangle dx, \tag{4.63}$$

which defines its central extension $\tilde{\mathscr{C}}_p((\mathfrak{g}))$. The Lie algebra $\tilde{\mathscr{C}}_p((\mathfrak{g}))$ is spanned by the generators $X_{a,k}(x)$ and I with the commutator

$$[X_{a,k}(x), X_{b,l}(y)]_p = C_{ab}^c X_{c,k+l} \delta(x-y) + \delta_{k+l,-p-1} K_{ab} \delta'(x-y) I, \tag{4.64}$$

$$[X_{a,k}(x), I]_p = 0. \tag{4.65}$$

The splitting (4.60) allows us to introduce a second structure of a Lie algebra on $\tilde{\mathscr{C}}((\mathfrak{g}))$ and a related *family of Lie-Poisson brackets* $\{,\}_p$ on the reduced dual space $\mathscr{C}^*((\mathfrak{g}))$ with coordinates $u_{a,k}(x)$:

$$\{u_{a,k}(x), u_{b,l}(y)\}_p = \begin{cases} -C_{ab}^c u_{c,k+l}(x)\delta(x-y) - K_{ab}\delta_{k+l,-p-1}\delta'(x-y) \\ \text{for} \quad k, l \geqslant 0, \\ C_{ab}^c u_{c,k+l}(x)\delta(x-y) + K_{ab}\delta_{k+l,-p-1}\delta'(x-y) \\ \text{for} \quad k, l < 0, \quad 0 \quad \text{otherwise}. \end{cases} \tag{4.66}$$

Of course, the non-ultralocal term $\pm K_{ab}\delta'(x-y)$ appears in only one line on the right hand side of (4.66), namely in the upper one for $p<0$ and in the lower one for $p>0$. For $p=0$ it does not appear at all, and so we recover the ultralocal Poisson brackets of § 1.

Thus, the discussion in § 1 is completely covered by the general scheme. It also becomes clear why it is natural to consider the differential operator L of the auxiliary linear problem and the invariants of the associated monodromy matrix.

In the general case, the coadjoint action $\tilde{\operatorname{ad}}_p^*$ of the Lie algebra $\mathscr{C}((\mathfrak{g}))$ on the reduced phase space $\mathscr{C}^*((\mathfrak{g}))$ is given by

$$\tilde{\operatorname{ad}}_p^* \xi \cdot U(x, \lambda) = \lambda^p \frac{d\xi}{dx}(x, \lambda) + [\xi(x, \lambda), U(x, \lambda)] \tag{4.67}$$

(cf. (4.39)). *Hence Hamilton's equations of motion with a Hamiltonian f from the Casimir algebra* $I(\mathscr{C}((\mathfrak{g})))$

$$\frac{\partial U(x, \lambda)}{\partial t} = \{f, U(x, \lambda)\}_p = \tilde{\operatorname{ad}}^* R \nabla f \cdot U(x, \lambda) \tag{4.68}$$

take the form of a zero curvature equation upon replacing U by

$$U_p(x,\lambda)=\lambda^{-p}\,U(x,\lambda). \tag{4.69}$$

Equations (4.68) for different p are essentially equivalent. More precisely, Hamilton's equation (4.68) for $p=p_1$ with the Hamiltonian f_1 can be written as an equation of the same form for $p=p_2$ with the Hamiltonian f_2 which is simply related to f_1. For instance, if f_1 is given by

$$f_1(U(x,\lambda))=\operatorname{Res}\lambda^N P(T(U(\cdot,\lambda))), \tag{4.70}$$

where P is some invariant of the monodromy matrix $T(U(\cdot,\lambda))$, then f_2 is given by

$$f_2(U(x,\lambda))=\operatorname{Res}\lambda^{N+p_2-p_1} P(T(U(\cdot,\lambda)\lambda^{p_1-p_2})). \tag{4.71}$$

For the proof we notice that since λ is a parameter in the definition of the monodromy matrix (4.44), there is a general expression for ∇f_1,

$$\nabla f_1(U(x,\lambda))=M(x,\lambda)\lambda^N, \tag{4.72}$$

where $M(x,\lambda)$ is a function of x and λ with values in $\mathfrak{g}$ which depends on λ only through $U(x,\lambda)$, $M(x,\lambda)=M(U(\cdot,\lambda),x)$. It follows that

$$\nabla f_2(U(x,\lambda)\lambda^{p_2-p_1})=\nabla f_1(U(x,\lambda)) \tag{4.73}$$

so that the zero curvature equations engendered by the corresponding Hamilton equations (4.68) coincide upon replacing $U(x,\lambda)\to U(x,\lambda)\lambda^{p_2-p_1}$.

The Poisson submanifolds $C^*_{N,M}$ for the Lie-Poisson bracket $\{,\}_0$ defined in § 1 are also Poisson submanifolds for the Lie-Poisson brackets $\{,\}_p$ for $p=-N,\dots,M$. The expressions for the corresponding Lie-Poisson brackets may be obtained from (4.66) by setting $u_{a,k}=0$ for $k\geqslant N$, $k<-M$. Formula (4.69) shows that Hamilton's equation (4.68) expressed as a zero curvature equation is more conveniently studied on the phase space $C^*_{N-p,M+p}$. The corresponding elements $U_p(x,\lambda)$ in this case all have the same structure

$$U_p(x,\lambda)=\sum_{k=-M}^{N-1} u^{(p)}_{a,k}(x)A^a\lambda^{-k-1} \tag{4.74}$$

and the phase spaces can be identified with one another in a natural way. In the next section we shall see for the NS model, where $N=0$, $M=2$, that the brackets $\{,\}_p$ give rise to the Λ-operator and the hierarchy of Poisson structures introduced in § III.5 of Part I.

4. Geometric interpretation of the dressing procedure

The dressing procedure for the general zero curvature equation

$$\frac{\partial U(\lambda)}{\partial t} - \frac{\partial V(\lambda)}{\partial x} + [U(\lambda), V(\lambda)] = 0 \tag{4.75}$$

was explained in § I.6, with the following result. Given the input data $U(x, t, \lambda)$ and $V(x, t, \lambda)$ which are rational functions of λ with values in the Lie algebra $\mathfrak{g}$ of the Lie group G and satisfy (4.75), we were able to construct a new solution, $U^g(\lambda)$ and $V^g(\lambda)$, of (4.75) parametrized by an element g of $C(G)$, the Lie group of functions $g(\lambda)$ on the contour Γ in $\mathbb{C}$ with values in G. To do so we took a solution of the compatible system of equations

$$\frac{\partial F}{\partial x} = U(x, t, \lambda) F, \tag{4.76}$$

$$\frac{\partial F}{\partial t} = V(x, t, \lambda) F \tag{4.77}$$

and solved for each x and t the factorization problem

$$F g F^{-1}(\lambda) = (F g F^{-1})_+(\lambda) (F g F^{-1})_-(\lambda), \tag{4.78}$$

where the functions $h_+(\lambda) = (F g F^{-1})_+(\lambda)$ and $h_-(\lambda) = (F g F^{-1})_-(\lambda)$ have an analytic continuation inside or outside the contour Γ, respectively. After that we set

$$F^g = h_+^{-1} F = h_- F g^{-1}, \tag{4.79}$$

and then $U^g(\lambda)$ and $V^g(\lambda)$ were determined by

$$U^g = \frac{\partial F^g}{\partial x} (F^g)^{-1}, \qquad V^g = \frac{\partial F^g}{\partial t} (F^g)^{-1}. \tag{4.80}$$

This operation preserves the pole divisors of $U(\lambda)$ and $V(\lambda)$ or, in our new terminology, the Poisson submanifolds that have finite dimension for fixed x (see § I.6 and § 1).

Here we shall elucidate the meaning of dressing transformations regarded as transformations in the phase space $\mathscr{C}^*((\mathfrak{g}))$ of functions $U(x, \lambda)$.

It is not hard to see that if two successive transformations are defined by the functions $g_1(\lambda)$ and $g_2(\lambda)$ then

$$F^{g_1 g_2}(x, \lambda) = (F^{g_2})^{g_1}(x, \lambda). \tag{4.81}$$

In fact, using both expressions (4.79) and the elementary relations

$$(g_+f)_-=f_-, \quad (gf_-)_-=g_-f_- \tag{4.82}$$

we find

$$\begin{aligned}(F^{g_2})^{g_1}&=(F^{g_2}g_1F^{g_2^{-1}})_-F^{g_2}g_1^{-1}\\&=((Fg_2F)_+^{-1}Fg_1g_2F^{-1}(Fg_2F^{-1})_-^{-1})_-\cdot(Fg_2F^{-1})_-Fg_2^{-1}g_1^{-1}\\&=(Fg_1g_2F^{-1})_-\cdot F(g_1g_2)^{-1}=F^{g_1g_2}.\end{aligned} \tag{4.83}$$

However, in general there is no such group property for elements U^g of phase space. The point is that the expression for U in terms of F given by (4.76) is invariant with respect to right multiplication $F\to Fg$ which does not commute with the above action of $C(G)$, $F\to F^g$. This may be remedied by fixing the value of $F(x,\lambda)$ at one point x, say, at $x=-L$,

$$F(x,\lambda)|_{x=-L}=I. \tag{4.84}$$

The modified dressing

$$F^g=h_+^{-1}Fg_+=h_-Fg_-^{-1} \tag{4.85}$$

preserves the boundary condition (4.84) but is no longer an action of the group $C(G)$.

Remarkably, *there exists a group such that (4.85) defines its group action. It coincides with $C(G)$ as a point set, but has a different multiplication law*

$$g\circ f=f_+g_+g_-f_-, \tag{4.86}$$

where $g=g_+g_-$ and $f=f_+f_-$ (under the assumption that the factorization problem in $C(G)$ has a unique solution). We shall denote this group by $C_0(G)$, and let $C_+(G)$ and $C_-(G)$ be the subgroups whose elements have the same analyticity properties as g_+ and g_-, respectively. These subgroups commute with one another in $C_0(G)$ and the multiplication law in each of them is given by

$$g_+\circ f_+=f_+g_+, \quad g_-\circ f_-=g_-f_-. \tag{4.87}$$

A comparison with (1.22)–(1.23) shows that the Lie algebras of $C_0(G)$ and $C_\pm(G)$ coincide (up to the sign of the commutator) respectively with the Lie algebras $C_0(\mathfrak{g})$ and $C_\pm(\mathfrak{g})$ introduced in § 1.

Let us now verify that (4.85) *defines an action of $C_0(G)$*. The verification is quite similar to (4.83):

$$\begin{aligned}(F^f)^g &= (F^f g F^{f^{-1}})_- F^f g_-^{-1} \\ &= ((FfF^{-1})_+^{-1} Ff_+ g f_- F^{-1}(FfF^{-1})_-^{-1})_- \cdot Ff_-^{-1} g_-^{-1} \\ &= (Ff_+ g_+ g_- f_- F^{-1})_- \cdot F(g_- f_-)^{-1} = F^{g\circ f}.\end{aligned} \tag{4.88}$$

The formula

$$U^g(x,\lambda) = \frac{d}{dx} F^g (F^g)^{-1} = \frac{d}{dx} h_- \cdot h_-^{-1} + h_- U h_-^{-1} = \tilde{\mathrm{A}}\mathrm{d}^* h_- \cdot \acute{U}(x,\lambda) \tag{4.89}$$

(cf. (4.41)) carries this action over to the phase space $\mathscr{C}^*((\mathfrak{g}))$. Moreover, the transformation $U \to U^g$ is only defined in a certain extension of $C^*((\mathfrak{g}))$ since in general it violates the periodicity condition. We will not go into details here.

It is appropriate to compare the dressing formulae (4.78), (4.89) with the formulae (4.16)–(4.17) for the solutions of the equations of motion within the general scheme of Subsection 1. They have practically the same structure but the matrices to be factorized in (4.78) and (4.16) differ at first sight: the dressing procedure involves an arbitrary matrix $g(\lambda)$ conjugated by the solution $F(x,\lambda)$ of the auxiliary linear problem, whereas in order to solve the equations of motion we must factorize the function $\exp\{-t\nabla f(U(x,\lambda))\}$, where $f(U)$ belongs to the Casimir algebra $I(\tilde{\mathscr{C}}((\mathfrak{g})))$. Nevertheless, it can be shown that

$$\exp\{-t\nabla f(U)\} = \hat{F} g_0(\lambda, t)\hat{F}^{-1}, \tag{4.90}$$

where $\hat{F}(x,\lambda)$ is a solution of the auxiliary linear equation for the initial value $U(x,\lambda)$ and $g_0(\lambda, t)$ takes values in a Cartan subgroup K of G that does not depend on t. In fact, consider a function $h(\lambda)$ which carries the monodromy matrix into a function with values in a fixed Cartan subgroup K,

$$T(U(\cdot,\lambda)) = h(\lambda)\hat{T}(U(\cdot,\lambda))h^{-1}(\lambda) = h(\lambda)\exp C(\lambda) h^{-1}(\lambda), \tag{4.91}$$

where $C(\lambda)$ takes values in the corresponding Cartan subalgebra $\mathfrak{k}$. It follows from (4.46)–(4.50) that

$$U(x,\lambda) = \tilde{\mathrm{A}}\mathrm{d}^* h(x,\lambda) \cdot \hat{U}_0(\lambda), \tag{4.92}$$

where

$$\hat{U}_0(\lambda) = \frac{1}{2L} C(\lambda), \qquad h(x,\lambda) = F(x,\lambda)h(\lambda)\exp\left(-\frac{x+L}{2L}C(\lambda)\right) \tag{4.93}$$

and $F(x, \lambda)$ satisfies the auxiliary linear equation with the boundary condition (4.84). Recalling that the gradient of an invariant function transforms by conjugations, we have

$$\nabla f(U(x,\lambda)) = F(x,\lambda) h(\lambda) \nabla f(\hat{U}_0) h^{-1}(\lambda) F^{-1}(x,\lambda). \tag{4.94}$$

Now, $\nabla f(\hat{U}_0)$ lies in the same Cartan subalgebra $\mathfrak{k}$ as $\hat{U}_0$. Hence (4.90) is an immediate consequence of (4.94). This observation shows that the dynamics flow is naturally incorporated in the general dressing group. In particular, this implies once again that the dressing procedure carries the set of solutions of the equations of motion into itself. However, unlike the dynamical transformations, the general dressing transformation is not a Hamiltonian one.

With this we close the description of a general geometric set-up for the inverse scattering method. Of course, our presentation was incomplete. The analytical justification of the formal infinite-dimensional constructions developed above is beyond the scope of the book. Nevertheless, we decided to outline the general scheme because it seems to be a faily elegant one and illuminates the main structures of the inverse scattering method used throughout the book in the study of particular models.

§ 5. The General Scheme as Illustrated with the NS Model

To close the book, we shall return to our basic example - the NS model - in order to show how the method for solving it developed in Part I agrees with the general geometric discussion of this chapter. More precisely, we will show in what sense the Riemann problem that was used to solve the initial value problem in Chapter III of Part I is interpreted as a factorization problem of § 4. Besides, we will relate the hierarchy of Poisson structures and the generating Λ-operator defined in § III.5, Part I, to the family of Poisson structures in Subsection 3 of § 4. This will provide a proof for the Jacobi identity as was promised in § III.5 of Part I.

When dealing with the Riemann problem we shall restrict our attention to the rapidly decreasing case and assume that there is no discrete spectrum. The solution of the initial problem for the NS equation via the Riemann problem given in § III.3, Part I, amounts to the following: given the initial data $\psi(x)$, $\bar{\psi}(x)$ one determines the transition coefficient $b(\lambda)$ and solves a family of regular Riemann problems

$$G(x,t,\lambda) = G_+(x,t,\lambda) G_-(x,t,\lambda), \tag{5.1}$$

with

$$G(x,t,\lambda)=\begin{pmatrix}1 & \varepsilon\bar{b}(\lambda)e^{-i\lambda x+i\lambda^2 t}\\ -b(\lambda)e^{i\lambda x-i\lambda^2 t} & 1\end{pmatrix}$$
$$=e^{\frac{i\lambda^2 t}{2}\sigma_3}G(x,\lambda)e^{-\frac{i\lambda^2 t}{2}\sigma_3},\quad \varepsilon=\operatorname{sign}\varkappa \tag{5.2}$$

and

$$G(x,\lambda)=G(x,0,\lambda). \tag{5.3}$$

The contour Γ is taken to be the real axis; the solution $G_\pm(x,t,\lambda)$ is required to have an analytic continuation into the half-plane $\pm\operatorname{Im}\lambda>0$ which is nondegenerate and normalized to I as $|\lambda|\to\infty$:

$$G_\pm(x,t,\infty)=I. \tag{5.4}$$

The matrix $U(x,t,\lambda)$ of the auxiliary linear problem is expressed in terms of $G_\pm(x,t,\lambda)$ as

$$U(x,t,\lambda)=-G_+^{-1}(x,t,\lambda)\frac{\partial G_+}{\partial x}(x,t,\lambda)+\frac{\lambda}{2i}G_+^{-1}(x,t,\lambda)\sigma_3 G_+(x,t,\lambda)$$
$$=\frac{\partial G_-}{\partial x}(x,t,\lambda)G_-^{-1}(x,t,\lambda)+\frac{\lambda}{2i}G_-(x,t,\lambda)\sigma_3 G_-^{-1}(x,t,\lambda). \tag{5.5}$$

With the notation of § 4 these formulae are compactly written as

$$U(t)=\widetilde{\mathrm{Ad}}^*G_+^{-1}(t)\left(\frac{\lambda\sigma_3}{2i}\right)=\widetilde{\mathrm{Ad}}^*G_-(t)\left(\frac{\lambda\sigma_3}{2i}\right), \tag{5.6}$$

where the dependence on x and λ is suppressed, as will often be done is what follows.

Using (5.6) and the group property of $\widetilde{\mathrm{Ad}}^*$, we find an expression for the solution $U(t)$ in terms of the initial data $U_0=U(t)|_{t=0}$,

$$U(t)=\widetilde{\mathrm{Ad}}^*h_+^{-1}(t)\,U_0=\widetilde{\mathrm{Ad}}^*h_-(t)\,U_0, \tag{5.7}$$

where the matrices $h_\pm(t)$ are given by

$$h_+(x,t,\lambda)=G_+^{-1}(x,0,\lambda)\,G_+(x,t,\lambda), \tag{5.8}$$

$$h_-(x,t,\lambda)=G_-(x,t,\lambda)\,G_-^{-1}(x,0,\lambda). \tag{5.9}$$

These matrices solve the factorization problem

$$h(t)=h_+(t)\,h_-(t), \tag{5.10}$$

where $h(t)$ is given by

$$h(x,t,\lambda)=G_{+}^{-1}(x,0,\lambda)e^{\frac{i\lambda^2 t}{2}\sigma_3}G_{+}(x,0,\lambda)$$
$$\times G_{-}(x,0,\lambda)e^{-\frac{i\lambda^2 t}{2}\sigma_3}G_{-}^{-1}(x,0,\lambda) \tag{5.11}$$

and so is expressed through the solution of the Riemann problem (5.1) for $t=0$ which is uniquely determined by the initial data U_0.

Expressions (5.7) coincide with the general formulae (4.17) from Subsection 1 of § 4 for the solution of the initial value problem for the abstract Hamilton equation (4.10), which is a zero curvature equation as shown in Subsection 2 of § 4. However, the general factorization problem (4.16) differs from the Riemann problem (5.10): the former deals with the factorization of a one-parameter matrix subgroup $g(t)=\exp\{-t\nabla H(U_0)\}$ whereas the matrices $h(t)$ involved in the latter problem do not make up a one-parameter subgroup.

To coordinate the two ways of solving the initial value problem we notice that (5.7) determines $h_{+}^{-1}(t)$ and $h_{-}(t)$ up to right factors from the centralizer of U_0 with respect to the action $\tilde{\mathrm{A}}\mathrm{d}^*$. Obviously, the matrices $F(x,\lambda)C(\lambda)F^{-1}(x,\lambda)$, where $F(x,\lambda)$ is a solution of the auxiliary linear problem with matrix $U_0(x,\lambda)$ and $C(\lambda)$ is an arbitrary matrix, lie in the centralizer. With this in mind we introduce the matrices

$$g_{+}(x,t,\lambda)=h_{+}(x,t,\lambda), \tag{5.12}$$

$$g_{-}(x,t,\lambda)=h_{-}(x,t,\lambda)G_{-}(x,0,\lambda)e^{\frac{i\lambda^2 t}{2}\sigma_3}G_{-}^{-1}(x,0,\lambda), \tag{5.13}$$

which satisfy

$$g_{+}(t)g_{-}(t)=g(t), \tag{5.14}$$

with

$$g(x,t,\lambda)=G_{+}^{-1}(x,0,\lambda)e^{\frac{i\lambda^2 t}{2}\sigma_3}G_{+}(x,0,\lambda). \tag{5.15}$$

The matrices $g(t)$ now form a one-parameter subgroup, and comparison with (4.90) shows that

$$g(t)=e^{-t\nabla H(U_0)}. \tag{5.16}$$

Thus (5.14) *gives a realization of the abstract factorization problem for the case of the NS model.* Expressions (5.12)–(5.13) indicate the functional classes that contain the required matrices $g_{\pm}(x,t,\lambda)$: these have an analytic continuation into the half-planes $\pm\operatorname{Im}\lambda>0$ with the following asymptotic behaviour, as $|\lambda|\to\infty$,

$$g_+(x,t,\lambda)=I+O\left(\frac{1}{|\lambda|}\right), \tag{5.17}$$

$$g_-(x,t,\lambda)=\left(I+O\left(\frac{1}{|\lambda|}\right)\right)e^{\frac{i\lambda^2 t}{2}\sigma_3}\left(I+O\left(\frac{1}{|\lambda|}\right)\right). \tag{5.18}$$

We thus have a formal agreement between the concrete Riemann problem for the NS model and the abstract factorization problem of Subsection 1, § 4. It should be noted, however, that we have not stated the abstract factorization problem in the infinite-dimensional group $\mathscr{C}((G))$ (or even in the corresponding Lie algebra) for the case of rapidly decreasing boundary conditions. Therefore the above reasoning should be regarded as defining such a problem on a particular orbit associated with the NS model. This example shows that the application of the general scheme of § 4 to a given nonlinear equation associated with a particular orbit requires additional analytical investigation of the corresponding auxiliary linear problem, which would give rise to a suitable Riemann problem. This concludes our discussion of the role of the factorization problem in solving the initial value problem for integrable nonlinear equations.

We shall now proceed to describe the geometric meaning of the Λ-operator defined in § III.5, Part I, and the associated hierarchy of Poisson structures. We recall the relevant definitions (for simplicity, we consider only the rapidly decreasing case and assume $\varkappa=-1$). The phase space $\mathscr{M}_0$ with coordinates $\psi(x)$, $\bar{\psi}(x)$ is equipped with the basic Poisson structure

$$\{f,g\}=\langle \operatorname{grad} f, \operatorname{grad} g\rangle = i\int_{-\infty}^{\infty} \operatorname{tr}(\operatorname{grad} f(x)\sigma_3 \operatorname{grad} g(x))\,dx, \tag{5.19}$$

where for any observable f

$$\operatorname{grad} f(x)=\frac{1}{i}\left(\frac{\delta f}{\delta\psi(x)}\sigma_+ + \frac{\delta f}{\delta\bar{\psi}(x)}\sigma_-\right). \tag{5.20}$$

In addition to (5.19) we have introduced a hierarchy of Poisson structures

$$\begin{aligned}\{f,g\}_{(k)} &= \langle \operatorname{grad} f, \Lambda^k \operatorname{grad} g\rangle \\ &= i\int_{-\infty}^{\infty} \operatorname{tr}(\operatorname{grad} f(x)\sigma_3\Lambda^k \operatorname{grad} g(x))\,dx,\end{aligned} \tag{5.21}$$

$k=-\infty,\dots,\infty$. Here Λ is an integro-differential operator acting on off-diagonal matrices $F(x)$ according to

$$\Lambda F(x)=i\sigma_3\left(\frac{dF}{dx}(x)-[U_0(x),d^{-1}([U_0(\cdot),F(\cdot)])(x)]\right), \tag{5.22}$$

where

$$U_0(x)=i(\psi(x)\sigma_-+\bar{\psi}(x)\sigma_+) \tag{5.23}$$

and $\langle,\rangle$ stands for the bilinear form given by the integral in (5.19). Obviously, $\{,\}=\{,\}_{(0)}$.

In § III.5, Part I, the Jacobi identity for the Poisson brackets $\{,\}_{(k)}$ was left unverified. *Here we shall indicate the geometric meaning of the Poisson structure* $\{,\}_{(1)}$ *and prove the Jacobi identity.*

In Subsection 3 of the preceding section we defined a family of Poisson structures $\{,\}_p$, $p=-N,\ldots,M$, on the phase space $C^*_{N,M}$. In particular, it was shown that the zero curvature equation for a matrix $U(x,\lambda)$ of the form

$$U(x,\lambda)=\lambda J+Q(x), \tag{5.24}$$

where for $\mathfrak{g}=su(2)$ we have

$$J=iJ_a\sigma_a,\qquad Q(x)=iQ_a(x)\sigma_a, \tag{5.25}$$

may be written in Hamiltonian form in three ways; in what follows we shall be interested in only two of them. The first one involves the phase space $C^*_{0,2}$ consisting of matrices $U(x,\lambda)$ of the form (5.24) with the Poisson bracket $\{,\}_0$:

$$\{J_a,J_b\}_0=0,\qquad \{J_a,Q_b(x)\}_0=0, \tag{5.26}$$

$$\{Q_a(x),Q_b(y)\}_0=\varepsilon_{abc}J_c\delta(x-y), \tag{5.27}$$

the second one makes use of the phase space $C^*_{1,1}$ consisting of matrices $\tilde{U}(x,\lambda)=J+\dfrac{Q(x)}{\lambda}$ with the Poisson bracket $\{,\}_{-1}$:

$$\{J_a,J_b\}_{-1}=0,\qquad \{J_a,Q_b(x)\}_{-1}=0, \tag{5.28}$$

$$\{Q_a(x),Q_b(y)\}_{-1}=-\varepsilon_{abc}Q_c(x)\delta(x-y)+\tfrac{1}{2}\delta_{ab}\delta'(x-y). \tag{5.29}$$

The zero curvature equation in the latter case is written for the matrix $U(x,\lambda)=\lambda\tilde{U}(x,\lambda)$ which has the same structure as (5.24).

The NS model is associated with a particular orbit in $C^*_{0,2}$ given by

$$J_1=J_2=0,\qquad J_3=-\tfrac{1}{2},\qquad Q_3(x)=0, \tag{5.30}$$

which is identified with the phase space $\mathcal{M}_0$ by setting

$$\psi(x)=Q_1(x)+iQ_2(x) \tag{5.31}$$

(see example 2 in § 1). However, $\mathscr{M}_0$ is not a Poisson submanifold with respect to the Poisson bracket $\{,\}_{-1}$.

We may, nevertheless, reduce the Poisson structure $\{,\}_{-1}$ to the manifold $\mathscr{M}_0$ if we regard the equations $Q_3(x)=0$ as constraints. To do so we shall calculate explicitly the corresponding Poisson-Dirac bracket given by

$$\{f,g\}^*_{-1}=\{f,g\}_{-1}+\int_{-\infty}^{\infty}\int_{-\infty}^{\infty}\{f,Q_3(x)\}_{-1}K^{-1}(x,y)\{Q_3(y),g\}_{-1}\,dx\,dy, \tag{5.32}$$

where $K^{-1}(x,y)$ is the kernel of the integral operator K^{-1} inverse to the operator K with the kernel $K(x,y)=\{Q_3(x),Q_3(y)\}_{-1}$. The right hand side of (5.32) must be restricted to the level surface $Q_3(x)=0$ of the constraints. From (5.29) we find

$$K(x,y)=\tfrac{1}{2}\delta'(x-y), \tag{5.33}$$

so that

$$K^{-1}(x,y)=\varepsilon(x-y), \tag{5.34}$$

where

$$\varepsilon(x)=\begin{cases} 1 & \text{for } x>0, \\ -1 & \text{for } x<0. \end{cases} \tag{5.35}$$

In particular, letting formally $f=\psi(x)$ and $g=\psi(y)$ or $\bar{\psi}(y)$ we get the Poisson-Dirac brackets of the coordinates $\psi(x)$, $\bar{\psi}(x)$

$$\{\psi(x),\psi(y)\}^*_{-1}=\psi(x)\psi(y)\varepsilon(x-y), \tag{5.36}$$

$$\{\psi(x),\bar{\psi}(y)\}^*_{-1}=\delta'(x-y)-\psi(x)\bar{\psi}(y)\varepsilon(x-y). \tag{5.37}$$

The same result is obtained by calculating the Poisson brackets of $\psi(x)$, $\bar{\psi}(x)$ according to (5.21) for $k=1$. *Thus we have established that the Poisson structures $\{,\}_{(1)}$ and $\{,\}^*_{-1}$ coincide. This implies, in particular, the Jacobi identity for the Poisson bracket $\{,\}_{(1)}$.*

To prove the Jacobi identity for all the Poisson structures $\{,\}_{(k)}$ we shall apply the following method. Observe that the Poisson brackets $\{,\}_0$ and $\{,\}_{-1}$ on $C^*_{0,2}$ *are compatible in the following sense: for any α the bracket $\{,\}^{(\alpha)}=\{,\}_{-1}+\alpha\{,\}_0$ satisfies the Jacobi identity.* This is best verified in the coordinates $Q_a(x)$, $a=1,2,3$. From (5.26)–(5.27) and (5.28)–(5.29) we derive

$$\begin{aligned}\{Q_a(x), Q_b(y)\}^{(\alpha)} &= \{Q_a(x), Q_b(y)\}_{-1} + \alpha\{Q_a(x), Q_b(y)\}_0 \\ &= -\varepsilon_{abc}\tilde{Q}_c(x)\delta(x-y) + \tfrac{1}{2}\delta_{ab}\delta'(x-y) \end{aligned} \tag{5.38}$$

where $\tilde{Q}_a(x) = Q_a(x) = \alpha J_a$, $a = 1, 2, 3$. Hence the bracket $\{,\}^{(\alpha)}$ results from the Poisson bracket $\{,\}_{-1}$ by a change of variables $Q_a(x) \to \tilde{Q}_a(x)$ in phase space and therefore satisfies the Jacobi identity. The reduced Poisson brackets $\{,\}_{(0)}$ and $\{,\}_{(1)}$ on the phase space $\mathcal{M}_0$ are compatible as well. Let us now consider the symplectic form Ω_α associated with the Poisson bracket $\{,\}^{(\alpha)}$. It is a bilinear form in off-diagonal matrices $\xi(x)$, $\eta(x)$ given by

$$\Omega_\alpha(\xi, \eta) = \Omega_{(0)}(\xi, (\Lambda + \alpha)^{-1}\eta), \tag{5.39}$$

where Ω_0 is the symplectic form of the Poisson bracket $\{,\}_{(0)}$. Since Ω_α is closed, it follows that all the forms

$$\Omega_{(k)}(\xi, \eta) = \Omega_{(0)}(\xi, \Lambda^{-k}\eta), \tag{5.40}$$

are closed, which can easily be seen by expanding $(\Lambda + \alpha)^{-1}$ in powers of α near $\alpha = 0$ and $\alpha = \infty$. The form $\Omega_{(k)}$ is associated with the Poisson bracket $\{,\}_{(k)}$ so that the fact that the former is closed implies the Jacobi identity for the latter.

Of course, these arguments also show *the validity of the Jacobi identity for the more general Poisson bracket*

$$\{f, g\}_\varphi = \langle \operatorname{grad} f, \varphi(\Lambda) \operatorname{grad} g \rangle, \tag{5.41}$$

where φ is an arbitrary smooth function. Therefore the Poisson brackets $\{,\}_\varphi$ and $\{,\}_\chi$ are compatible for any functions φ and χ. The formal proof given above can be made precise by assuming that Λ has an inverse; if Λ has a nontrivial null-space (this is the case for the Λ-operator (5.22)), then the phase space $\mathcal{M}_0$ must be reduced by fixing the values of the functionals from the annihilator.

In § III.5 of Part I, we also indicated another role of Λ, that of a generating operator for a family of integrals of the motion I_n in involution related by

$$\operatorname{grad} I_n(x) = \Lambda \operatorname{grad} I_{n-1}(x). \tag{5.42}$$

Here we will show how this formula, which was obtained by a straightforward calculation in § III.5, Part I, results from simple geometric considerations and may serve to define the family I_n.

Suppose we are given two functionals I_1 and I_2 such that the associated Hamilton equations of motion on the phase space $\mathcal{M}_0$ relative to the Poisson *brackets $\{,\}_{(1)}$ and $\{,\}_{(0)}$, respectively, coincide with one another, i.e. for any observable f we have*

$$\{I_1, f\}_{(1)} = \{I_2, f\}_{(0)}. \tag{5.43}$$

Using the compatibility of the Poisson brackets $\{,\}_{(0)}$ and $\{,\}_{(1)}$ we will show that this equation implies the existence of a family of functionals I_n that are in involution with respect to both Poisson brackets $\{,\}_{(0)}$ and $\{,\}_{(1)}$ and satisfy

$$\{I_n, f\}_{(1)} = \{I_{n+1}, f\}_{(0)}. \tag{5.44}$$

For the proof it suffices to show the existence of a functional I_3 such that

$$\{I_2, f\}_{(1)} = \{I_3, f\}_{(0)}. \tag{5.45}$$

For that purpose we will show that the vector field

$$Xf = \{I_2, f\}_{(1)} \tag{5.46}$$

is (locally) Hamiltonian with respect to the Poisson bracket $\{,\}_{(0)}$ i.e.

$$X\{f, g\}_{(0)} = \{Xf, g\}_{(0)} + \{f, Xg\}_{(0)}. \tag{5.47}$$

The last formula can be written as

$$\{I_2, \{f, g\}_{(0)}\}_{(1)} = \{\{I_2, f\}_{(1)}, g\}_{(0)} + \{f, \{I_2, g\}_{(1)}\}_{(0)} \tag{5.48}$$

and follows from the Jacobi identity for the Poisson bracket $\{,\}_{(0)} + \{,\}_{(1)}$ and the equation

$$\{I_2, \{f, g\}_{(1)}\}_{(0)} = \{\{I_2, f\}_{(0)}, g\}_{(1)} + \{f, \{I_2, g\}_{(0)}\}_{(1)}, \tag{5.49}$$

which is derived from (5.43) in a similar way to (5.48).

The explicit expressions (5.19) and (5.21) for $\{,\}_{(0)}$ and $\{,\}_{(1)}$ yield that the functional I_3 can be determined from

$$\operatorname{grad} I_3(x) = \Lambda \operatorname{grad} I_2(x). \tag{5.50}$$

The above argument may be regarded as an existence proof for this equation (in a simply connected phase space).

The final expression for I_n is

$$I_n = \operatorname{grad}^{-1} \Lambda^{n-1} \operatorname{grad} I_1 = \operatorname{grad}^{-1} \Lambda^{n-m} \operatorname{grad} I_m. \tag{5.51}$$

Together with the definition (5.21) of the Poisson brackets $\{,\}_{(k)}$ this implies that the functionals I_n are in involution with respect to all these Poisson structures; we also have a relation generalizing (5.44),

$$\{I_k, f\}_{(l)} = \{I_m, f\}_{(n)}, \tag{5.52}$$

for $k+l=m+n$.

For the NS model, I_1 and I_2 may be chosen to be the number of particles N

$$N = \int_{-\infty}^{\infty} |\psi(x)|^2 \, dx \tag{5.53}$$

and the momentum P

$$P = \frac{1}{2i} \int_{-\infty}^{\infty} \left(\bar{\psi} \frac{d\psi}{dx} - \psi \frac{d\bar{\psi}}{dx} \right) dx. \tag{5.54}$$

Relation (5.43) for the Poisson brackets $\{,\}_{(0)}$ and $\{,\}_{(1)}$ given by (5.19) and (5.36)–(5.37) is easily verified. *Thus the above discussion yields the existence of a Λ-operator, a hierarchy of Poisson structures $\{,\}_{(k)}$, and a family of functionals I_n which are in involution with respect to all these Poisson brackets.*

So, coming back to the NS model we have cast a new light on the related structures. The chain of ideas developed in the text is now closed, and this brings the book to its natural end.

§ 6. Notes and References

1. The Lie-Poisson bracket of the form (1.3) on the phase space $\mathfrak{g}^*$ was introduced and studied by S. Lie [L 1888]. This Poisson structure and the corresponding symplectic structure on the orbits of the coadjoint action of the Lie algebra $\mathfrak{g}$ was subsequently rediscovered by different people [B 1967], [K 1970], [S 1970], [Ki 1972]. For an up-dated treatment of the properties of the Lie-Poisson bracket and the theory of Poisson manifolds see [W 1983].

2. The term current algebra for the infinite-dimensional Lie algebra $C(\mathfrak{g}) = \mathfrak{g} \otimes \mathbb{C}[[\lambda, \lambda^{-1}]]$ is borrowed from quantum field theory. In the mathematical literature, the Lie algebra $C(\mathfrak{g})$ is called a loop algebra.

3. The construction of integrable systems using the decomposition of a Lie algebra into a linear sum of two subalgebras was proposed by B. Kostant for the case of the Toda model with free ends [K 1979a]. In [A 1979],

[RSF 1979], [RS 1979], [MM 1979], [S 1980], [RS 1981] the scheme was developed further and applied to a wide class of Lie algebras including infinite dimensional ones. The second Lie algebra structure with commutator $[\,,]_0$ was introduced in [MM 1979], [R 1980].

4. The connection between the r-matrix formulation and the Lie algebra $C_0(\mathfrak{g})$ was found in [RF 1983].

5. Formula (1.25) suggests an abstract definition of the r-matrix as given in [S 1983]: an r-matrix is an operator R in a Lie algebra $\mathfrak{g}$ such that the commutator defined by (1.25) satisfies the Jacobi identity. An important class of r-matrices consists of those operators R that satisfy the equation

$$[R\xi, R\eta]-R([R\xi,\eta]+[\xi,R\eta])=-[\xi,\eta] \tag{6.1}$$

for any ξ,η in $\mathfrak{g}$, which is the so-called modified Yang-Baxter equation (see [S 1983]). This is precisely the equation that holds for the integral operator R in $C(\mathfrak{g})$ with the kernel $r(\lambda-\mu)$ of the form (1.31), the integral being taken in the sense of principal value; for $\lambda\neq\mu$ it coincides with the usual Yang-Baxter equation (2.3).

6. In order to attach precise meaning to the symbols $A\otimes I$ and $I\otimes A$ in § 1, the Lie algebra $\mathfrak{g}$ must be assumed embedded into an associative algebra with unity (for instance, into the universal enveloping algebra $U(\mathfrak{g})$); clearly, this notation is "functorial".

7. A description of the algebra of Casimir functions of the Lie algebra $C_{N,M}(\mathfrak{g})$ is given in [T 1982], [KR 1983].

8. Multi-pole Poisson submanifolds discussed in § 1 have a simple interpretation in terms of the Lie algebra $\mathfrak{g}_{\mathbb{A}}$ over the adele ring $\mathbb{A}$ of the field $\mathbb{C}(\lambda)$ of rational functions on $\mathbb{C}$ (see [C 1983]). The analogue of the decomposition (1.16) is given by splitting $\mathfrak{g}_{\mathbb{A}}$ into the sum of the subalgebra of principal adeles $\mathfrak{g}(\lambda)$ consisting of rational functions in λ with values in $\mathfrak{g}$, and the subalgebra of integral adeles $\mathfrak{g}_{\mathbb{A}}^0$ (a reformulation of the resolution of a rational function into partial fractions). The elements $U(\lambda)$ of the form (1.55) make up a Poisson submanifold of $\mathfrak{g}(\lambda)$ with respect to the action of the subalgebra $\mathfrak{g}_{\mathbb{A}}^0$.

9. The classification of solutions of (2.2)-(2.3) with values in a simple algebra $\mathfrak{g}$ is given in [BD 1982]. This paper describes all trigonometric and elliptic r-matrices of the form $r(u)=X^{ab}(u)\,t_a\otimes t_b$ where the matrix $X^{ab}(u)$ is not identically degenerate and satisfies $X^{ab}(u)=\dfrac{\delta^{ab}}{u}+O(1)$ as $u\to 0$, and $\{t_a\}$ is an orthonormal basis in $\mathfrak{g}$ with respect to the Killing form; also, a large family of rational r-matrices is constructed. There turned out to be a close connection between the problem of classification of trigonometric r-matrices and the structure theory of affine Lie algebras.

10. A method for constructing trigonometric and elliptic r-matrices and the associated fundamental Poisson brackets by means of the averaging pro-

cedure was proposed in [RF 1983]. In fact, one does not get all the trigonometric r-matrices in this way; nevertheless, it may be shown that any such matrix is obtained by combining the averaging procedure with extension theory of linear operators due to von Neumann [BD 1982], [S 1983].

11. We point out that if two automorphisms of finite order, θ and θ', of the Lie algebra $\mathfrak{g}$ with abelian fixed point subalgebras differ by an inner automorphism, then the associated trigonometric r-matrices are equivalent [BD 1982].

12. The elliptic r-matrix for $n=2$ was introduced by E. K. Sklyanin in [S 1979] and extended to higher dimensions in [Be 1980]. In [T 1984] it was interpreted as a matrix analogue of the Weierstrass zeta function.

13. There is an analogue of the Lie algebra $C_0(\mathfrak{g})$ for the trigonometric and elliptic cases. The algebra $C(\mathfrak{g})$ and its subalgebra $C_+(\mathfrak{g})$ are the same as in the rational case, while the analogue of $C_-(\mathfrak{g})$ is defined by using the lattice Λ_1 or Λ_2. For instance, in the elliptic case it is defined as follows. Let $\mathscr{E}(\mathfrak{g})$ be the algebra of meromorphic functions $\xi(\lambda)$ on $\mathbb{C}$ with values in $\mathfrak{g}$ that satisfy the quasi-periodicity conditions

$$\xi(\lambda+\omega_i)=\theta_i\,\xi(\lambda), \quad i=1,2, \tag{6.2}$$

and whose poles are at the points of the lattice Λ_2. We identify $\mathscr{E}(\mathfrak{g})$ with a subalgebra of $C(\mathfrak{g})$ by assigning to a function in $\mathscr{E}(\mathfrak{g})$ its Laurent series at $\lambda=0$. A function in $\mathscr{E}(\mathfrak{g})$ is uniquely determined by its principal part at $\lambda=0$, so that we have the decomposition

$$C(\mathfrak{g})=C_+(\mathfrak{g})+\mathscr{E}(\mathfrak{g}) \tag{6.3}$$

which defines the Lie algebra $C_0(\mathfrak{g})$. The operator $R=\frac{1}{2}(P_+-P_-)$ (see (1.24)) gives the elliptic r-matrix $r^{\Lambda_2}(\lambda)$. The linear space $\mathscr{E}(\mathfrak{g})$ is dual to the Lie algebra $C_+(\mathfrak{g})$ relative to the pairing (1.13), so the orbits of the coadjoint action of the subalgebra $C_+(\mathfrak{g})$ described in § 1 acquire a new functional realization in the space $\mathscr{E}(\mathfrak{g})$. In particular, the simplest orbit associated with the HM model turns into the one associated with the LL model. This interpretation is due to A. G. Reyman and M. A. Semenov-Tian-Shansky [RS 1986].

14. The simplest matrices $L(\lambda)$ that satisfy the fundamental Poisson brackets with a rational r-matrix for classical Lie algebras are given in [RF 1983].

15. An analytical justification of the multiplicative averaging (3.12) for the case of $\mathfrak{g}=sl(2)$ and the matrix $L(\lambda)$ of the LHM model was carried out in [RF 1983]. The infinite product (3.12) can be computed explicitly and gives the matrix $L(\lambda)$ for the LSG model discussed in § III.5.

16. The fundamental Poisson brackets in one site define a Poisson structure on the Lie group $C(G)$: the elements $L(\lambda)$ for all λ may be thought of as generators of the ring of functions on $C(G)$, and the Poisson bracket may be

extended by the "Leibnitz rule" to the whole ring of functions. The resulting Poisson bracket is one example in the class of Poisson brackets introduced in [D 1983]. The key property of such Poisson structures is that the operation of group multiplication is a Poisson mapping. This property formalizes the fact that the monodromy matrix $T_N(\lambda)$ for lattice models has the same Poisson brackets as the matrices $L_n(\lambda)$ (see § III.1). A Lie group with such a Poisson structure is called a Poisson-Lie group (or Hamilton-Lie group in [D 1983]). The quadratic Poisson bracket defined in [GD 1978] also provides an example of a Poisson-Lie group [S 1983].

17. A geometric theory of integrable lattice models is developed in [S 1983], [S 1985b]; the corresponding Poisson submanifolds and orbits have been described by V. G. Drinfeld (see [S 1985b]).

18. The fact that Casimir functions lying in $I(\mathfrak{g})$ are in involution with respect to the Poisson bracket $\{,\}_0$ – the "involutivity theorem" – is essentially contained in [K 1979a]. Its r-matrix formulation is given in [S 1983].

19. A method for solving the Hamilton equations of motion (4.10) induced by a Casimir function relative to the Poisson bracked $\{,\}_0$ by means of the factorization problem (4.16) in the Lie group G (the "factorization theorem") was proposed in [RS 1979], [RS 1981]. The idea of the method goes back to the work of V. E. Zakharov and A. B. Shabat [ZS 1979].

20. Finite-dimensional simple Lie algebras lead to integrable systems that generalize the Toda model with free ends to the case of an arbitrary root system; these models were introduced in [B 1976], where a Lax representation was found. The solution of the corresponding equations of motion and the study of the asymptotic dynamics was carried out in [K 1979b].

21. The central extension $\mathscr{C}(\mathfrak{g})$ of the current algebra $C(\mathfrak{g})$ (when $\mathfrak{g}$ is a simple Lie algebra) is an example of a Kac-Moody algebra – an affine Lie algebra of height 1 [K 1983]. The reason for introducing it is that the coadjoint action of the Lie group $\mathscr{C}(G)$ is given by gauge transformations and leads to the equations of motion in the form of a zero curvature condition. This was first observed in [RS 1980a] where these algebras were used for constructing integrable equations.

22. The introduction of the spectral parameter λ leads to infinitely many Casimir functions and hence enables one to construct meaningful examples of nonlinear equations which are completely integrable Hamiltonian systems. If $\mathfrak{g}$ is a finite-dimensional Lie algebra, then associated with the Lie algebra $C(\mathfrak{g})$ (no x-dependence) there is a series of interesting finite-dimensional integrable systems: the generalized periodic Toda models, multi-dimensional tops in potential fields, and systems of interacting tops [RS 1979], [R 1980], [AM 1980], [RS 1981], [B 1984], [RS 1986]. In this case the factorization theorem immediately yields the linearization of the equations of motion on the Jacobian of the spectral curve, an algebraic curve defined by the equation $\det(U(\lambda)-\mu)=0$ [RS 1981]. This links the Lie-algebraic approach to integrable systems with the theory of finite-gap integration and Novikov's equations [N 1974], [DMN 1976].

23. "Switching on" the x-dependence may be regarded as the "two-dimensional extension" of finite-dimensional integrable systems. Thus, for instance, there is a natural two-dimensional extension of the periodic Toda models [Mi 1979]; the two-dimensional extension of tops leads to systems such as the matrix sine-Gordon model (see § I.8).

24. If in the zero curvature equation

$$\frac{\partial U}{\partial t} - \frac{\partial V}{\partial x} + [U, V] = 0 \tag{6.4}$$

the dependence on t disappears (the stationary case), then (6.4) takes the Lax form with respect to the matrix $V(x, \lambda)$,

$$\frac{dV}{dx} = [U, V]. \tag{6.5}$$

The Lie-algebraic scheme provides a simple interpretation for the Poisson structure of these stationary equations determined in [GD 1975], [BN 1976]: it coincides with the Lie-Poisson bracket for the finite-dimensional orbit passing through $V(x, \lambda)$ [R 1983], [FNR 1983b]. In addition, there is a compatible system of higher equations (6.4),

$$\frac{\partial U}{\partial t_n} - \frac{\partial V_n}{\partial x} + [U, V_n] = 0, \tag{6.6}$$

where each equation is equipped with its own time variable t_n, so that $t = t_N$, $V = V_N$; the NS model corresponds to $N = 3$. Since (6.6) is a compatible system, the matrices $V_n(x, t_1, \ldots; \lambda)$ also satisfy the equations

$$\frac{\partial V_n}{\partial t_k} - \frac{\partial V_k}{\partial t_n} + [V_n, V_k] = 0. \tag{6.7}$$

On the manifold of stationary solutions $\left(\frac{\partial}{\partial t_k} = 0\right)$ this gives rise to a compatible system of Novikov's equations

$$\frac{\partial V}{\partial t_n} = [V_n, V] \tag{6.8}$$

which also have Hamiltonian form with respect to the same Lie-Poisson bracket on the orbit [R 1983], [FNR 1983b].

25. The observation that different cocycles ω_p give different Poisson structures $\{,\}_p$ which produce the same family of integrable equations is due to [KR 1983].

26. Equations (6.6) may be put in Hamiltonian form, so that the densities $h_n(x, t_1, \ldots)$ of their Hamiltonians restricted to solutions of the equations of motion satisfy

$$\frac{\partial h_n}{\partial t_k} = \frac{\partial p_n^k}{\partial x}, \tag{6.9}$$

where the $p_n^k(x, t_1, \ldots)$ are the densities of the Hamiltonians for (6.7). The densities h_n can be chosen in such a way (by adding perfect derivatives, if necessary) that they are given, as well as the p_n^k, by a generating function $\tau(x, t_1, \ldots)$,

$$h_n = \frac{\partial^2}{\partial t_n \partial x} \log \tau, \quad p_n^k = \frac{\partial^2}{\partial t_n \partial t_k} \log \tau \tag{6.10}$$

[R 1983], [FNR 1983 b, c]. (In the case of the NS model, h_n should be taken as in (III.5.42), Part I). The function τ is the well-known τ-function of the Japanese authors [DJKM 1981] who introduced it on the basis of representation theory for affine Lie algebras. Originally the τ-function was defined by R. Hirota [H 1976] in a special setting, as a solution of a system of bilinear equations (Hirota's equations). These equations have subsequently got a natural interpretation in terms of representations of affine Lie algebras [DJKM 1981], [K 1983]. A different approach to the τ-function based on the analyticity properties of solutions of the auxiliary linear problem has been developed in [SW 1985].

27. The general Lie-algebraic scheme of § 4 applies not only to the current algebra but also the algebra of formal pseudo-differential operators [A 1979], [LM 1979]. This provides a natural interpretation for the Poisson structures introduced in [GD 1975]. The two-dimensional extension that uses the Maurer-Cartan 2-cocycle leads in this case to equations such as the Kadomtsev-Petviashvili equation [RS 1984].

28. The behaviour of Poisson brackets under dressing transformations has a natural explanation based on the theory of Poisson-Lie groups. Specifically, dressing transformations define a Poisson action with respect to a certain Poisson structure on the Lie group $C_0(G)$ [S 1985 b]. Infinitesimal dressing transformations may also be defined by direct methods. The vector fields that realize them are often associated with "hidden symmetries"; for more details see the survey [D 1984]. We notice that in general these vector fields are not Hamiltonian [S 1985 b].

29. An important class of dressing transformations which can be given explicitly are the so-called Bäcklund transformations defined by A. Bäck-

lund for the sine-Gordon equation [B 1882]; they are frequently met in modern literature on the inverse scattering method and soliton theory [M 1976].

30. Expression (5.16) can be made quite rigorous if, under the rapidly decreasing boundary conditions, one considers the current group $\mathscr{C}((G))$ with triangular asymptotics as $x \to \pm\infty$ (cf. the corresponding asymptotic formulae for the matrices $G_\pm(x, t, \lambda)$ in § II.1 of Part I). The invariants of the coadjoint action in this case are the minors of the reduced monodromy matrix $T(\lambda)$ (i.e. the coefficient $a(\lambda)$ for the NS model).

31. The fact that the Poisson bracket $\{,\}_{(1)}$ is a reduction in the sense of Dirac [D 1964] of the Lie-Poisson bracket $\{,\}_{-1}$ was observed in [RS 1980b].

32. The notion of compatible Poisson brackets was introduced in [Ma 1978] and elaborated in [GD 1979], [GD 1981] (we point out that the Jacobi identity for the Poisson bracket $\{,\}^{(\alpha)}$ for every α is a consequence of the Jacobi identity for a single value of $\alpha \neq 0$). The derivation of the Jacobi identity for the Poisson brackets $\{,\}_{(k)}$ given in § 5 follows [GD 1979].

33. The construction of the family of functionals I_n and the hierarchy of compatible Poisson brackets $\{,\}_{(k)}$ given in § 5 goes back to [Ma 1978] (see also [KR 1978]).

34. The HM model also possesses a Λ-operator and a hierarchy of Poisson structures. The corresponding second Poisson bracket $\{,\}_{(1)}$ is obtained from the Lie-Poisson bracket

$$\{S_a(x), S_b(y)\}_{(1)} = \tfrac{1}{2}\,\delta_{ab}\,\delta'(x-y) \tag{6.11}$$

by reducing to the orbit $\vec{S}^2 = 1$; the Λ-operator that arises coincides with the one given in [GY 1984].

35. The general Lie-algebraic scheme of § 4 also leads to interesting results if the Lie algebra $\mathfrak{g}$ is the algebra of vector fields on the circle, and the central extension of $C(\mathfrak{g})$ is determined by the Gelfand-Fuks 2-cocycle [GF 1968], which gives the Virasoro algebra. In particular, the second Poisson structure for the KdV equation associated with the Λ-operator coincides with the corresponding Lie-Poisson bracket [S 1985a].

36. The connection between the basic and the second Poisson structures for the KdV equation is established by the well-known Miura transformation [M 1968]. This was generalized in [DS 1981] to equations in the algebra of formal pseudo-differential operators, associated with higher order differential operators. More precisely, any affine Lie algebra gives rise to a sequence of the KdV-type equations and to several sequences of the modified KdV-type equations, their solutions being related by a generalized Miura transformation. The structure of this Miura transformation is determined by the Dynkin diagram of the affine Lie algebra [DS 1984].

37. A different approach to the classification of integrable equations having an a priori assigned functional form was proposed in [MS 1985] relying on classical methods of the theory of Lie-Bäcklund transformations.

References

[A 1979] Adler, M.: On a trace functional for formal pseudodifferential operators and symplectic structure of the Korteweg-de Vries type equations. Invent. Math. *50*, 219–248 (1979)

[AM 1980] Adler, M., van Moerbeke, P.: Completely integrable systems, Euclidean Lie algebras and curves. Adv. Math. *38*, 267–317 (1980)

[B 1882] Bäcklund, A. V.: Zur Theorie der Flächentransformationen. Math. Ann. *19*, 387–422 (1882)

[B 1967] Beresin, F. A.: Several remarks on the associative envelope of a Lie algebra. Funk. Anal. Prilož. *1* (2), 1–14 (1967) [Russian]

[B 1976] Bogoyavlensky, O. I.: On perturbations of the periodic Toda lattice. Comm. Math. Phys. *51*, 201–209 (1976)

[B 1984] Bogoyavlensky, O. I.: Integrable equations on Lie algebras arising in problems of mathematical physics. Izv. Akad. Nauk SSSR, Ser. Mat. *48*, 883–938 (1984) [Russian]

[BD 1982] Belavin, A. A., Drinfeld, V. G.: Solutions of the classical Yang-Baxter equation for simple Lie algebras. Funk. Anal. Prilož. *16* (3), 1–29 (1982) [Russian]; English transl. in Funct. Anal. Appl. *16*, 159–180 (1982)

[Be 1980] Belavin, A. A.: Discrete groups and the integrability of quantum systems. Funk. Anal. Prilož. *14* (4), 18–26 (1980) [Russian]; English transl. in Funct. Anal. Appl. *14*, 260–267 (1980)

[BN 1976] Bogoyavlensky, O. I., Novikov, S. P.: The connection between the Hamiltonian formalisms of stationary and non-stationary problems. Funk. Anal. Prilož. *10* (1), 9–13 (1976) [Russian]; English transl. in Funct. Anal. Appl. *10*, 8–11 (1976)

[C 1983] Cherednik, I. V.: On the definition of τ-function for generalized affine Lie algebras. Funk. Anal. Prilož. *17* (3), 93–95 (1983) [Russian]

[D 1964] Dirac, P. A. M.: Lectures on quantum mechanics. Belfer Grad. School of Science, Yeshiva University, N-Y 1964

[D 1983] Drinfeld, V. G.: Hamiltonian structures on Lie groups, Lie bialgebras and the geometrical meaning of classical Yang-Baxter equations. Dokl. Akad. Nauk SSSR *268*, 285–287 (1983) [Russian]; English transl. in Sov. Math. Dokl. *27*, 68–71 (1983)

[D 1984] Dolan, L.: Kac-Moody algebras and exact solvability in hadronic physics. Phys. Rep. *109* (1), 1–94 (1984)

[DJKM 1981] Date, E., Jimbo, M., Kashiwara, M., Miwa, T.: Operator approach to the Kadomtsev-Petviashvili equation. Transformation groups for soliton equations. III. J. Phys. Soc. Japan *50*, 3806–3812 (1981)

[DMN 1976] Dubrovin, B. A., Matveev, V. B., Novikov, S. P.: Nonlinear equations of Korteweg-de Vries type, finite-zone linear operators and abelian varieties. Uspekhi Mat. Nauk *31* (1), 55–136 (1976) [Russian]; English transl. in Russian Math. Surveys *31* (1), 59–146 (1976)

[DS 1981] Drinfeld, V. G., Sokolov, V. V.: Equations of Korteweg-de Vries type and simple Lie algebras. Dokl. Akad. Nauk SSSR *258*, 11–16 (1981) [Russian]; English transl. in Sov. Math. Dokl. *23*, 457–461 (1981)

[DS 1984] Drinfeld, V. G., Sokolov, V. V.: Lie algebras and equations of Korteweg-de Vries type. Itogi Nauki Tekh., Ser. Sovrem. Probl. Mat. *24*, 81–180 Moscow, VINITI 1984 [Russian]

[FNR 1983 a] Flaschka, H., Newell, A. C., Ratiu, T.: Kac-Moody Lie algebras and soliton equations. II. Lax equations associated with $A_1^{(1)}$. Physica *D 9*, 303–323 (1983)

[FNR 1983 b] Flaschka, H., Newell, A. C., Ratiu, T.: Kac-Moody Lie algebras and soliton equations. III. Stationary equations associated with $A_1^{(1)}$. Physica *D 9*, 324–332 (1983)

[FNR 1983c] Flaschka, H., Newell, A. C., Ratiu, T.: Kac-Moody Lie algebras and soliton equations. IV. Physica *D9*, 333-345 (1983)

[GD 1975] Gelfand, I. M., Dikiĭ, L. A.: Asymptotic behaviour of the resolvent of Sturm-Liouville equations and the algebra of the Korteweg-de Vries equations. Usp. Mat. Nauk *30* (5), 67-100 (1975) [Russian]; English transl. in Russian Math. Surveys *30* (5), 77-113 (1975)

[GD 1978] Gelfand, I. M., Dikiĭ, L. A.: A family of Hamiltonian structures connected with integrable nonlinear differential equations. Preprint No 136 Inst. Applied. Math., Moscow 1978

[GD 1979] Gelfand, I. M., Dorfman, I. Ya.: Hamiltonian operators and algebraic structures associated with them. Funk. Anal. Prilož. *13* (4), 13-30 (1979) [Russian]; English transl. in Funct. Anal. Appl. *13*, 248-262 (1979)

[GD 1981] Gelfand, I. M., Dorfman, I. Ya.: Hamiltonian operators and infinite-dimensional Lie algebras. Funk. Anal. Prilož. *15* (3), 23-40 (1981) [Russian]; English transl. in Funct. Anal. Appl. *15*, 173-187 (1981)

[GF 1968] Gelfand, I. M., Fuks, D. B.: Cohomology of the Lie algebra of vector fields on the circle. Funk. Anal. Prilož. *2* (4), 92-93 (1968) [Russian]

[GY 1984] Gerdjikov, V. S., Yanovski, A. B.: Gauge covariant formulation of the generating operator. 1. The Zakharov-Shabat system. Phys. Lett. *103A*, 232-236 (1984)

[H 1976] Hirota, R.: Direct method of finding exact solutions of nonlinear evolution equations. In: Miura R. (ed.) Bäcklund transformations. Lecture Notes in Mathematics, Vol. 515. Berlin-Heidelberg-New York, Springer 1976

[K 1970] Kostant, B.: Quantization and unitary representations. I. Prequantization. Lecture Notes in Mathematics, Vol. 170, 87-208. Berlin-Heidelberg-New York, Springer 1970

[K 1979a] Kostant, B.: Quantization and representation theory. Proc. of Symposium on Representations of Lie groups. Oxford 1977, London Math. Soc. Lect. Notes Ser. *34*, 287-316 (1979)

[K 1979b] Kostant, B.: The solution to a generalized Toda lattice and representation theory. Adv. Math. *34*, 195-338 (1979)

[K 1983] Kac, V. G.: Infinite dimensional Lie algebras. Progress in Mathematics, v. 44. Boston, Birkhäuser 1983

[Ki 1972] Kirillov, A. A.: Elements of the Theory of Representations. Moscow: Nauka 1972 [Russian]; English transl. Berlin-Heidelberg-New York, Springer 1976

[KR 1978] Kulish, P. P., Reyman, A. G.: A hierarchy of symplectic forms for the Schrödinger and the Dirac equations on the line. In: Problems in quantum field theory and statistical physics. I. Zapiski Nauchn. Semin. LOMI *77*, 134-147 (1978) [Russian]; English transl. in J. Sov. Math. *22*, 1627-1637 (1983)

[KR 1983] Kulish, P. P., Reyman, A. G.: Hamiltonian structure of polynomial bundles. In: Differential geometry, Lie groups and mechanics. V. Zap. Nauchn. Semin. LOMI *123*, 67-76 (1983) [Russian]

[L 1888] Lie, S. (unter Mitwirkung von F. Engel): Theorie der Transformationsgruppen. Bd. 1-3. Leipzig, Teubner, 1888, 1890, 1893

[LM 1979] Lebedev, D. R., Manin, Yu. I.: Gelfand-Dikiĭ Hamiltonian operator and coadjoint representation of Volterra group. Funk. Anal. Prilož. *13* (4), 40-46 (1979) [Russian]

[M 1968] Miura, R. M.: Korteweg-de Vries equation and generalizations. I. A remarkable explicit nonlinear transformation. J. Math. Phys. *9*, 1202-1204 (1968)

[M 1976] Miura, R. (ed.): Bäcklund transformations. Lecture Notes in Mathematics, Vol. 515. Berlin-Heidelberg-New York, Springer 1976

[Ma 1978] Magri, F.: A simple model of the integrable Hamiltonian equation. J. Math. Phys. *19*, 1156–1162 (1978)

[Mi 1979] Mikhailov, A. V.: Integrability of the two-dimensional generalization of the Toda chain. Pisma Zh. Exp. Teor. Fiz. *30*, 443–448 (1979) [Russian]; English transl. in Sov. Phys. JETP Letters *30*, 414–418 (1979)

[MM 1979] Moerbeke, P. van, Mumford, D.: The spectrum of difference operators and algebraic curves. Acta Math. *143*, 93–154 (1979)

[MS 1985] Mikhailov, A. V., Shabat, A. B.: Integrability conditions for a system of two equations of the form $u_t = A(u)u_{xx} + F(u, u_x)$. I. Teor. Mat. Fiz. *62*, 163–185 (1985) [Russian]

[N 1974] Novikov, S. P.: The periodic problem for the Korteweg-de Vries equation. Funk. Anal. Prilož. *8* (3), 54–66 (1974) [Russian]; English transl. in Funct. Anal. Appl. *8*, 236–246 (1974)

[R 1980] Reyman, A. G.: Integrable Hamiltonian systems connected with graded Lie algebras. In: Differential geometry, Lie groups and mechanics. III. Zap. Nauchn. Semin. LOMI *95*, 3–54 (1980) [Russian]; English transl. in J. Sov. Math. *19*, 1507–1545 (1982)

[R 1983] Reyman, A. G.: A unified Hamiltonian system on polynomial bundles and the structure of stationary problems. In: Problems in quantum field theory and statistical physics. 4. Zap. Nauchn. Semin. LOMI *131*, 118–127 (1983) [Russian]; English transl. in J. Sov. Math. *30*, 2319–2325 (1985)

[RF 1983] Reshetikhin, N. Yu., Faddeev, L. D.: Hamiltonian structures for integrable models of field theory. Teor. Mat. Fiz. *56*, 323–343 (1983) [Russian]; English transl. in Theor. Math. Phys. *56*, 847–862 (1983)

[RS 1979] Reyman, A. G., Semenov-Tian-Shansky, M. A.: Reduction of Hamiltonian systems, affine Lie algebras and Lax equations. I. Invent. Math. *54*, 81–100 (1979)

[RS 1980a] Reyman, A. G., Semenov-Tian-Shansky, M. A.: Current algebras and nonlinear partial differential equations. Dokl. Akad. Nauk SSSR *251*, 1310–1314 (1980) [Russian]; English transl. in Sov. Math. Dokl. *21*, 630–634 (1980)

[RS 1980b] Reyman, A. G., Semenov-Tian-Shansky, M. A.: A family of Hamiltonian structures, a hierarchy of Hamiltonians and reduction for first order matrix differential operators. Funk. Anal. Prilož. *14* (2), 77–78 (1980) [Russian]; English transl. in Funct. Anal. Appl. *14*, 146–148 (1980)

[RS 1981] Reyman, A. G., Semenov-Tian-Shansky, M. A.: Reduction of Hamiltonian systems, affine Lie algebras and Lax equations. II. Invent. Math. *63*, 423–432 (1981)

[RS 1984] Reyman, A. G., Semenov-Tian-Shansky, M. A.: Hamiltonian structure of equations of the Kadomtsev-Petviashvili type. In: Differential geometry, Lie groups and mechanics. VI. Zap. Nauchn. Semin. LOMI *133*, 212–227 (1984) [Russian]; English transl. in J. Sov. Math. *31*, 3399–3410 (1985)

[RS 1986] Reyman, A. G., Semenov-Tian-Shansky, M. A.: Lie algebras and Lax equations with spectral parameter on an elliptic curve. In: Problems of quantum field theory and statistical physics 6. Zap. Nauchn. Semin. LOMI *150*, 104–118 (1986) [Russian]

[RSF 1979] Reyman, A. G., Semenov-Tian-Shansky, M. A., Frenkel, I. B.: Graded Lie algebras and completely integrable systems. Dokl. Akad. Nauk SSSR *247*, 802–805 (1979) [Russian]; English transl. in Sov. Math. Dokl. *20*, 811–814 (1979)

[S 1970] Souriau, J.-M.: Structure des systèmes dynamiques. Paris, Dunod 1970

[S 1979] Sklyanin, E. K.: On complete integrability of the Landau-Lifshitz equation. Preprint LOMI E-3-79, Leningrad 1979

[S 1980] Symes, W.: Systems of Toda type, inverse spectral problems and representation theory. Invent. math. *59*, 13–53 (1980)

[S 1983] Semenov-Tian-Shansky, M. A.: What is a classical *r*-matrix. Funk. Anal. Prilož. *17* (4), 17–33 (1983) [Russian]; English transl. in Funct. Anal. Appl. *17*, 259–272 (1983)

[S 1985 a] Semenov-Tian-Shansky, M. A.: Group-theoretic methods in the theory of integrable systems. Doctoral thesis, Leningrad 1985 [Russian]

[S 1985 b] Semenov-Tian-Shansky, M. A.: Dressing transformations and Poisson group actions. Publ. RIMS *21*, 1237–1260 (1985)

[SW 1985] Segal, G., Wilson, G.: Loop groups and equations of KdV type. Publ. Math. IHES *61*, 5–65 (1985)

[T 1982] Trofimov, V. V.: Completely integrable geodesic flows of left-invariant metries on Lie groups connected with commutative graded algebras with Poincare duality. Dokl. Akad. Nauk SSSR *263*, 812–816 (1982) [Russian]; English transl. in Sov. Math. Dokl. *25*, 449–453 (1982)

[T 1984] Takhtajan, L. A.: Solutions of the triangle equations with $\mathbb{Z}_N \times \mathbb{Z}_N$ symmetry as matrix analogues of the Weierstrass zeta and sigma functions. In: Differential geometry, Lie groups and mechanics. VI. Zap. Nauchn. Semin. LOMI *133*, 258–276 (1984) [Russian]

[W 1983] Weinstein, A.: The local structure of Poisson manifolds. J. Diff. Geometry *18*, 523–557 (1983)

[ZMNP 1980] Zakharov, V. E., Manakov, S. V., Novikov, S. P., Pitaievski, L. P.: Theory of Solitons. The Inverse Problem Method. Moscow, Nauka 1980 [Russian]; English transl.: New York, Plenum 1984

[ZS 1979] Zakharov, V. E., Shabat, A. B.: Integration of the nonlinear equations of mathematical physics by the method of the inverse scattering problem. II. Funk. Anal. Prilož. *13* (3), 13–22 (1979) [Russian]; English transl. in Funct. Anal. Appl. *13*, 166–174 (1979)

Conclusion

This conclusion is intended for those who have read the book to the end. We hope that the main text and the notes to separate chapters have furnished a convincing evidence of how rich from the mathematical point of view, both conceptually and technically, is the subject of solitons and integrable partial differential equations. In fact, the inverse scattering method naturally intertwines various branches of mathematics: differential geometry, the theory of Lie groups and Lie algebras and their representations, complex and functional analysis. All of them serve one common purpose, to classify integrable equations and describe their solutions. As a result, the traditional parts of these branches, such as Hamiltonian formalism, affine Lie algebras, or the Riemann problem are seen in a new light.

Moreover, the progress of the inverse scattering method has given rise to new problems and structures in these domains. It will suffice to remind of the general notion of r-matrix and its interpretation from the Hamiltonian, group-theoretic, and analytical points of view. This is what reflects the modern trend in mathematics when theoretical disciplines that at first sight seem unrelated interact and draw from one another in the joint effort to solve concrete problems having important physical applications.

To a still greater extent this trend persists when the methods developed in our book are generalized to models of quantum mechanics and field theory. Research in this direction has been very active in the last years. And again, the unifying concept proves to be the (quantum) R-matrix. We hope that this line of study will soon be summarized in yet another monograph like the one we present.

List of Symbols

Symbols are listed in alphabetical order (first Latin, then Greek), with an indication of the number of the page and formula where they are defined. A symbol is repeated if it has more than one meaning in the text.

Index

Lars Hörmander

The Analysis of Linear Partial Differential Operators

Volume I

Distribution Theory and Fourier Analysis

1983. 5 figures. IX, 391 pages. (Grundlehren der mathematischen Wissenschaften, Volume 256). ISBN 3-540-12104-8

Volume I is primarily directed to a broad range of students with an interest in analysis, assuming a basic knowledge of advanced calculus, integration theory, and some functional analysis. Going beyond previous books on distribution theory, Hörmander places a special emphasis on Fourier analysis, particularly its important microlocal aspects. These are also discussed in the wider framework of hyperfunction theory, providing a useful introduction (in the spirit of Schwartz distributions) to harmonic analysis and to the analytic theory of partial differential equations.

Volume II

Differential Operators with Constant Coefficients

1983. 7 figures. VIII, 391 pages. (Grundlehren der mathematischen Wissenschaften, Volume 257). ISBN 3-540-12139-0

Volume II is a systematic study of partial differential operators with constant coefficients, and some of their perturbations. New chapters cover convolution operators, scattering theory, and methods from the theory of analytic functions of several complex variables. This volume is more advanced than the first one, and directed to research workers and graduate students in mathematics and mathematical physics.

Printed in the United States
By Bookmasters